D0764353

IUPAC Periodic Table of the Elements

Key:

atomic number **Symbol** name standard atomic weight

1	2	3	4	5	6	7	8	9	10	11	12	13	14	15	16	17	18
1 **H** hydrogen [1.007; 1.009]																	2 **He** helium 4.003
3 **Li** lithium [6.938; 6.997]	4 **Be** beryllium 9.012											5 **B** boron [10.80; 10.83]	6 **C** carbon [12.00; 12.02]	7 **N** nitrogen [14.00; 14.01]	8 **O** oxygen [15.99; 16.00]	9 **F** fluorine 19.00	10 **Ne** neon 20.18
11 **Na** sodium 22.99	12 **Mg** magnesium 24.31											13 **Al** aluminium 26.98	14 **Si** silicon [28.08; 28.09]	15 **P** phosphorus 30.97	16 **S** sulfur [32.05; 32.08]	17 **Cl** chlorine [35.44; 35.46]	18 **Ar** argon 39.95
19 **K** potassium 39.10	20 **Ca** calcium 40.08	21 **Sc** scandium 44.96	22 **Ti** titanium 47.87	23 **V** vanadium 50.94	24 **Cr** chromium 52.00	25 **Mn** manganese 54.94	26 **Fe** iron 55.85	27 **Co** cobalt 58.93	28 **Ni** nickel 58.69	29 **Cu** copper 63.55	30 **Zn** zinc 65.38(2)	31 **Ga** gallium 69.72	32 **Ge** germanium 72.63	33 **As** arsenic 74.92	34 **Se** selenium 78.96(3)	35 **Br** bromine 79.90	36 **Kr** krypton 83.80
37 **Rb** rubidium 85.47	38 **Sr** strontium 87.62	39 **Y** yttrium 88.91	40 **Zr** zirconium 91.22	41 **Nb** niobium 92.91	42 **Mo** molybdenum 95.96(2)	43 **Tc** technetium	44 **Ru** ruthenium 101.1	45 **Rh** rhodium 102.9	46 **Pd** palladium 106.4	47 **Ag** silver 107.9	48 **Cd** cadmium 112.4	49 **In** indium 114.8	50 **Sn** tin 118.7	51 **Sb** antimony 121.8	52 **Te** tellurium 127.6	53 **I** iodine 126.9	54 **Xe** xenon 131.3
55 **Cs** caesium 132.9	56 **Ba** barium 137.3	57-71 lanthanoids	72 **Hf** hafnium 178.5	73 **Ta** tantalum 180.9	74 **W** tungsten 183.8	75 **Re** rhenium 186.2	76 **Os** osmium 190.2	77 **Ir** iridium 192.2	78 **Pt** platinum 195.1	79 **Au** gold 197.0	80 **Hg** mercury 200.6	81 **Tl** thallium [204.3; 204.4]	82 **Pb** lead 207.2	83 **Bi** bismuth 209.0	84 **Po** polonium	85 **At** astatine	86 **Rn** radon
87 **Fr** francium	88 **Ra** radium	89-103 actinoids	104 **Rf** rutherfordium	105 **Db** dubnium	106 **Sg** seaborgium	107 **Bh** bohrium	108 **Hs** hassium	109 **Mt** meitnerium	110 **Ds** darmstadtium	111 **Rg** roentgenium	112 **Cn** copernicium		114 **Fl** flerovium		116 **Lv** livermorium		

57 **La** lanthanum 138.9	58 **Ce** cerium 140.1	59 **Pr** praseodymium 140.9	60 **Nd** neodymium 144.2	61 **Pm** promethium	62 **Sm** samarium 150.4	63 **Eu** europium 152.0	64 **Gd** gadolinium 157.3	65 **Tb** terbium 158.9	66 **Dy** dysprosium 162.5	67 **Ho** holmium 164.9	68 **Er** erbium 167.3	69 **Tm** thulium 168.9	70 **Yb** ytterbium 173.1	71 **Lu** lutetium 175.0
89 **Ac** actinium	90 **Th** thorium 232.0	91 **Pa** protactinium 231.0	92 **U** uranium 238.0	93 **Np** neptunium	94 **Pu** plutonium	95 **Am** americium	96 **Cm** curium	97 **Bk** berkelium	98 **Cf** californium	99 **Es** einsteinium	100 **Fm** fermium	101 **Md** mendelevium	102 **No** nobelium	103 **Lr** lawrencium

The chemical elements, IUPAC (2007)

Name	Symbol	Atomic number	Atomic weight	Name	Symbol	Atomic number	Atomic weight
Actinium	Ac	89	227	Neodymium	Nd	60	144.242
Aluminum	Al	13	26.982	Neon	Ne	10	20.180
Americium	Am	95	243	Neptunium	Np	93	237
Antimony	Sb	51	121.76	Nickel	Ni	28	58.693
Argon	Ar	18	39.948	Niobium	Nb	41	92.906
Arsenic	As	33	74.922	Nitrogen	N	7	14.0067
Astatine	At	85	210	Osmium	Os	76	190.23
Berkelium	Bk	97	247	Oxygen	O	8	15.999
Beryllium	Be	4	9.012	Palladium	Pd	46	106.42
Bismuth	Bi	83	208.980	Phosphorus	P	15	30.974
Boron	B	5	10.811	Platinum	Pt	78	195.084
Bromine	Br	35	79.904	Plutonium	Pu	94	244
Cadmium	Cd	48	112.411	Polonium	Po	84	209
Caesium	Cs	55	132.905	Potassium	K	19	39.098
Calcium	Ca	20	40.078	Praseodymium	Pr	59	140.908
Californium	Cf	98	251	Promethium	Pm	61	145
Carbon	C	6	12.0107	Protactinium	Pa	91	231.036
Cerium	Ce	58	140.116	Radium	Ra	88	226
Chlorine	Cl	17	35.453	Radon	Rn	86	222
Chromium	Cr	24	51.996	Rhenium	Re	75	186.207
Cobalt	Co	27	58.933	Rhodium	Rh	45	102.906
Copper	Cu	29	63.546	Rubidium	Rb	37	85.468
Curium	Cm	96	247	Ruthenium	Ru	44	101.07
Dysprosium	Dy	66	162.5	Samarium	Sm	62	150.36
Erbium	Er	68	167.259	Scandium	Sc	21	44.956
Europium	Eu	63	151.964	Selenium	Se	34	78.96
Fluorine	F	9	18.998	Silicon	Si	14	28.0855
Francium	Fr	87	223	Silver	Ag	47	107.868
Gadolinium	Gd	64	157.25	Sodium	Na	11	22.990
Gallium	Ga	31	69.723	Strontium	Sr	38	87.62
Germanium	Ge	32	72.64	Sulfur	S	16	32.07
Gold	Au	79	196.967	Tantalum	Ta	73	180.948
Hafnium	Hf	72	178.49	Technetium	Tc	43	98
Helium	He	2	4.0026	Tellurium	Te	52	127.6
Holmium	Ho	67	164.930	Terbium	Tb	65	158.925
Hydrogen	H	1	1.00794	Thallium	Tl	81	204.383
Indium	In	49	114.818	Thorium	Th	90	232.038
Iodine	I	53	126.904	Thulium	Tm	69	168.934
Iridium	Ir	77	192.217	Tin	Sn	50	118.71
Iron	Fe	26	55.845	Titanium	Ti	22	47.867
Krypton	Kr	36	83.798	Tungsten	W	74	183.84
Lanthanum	La	57	138.905	Uranium	U	92	238.029
Lead	Pb	82	207.2	Vanadium	V	23	50.942
Lithium	Li	3	6.941	Xenon	Xe	54	131.293
Lutetium	Lu	71	174.967	Ytterbium	Yb	70	173.054
Magnesium	Mg	12	24.305	Yttrium	Y	39	88.906
Manganese	Mn	25	54.938	Zinc	Zn	30	65.38
Mercury	Hg	80	200.59	Zirconium	Zr	40	91.224
Molybdenum	Mo	42	95.96				

Data from Wieser (2006), updates from IUPAC website. Where there is no figure after the decimal point the element is man-made or the isotopic composition is variable so only the approximate weight of the most stable or most important isotope is listed.

An Introduction to the Chemistry of the Sea

Engaging and clearly written, this new edition is an accessible introduction that provides students in oceanography, marine chemistry and biogeochemistry with the fundamental tools they need. It highlights geochemical interactions between the ocean, solid earth, atmosphere, and climate, enabling students to appreciate the interconnectedness of Earth's processes and systems, and elucidates the huge variations in the oceans' chemical environment, from surface waters to deep water.

- Fully updated to cover exciting recent developments in the field, including all-new sections and expanded material ranging from estuaries and sediments to methane hydrates and ligands.
- Now with a glossary, new end-of-chapter summaries, and questions for students to review their learning and put the theory into practice.
- Appendices provide a useful detailed reference for students, and include seawater properties, and key equations and constants for calculating oceanographic processes.

"This second edition is a welcome updating of Pilson's classic text. The book treats the broad range of aspects of marine chemistry for advanced undergraduates and graduate students, but even seasoned researchers will find it a valuable resource. The inclusion of a historical perspective throughout the book is particularly useful."

– Professor J. Kirk Cochran

School of Marine and Atmospheric Sciences, Stony Brook University

"Contains an enormous wealth of information, presenting marine chemistry in a way that even non-chemists will be able to grasp. Pilson has taken great care to explain the basics as well as the state-of-the-art. It is a book that any marine scientist (from undergraduate through to professor) will benefit from having close to hand."

– Professor David Thomas

School of Ocean Sciences, Bangor University;
UK & Finnish Environment Institute, Marine Research Centre, Helsinki;
Finland & Arctic Research Centre, Aarhus University, Denmark

"This book is unique and clearly presented: not just a textbook for courses on ocean chemistry, but definitely a great reference book for coastal ecosystems and freshwater bodies as well."

– Dr. Yushun Chen

Director of Environmental Quality and Ecosystems Health Research
Laboratory Aquaculture and Fisheries Center, University of Arkansas

"This new edition of Pilson's classic textbook has been brought up to date with regard to the latest analytical techniques and scientific insights of modern marine chemistry. With its broad scope and light narrative style it will likely remain the foundation of many excellent graduate and advanced undergraduate courses for years to come."

– Dr. Johan Schijf
University of Maryland Center for Environmental Science | Chesapeake Biological Laboratory

Michael E. Q. Pilson is Emeritus Professor of Oceanography at the Graduate School of Oceanography, University of Rhode Island, and has taught a course on chemical oceanography for most of the past 40 years. For some years, he directed the Marine Ecosystems Research Laboratory at Rhode Island, conducting experimental study of biogeochemistry in shallow coastal waters. Professor Pilson has published around 90 papers on chemical, physical, biological, ecological, and geological aspects of oceanography and has advised students in chemical and biological areas. His broad range of experience has shaped his oceanography course to be accessible and interesting to students with diverse backgrounds, interests and professional goals.

An Introduction to the Chemistry of the Sea

Second Edition

MICHAEL E. Q. PILSON
University of Rhode Island

CAMBRIDGE
UNIVERSITY PRESS

University Printing House, Cambridge CB2 8BS, United Kingdom

One Liberty Plaza, 20th Floor, New York, NY 10006, USA

477 Williamstown Road, Port Melbourne, VIC 3207, Australia

4843/24, 2nd Floor, Ansari Road, Daryaganj, Delhi - 110002, India

79 Anson Road, #06-04/06, Singapore 079906

Cambridge University Press is part of the University of Cambridge.

It furthers the University's mission by disseminating knowledge in the pursuit of education, learning and research at the highest international levels of excellence.

www.cambridge.org
Information on this title: www.cambridge.org/9780521887076

First edition published by Prentice Hall (Pearson Education Inc.), UK, 1998, as *An Introduction to the Chemistry of the Sea*
Second edition published 2013

A catalogue record for this publication is available from the British Library

Library of Congress Cataloging in Publication data
Pilson, Michael E. Q.
An introduction to the chemistry of the sea / Michael E. Q. Pilson, University of Rhode Island. – Second edition.
pages cm
Includes bibliographical references and index.
ISBN 978-0-521-88707-6
1. Chemical oceanography. I. Title.
GC111.2.P55 2013
551.46′6–dc23 2012028896

ISBN 978-0-521-88707-6 Hardback

Additional resources for this publication at www.cambridge.org/pilson

CONTENTS

PREFACE

This textbook grew out of a set of class notes developed for teaching an introductory course on the chemistry of the sea to successive classes of beginning graduate students in oceanography, and to a few undergraduates. Students enter marine science from a wide variety of disciplines; I have not assumed an advanced knowledge of chemistry nor, indeed, of any single science. The book is intended to be accessible to anyone with some exposure to chemistry, physics, and biology (such as would be the case for most people with a degree in a natural science), as well as a little geology. It might be helpful if the reader has previously seen one or more of the many excellent books that provide a general description of the oceans or has otherwise acquired a similar overview.

The coverage is intended to provide an introduction appropriate for most students of marine science, as well as a general base from which the professional marine chemist can go forward to specialized studies. A rather large number of references provides an entrée to reviews, textbooks, and more specialized literature. Tables of numerical values and constants, provided in the Appendix, will be useful to oceanographers and other environmental scientists.

Every branch of oceanography draws to some extent upon the others, and this interaction often passes through a common link in the chemistry of seawater. The chemistry of the sea affects and is affected by numerous physical, biological, and geological processes. Much of the knowledge of physical processes comes from chemical measurement, and in turn the great flows and mixings in the sea influence the distribution of chemical substances. The biota influence many aspects of the chemistry of the sea and in turn are controlled in part by the chemistry of the medium in which they float or swim. The sediments of the sea floor are partly the product of chemical and biological processes, and finally the chemistry of the sea is ultimately controlled in large part by geological processes. I have endeavored to point out or draw upon these linkages, when I have had the strength to do so.

Some subjects are covered more completely than others. For example, in Chapter 7, the subject of carbon dioxide is presented in considerable detail, for several reasons. The carbon dioxide system itself touches on or is deeply intertwined with most areas of marine science. Air–sea gas exchange, solubility of gases, physical chemistry of seawater, ionic reactions, biological production and respiration, precipitation and dissolution of minerals in seawater, ocean circulation, isotopic tracers and the dating of seawater, and the history of ancient climates are all subjects where knowledge of the carbon dioxide system is important as a base or contributes to the analysis. Furthermore, one of the major impacts by humans on the biogeochemistry of Earth is caused by the discharge of carbon dioxide. The role of CO_2 in climate change and the role of the oceans in modulating the concentration of CO_2 in the atmosphere are so important that it seemed useful to make this section as thorough as a reasonable balance of space would allow.

When it seemed interesting or illuminating, I have provided a brief recounting of the history of a subject, for I think that science textbooks too often present the state of a field as if it is rather fixed, with too little appreciation of how we got to where we are. Sometimes the sense of where we are now in the development of our understanding can only be gained by a little familiarity with the rhythm and pace of scientific discovery. In a few cases I have included early figures from the literature, rather than figures derived from better modern data, because often the general principle involved in some process was initially fixed into our minds by a glance at that early figure.

Many subjects touched on in this book are active fields of research. Every month the journals coming into the library bring relevant facts to enhance our knowledge or new insights to change in some detail the way we view the chemistry of the sea. Those who would maintain a current knowledge of the field must keep reading the journal literature. The material covered here will help to prepare the reader to read those journals. I will be most grateful to those who make substantive suggestions or report the inevitable errors.

ACKNOWLEDGMENTS

I thank all the many students, colleagues, and others (some of whom I know only from what they have written) who have helped or stimulated me over many years. I especially acknowledge Dana Kester, John Knauss, Jim Quinn, and Dave Schink. A great many helpful comments on the manuscript were received from David R. Schink, Texas A & M University; Douglas E. Hammond, University of Southern California; James G. Quinn, University of Rhode Island; Theodore C. Loder III, University of New Hampshire; and Kent A. Fanning, University of South Florida. The organizers of each of the Gordon Conferences in Chemical Oceanography are also due special thanks, beginning with Norris Rakestraw and Victor Linnenbom who initiated the series. Without these excellent and stimulating conferences, held at intervals since 1969, this book might not have been written. Early drafts of the first edition were typed by my wonderfully competent secretary, Evelyn Dyer.

The preparation of the second edition has been substantially aided by the many suggestions from several dozen students and other colleagues of several institutions. I am grateful to all.

Lastly, our present understanding of the chemistry of the sea has been won by the cleverness and dedication of only a few hundred investigators inquiring into the processes at work in Earth's environment, especially the oceans. As sciences go, it is a relatively small, but, for me, a very special group of people.

Michael E. Q. Pilson
Graduate School of Oceanography
University of Rhode Island

1 Introduction

A complete working model of the earth is still a rather distant goal. HOLLAND 1978

Those magnificent pictures of Earth from space, among the most humanly important and evocative results from the placing of manned and unmanned satellites in orbit during the last several decades, have shown us that underneath the clouds most of the world is blue. Viewed in this way from space, Earth is seen as a planet covered mostly by water, and most of that water is seawater.

This thin layer of water, covering 71% of Earth, affects or controls much of its climate and chemistry. The blue color of the sea tells us that most sunlight on the ocean is absorbed and not reflected. The absorption of sunlight warms the planet. The warmth evaporates water, especially from the tropical ocean. Water in the atmosphere is a greenhouse gas, and this also helps to warm Earth. Some atmospheric water forms clouds; by reflecting sunlight clouds help to cool Earth. The balance between heating and cooling is always changing as clouds form and dissipate, so exact calculation of the balance is difficult. The water in the ocean is a vast reservoir of heat, which buffers and slows global change. Both the currents in the ocean and the winds in the atmosphere carry heat from low latitudes, where there is a net input, towards the poles, where there is a net loss of heat to space. These vast currents, simple in general concept, immensely complex in detail, profoundly affect both the local distribution of climate and the overall climate of Earth. The ocean contains an enormous reservoir of carbon dioxide; it both absorbs and releases this greenhouse gas from and to the atmosphere. Living things in the sea influence both the release and the uptake. Physical, chemical, biological, and geological processes interact with each other in the ocean and with the atmosphere to create the environment of the planet we live on.

Table 1.1 Volumes of water on Earth

	Volume; units of 10^3 km^3	%
Seawater	1 332 000	97.4
Sea ice	20	—
Continental ice	27 800	2.0
Lakes and rivers	225	—
Groundwater	8062	0.6
Vapor (liquid volume)	13	—
TOTAL	1 386 120	100

From Baumgartner and Reichel 1975, Menard and Smith 1966, and Table J-6. The water chemically combined in crustal rocks and in the mantle is not included in these estimates.

Table 1.2 Miscellaneous data on the Earth and its oceans

Earth surface area	510×10^6 km^2
Land surface area	148×10^6 km^2
Ocean surface area	362×10^6 km^2
Ocean surface area, % of Earth area	71%
Ocean: total volume	1.33×10^9 km^3
	1.33×10^{18} m^3
Average depth	3680 m
Temperature range	~ -2 to ~40 °C
Pressure range	~1 to ~1000 atm
	~1 to ~1000 kg cm^{-2}
	~1 to ~1000 bar
	~100 to ~10^5 kPa

Menard and Smith 1966, Charette and Smith 2010. For more detail, see Appendix J.

1.1 Scope of chemical oceanography

Most of the water on Earth is within the province of the marine chemist, because over 97% of all (non-mineral-bound) water on Earth is in the sea, and most of the rest is frozen (Table 1.1). The maximum depth is a little over 10 km, so the pressures encountered in the ocean range from 1 atmosphere to about 1000 atmospheres, or roughly 100 to 100 000 kPa (Table 1.2). The marine chemist must therefore consider the effects of pressures within this range on chemical reactions within the sea.

The temperature of the open sea ranges normally from about –2 °C to about 30 °C, although in tropical bays and tide pools the temperature may reach or exceed 40 °C, so most of the time a marine chemist must be concerned with temperature effects only within this range. The salt content varies from 0 (when, of course, it is not really seawater) to 41 g/kg (grams per kilogram = parts per thousand by weight, usually

Table 1.3 Mean and percentile distribution[a] of potential temperature[b] and salinity in the major oceans of the world

Ocean	Mean	$P_{5\%}$	$P_{25\%}$	$P_{50\%}$	$P_{75\%}$	$P_{95\%}$
			Potential temperature, °C			
Pacific	3.36	0.8	1.3	1.9	3.4	11.1
Indian	3.72	−0.2	1.0	1.9	4.4	12.7
Atlantic	3.73	−0.6	1.7	2.6	3.9	13.7
World	3.52	0.0	1.3	2.1	3.8	12.6
			Salinity, *S*			
Pacific	34.62	34.27	34.57	34.65	34.70	34.79
Indian	34.76	34.44	34.66	34.73	34.79	35.19
Atlantic	34.90	34.33	34.61	34.90	34.97	35.73
World	34.72	34.33	34.61	34.69	34.79	35.10

[a] Percent of the total volume of the ocean with temperature and salinity characteristics falling below the indicated values ($P_{5\%}$, $P_{25\%}$, etc.). From Montgomery 1958.

[b] When water is placed under pressure it is slightly compressed and its temperature rises; conversely, when it is decompressed its temperature drops. The temperature of seawater from various depths is usually expressed as the "potential temperature." This is the temperature it would have if brought to the surface without gain or loss of heat (i.e. it is decompressed adiabatically). Similarly, the potential density is its density when brought to the surface. (For special purposes the potential density is sometimes specified as that at some other standard depth.)

abbreviated as ppt or ‰; later we will learn a more precise definition of salinity); sometimes one may have to consider the more saline products of evaporation in closed basins. Other special circumstances are occasionally encountered, the most interesting and important being the occurrence of extremely hot seawater (sometimes greater than 400 °C) circulating through hot basaltic rocks exposed at the oceanic spreading centers and other regions of submarine volcanism.

Even the narrow temperature ranges, noted above as characteristic of the ocean, may give a misleading impression about the relative constancy of typical seawater (Table 1.3). Some 90% of the entire ocean volume is found between 0 and 12.6 °C, with a salt content ranging from 34.33 to 35.10‰. One-half of the entire volume of the ocean is characterized by temperatures between 1.3 and 3.8 °C and concentrations of salt between 34.61 to 34.79‰. The mean values for the world ocean are 3.52 °C and 34.72‰. The observation that 90% of the total volume of seawater is found within a range in the concentration of salt only ±1.1% from the overall mean value posed an extraordinary challenge to analytical chemists during the early days of oceanography. A great wealth of information resides in the real patterns of differences in the content of salt and other substances from place to place throughout the mass of seawater, but that information could only be revealed by chemical measurements of extraordinary sensitivity and precision. The majority of analysts in most other fields are satisfied to achieve a two-standard-deviation accuracy and precision of ±1% of the measured value. However, in evaluating distributions of salt throughout the water masses of the oceans, data with only this precision would be essentially useless. In the ocean we need measurements of salinity that are at least 100 times better than this, with respect to both precision and accuracy.

The field of chemical oceanography is primarily concerned with the chemical and physical nature of seawater under the conditions and within the limits described above, with the chemical reactions which proceed in the medium, with the processes which control the composition of seawater, and with the processes by which the sea affects the atmosphere above and the solid earth below. To this end it is important to study the chemical nature of seawater itself, and the interactions of seawater with the atmosphere, with the sediments, with hot basaltic rocks, and with the organisms that live in the sea and to a great extent also control its chemistry.

We may look at the ocean in a variety of ways, and ask a variety of questions:

- Why is the ocean full of seawater? Where did it come from? How long will it last?
- Is the ocean the wastewater and salt of the whole world, accumulated for millions or billions of years, and is this question a useful approach to understanding the sea?
- Seawater is a very concentrated solution of some substances, and an extremely dilute solution of some others; in each case, what are the chemical consequences?
- The ocean is the environment for much of the life on Earth, and is probably where life began; to what extent is its chemistry today the product of life processes, and to what extent does the chemical environment in the sea control these processes?
- The oceans are the chemical environment through which and under which most of the world's sediment accumulation takes place. In addition, there are very important exchanges between hot basaltic rocks and seawater at the undersea spreading centers. To what extent are the chemistry of the sea, the nature of the sediments, and even the chemistry of the mantle, influenced or controlled by these chemical exchanges?
- The ocean is an integral part of the processes affecting the climate on Earth; to what extent does it both record and modulate those processes which tend to change the climate?

While, as Holland suggested, we may never have a complete working model of the Earth, it is nevertheless clear that many pieces of such a model are in place and pretty well understood, and the same is true of the sea. We know the sizes of some geochemical reservoirs very well, and most of the rest well enough. Many of the exchange processes are known, but in many cases the rates and what controls them are poorly known. We have little knowledge of changes in reservoir sizes through time, even less do we understand changes in the rates of important processes through time and with changing climate.

Our drive to study these matters is fueled in part by the necessity to examine the extent to which humans have caused, and continue to cause, changes in the geochemical transport rates of many substances and consequent changes in our environment. The waters of the ocean contain detectable records of some human activities over several decades, and perhaps longer. The sediments on the sea floor contain magnificent records of changing climate and other conditions on Earth over many tens of millions of years. We can test our understanding of natural processes by attempting to understand these records.

Because the chemistry of seawater plays such a central role in atmospheric, biological, and geochemical processes, the field of chemical oceanography overlaps many scientific disciplines and draws upon them for a variety of experimental and theoretical approaches to the study of the natural world.

1.2 History of chemical oceanography

What do we know, and when did we know it? *After Howard Baker*

In reading through some of the early records of observations on the chemical nature of seawater, a striking feature becomes apparent: Most of the names one comes across are already familiar because of their contributions to the foundation of the natural sciences or to the groundwork on which we base the modern study of chemistry.

For example, in the fourth century BCE, Aristotle (384–322 BCE) was interested in both the origin and the nature of seawater: "We must now discuss the origin of the sea and the cause of its salt and bitter taste." (Aristotle 1931, vol. III, p. 354[b]). As we might expect, his explanations were not entirely clear. While he considered the words of Empedocles, that the sea is "the sweat of the earth," to be absurdly metaphorical, suitable for the requirements of a poem but unsatisfactory as a scientific theory, today we might still appreciate and use the metaphor, because at least some of the water in the sea must have been driven by heat out of the interior of the Earth. Aristotle did consider and use the analogy that the sweet water we drink becomes both salt and bitter as it gathers in the bladder as urine. He seemed convinced that water became salt through admixture with some earthy substance (vol III, p. 357[a]). The bitter taste of sea salt or of seawater was noted again and again for well over 2000 years before the explanation was discovered. Aristotle was also aware of the relative density of seawater and fresh water. He reported a simple experiment showing that an egg would float in brine and sink in fresh water. Sailors of the time knew that ships floated higher in seawater than in fresh water, and indeed Aristotle reported that ships had for this reason sometimes come to grief upon sailing from salt into fresh water. The first recorded observations on the organic chemistry of seawater are also due to Aristotle. He noted that, especially in warm weather, "a fatty oily substance formed on the surface ... having been excreted by the sea which has fat in it." (Aristotle 1931, vol. VII, p. 932[b]).

Aristotle's ideas on the hydrological and geochemical cycles were rather vague. He did, however, believe, apparently based on experimental evidence, that the vapors of sweet water arose from the sea, and fell again as rain, and he disputed the notion common at the time that there were subterranean passages by which seawater rose through the earth, perhaps straining out the salt by passing through soil, and discharging from springs over the land. In regard to the latter notion, it appears that a belief then current was that salt could be removed from seawater by passage through soil (we might now invoke the concepts of ion exchange). The Roman writer Pliny the Elder also stated that salt could be removed from seawater in this way. I have never seen any modern test of the suggestion, however. The notion of a subterranean return of water from sea to land was supported by the observation that in some places the springs from the ground yielded salty water. Aristotle did not dismiss the somewhat vague current notion that salt evaporated from the sea as a dry exhalation and mixed with rain to come down again. In fact, there is sea salt in rain, and one wonders if Aristotle could have had evidence of it.

Aristotle's largest contributions to marine science were his investigations in marine biology. Taking his contributions in total, however, Margaret Deacon (1971) wrote that Aristotle has a legitimate claim to the title "father of oceanography," although I am inclined to doubt that such a term has any real meaning. While certain additional facts about the chemistry of seawater were recorded by other ancient authors (such as Pliny the Elder), there were no real advances made for some 2000 years following Aristotle.

The next prominent person associated with the development of knowledge in this field was Robert Boyle (1627–1691). While Robert Boyle has been called the "father of chemical oceanography," he was also one of the founders of the Royal Society, the first scientific society, and he is more commonly called the "father of modern chemistry." Boyle added considerable weight to the notion that an element is "that which has not been decomposed." Robert Boyle is best known for his work in pneumatics and for the simple law relating the pressures and volumes of a gas at constant temperature ($P_1V_1 = P_2V_2$), known as Boyle's law.

In 1673, Robert Boyle published a tract entitled *Observations and Experiments about the Saltness of the Sea*, in which he related experiments on the chemistry of seawater. As a result of some of the first real chemical tests, Boyle learned that even so-called fresh waters – river and lake waters – contain small amounts of salt, and therefore the salt in the oceans must be, at least in part, the accumulated washings from the land. In this he displayed an elementary perception of the geochemical cycle. No report on the matter having been published for 2000 years, Boyle demonstrated once again that sweet water could be obtained by evaporating seawater, and suggested that this could be of use to mariners who are short of water while at sea.

Boyle reported that seawater was of similar saltness at the surface and at depth. He recorded this observation after hearing reports from divers that the water tasted just as salty at depth as it did on the surface. He also analyzed "deep" samples collected in the English Channel with what must have been the first oceanographic sampling bottle, devised by Robert Hooke. Boyle was the first perceptive analyst of seawater, but he could not get reliable estimates of the salt in seawater by drying it and weighing the residue. As we will later see, this fact of nature is still true, and has caused an endless amount of trouble for oceanographers. He did report an approximate value of 30 g in a kilogram of seawater (in our units), nearly correct for that location. Some years after, Boyle was responsible for introducing the silver nitrate test for the saltness of water and we will later see what an important role that this chemical reaction has had in the development of modern oceanography. Boyle found that direct measurements of specific gravity were more reliable and consistent than measurements of the total salt content, by whatever method. During Boyle's time chemical knowledge was not sufficiently advanced that it was possible to determine the individual salts in seawater. Boyle used the word "saltness" to mean the concentration of salt in seawater, and this may be the first use of one word for that purpose.

Edmond Halley (1656–1742) was an English natural scientist of remarkable accomplishments. He was a mathematician, scientist, and navigator, and he became the Astronomer Royal of Great Britain. A friend of Isaac Newton, he urged Newton to publish the *Principia Mathematica*, edited the manuscript, and paid for its publication. His name lives on attached to "Halley's Comet," which he predicted would return each 76 years, and it did most recently in 1985–1986. Halley (1687) was the first to attempt a calculation of the hydrological balance for a part of the ocean. He measured the rate of evaporation of water from open pans and estimated that the Mediterranean must lose 53×10^9 tons of water per day in the summer and gain 1.8×10^9 tons per day from the inflowing rivers and rain. The difference between these numbers led him to comment that he might pursue further the question of what happens to the vapor, and the relation of this to the fact, known to mariners, that the current always sets inward through the Strait of Gibraltar; however, he never returned to the topic. His comments on this and other topics stimulated others to think about the world in rational and quantitative ways. Halley was perhaps the first person to think that the concentration of salt in the sea might be increasing with time, and that this increase might eventually be measurable. He urged that samples of seawater from various places be analyzed and the results recorded for comparison with analyses in future years (Halley 1715). Halley was concerned with the terribly difficult problem of determining the longitude of ships at sea, and commanded a vessel, the *Paramore*, on three voyages between 1698 and 1701 to plot the variations in the magnetic field over much of the North Atlantic. He wished to investigate whether the observed variations of magnetic north relative to the bearing to the pole star might provide a suitable method for determining longitude. These voyages were the first carried out purely for scientific studies of Earth.

Halley's influence extends on in other ways through more or less vague historical connections. He was concerned with making accurate measurements of the distance from Earth to the Sun, and in 1716 he wrote a paper calling attention to the value of making observations, at several locations around Earth, of the transits of Venus across the Sun, which he calculated were due in 1761 and 1769, long after he would have died. The British Admiralty fitted out the Transit of Venus Expedition in 1769 and sent it to the Pacific to make observations in the area around Tahiti. The name of the ship was the *Endeavour*, and the captain was James Cook. Some years later the *Endeavour* was sold into commercial service, and was sunk in the harbor of Newport, RI. The University of Rhode Island's research vessel, *RV Endeavor*, is named after Captain Cook's ship and carries as a talisman a piece of the wood originally believed to be from the original *Endeavour* (though now believed to be from Cook's other ship, *Resolution*, also sunk in Narragansett Bay [Abbass 1999]), and a sliver from *that* piece was carried by the space shuttle *Endeavour*.

It may not be possible to agree on who was the first true oceanographer, but one person who could claim the title was Count Luigi Ferdinando Marsigli (*aka* Louis

Ferdinand Marsilli, 1658–1730). In 1725, he published the first book entirely devoted to oceanography: *Histoire Physique de la Mer*. This book was the first to include sections on the sea's basin, on the water in it, on the movements of the water, and on marine plants and animals. He had often measured the saltness of the water from various locations, and this book contains the first recorded tabulated data of such characteristics. He collected water from various depths to determine the saltness (he used the word "salure"), and also found a way to measure the temperature down to some tens of meters. His was the first evidence for a seasonal thermocline. Marsigli was the first person to apply color tests to determine the acid or basic quality of seawater, and reported that seawater was basic. He also extended Boyle's observations on the unreliability of measurements of the salt content of seawater made by weighing the residue left after evaporation. Marsigli was concerned about the problem of the bitter taste of seawater and sea salt and, like many at that time, believed it was due to the "bitumin" material present in seawater.

As a young man Marsigli had accompanied the Venetian envoy to the Ottoman court at Constantinople. There he interested himself in many aspects of that exotic place, including the two-layer flow of water in the Bosporus, which he learned about from talking to fishermen and ship captains. In 1681, he produced the first clear explanation of this important phenomenon (Soffientino and Pilson 2005, 2009). Interesting aspects of his life are given in Stoye (1994).

Antoine Laurent Lavoisier (1743–1794) was the first chemist to discard the phlogiston theory and to substitute for it the concepts of oxidation and reduction; he was the first chemist to write real chemical equations, and his experiments were superbly quantitative. Lavoisier can be regarded as the person chiefly responsible for setting chemistry on its modern path. About a hundred years after Boyle's pioneering work, Lavoisier (1772) attempted to analyze the major components of sea salt. He evaporated seawater, and extracted the dried residue with alcohol and with alcohol–water mixtures and recrystallized the salts from the resulting solutions. By this approach he was able to identify several products (expressed here in modern terms):

$CaCO_3$	calcium carbonate
$CaSO_4{\cdot}2H_2O$	gypsum, calcium sulfate
$NaCl$	sodium chloride
$Na_2SO_4{\cdot}xH_2O$	Glauber's salt, sodium sulfate
$MgSO_4$	Epsom salt, magnesium sulfate
$MgCl_2$	magnesium chloride

This early attempt to analyze such a mixture of salts showed the complexity of seawater as a chemical system and defined the approximate amounts of the major components. The following quotation (Lavoisier 1772) suggests both his sense of the geochemical cycle and the complexity of seawater as a chemical system:

> The water of the sea results from a washing of the entire surface of the globe; these are in some ways the rinsing of the grand laboratory of nature; one therefore expects to find mixed together in this water all the salts which can be encountered in the mineral kingdom, and that is indeed what is found.

Lavoisier was the first person to realize clearly that the bitter taste of seawater and of sea salt was not due to some bituminous substance, but rather to the presence of $MgSO_4$ or $MgCl_2$ ("sel d'Epsom" or "sel marin a base de sel d'Epsom").

Lavoisier also made significant contributions to geology, agriculture, public administration, and the creation of the metric system, and was perhaps the greatest scientist of the time. He was a political moderate, and supported the early goals of the French Revolution. Unfortunately, he was born to wealth, he had inherited a minor noble title purchased by his father, and he was a partner in the Ferme Général, a private company that had collected certain taxes for the government under the regime of Louis XVI. During the last excesses of the Revolution he was guillotined. At the time his friend, Joseph-Louis Lagrange, said: "It took but a moment to cut off that head, though a hundred years perhaps will be required to produce another like it."

Joseph Louis Gay-Lussac (1778–1850) is perhaps best known for Gay-Lussac's law (for every degree temperature increase, gases expand 1/273 of their volume at 0 °C), which made possible the early attempts at devising an absolute scale of temperature. Gay-Lussac is responsible for the observation that oxygen and hydrogen react to form water exactly in the proportions of one volume of oxygen to two volumes of hydrogen; careful measurements of this sort were fundamental to the development of the atomic and molecular theories. He introduced the verb "to titrate" and developed volumetric methods of standardizing acids and titrating solutions of silver salts. This latter development was of exceedingly great importance in subsequent analyses of seawater.

Gay-Lussac investigated the concentration of total salt in the ocean, and showed that, while variations may exist in near-shore waters, and small variations in the open ocean, in general the concentration of total salt is remarkably constant throughout the Atlantic. He understood that such small variations as there were must be due to variations in river runoff and to differences between evaporation and precipitation. He also understood that the ocean currents would tend to mix the water and minimize differences.

The idea of a "column" of seawater, so universally used today in the phrase "water column," seems to have been his: "la salure d'une colunne d'eau de l'Océan" (Gay-Lussac 1817). He also considered whether settling of the salts might cause deep water to be more saline than surface water, and addressed the matter experimentally by keeping salt solutions in 2-meter-high glass tubes for 6 to 20 months, and then analyzing the top and bottom (Gay-Lussac 1817, 1819). He found no difference.[1]

[1] It is a minor curiosity of nature that salts could indeed be slightly more concentrated at the bottom of such a tube, where at equilibrium the gravitational tendency to settle is balanced against the effects of molecular diffusion. Pytkowicz (1962) calculated that the effect is about 0.18 km^{-1}, or 1.8×10^{-4} m^{-1}. If Gay-Lussac had started with seawater having a salinity of 35.000, and if he had been able to keep the tubes absolutely unstirred and with no convection for several years, the top of the tube might have had a salinity of 34.994 and the bottom a salinity of 35.006, a difference he could not have been able to measure. In the ocean the general water motions are much too great to allow the effects of molecular settling to become apparent, even in the most isolated basins.

During the middle of the nineteenth century there was a remarkable development of analytical technique and of chemical understanding. For some investigators, seawater, because of its complexity, was even then regarded as a challenge to the analyst, so a considerable number of analyses were published. Alexander Marcet (1819), having analyzed samples from the North and South Atlantic, and the Mediterranean and other arms of the Atlantic, reported: "... all specimens of seawater which I have examined, however different in their strength, contain the same ingredients all over the world, these bearing very nearly the same proportions to each other; so that they differ only as to the total amount of their saline contents." This statement was "probably the first suggestion of the relative constancy of composition of seawater" (Wallace 1974), and the so-called Principle of Constant Composition is often called Marcet's Principle. In 1851, the German chemist Ernst von Bibra published analyses of five samples from the Atlantic, three from the Pacific and one from the North Sea (von Bibra 1851). All samples showed essentially the same relative proportions of the component substances.

The first book describing the physical characteristics of the ocean that achieved a large popular audience was Matthew Fontaine Maury's *Physical Geography of the Sea*, first published in 1855, widely read at the time, and influential in spreading oceanographic knowledge. The book was influential in part because of the fame of the author, who was well known and honored for his pioneering work (while superintendent of the US National Observatory) in compiling and publishing his wind and current charts of the oceans and accompanying sailing directions. These were of immense benefit to seamen of all nations, and to all sea-going commerce.

In his book Maury stated, probably on the evidence from von Bibra, that "if we take a sample of ... the average water of the Pacific Ocean, and analyze it, and if we do the same by a similar sample from the Atlantic, we shall find the analysis of the one to resemble that from the other as closely as though the two samples had been from the same bottle after having been well shaken." (Maury 1855, p. 152). He drew from this evidence the conclusion that the ocean currents must be sufficient to move and mix the whole ocean throughout the world, and expressed the idea succinctly as: "seawater is, with the exceptions above stated, everywhere and always the same, and ... it can only be made so by being well shaken together." (Maury 1855, p. 151.)

Since that time the notions of the general near-constancy of composition of seawater, of ocean mixing, and of the importance of ocean currents in carrying water around the world, have become generally appreciated. The total salt content and specific gravity were also known to vary regionally, but only to a very small extent. This principle of constant composition, however, was really established on a secure foundation by Johan Georg Forchhammer, a prominent Danish geologist and a superb chemical analyst (Forchhammer 1865). He believed that to understand the formation of sediments on the sea floor it was necessary to have an accurate knowledge of the chemical components in the water. He analyzed about 260 samples of seawater from all the major regions of the ocean and even the Caspian Sea. He introduced the practice of reporting a ratio of the concentration of each of the major elements to that of chloride, and demonstrated the remarkable constancy of these ratios. He introduced the use of the word "salinity" to characterize the total salt content of a sample of seawater.[2] He also reported a semi-quantitative test for the organic matter in seawater.

Following this work, the major highlight of chemical oceanography was the *Challenger* expedition. This was not the first oceanographic expedition, but it was the biggest and most comprehensive up until that time. It was the most scientifically well conceived and it is often held to have marked the beginning of oceanography as a scientific discipline. In 1871, a committee of the Royal Society formulated the goals of what was to become the *Challenger* expedition, as follows (Yonge 1972):

(1) To investigate the *Physical Conditions* of the *Deep Sea*, in the Great Ocean-basins . . . in regard to Depth, Temperature, Circulation, Specific Gravity and Penetration of Light . . . at various ranges of depth from the surface to the bottom.
(2) To determine the *Chemical Composition* of *Sea Water* . . . at surface and bottom . . . [and] various intermediate depths . . . to include the Saline Constituents, the Gases, and the Organic Matter in *solution*, and the nature of any particles found in *suspension*.
(3) To ascertain the *Physical* and *Chemical* characters of the *Deposits* everywhere in progress on the Sea-bottom, and to trace, so far as may be possible, the sources of these deposits.
(4) To examine the Distribution of *Organic Life* throughout the areas traversed, especially in the *deep* Ocean-bottoms and at different depths; with especial reference to the Physical and Chemical conditions already referred to, and to the connection of the present with the past condition of the Globe.

The sense that these studies are properly pursued together and at the same time, and that there is some unity to the discipline of oceanography, received here its first formal expression. In modern evolving formulations, which have greater explicit emphasis on subsequently discovered processes and global changes, these still remain the goals of modern oceanographic institutions. The idea that those who study the ocean should know something of all these aspects of the subject is perhaps traceable to that committee of the Royal Society.

In 1872, the British Government outfitted the *HMS Challenger* to sail around the world investigating the sea. During this expedition, which lasted from 1873 to 1876, people on the *Challenger* collected samples that laid the groundwork for the modern study of physical, chemical, biological, and geological oceanography. Many of the results of the *Challenger* expedition are still referred to.

Many of the seawater samples collected by the *Challenger* were analyzed by the German chemist William Dittmar, working in Glasgow. By this time, chemical analytical technique had sufficiently advanced that Dittmar (1884) was able to report such complete and well-regarded analyses of seawater that not until the 1960s were any subsequent samples of seawater so thoroughly analyzed. Since the

[2] The word *salinity* was actually used by Franck (1694, p. 181) in an obscure and idiosyncratic publication, so it may have had some use in conversation in England; I have not come across it in technical publications. Forchhammer would have been familiar with the use of the word *salure* in French, and seems to have deliberately and independently introduced the word *salinity* into what became an important and widely read paper; the word was used universally thereafter, at least in English.

publication of Dittmar's work we have known reasonably well the ratios, one to another, of the major constituents of seawater, and the adjustments since have been relatively minor.

After the *Challenger* expedition and the work of Dittmar, the history of oceanography is essentially modern. Many of the results since then can be considered as modern chemistry, or at least as more or less valid data. In the chapters that follow we will have occasion to refer to information collected during approximately the last hundred years, although, of course, by far the greatest proportion of our knowledge is from the last few decades.

1.3 Major features of ocean circulation

The distribution of substances within the sea is controlled in part by the general features of ocean circulation, and can be understood only with some knowledge of these in mind. Conversely, much of what we know about the deep circulation of the oceans has come from examining and interpreting the distributions of various physical and chemical properties. It is beyond our scope here to discuss in detail the currents in the sea, or how our knowledge of them has been won; for this information the reader should consult books of general oceanography or texts of physical oceanography. Here we can present only a general outline of the currents in the ocean, and the forces that cause them.

Early observations of ocean currents by fishermen and other sailors provided a general appreciation of some of the major surface currents, so that even several hundred years ago the better navigators routinely took advantage of them. For example, the Florida Current, as the Gulf Stream between Florida and the Bahamas is called, was first described by Juan Ponce de León in 1513, and by 1519 Spanish galleons commonly rode the Gulf Stream north and northeast before turning east on their way back to Spain. Later observations by oceanographers using increasingly sophisticated procedures have led to a quite detailed knowledge of the ocean surface currents throughout the world (Figure 1.1), and their variability. Some currents are narrow and reasonably well defined, others are broad general drifts; some are relatively shallow, reaching only a few hundred meters in depth, others may extend to the bottom of the sea. All can vary in strength and location and some show convoluted meanders. Velocities may vary from a few centimeters per second up to perhaps 250 cm per second (about five knots). Because these currents may be hundreds of meters deep and many kilometers wide, they transport immense volumes of water. A unit introduced by Max Dunbar (1967) and commonly used to express these transports is the Sverdrup (Sv), equal to 1 000 000 m^3 per second. The flow of all the world's rivers combined is about 1.2 Sv, while the flow of the Gulf Stream, for example, varies according to location from about 30 to about 100 Sv (Knauss 1997). The flow of possibly the largest ocean current, the Antarctic Circumpolar Current, may be 180 Sv through the Drake Passage (Grose *et al.* 1995). Many currents are much smaller.

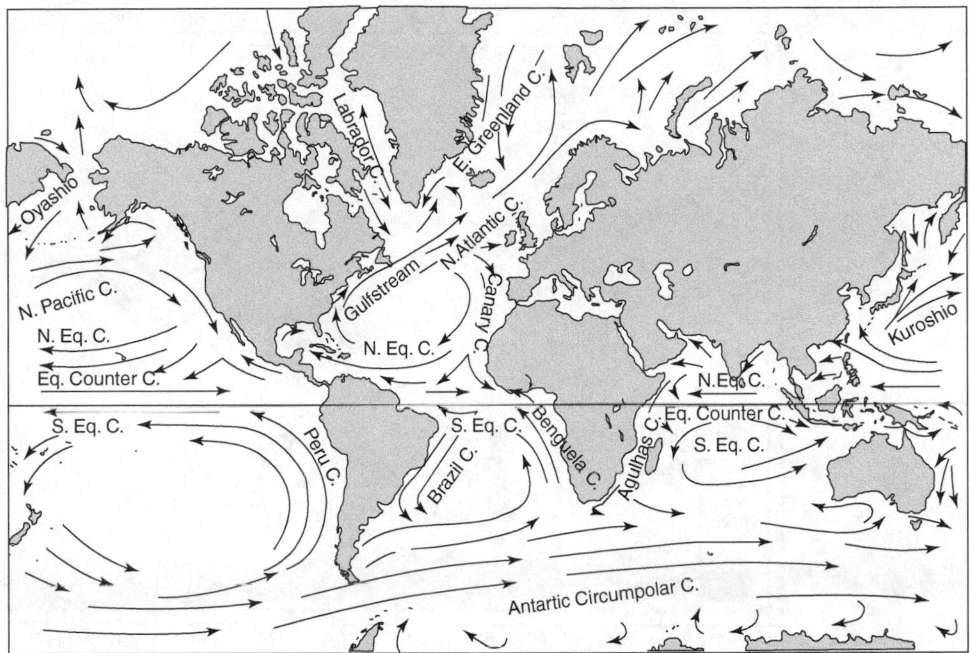

Figure 1.1 Simplified presentation of the major surface currents in the world ocean. Redrawn from Knauss (1997).

While the ocean currents transport great quantities of water from place to place, transport of water through the atmosphere has important consequences in addition to the obvious one of providing rain for land areas. Consider the distribution of salinity. Looking at Table 1.3, you may have wondered why the Atlantic is saltier than the Pacific. The oceans have been mixing since the world began. What now keeps the Atlantic saltier? This can only happen if there is a continuing input of salt to the Atlantic, or a continuing removal of fresh water. The latter appears to be the cause. Examination of a map of the world to discover where the fresh water might be removed from the Atlantic reveals few possibilities, and some surprises.

Most of the land on Earth drains via rivers into the Atlantic Ocean (Figure 1.2). The distribution of drainage areas is controlled by the distribution of mountain ranges which (because of the present arrangement of the motions of Earth's crustal plates) are characteristically close to the ocean around the Pacific, so that warm moist air from the Pacific normally cannot penetrate very far over land before being forced to rise, cool, and lose its water as rain, whereas humid Atlantic air can go farther before losing its water. In both cases, much of the water evaporated from the ocean returns to the same ocean. How then can the Atlantic be saltier? There are two places where the distribution of mountains and winds allows warm humid air to leave the drainage basin of the Atlantic. The mountains of Panama and Nicaragua are low, so in that region and nearby areas the prevailing easterlies can carry humid air from the Atlantic to the Pacific, bearing water at an estimated

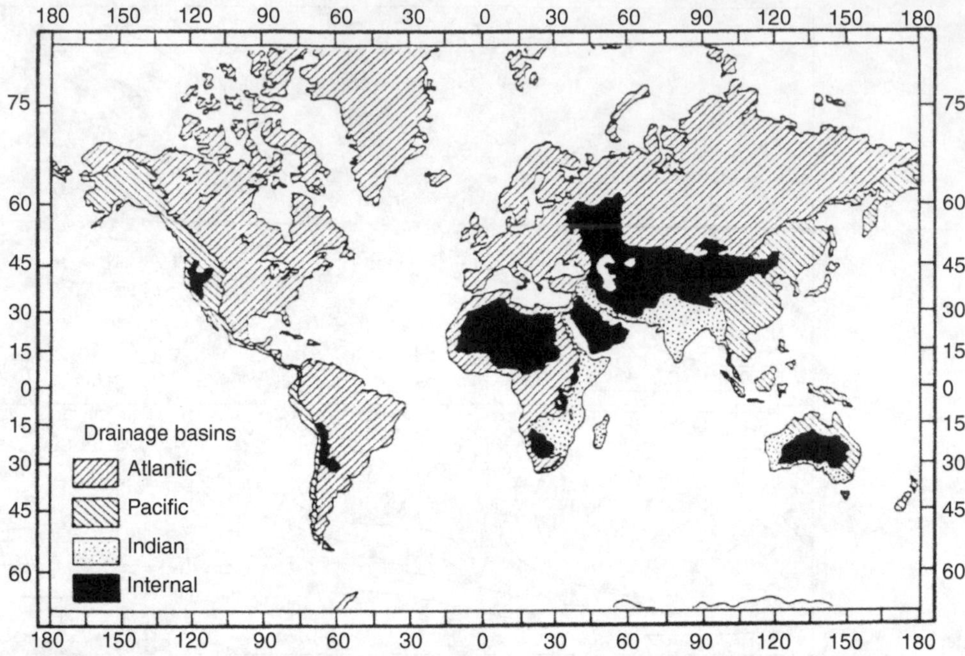

Figure 1.2 Drainage areas of the continental regions, showing discharge by ocean basin (Broecker 1989). Areas labeled "internal" have no drainage to the ocean; these may drain to internal seas or salt lakes, if they have enough water to drain at all. Areas draining to the Arctic Ocean are included in the Atlantic drainage because seawater flows out of the Pacific into the Arctic, and out of the Arctic into the Atlantic.

average rate of 0.36 Sv (Broecker 1991). Similarly, the prevailing westerlies carrying evaporated water from the Mediterranean can cross into the Pacific drainage basin before losing their water; this transport of water from the Mediterranean is about 0.075 Sv (Sverdrup *et al.* 1942), and the total loss across Europe to the Pacific drainage basin is about 0.25 Sv (Broecker 1991). Other atmospheric transports bring small amounts of water from the Pacific into the Atlantic, so that the net loss from the Atlantic amounts to about 0.32 Sv. The ocean currents also play an important role, but calculation shows that atmospheric transport alone is sufficient to maintain the Atlantic as a whole in a slightly saltier state than the Pacific. The overall balance is maintained by large surface flows of slightly diluted salt water from the Pacific through the Bering Straits to the Arctic and thence to the Atlantic, from the Pacific around Cape Horn, and from the Indian Ocean around the Cape of Good Hope.

Living organisms are present at all depths everywhere in the ocean, and most of these require oxygen. The only source of this oxygen is at or near the surface, but oxygen is present nearly everywhere in the ocean, so we know that the deep water must be renewed from time to time with water from the surface. Tropical surface water is warm, and in temperate regions the surface is moderately warm, but deep water everywhere is cold, only a few degrees above freezing. It follows that most deep water must have entered the depths by sinking in polar regions.

The deepest water is the most dense, so only in those regions where surface water becomes exceptionally dense can this water sink all the way to the bottom. Other regions may form water slightly less dense; this water can sink to intermediate depths. The density of seawater increases as the temperature drops and as the content of salt increases (the influence of salt is relatively more important at low temperatures). Therefore, those regions where the water is both cold and salty will make the most dense water. At the present time there are only two regions where large volumes of water become dense enough to sink to the bottom: the northern parts of the North Atlantic Ocean and the Weddell and Ross seas near the coast of Antarctica.[3] In the north, the Norwegian, Greenland, and Irminger seas east of Greenland, and the Labrador Sea west of Greenland are regions where large amounts of cold dense water are made in the winter. Relatively salty surface water originating in the tropical Atlantic and transported north by the Gulf Stream makes its way into these northern regions. When it cools in the winter it becomes as dense as water anywhere in the major ocean basins, and it sinks there and makes its way south, mixing with somewhat lighter water on the way. This mixture is called North Atlantic Deep Water (NADW). As it moves south it encounters and mixes with several other sources of deep water (Figure 1.3). Antarctic Bottom Water (AABW), colder and denser there because it has not yet been so much diluted, lies under and mixes upwards into NADW. Antarctic Intermediate Water (AIW), formed by cooling of surface water but not so dense as NADW, lies above the latter water mass and mixes down. The resulting modified NADW mixes into the Antarctic Circumpolar Current at mid depths.

Very little deep water is currently made in the North Pacific.[4] The reason is that surface water in the Atlantic is saltier, and therefore more dense when it gets cold, than surface water in the Pacific. The deep Pacific is filled with water that originates in the North Atlantic, mixes with water sinking around Antarctica, and then flows along the bottom into the Pacific and Indian oceans, and this water is denser than the less saline surface water in the North Pacific can become when it cools in the winter.

A very general schematic picture of the major circulation pathways of surface to deep water and back to the surface again is shown in Figure 1.4. This generalized pathway was drawn to show the very broadest features of oceanic water movements and to show how this circulation might lead to the various observed gradients in the concentrations of substances that are observed in the ocean.

Many intermediate-density waters are formed in different places, sinking to intermediate depths. Here we mention only three. Mediterranean water is not very cold, but it is salty enough to be denser than water in the Atlantic. As Mediterranean water

[3] For additional perspective, however, see Appendix Figure C.1.1

[4] It is conventional to state that no deep water is made there. In fact, however, there is some evidence that very small amounts of water dense enough to flow all the way to the bottom may be made in the winter by freezing of sea ice over the Bering Sea shelf. This water has been traced by the presence of Freon (CFC-11) in the deepest bottom water of the Aleutian Basin (Warner and Roden 1995). This chemical has been introduced much too recently to have otherwise found its way into the deep water of the Pacific and thence to the Aleutian Basin.

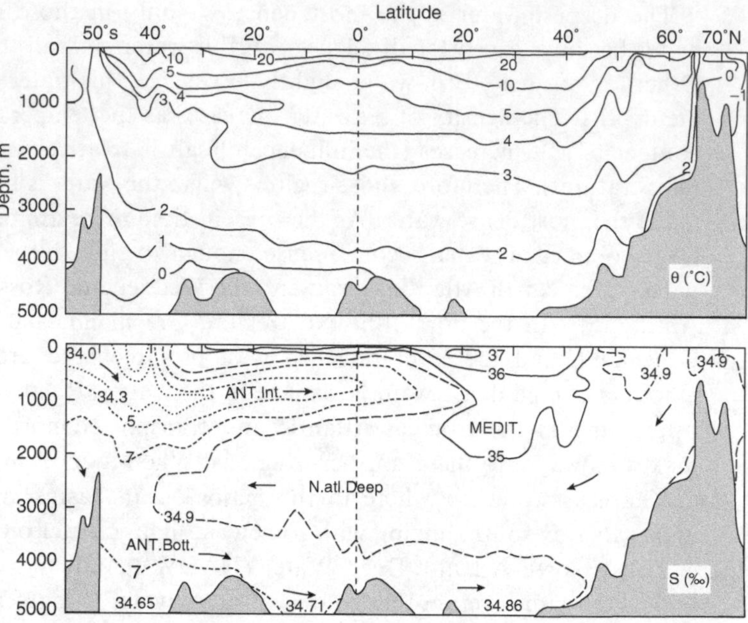

Figure 1.3 Cross section of the Atlantic Ocean from south to north, showing some of the general features of the deep circulation. The top panel is temperature, the bottom panel is salinity. Cold water formed east and west of Greenland sinks and flows along the western slope into the basin of the North Atlantic. The deep water mass filling much of the basin of the North Atlantic is known as North Atlantic Deep Water (NADW); at depths of 2 to 4 km it flows towards the south. Cold water formed under the ice near Antarctica sinks and flows to the north; this is called Antarctic Bottom Water (AABW). Being denser than NADW, the AABW flows underneath NADW, gradually mixing with it and eventually losing its identity. Less-dense water formed by cooling around 50° S called Antarctic Intermediate Water (AAIW) sinks and moves north overlying and also mixing with the NADW. Modified from Pickard and Emery (1990).

pours out through the Straits of Gibraltar it cannot, however, make it all the way to the floor of the Atlantic because it mixes with Atlantic water as it flows down the slope, so it ends up as water of an intermediate density that spreads out across the Atlantic at depths between about 1000 to 2000 m. The line identifying the salinity of 35‰ (Figure 1.3) is rather convoluted in the region west of Gibraltar, due to the influence of Mediterranean water there. As this water mixes gradually into the rest of the Atlantic it contributes to the saltiness of this ocean. Antarctic Intermediate Water (Figure 1.3) is formed in the South Atlantic, and is slightly less dense than water formed in the North Atlantic, so, as mentioned already, it lies over the water coming down from the north. A similar body of water, North Pacific Intermediate Water, is formed in the North Pacific in the wintertime, with a density sufficient to allow it to reach depths of about 800 m.

Figures such as 1.3 and 1.4 might give the misleading impression that these deep flows are smooth, regular, and slow. In fact, the sources are sporadic, the flows often channeled, some close to the sea floor, as deep currents that waver and snake around

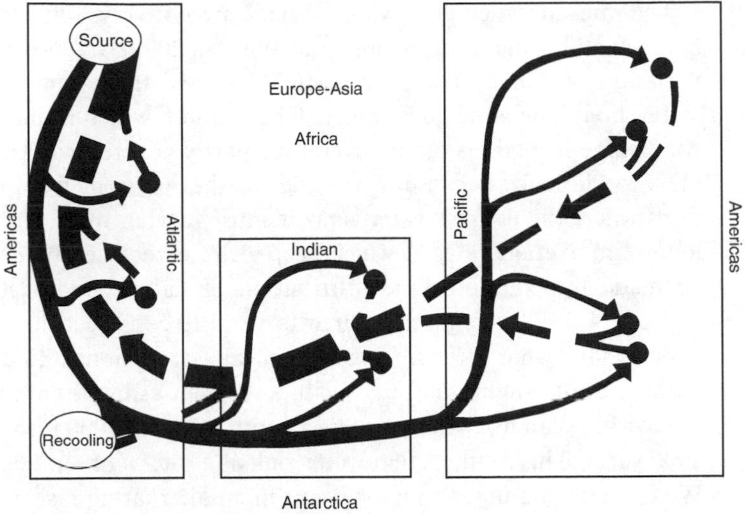

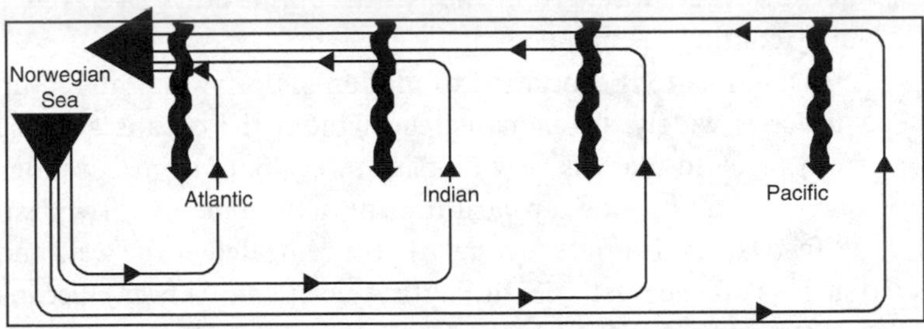

Figure 1.4 This figure, from Broecker and Peng (1982), is a simple and generalized presentation of the circulation of water in the world ocean and the resulting redistribution within the ocean of many biologically transported substances. Water sinks in the North Atlantic and flows south in the western Atlantic at a rate of about 15 Sv. This NADW mixes into the vigorous circulation around Antarctica and along with very cold AABW formed under the ice shelves contributes to the formation of a further 15 to 20 Sv of dense bottom water. The return flow comes from upwelling throughout the ocean basins and especially in several regions on the eastern sides of the basins (small circles in the upper panel) and along the equator. The return flow (dashed lines) eventually returns water on the surface to those places where it sinks again. During its long path the deep water continually receives falling debris from the surface, and thus is continually enriched with nutrients and other substances. This diagram captured in a visual presentation the gross features of the ocean circulation and material transports that lead to major chemical differences between the ocean basins. The upper part of the diagram has been redrawn, often in much more elaborate form, and reused dozens of times in professional and popular publications. The details of the flows and processes involved are, however, immensely complex; they continue to be the subject of intense investigation.

over the bottom, and there are large eddies that circle and move through the water column at all depths. These all make it difficult to gain a detailed knowledge of movements in the deep water, though the general features such as shown in the figures are reasonably well known.

The rates at which deep water is made are not very well known, and the estimates here are still rough (Ganachaud and Wunsch 2000). Inter-annual- and decadal-scale variations are likely (Dickson *et al.* 1988), and the common assumption of steady state should be used judiciously. The densest water is made sporadically in the wintertime in regions often covered or partly covered with ice, where observations are very difficult. Because of varying weather patterns from year to year, the locations where the densest water is made must vary, as must the rates. Estimates of the long-term average rates at which deep water is replaced by newly sinking water can be made by examining the distribution of natural radioactive tracers, especially carbon-14. It is not entirely straightforward to construct these estimates, and results depend somewhat on various geochemical assumptions and definitions, so different authors quote slightly different values. Present estimates are that the deep water in the world ocean as a whole, below a depth of about 1000 meters, is replaced in about 1000 years. This requires that water sink at a rate of about 30 Sv, annually averaged. We can imagine that this must occur in an alternating fashion, mostly in the North Atlantic in the northern winter, and around Antarctica in the southern winter. In fact, however, water from the Nordic Basin flows over sills into the Atlantic in a fairly continuous manner.

In contrast to the localized formation of deep water, the return of deep water to the surface is widely disseminated throughout the oceans and the variation of its rate from place to place is very difficult to evaluate from oceanographic measurements. Some occurs by slow upward transport combined with widespread mixing with the surface layers. It is not particularly concentrated at the well-known upwelling regions (such as off the west sides of South America and Africa), because the water upwelling in these regions comes mostly from relatively shallow depths. Possibly some return of deep water occurs by mixing towards the surface in the Pacific and Indian oceans, but it is known that a great deal of deep water from the Pacific returns to the south and is entrained into the Antarctic Circumpolar Current where it encounters vigorous vertical mixing during the circuit around Antarctica.

Summary

Most of the water on Earth is seawater, and this is what marine chemists are concerned with. The properties of this water vary only over very narrow ranges in the open ocean (temperature –2 °C to 30 °C and salt content of 34.3 to 35.1 grams per kilogram), so measurements to distinguish different water masses must be extremely precise. There are a few exceptional circumstances where the temperature may exceed 400 °C, and here the chemistry may be quite exotic.

The composition of seawater was considered by the ancient Greeks, but recognizably scientific studies began with Robert Boyle in the seventeenth century. Early contributors included Edmond Halley, Count Luigi Ferdinando Marsigli, Antoine Laurent Lavoisier, Joseph Louis Gay-Lussac, Alexander Marcet, and William Dittmar. After Dittmar, in the late nineteenth century, the gross composition of seawater may be considered as quantitatively established.

The Atlantic Ocean is saltier than the Pacific (and Indian), due to the net removal of fresh water from the Atlantic by evaporation and atmospheric transport to the Pacific. This circumstance is responsible for much of the large-scale hydrographic structure and water transport within the oceans, and consequently for the distribution patterns of many substances dissolved in the water.

The large-scale circulation of the ocean is characterized by the sinking of relatively salty and cold water in the Nordic seas that contributes to North Atlantic Deep Water, which then spreads down and is entrained into the great Antarctic Circumpolar Current and contributes to the deep water of the Pacific. A roughly equal quantity of deep water is contributed by complex mixing and re-cooling in the Antarctic currents combined with very cold water influenced by freezing under the ice shelves there. Antarctic Bottom Water is made there and flows along the bottom into both the Atlantic and Pacific. The deep water of the world ocean is thus replaced in approximately 1000 years, at a rate of about 30 Sv.

The great surface currents flowing around the central gyres in the major ocean basins transport immense quantities of water, heat, salt, and other substances.

SUGGESTIONS FOR FURTHER READING

Deacon, Margaret B., ed. 1978. *Oceanography, Concepts and History*. Dowden, Hutchinson & Ross, Stroudsburg, PA.

Deacon, Margaret B. 1971. *Scientists and the Sea: A Study of Marine Science*. Academic Press, New York.

These two books by Margaret Deacon are excellent sources for much of the early history of oceanography. The latter text, a real classic, was re-issued, with some corrections and addenda, by Scholar Press in 1997.

Knauss, John A. 2005 *Introduction to Physical Oceanography*, 2nd edn. Waveland Press, Long Grove, IL.

An excellent general text in the field.

Riley, J. P. 1965. Historical introduction. In *Chemical Oceanography*, 1st edn., vol. **1**, ed. Riley, J. P. and G. Skirrow. Academic Press, New York, ch. 1, pp. 1–41.

One of very few places to find information on the history specifically of chemical oceanography.

Sverdrup, Keith A., Alyn C. Duxbury, and Alison B. Duxbury. 2005. *Introduction to the World's Oceans*, 8th edn. McGraw-Hill, New York.

Everyone beginning in oceanography is advised to read one of several good general undergraduate-level texts. This one is quite satisfactory.

2 The water in seawater

. . . the earth and its atmosphere constitute a vast distilling apparatus in which the equatorial ocean plays the part of the boiler, and the chill regions of the poles the part of the condenser. TYNDALL 1896

Water makes up more than 96% of the total mass of seawater, so it is appropriate to begin with a short discussion of the properties of water itself. The weather and climate on Earth are in several ways controlled by the physical properties of water. Factors of special importance are its high heat capacity, its high heat of evaporation and condensation, its high heat of freezing and thawing, the molecular structuring associated with its expansion when freezing, and the relationship of its vapor pressure to temperature. In addition, its infrared absorption spectrum causes water vapor to be the most important greenhouse gas, and its radiative properties cause water to be important in the radiation of heat away from Earth. Variations from place to place in its isotopic composition provide insights into several aspects of Earth science. The chemical property of water that is perhaps most important to the marine chemist is its high solvent power for polar substances and substances that form charged ions in solution.

2.1 Physical properties of water

We often think of water as a typical liquid. It is certainly the most common liquid around, but it is not typical, as far as its chemical and physical properties are concerned, compared to other liquids.

Most of the properties of water could not have been easily predicted from the known relationships between the properties of related liquids. For example, Figure 2.1 shows that the boiling point of water is unusually high. This very high boiling point suggests that the water molecules must be exceptionally attracted to each other, and this strong tendency for the molecules to hang together is also advanced to explain many of the other anomalous properties of water (Table 2.1).

Table 2.1 Unusual properties of water

Property	Comments
High boiling point High melting point High specific heat High heat conductivity High heat of evaporation High heat of melting	All these properties relating to heat cause water to be important in moderating temperature extremes, and in transporting heat from place to place around Earth.
Maximum density at 4 °C	Contrast between oceans and lakes
High surface tension	Droplet formation in clouds, breaking waves
High viscosity	Biologically important
High dielectric constant	Makes a good solvent for ionized substances

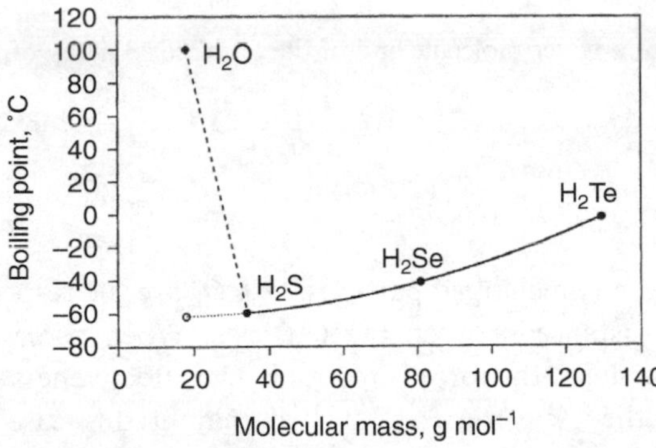

Figure 2.1 Relationship of boiling point to molecular weight for the hydrides of the elements (tellurium, selenium, sulfur, and oxygen) belonging to Group 16 in the periodic table. By extrapolation from the first three, one might plausibly predict the boiling point of water to be about –62 °C, whereas it is really 100 °C. ■

2.1.1 Water molecules

These unusual properties ultimately result from the shape of the molecule itself, as shown schematically in Figure 2.2a. In a water molecule, the two hydrogen atoms are arranged at an angle of about 104.5° to each other. There is a small tendency for the electrons to be concentrated towards the oxygen and away from the hydrogen atoms, so that the weighted average locations of negative and positive charges are separated, the molecule is a dipole, and the molecules orient in an electric field. This feature of the molecule also leads to the rather strong tendency for hydrogen bonds to form, as shown in the schematic of a group of molecules. This results in a relatively high freezing point, and in strong associations between the molecules in the liquid state as well.

The separation of the positive and negative charges results in water having a very high dielectric constant. The high dielectric constant is partly responsible for the strong solvent power that water has for ionized substances. This may be understood by noting that the force between two charged particles in a vacuum is given by

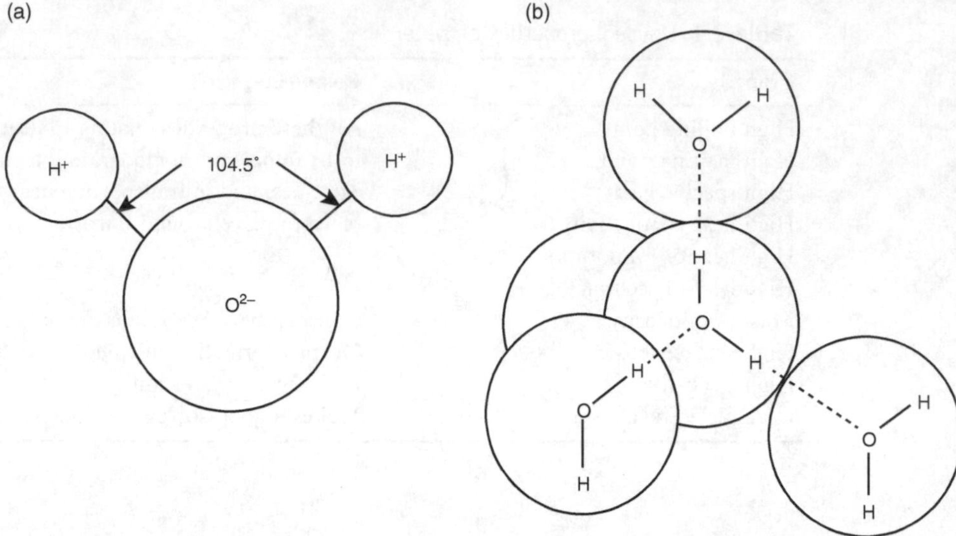

Figure 2.2 Schematic views of a water molecule and of the tetrahedral array of water molecules in ice.

$$F \propto \frac{Q^+ Q^-}{r^2}, \qquad (2.1)$$

where F is the force between two charged particles, Q^+, Q^- are the respective charges and r is the separation distance between the particles. When some substance is interposed between the particles, the force is reduced. The effectiveness of the reduction is designated by ε, the dielectric constant, so that in this case the force is expressed as

$$F \propto \frac{Q^+ Q^-}{r^2 \varepsilon}. \qquad (2.2)$$

Since water has an extremely high dielectric constant, it quite effectively shields oppositely charged particles from each other, thus reducing the attractive force between them. This is one reason why water is such a good solvent for substances that dissociate into oppositely charged ions.

Oxygen atoms in the water molecule have unpaired electrons; their presence enables a water molecule to coordinate with the hydrogens of the other water molecules. This, and the high dipole moment, leads to the strong association of the water molecules with each other. In turn, this leads to the low vapor pressure and the relatively high boiling point.

The angle between the hydrogens on the water molecule leads to a three-dimensional arrangement of the molecules in an especially open, loosely packed structure, whose nature in the liquid state is not completely understood. Whatever the exact arrangements, the molecules take up more space than would be predicted from a knowledge of their size and the assumption of a random array.

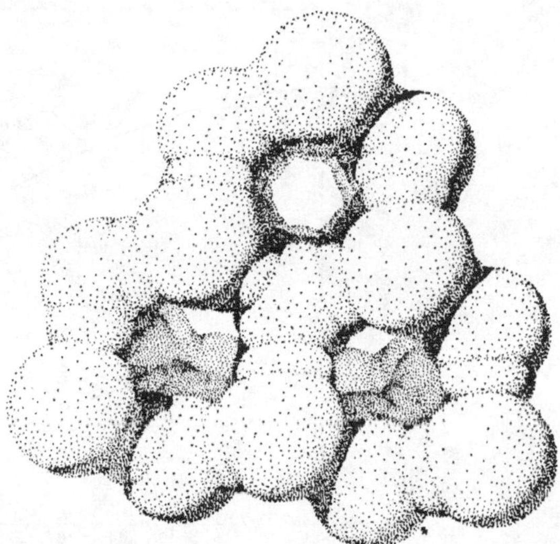

Figure 2.3 Diagram of the arrangement of the water molecules in ordinary ice. Modified from Pimentel and McClellan (1960). In this view (down the *c*-axis of the crystal) the open structure, and thus the low density, of the ice is emphasized. The open spaces within the ice appear to be large enough that the smallest noble gases, helium and neon, might be accommodated without distorting the crystal lattice (Kahane *et al.* 1969). Thus, helium at least, and possibly neon, may be more "soluble" in ice than in liquid water. I have not found a recent evaluation of this suggestion.

2.1.2 Density

When ice is formed, the water molecules become oriented into a regular tetrahedral structure, shown in Figures 2.2b and 2.3, which occupies even more volume than in the liquid state. Water is one of the very few substances that increase in volume upon freezing. The increase is about 9% (thus reducing the density by about 90 kg m^{-3}), so ice floats. Were this not the case, the climate and other aspects of the physical nature of the Earth would be vastly different.

Figure 2.4 shows the relationship between density and temperature for both pure water and seawater. A peculiarity of pure water is that as it is cooled the density reaches a maximum at 3.98 °C, well above the freezing point, and then decreases slightly towards the freezing point. This phenomenon has led to the speculation that even in liquid water some of the molecules are arranged in an ice-like structure, thought of as miniature "icebergs," and that as water is cooled the fraction of the molecules arranged in this way increases, thus causing a decrease in density. The most common arrangement is thought to be as regular pentamers (Liu *et al.* 1996). These icebergs cannot be permanent. They have been called "flickering clusters," and they are believed to be transient associations which, as temperature and pressure go up or down, change in statistical frequency of occurrence, and thus proportionately change the overall properties of the water. One result of this

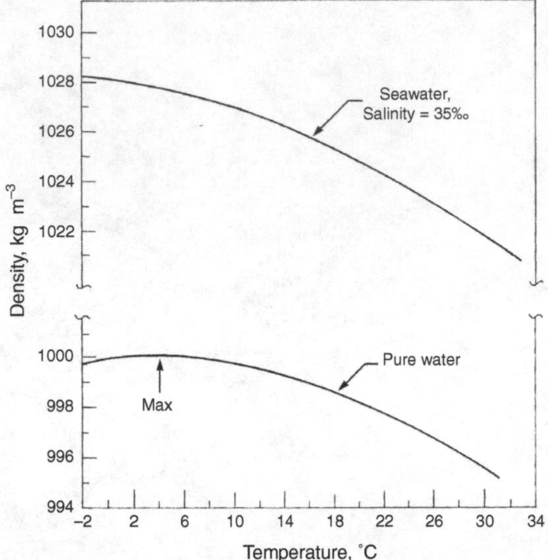

Figure 2.4 Relationship between density and temperature at 1 bar total pressure for pure water and for seawater at $S = 35‰$. In fresh water the maximum density is at 3.98 °C. When a lake, for example, cools to this point the densest water fills the lake at this temperature from top to bottom. With further cooling the water becomes less dense and floats on the top. Ice is even less dense and floats above this. Deep lakes therefore cannot get colder than 3.98 °C on the bottom, unless vigorously mixed by the wind. In contrast, seawater continues to become denser down to its freezing point, so water just at the freezing point may, if the circumstances allow, sink all the way to the bottom.

phenomenon is that after a fresh-water lake cools to 4 °C in the wintertime, progressively colder water is less dense and floats on the surface, where it may eventually freeze, and the deep water cannot not cool below 4 °C unless the lake is shallow and physically mixed.

The presence of salt in the water must disrupt this tendency toward an ordered arrangement; the more salt the lower the temperature of maximum density. Figures 2.4 and 2.5 show that in full-strength seawater this maximum is below the freezing point. This phenomenon is important in the ocean because it leads to the necessity for the entire water column to cool to the freezing point before freezing begins (provided the distribution of salinity allows this).

2.1.3 The vapor phase

In our common experience water evaporates and ice sublimes, so we know that both forms of water have an appreciable vapor pressure. As we see in Figure 2.6, the vapor pressure increases with temperature, and equals the pressure of the standard atmosphere at sea level (101.325 kPa = 1.01325 bar, or 760 mm of Hg) at 100 °C, which with the freezing point forms the basis of the centigrade or Celsius thermometric scale.

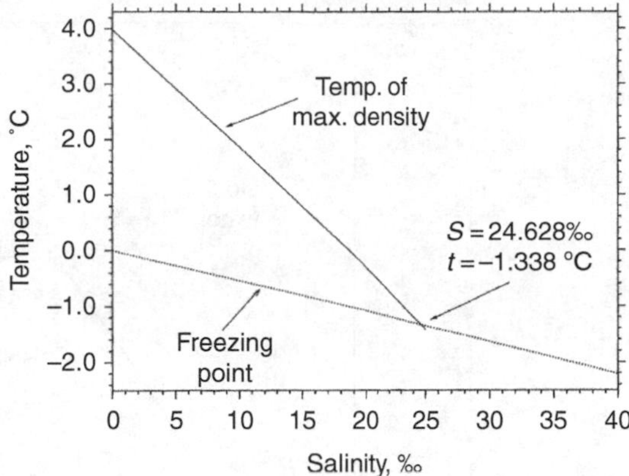

Figure 2.5 The temperature of maximum density and the freezing point are each plotted as a function of the salinity of seawater. The temperature of maximum density is calculated from the relationship in Box 3.3, and the freezing point as in Appendix Table C.2. The presence of salts in the water disrupts the tendency of water molecules to associate in low-density arrangements; as salt concentration increases, the temperature of maximum density moves to colder and colder temperatures. The graph shows that at all salinities greater than 24.63‰ there is no maximum in the density before the water freezes. This situation applies throughout nearly all the surface of the ocean, except for low-salinity or brackish regions such as coastal ponds and estuaries with salinities less than 24.63‰.

If pure water is heated above 100 °C, with just sufficient pressure maintained to prevent it all turning to vapor, the liquid continues to expand, and any vapor in equilibrium with it becomes more dense. At 100 °C water has a density of 958.39 kg m^{-3} (specific volume 1.0434 m^3 t^{-1}). Figure 2.7 shows that as the temperature increases the density changes become greater and greater. For example, at a temperature of 373 °C the density drops to approximately 402 kg m^{-3} (specific volume about 2.49 m^3 t^{-1}). Maintaining water in the liquid state at this temperature requires a pressure of about 218 bar (about 215 atmospheres). With a further 0.98 °C rise in temperature, the density, already dramatically low, drops further, to 322 kg m^{-3}, and the pressure is 220.6 bar.

At this point, called the *critical point* or *critical temperature*, a remarkable change takes place. The liquid and the vapor now have exactly the same density, and are completely miscible. The system has only one phase, and as temperature increases it behaves like a gas. Above the critical temperature of 373.98 °C, no matter what the pressure, liquid water does not exist; this is the *super-critical* state, and the water is called "super-critical water." This transition has been very well studied with pure water, due to the importance of understanding the properties of steam during the transmission of heat and the generation of power. Less is known about the behavior of salt solutions in the super-critical region, because it is both theoretically and experimentally difficult to evaluate the solubility of salts and the thermodynamic

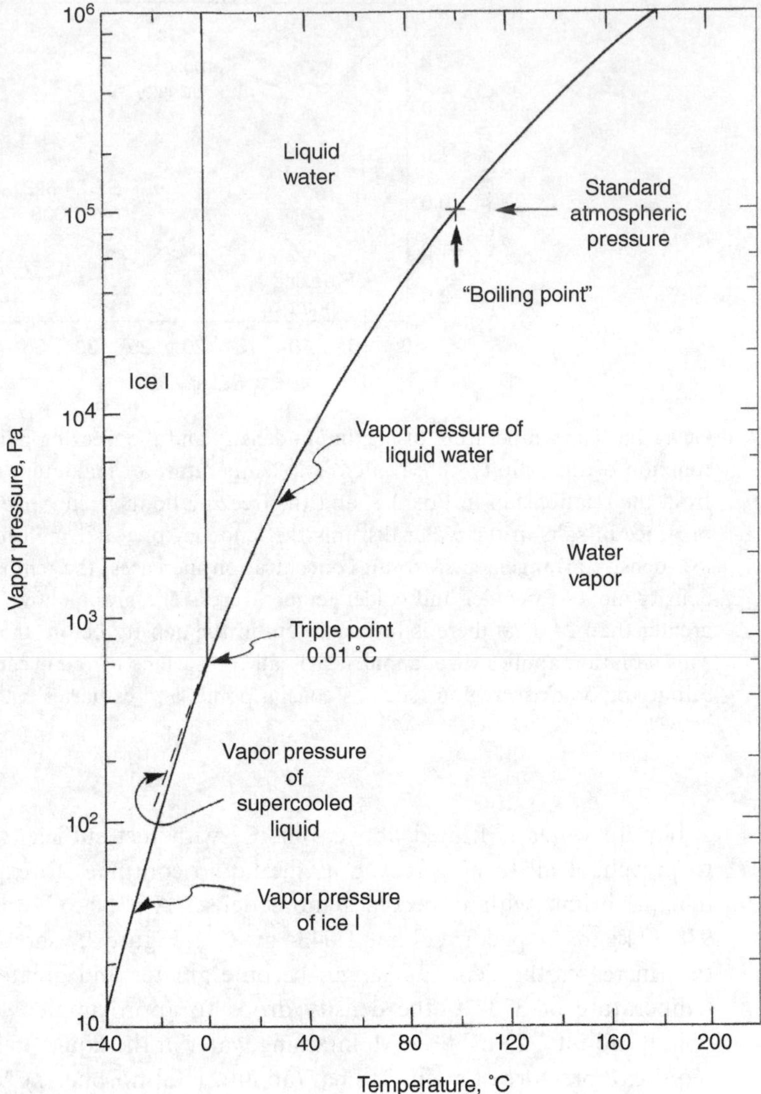

Figure 2.6 Pressure–temperature relationships of water in the vicinity of the triple point, which is a unique point where ice, water, and vapor can co-exist in equilibrium. Measurements have placed the triple point at +0.01 °C. The ice of common experience is ice I; under more extreme conditions several other forms may exist. (From data in the *CRC Handbook of Chemistry and Physics* (Haynes 2011).)

properties of solutions at such elevated temperatures and pressures, but there are good data for solutions of sodium chloride.

The behavior of NaCl and water at high temperatures and pressures was extensively investigated by Sourirajan and Kennedy (1962), who examined the concentration range from pure water to saturated solutions, and temperatures up

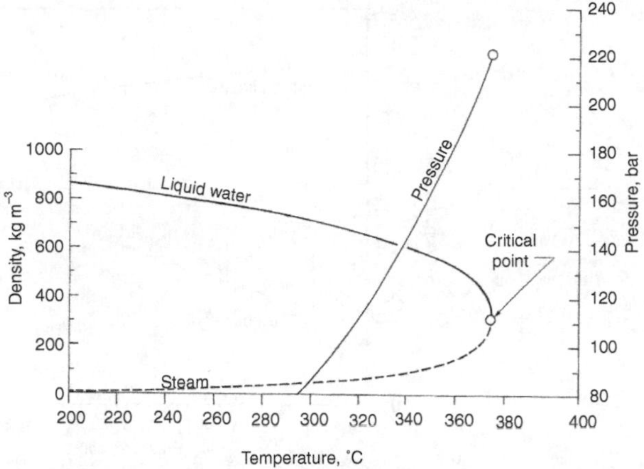

Figure 2.7 Relationship of density to temperature for liquid water and steam between 200 °C and the critical point at 374 °C, when the pressure at each point is maintained just equal to the vapor pressure of the water for each temperature. Plotted from data in Haar *et al.* (1984). The required pressure, so that liquid and vapor phases are both present, is plotted only for the region above 300 °C up to the critical pressure of 221 bar (= 22.1 MPa). Note that the densities of steam and liquid water become equal (= 322 kg m^{-3}) at the critical point, and the two phases become one. At higher temperatures there is only super-critical water, and liquid water does not exist, no matter what the pressure.

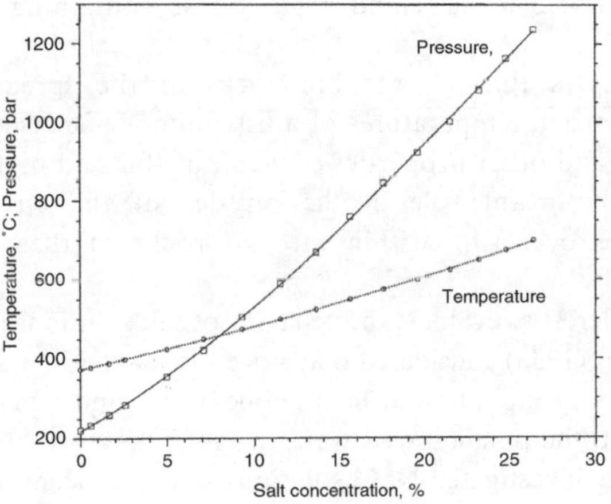

Figure 2.8 Effect of NaCl in water on the critical temperature and critical pressure of the solution. Plotted from data in Sourirajan and Kennedy (1962). The points yield the following equations: $T = 0.093x^2 + 9.794x + 374.246$; $P = -0.010x^3 + 0.801x^2 + 24.193x + 218.223$.

to 700 °C. In the presence of NaCl the critical temperature and pressure are elevated, so that, for example, at a concentration of 5% salt the critical temperature is 425 °C and the critical pressure about 356 bar (Figure 2.8). Above this point there is only a super-critical substance that behaves like a gas and contains 5% NaCl and 95% water. The presence of salt introduces additional behavior not found in pure water. In order to understand the situation around the critical point in salt solutions, imagine a super-critical solution with ~5% salt

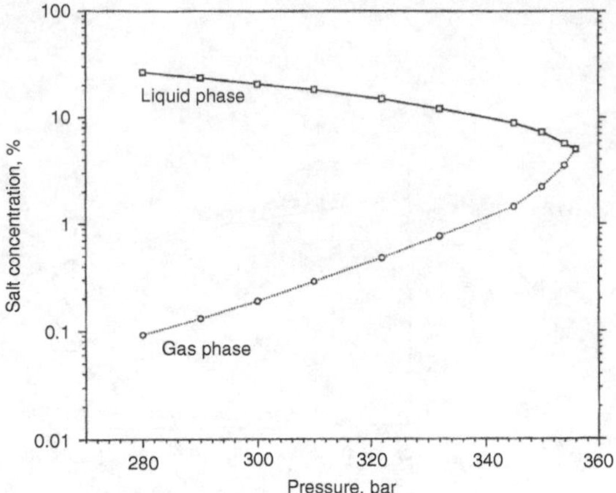

Figure 2.9 Concentrations of salt in liquid and gas phases. Plotted from data in Sourirajan and Kennedy (1962). In this case the total concentration of NaCl in the water was 5.0 weight percent, the temperature was held constant at the critical value of 425 °C, and the pressure was varied. Above the critical pressure of 356 bar there is only one phase; this behaves as a gas and has a salt concentration of 5.0%. Below the critical pressure of 356 bar there are two phases with salt concentrations that vary with pressure.

just at the critical point (Figure 2.9). The temperature is 425 °C. If the pressure is now reduced to 350 bar *with no change in temperature*, two phases will appear, a liquid phase with about 7.2% NaCl and a gas phase containing about 2.2% NaCl.

Seawater that circulates through the hot rocks at the spreading centers is known to sometimes reach temperatures of a little above 400 °C. The dramatically changing density and other properties of water at this and higher temperatures must play an important role in the physics of the circulation and the chemical exchanges occurring within the hot rocks in these remarkable regions.

It is of considerable interest to evaluate the behavior of salt solutions in this region. Bischoff and Rosenbauer (1988) considered that since the magnesium and sulfate in seawater are lost both by precipitation at high temperatures and by chemical reactions with the hot basalt, the remaining seawater would behave rather like a 3.2% solution of NaCl, so they investigated NaCl solutions at this concentration in some detail. They established that for such a solution the critical point is at 407 °C and 298.5 bar. In considering the thermodynamic properties of seawater circulating past hot basalt deep below the sea floor they concluded that seawater must approach the critical point, and also that separation of phases could well occur deep within the conduits in the rocks. This may account for the observation that in some samples of the hot water exiting at the spreading centers the concentrations of salt are very different from that of the ambient seawater in those regions (see further discussion in Chapter 13).

2.2 Isotopes of hydrogen and oxygen

The heaviest water must fall first. *Aristotle*

In ordinary natural water the hydrogen and oxygen each consist of three different isotopic forms with different masses; this fact is of some geochemical significance. The isotopes present in water are:

$$^{1}H, ^{2}H \text{ or } D, ^{3}H \text{ or } T, ^{16}O, ^{17}O, ^{18}O.$$

The first isotope of hydrogen has a nucleus consisting of only a single proton, and accordingly the mass number is 1. The second isotope has both one proton and one neutron in the nucleus, and the mass number is 2; this isotope has a special name: "deuterium," hence the common use of the letter "D" to represent the isotope. The third isotope has one proton and two neutrons, a mass number of 3, and also has a special name: "tritium." Tritium is radioactive, and its discussion is deferred to Chapter 10. The nuclei of oxygen have 8 protons, and 8, 9, or 10 neutrons. These isotopes are present in all possible combinations, so that (not including molecules with tritium) all natural water contains nine kinds of water molecules (Table 2.2). The ratios of the different isotopes, one to another, can be measured with remarkably high precision using an isotope ratio mass spectrometer.

The different isotopic forms of water each have different vapor pressures and freezing points (Table 2.3); these physical differences are important because they make it possible to use the isotopes as tracers of geochemical processes. The differences in vapor pressure lead to a fractionation of the isotopes whenever water evaporates or condenses. The heavier isotopes are, in every case, concentrated into the liquid phase. This leads to differing isotopic compositions in water from different sources. When water freezes, there is also a fractionation such that the heavier isotopes are concentrated into the solid phase, although the effect is less than in the gas–liquid exchange processes.

2.2.1 Isotope standards

The differing isotopic composition of "tap water" in different parts of the world, as well as the changes caused by distillation, led to a need for a standard water to which all measurements could be related. This is especially important because, for technical reasons in mass spectrometry, it is possible to determine relative isotopic ratios much more accurately than absolute isotopic concentrations. To establish the needed standard, investigators at the National Bureau of Standards in Washington distilled a large sample of water from the Potomac River. This distilled water is distributed in small volumes as the primary standard termed NBS-1. This standard is the ultimate reference for the mass spectrometer determination of both hydrogen and oxygen isotopes in water. It is quite different in composition from the water in seawater, however, so it is inconvenient to use as a reference when presenting oceanographic data.

Table 2.2 Isotopic composition of VSMOW

	Number of molecules in a total of 10^{12} molecules
$H_2{}^{16}O$	997 318 000 000
$H_2{}^{18}O$	1 999 800 000
$H_2{}^{17}O$	370 885 000
$HD^{16}O$	310 685 000
$HD^{18}O$	622 985
$HD^{17}O$	115 379
$D_2{}^{16}O$	24 196
$D_2{}^{18}O$	49
$D_2{}^{17}O$	9

The absolute composition was measured by Hagemann *et al.* (1970) to give a D/H ratio of 155.76×10^{-6}, and by Baertschi (1976) to give a $^{18}O/^{16}O$ ratio of 2005.2×10^{-6}. In each case the last significant figure is quite uncertain. Together with an estimate that ^{17}O is about 0.0371% of the oxygen in VSMOW, the approximate concentrations of the different molecules in typical seawater may be calculated. (Note that a kilogram of seawater at $S =$ 35‰ contains about 3.225×10^{25} molecules of water.)

Table 2.3 Some characteristics of the different isotopic forms of "pure" water

Form	FP, °C	BP, °C	Temp. of max. density, °C	Max. density, kg m^{-3}	VP at 20 °C, Pa
"Ordinary water"	0.00	100.00	3.98	999.975	2338
D_2O	3.81	101.40	11.2	1106.0	2140
$D_2{}^{18}O$	–	–	11.45	1216.88	–
T_2O	–	–	13.40	1215.01	–
$HD^{16}O$	–	–	–	–	2170
$H_2{}^{18}O$	–	–	4.21	1112.49	2316

Data from Kell (1972) and Haynes (2011). Included are the freezing point (FP), boiling point (BP), temperature of maximum density, the maximum density, and the vapor pressure (VP) in pascals. In the cases of "ordinary water," D_2O, and T_2O, the complete isotopic composition was not specified in the sources, but one might assume that ordinary water is more or less typical distilled tap water.

The standard originally chosen for oceanographic use (Craig 1961b) was close to a volume-weighted average isotopic composition for deep water from the Atlantic, Pacific, and Indian oceans, and was called:

"Standard Mean Ocean Water," or SMOW.

No actual sample of SMOW ever existed; Harmon Craig simply calculated the numerical values from early data characterizing the major oceans. He defined

SMOW relative to the distilled water standard (NBS-1), according to the following relationships:

$$\left(\frac{D}{H}\right)_{SMOW} = 1.050\left(\frac{D}{H}\right)_{NBS-1}, \tag{2.3}$$

$$\left(\frac{^{18}O}{^{16}O}\right)_{SMOW} = 1.008\left(\frac{^{18}O}{^{16}O}\right)_{NBS-1}. \tag{2.4}$$

In order to provide a practical standard for use in mass spectrometric measurements, the International Atomic Energy Agency (IAEA) in Vienna prepared a considerable volume of distilled water adjusted to have an isotopic composition as close as possible to that of SMOW. This is distributed as VSMOW (Vienna-SMOW), and is now the international standard (Gonfiantini 1978, Coplen 1994). Estimates of the absolute ratios of the isotopes in VSMOW are given in Table 2.2. There is, for example, one atom of D for each 6420 atoms of H. Another standard is "Standard Light Antarctic Precipitation," or SLAP, prepared from ice in central Antarctica, in which there is one atom of D for each 12 192 atoms of H, or an absolute D/H ratio of 82.02×10^{-6} (Hagemann *et al.* 1970), not much more than one-half that in VSMOW. Isotopic abundances of D and ^{18}O in natural waters are now reported relative to VSMOW, and measured with mass spectrometers that should also be calibrated with SLAP (Coplen 1994).

2.2.2 Fractionation

When water evaporates from the ocean, the vapor is commonly depleted by about 0.8% in ^{18}O relative to the water left behind (Craig and Gordon 1965). Suppose the starting ratio of ^{18}O to ^{16}O in the seawater is the same as in VSMOW, namely 0.0020052, or 1/498.70, then the value in the vapor would be 0.0019892 or 0.992/498.70. These numbers are awkward to use, and it is more convenient to express the concentration as a difference from the arbitrary standard, VSMOW. Since the differences are small, it is usual to express them in parts per thousand, as in the following example:

$$\frac{R_{sample} - R_{VSMOW}}{R_{VSMOW}} \times 1000 = \delta \tag{2.5}$$

where R is the ratio of D/H or $^{18}O/^{16}O$, and δ (delta or "del") is a conventional notation to express the differences. In the example given above, the ^{18}O concentration in the first vapor appearing when the water evaporates would be expressed as:

$$\frac{\frac{0.992}{498.7} - \frac{1}{498.7}}{\frac{1}{498.7}} \times 1000 = -8.00‰.$$

We would say that the vapor is "0.8%," or "8‰" or "8 per mil" depleted in ^{18}O relative to the liquid water, or, in the usual phrasing, that it is characterized by $\delta^{18}O = -8.00$.

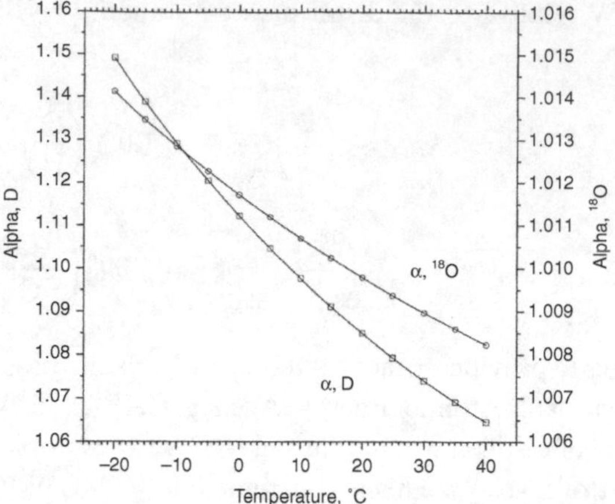

Figure 2.10 Effect of temperature on the fractionation factors which are observed for the isotopes of hydrogen and oxygen when water vapor and liquid water are in equilibrium, plotted from relationships developed by Majoube (1971). Here the fractionation factors are defined as: $\alpha_D = (D/H)_{liquid}/(D/H)_{vapor}$ and $\alpha_{O-18} = (^{18}O/^{16}O)_{liquid}/(^{18}O/^{16}O)_{vapor}$ and the algorithms are: for D, $\ln \alpha = 24\,844/(T^2) - 76.248/T - 0.052612$; for ^{18}O, $\ln \alpha = 1137/(T^2)$ $0.4156/T - 0.0020667$; where T is in kelvin (= °C + 273.15). Note the different scales; the fractionation factor for D/H is roughly ten times greater than that for $^{18}O/^{16}O$. During the actual processes of evaporation or condensation in nature the vapor and liquid will often not be in equilibrium, so additional differences due to kinetic effects are also expected and observed.

The fractionation that occurs during evaporation or condensation is different for each isotope, and depends on temperature, and even the temperature dependence is somewhat different for each isotope (Figure 2.10). In addition to the equilibrium relationships shown in Figure 2.10, a kinetic effect due to the higher mobility (leading to faster diffusion) of the lighter isotopes is also important. Because the latter effect cannot be known *a priori*, and must vary, it is necessary to measure the actual resulting empirical effect. Despite potential sources of variability, the fractionations of the oxygen and hydrogen isotopes are strongly correlated in ordinary atmospheric processes, and the relationship

$$\delta D = 8 \times \delta^{18}O + 10 \tag{2.6}$$

holds for most samples (Figure 2.11). Where water is fractionated by geological processes, however, the fractionations may be different, as noted in Box 2.1

The distillation apparatus referred to by Tyndall in the opening epigraph causes a considerable latitudinal fractionation of the isotopes in water. The tropical regions of the ocean are generally characterized by net evaporation relative to precipitation. The vapor which leaves the surface is initially depleted by about 8‰ in ^{18}O, relative to ocean surface water. The vapor is transported away from the tropical areas in the atmosphere. If the air cools and rain falls out, the rain will be enriched in ^{18}O relative to the residual vapor, and consequently the vapor will be depleted in this heavy

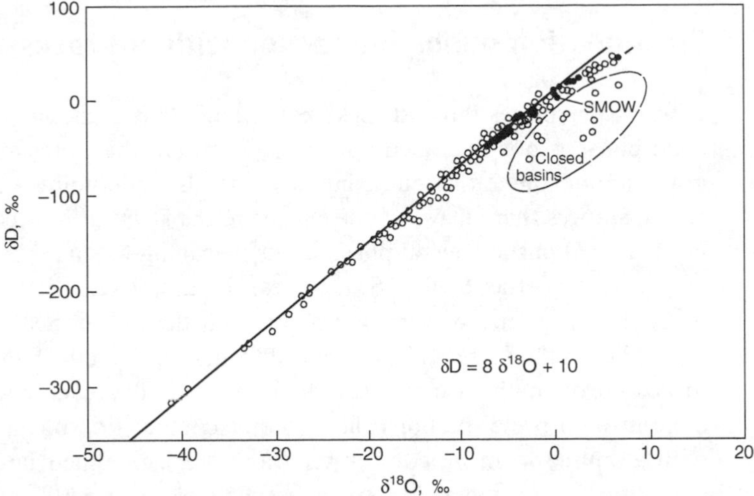

Figure 2.11 Relationship between δD and $\delta^{18}O$ in a wide range of precipitation, lake, and river waters. These samples were collected from many parts of the world and at different seasons of the year. Extreme evaporation products from places like the Dead Sea or Great Salt Lake were believed to be exceptions, but subsequent investigation showed that such waters pose analytical problems, and also that the thermodynamic activities of the water molecules are sufficiently modified that this effect must be taken into account in evaluating the fractionations. When these problems are addressed, and corrections applied, the concentrations fall back into line (Horita and Gat 1989). There are some exceptions to the simple straight line relationship; the points on the dashed line at the upper end of the samples are from rivers and lakes in East Africa. The relationship is robust enough, however, that it has been generally used ever since this graph was published by Craig (1961a). The total range shown here (about 50‰ in $\delta^{18}O$ and 400‰ in δD) approximates the maximum found in ordinary natural waters. The highest values tend to be found in warm, saline ocean water and in salt lakes. The lowest values are observed in snow falling on the coldest regions of Antarctica.

isotope. If the air mass then strikes a mountain and rises and cools, losing more rain, the vapor will be further depleted in ^{18}O. The same will happen if it is carried to the polar regions. The result of this is that there is a remarkable relationship between the average temperature of a location and the average ^{18}O content of the local precipitation (Figure 2.13). The relationship is good enough that it is possible to assign ancient temperature values to the snow that eventually made ice recovered from cores deep in the Greenland and Antarctic ice sheets. Additional factors and local relationships may have to be accounted for in dealing with conditions on these ice caps; a further uncertainty is introduced because it is likely that the pathways and fractionations that operate today may not have been exactly the same during earlier and different climate regimes. It is evident in Figure 2.13 that there is also a strong correlation with latitude, and in fact precipitation in all far northern and far southern regions is greatly depleted in ^{18}O.

The continuous atmospheric transport of water vapor depleted in ^{18}O, from the tropics towards the poles, results in a general tendency for seawater also to show a

Box 2.1 Fractionation during interaction with hot rocks

If water percolates through rocks hot enough that exchange with the oxygen in the solid phase is possible, then the $^{18}O/^{16}O$ ratio in the water may change, ultimately approaching the ratio found in the minerals undergoing exchange. Under these circumstances there may be little change in the D/H ratio, because there is relatively little H or D in the mineral phases. This reasoning was used by Craig (1966) to trace the origin of the Salton Sea brines. Figure 2.12, shows a plot of the δD, $\delta^{18}O$ characteristics of various waters from the Salton Sea geothermal area. This region lies in the Imperial Valley of California, right on the San Andreas Fault. Wells drilled through the sedimentary floor of the valley encounter extremely hot and concentrated brine solutions. The isotopic composition of the water was consistent with a source from meteoric water (rainwater), modified under the extremely hot conditions by an increasing exchange of the oxygen with sedimentary or intrusive rocks so that the water is enriched with ^{18}O. Most rocks have a $\delta^{18}O$ in the range of +5 to +12, relative to VSMOW (Faure 1986). The D/H ratio appears unaffected because of the comparatively small amount of hydrogen in typical rocks. Similar exchanges must take place as seawater circulates through hot rocks at the ocean spreading centers, and this should have some effect on the composition of ocean water.

slight enrichment of ^{18}O in tropical regions (to the extent that salinity is enhanced by evaporation) and a depletion of ^{18}O in polar regions (to the extent that it is diluted with precipitation water), and consequently there is, generally, a relationship to salinity for each region considered (Figure 2.14).

2.2.3 Isotopes as tracers

The variation of isotopic composition from place to place suggests that this parameter could be useful as a tracer for ocean mixing processes. In principle it can be, but the variations are generally small within the ocean itself, and co-vary with salinity. Figure 2.15 shows that different major oceanic regions do show different trends in the relationship between salinity and ^{18}O, due to regional differences in various factors, of which the most important may be the pathways by which evaporated water is carried from and back to the ocean. Temperature and salinity can be determined very precisely and much more easily than the isotopic ratios, so in many circumstances the isotopic composition is not a useful tracer.

There is, however, one circumstance in which the ^{18}O to salinity relationship can be a sensitive and important tracer. When water freezes, the heavier isotopes are slightly concentrated in the ice. The equilibrium fractionation factor is about 0.9970, but in practice it appears that in ice that is freezing out of seawater there is a kinetic limitation to the fractionation, so that a fractionation factor of about 0.9974 is observed (Macdonald *et al.* 1995). This fractionation is relatively small, but in some

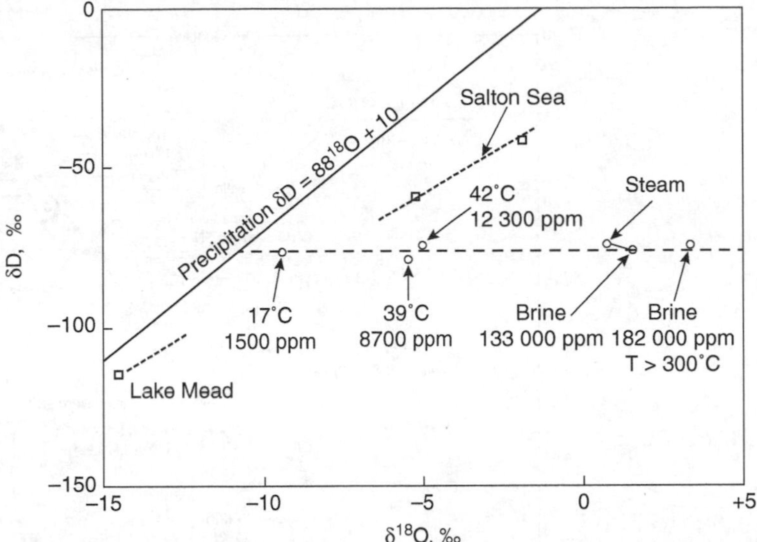

Figure 2.12 Isotopic composition of waters in the Salton Sea geothermal area: temperatures and chloride concentrations (parts per million) shown for subsurface waters (open circles). Modified from Craig (1966). The water of the Salton Sea itself (upper two squares; the composition varies with distance from inlet) comes from the Colorado River at Lake Mead via irrigation ditches; the short dashed line for this system has a slope of 5, possibly because of kinetic effects in evaporation and molecular exchange with atmospheric water vapor in this salty enclosed basin. The interpretation of the data along the horizontal dashed line is that the geothermal waters are derived from local precipitation with $\delta^{18}O$ of about −11‰, which is then modified by exchange of oxygen isotopes with the local rocks; in order of increasing oxygen-18 enrichment they are: a spring at the base of the Chocolate Mountains east of the Salton Sea, water issuing from "mud volcanoes" on the east shore, water from old CO_2 production wells about 300 m deep, and two geothermal wells. (See Box 2.1.) ■

circumstances it may provide a unique tracer, because it is opposite to the effect of evaporation and causes the water left behind to depart from the general salinity to ^{18}O relationship (Figure 2.15). An important use of this isotopic tracer is to examine the extent to which a water mass may contain contributions of brine left when ice has been frozen out of seawater.

Consider the example shown in Figure 2.16. The $\delta^{18}O$ vs. salinity values for North Atlantic Deep Water (NADW) all fall close to a single point in Figure 2.14. Neither the salinity nor the $\delta^{18}O$ show much variation, at least on the scale used here. This is consistent with the view that it is possible to form NADW just by cooling North Atlantic surface water. In fact, the greatest influence on NADW is from surface water cooled in the Norwegian Sea, subsequently entering the North Atlantic over the sill in the Denmark Straits, and additional water formed in the Labrador Sea. The $\delta^{18}O$ vs. salinity data for the greatest mass of water in the ocean, the Pacific and Indian Ocean Deep Water (P&IDW), all fall well away from the surface-water relationships, so it appears impossible just by cooling to make P&IDW from either

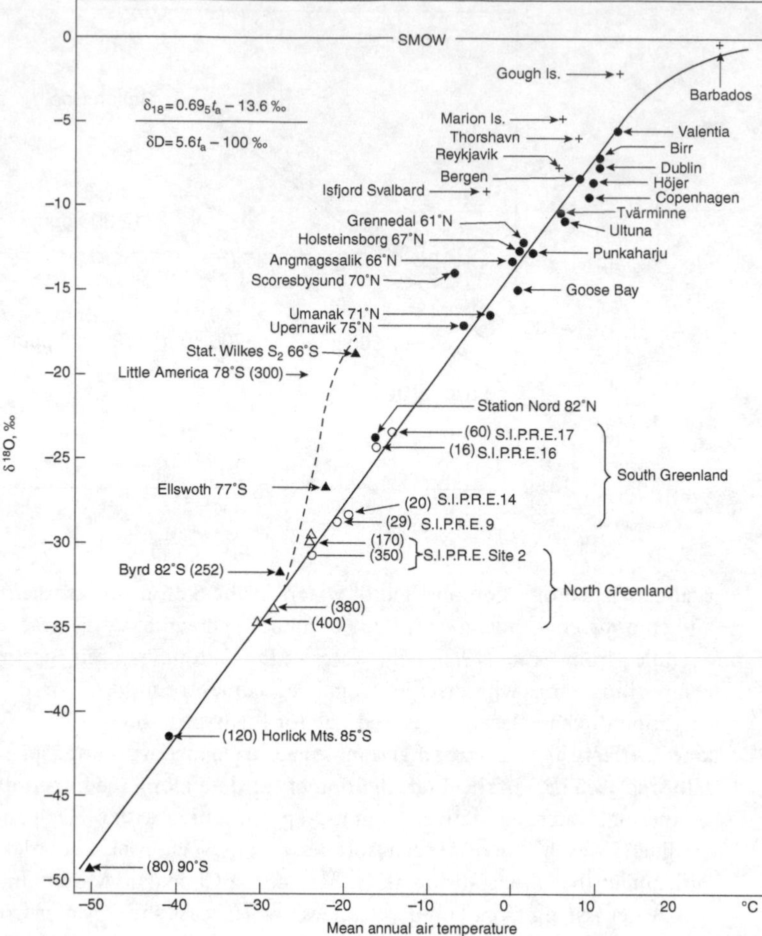

Figure 2.13 Annual mean $\delta^{18}O$ in precipitation as a function of the annual mean air temperature at the location where the precipitation (rain or snow) was collected. The ordinate is $\delta^{18}O$, the abscissa is mean air temperature in °C. The straight part of the line has the equation: $\delta^{18}O = 0.695T - 13.6$, where T is in °C. The comparable equation for δD is also given. The numbers in brackets at some of the Greenland and Antarctic stations refer to the thickness in centimeters of the snow layer collected for analysis. When this graph (redrawn from the original) was first published by Dansgaard (1964) it became evident that it is possible to establish the approximate temperature of a location from a measurement of the isotopic composition of the water falling there. Of greatest importance, it became possible to estimate the temperature of regions where ice accumulates, as far back in time as ancient ice can be dated. The relationship is evidently to some extent characteristic of a particular location, as can be seen by the samples from parts of Antarctica, so for any detailed evaluation at a particular location the local relationship needs to be established. It is still somewhat of a question how closely the relationships may hold back through time. ■

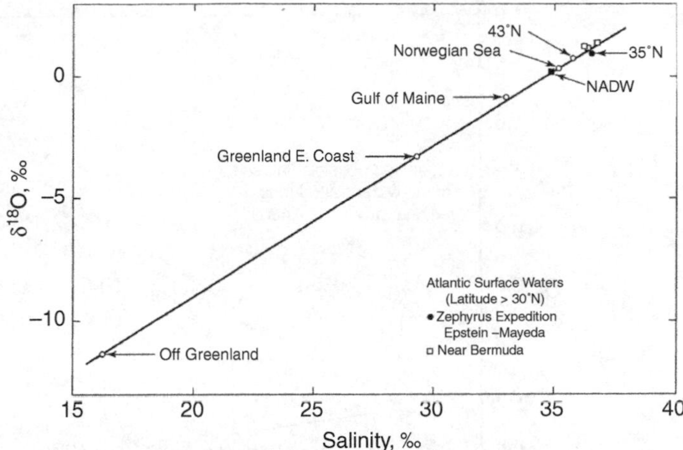

Figure 2.14 Relationship between $\delta^{18}O$ and salinity for selected samples of North Atlantic surface waters and the North Atlantic Deep Water (NADW). From Craig and Gordon (1965).

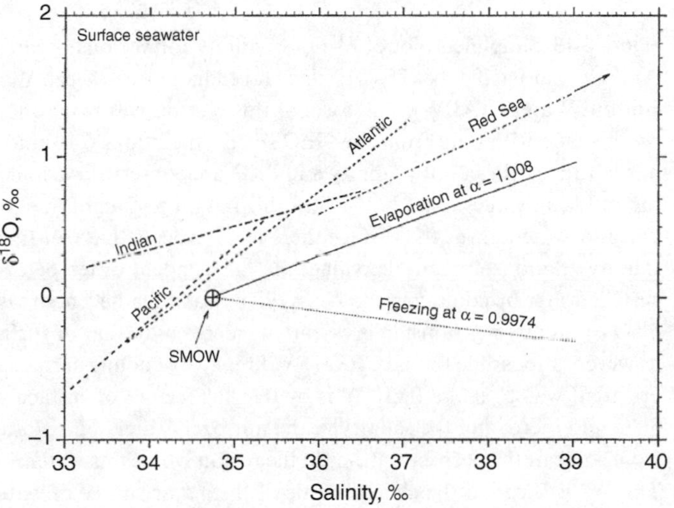

Figure 2.15 Relationship between $\delta^{18}O$ in seawater during evaporation or freezing. The world average salinity, appropriate for the water originally defined as SMOW, was taken as a convenient starting point for calculating the change of $\delta^{18}O$ and salinity in the liquid phase during progressive evaporation and freezing of seawater. An α value of 1.008 during evaporation was used, as this is thought typical of water vapor coming from warm seawater (Dansgaard 1964); depending on temperature and kinetics a range of values must occur in nature. The α value of 0.9974 during freezing was the average difference in $\delta^{18}O$ observed between seawater and ice forming on the Beaufort shelf and in the Mackenzie estuary (Macdonald *et al.* 1995). The four dashed lines are linear fits to the observations from various regions. Data for the Red Sea are from Craig (1966); observations there extend nearly to a salinity of 41‰. Surface-water observations for the Indian, Pacific, and Atlantic oceans, omitting some high-latitude values, are from the GEOSECS survey (Ostlund *et al.* 1987).

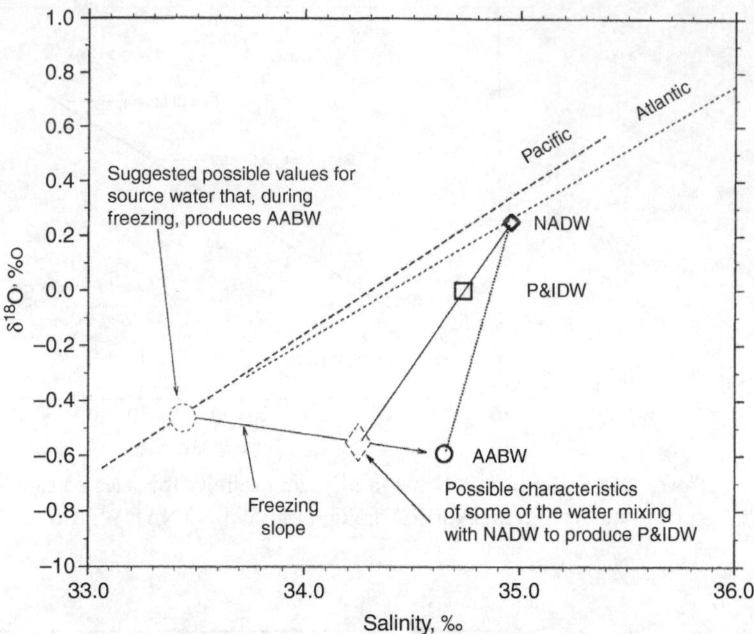

Figure 2.16 Simplified plot of $\delta^{18}O$ vs. salinity for various oceanic areas: surface water in the Atlantic and Pacific, NADW, Pacific and Indian Ocean Deep Water (P&IDW), and Antarctic Bottom Water (AABW). The form of this plot derives from one in Craig and Gordon (1965) but is made with data from the GEOSECS expeditions (Ostlund *et al.* 1987). In water which has left the surface, both salinity and $\delta^{18}O$ are conservative, and when two water masses mix, the resulting values in a plot such as this fall on a straight line joining the two end members. Once seawater leaves the surface there is no process that can significantly change the $\delta^{18}O$ to salinity relationship, and the values are independent of temperature. NADW is made from surface water by cooling in the Norwegian Sea and adjacent areas (see also Figure 2.14), so its $\delta^{18}O$ to salinity relationship is essentially identical to that of surface water in that region. It is, however, impossible to make P&IDW directly by cooling surface water in any ocean. The only apparent way to make P&IDW is by freezing ice out of surface water and thus decreasing $\delta^{18}O$ and increasing the salinity in the unfrozen water. The plotted values for AABW were from some of the deepest stations in the region where this water mass has fairly extreme values. This AABW cannot be characteristic of the major source of water forming P&IDW; rather, this source must be a mixture of AABW and some intermediate-depth water.

Pacific or North Atlantic surface water. The P&IDW cannot be formed from a mixture of NADW and AABW, because the salinities are too high. However, if some of the water formed by freezing of ice in the Weddell Sea is not quite saline enough (i.e. dense enough) to sink all the way to the bottom, but sinks partway down while mixing with NADW, this could help generate Pacific Deep Water. The situation is known to be more complicated than this, with intermediate waters from the Pacific and even glacial meltwater apparently also contributing, but this simplified example illustrates the principle involved.

The isotope ratio can be an even more sensitive method of distinguishing two very different sources of water diluting surface water. Rivers draining into the Arctic

Ocean are very "light," carrying a $\delta^{18}O$ of –18 to –20‰, whereas the seawater there is around +3‰. In the regions within perhaps 100 km of the coast several meters of ice freeze and melt each year; if the ice freezes from seawater it will have a $\delta^{18}O$ of close to 0, whereas if it freezes from river water spreading out below the ice it will have a $\delta^{18}O$ of about –17. Both situations exist; the sources of water for making the ice can be distinguished using the oxygen isotopes as a tracer, the geographic extent of the river plume under the ice can be established, and the resulting meltwater in the spring can be similarly characterized. In this way, one can also learn the extent to which the formation of sea ice generates brine that might contribute to deeper, more saline water in the Arctic Ocean, and how much is just the freezing of river water (Macdonald *et al.* 1995).

2.2.4 Chemical equilibrium

Measurement of the isotopic composition of various substances makes it possible to evaluate the degree to which they are in chemical equilibrium. For example, in atmospheric oxygen $\delta^{18}O = +23.5$, while if the O_2 was in equilibrium with SMOW the value would be near zero. The isotopic composition of the O_2 dissolved in seawater is close to that in the air (allowing for a small effect due to the different solubilities of the heavy and light isotopes.[1] Therefore, we know that the O_2 in air and the O_2 dissolved in seawater do not equilibrate rapidly with the oxygen in water molecules. The isotopically heavy character of atmospheric O_2 is known as the *Dole effect* (see Bender *et al.* 1994 for a discussion of this phenomenon) and appears to be due mostly to a combination of the isotopic effects associated with photosynthesis on land and the preferential uptake of $^{16}O_2$ during respiration everywhere. In contrast, the $\delta^{18}O$ of atmospheric CO_2 is very close to that expected if the oxygen in CO_2 were freely exchangeable with that of water, after allowance (in this case) is made for a significant isotopic fractionation noted at equilibrium.

2.2.5 Ancient climates

Through geological time the isotopic composition of the ocean has varied, and this provides evidence concerning the climate in earlier ages. Continental ice has a different isotopic composition than the water in seawater, so melting of the ice and mixing of the resulting meltwater into the sea will change the isotopic composition of the ocean. Present continental ice totals about 28×10^6 km^3 (Appendix J, Table J.3), of which about 10% is in Greenland and most of the rest in Antarctica.

[1] Gases dissolved in water are enriched in their heavier isotopes, relative to the gas phase. The effect is a function of temperature; for oxygen the empirical equation is:

$$\Delta = (427/T) - 0.730 \qquad (2.7)$$

where temperature is in kelvin, and Δ is ($\delta^{18}O$ of dissolved gas – $\delta^{18}O$ of gas in air). Thus, at 10 °C, dissolved oxygen is characterized by a $\delta^{18}O$ greater than that in air by 0.78‰, provided the dissolved and gas phases are in equilibrium. The equation is from measurements by Benson and Krause (1980).

The volume-weighted isotopic composition of this ice is not precisely known, due to limited sampling and variations with age and location of the ice. Estimates from Lhomme *et al.* (2005) that Greenland ice is characterized by an average value of $\delta^{18}O \approx -34.2$ and Antarctic ice by $\delta^{18}O \approx -52$ lead to a present world average of about –50.2. The corresponding δD would be about –284.

During the last glaciation the amount of water stored in continental ice sheets may have amounted to an additional 43×10^6 km^3 (equivalent to a fall in sea level of about 125 m). Making the (possibly risky) assumption that the isotopic composition of the additional glacial ice was the same as that estimated for present Greenland ice, the following changes from present values of the mean oceanic isotopic composition may be calculated (from above estimates and data in Tables J.3, J.6):

	Change in oceanic values	
	$\delta^{18}O$	δD
Melt all present continental ice	−1.03	−8.03
Return to glacial age	+1.14	+9.43

Another piece of evidence concerning the ^{18}O concentration that might characterize the water in the glacial ocean comes through analyzing the pore water extracted at intervals down a 600-m-long core of ocean sediment collected at a depth of 3041 m in the tropical Atlantic. Schrag *et al.* (1996) modeled the observed distribution gradients and concluded that during the last glacial maximum, about 20 000 years ago, the deep water in that region was characterized by $\delta^{18}O \approx +0.8‰$ greater than the value today. Either this pore-water record is not sufficient to establish the difference, or the additional glacial ice had more ^{18}O in it than was assumed for the calculation above. The discrepancy is yet unresolved.

In the geological record we have no clear evidence of the deuterium concentration in the ocean, but we do have a considerable record of the ^{18}O. When organisms make shells or skeletons of calcium carbonate, the oxygen in the $CaCO_3$ has an $^{18}O/^{16}O$ ratio that is related to that of the water in which they are living. Analysis of the $\delta^{18}O$ in ancient marine shells therefore provides evidence on the volume of continental ice at the time. A considerable complication, however, is that the ratio $(\delta^{18}O_{shell})/(\delta^{18}O_{water})$ is a function of temperature; indeed, such measurements were first made to estimate ancient temperatures.

The temperature sensitivity is such that an increase of one unit (1‰) in the $\delta^{18}O$ characterizing biologically formed marine calcite in bottom waters corresponds to a decrease in water temperature of about 4.8 °C (Bemis *et al.* 1998), varying somewhat with conditions and with species. During glacial times it is possible and even likely that the deep water was colder than today, so the two effects are somewhat parallel and must be allowed for in either paleotemperature determinations or estimates of the volume and isotopic composition of continental ice. This limits the accuracy with which one can know one or the other, but the fossil isotopic record nevertheless has provided an exceedingly valuable assessment of past changes.

2.3 Clathrate compounds

A remarkable consequence of the tendency of water to form a rather open structure when it freezes is that under the right conditions it can form a lattice-like cage around certain other small molecules. Humphry Davy (1811) made the first observation that led to a knowledge of this phenomenon when he noted that a solution of chlorine gas in water "freezes more readily than pure water." Subsequent studies led to an understanding that a solid ice-like compound, containing gas and water, can form at a temperature above the freezing point of water. Michael Faraday (1823) reported that the chlorine hydrate would form at a temperature as high as about 9 °C, and had the composition of 10 H_2O for each molecule of chlorine gas. (Later researches revised this formula to contain less water.) During the next hundred years studies of this laboratory curiosity (Sloan and Koh 2008) led to the discovery that numerous other small gas molecules (including methane, CH_4) will form hydrates (now commonly called **clathrate** structures or *clathrates*), though usually this requires a higher gas pressure than was the case with chlorine, which forms hydrates at atmospheric pressure. Methane clathrate then achieved some prominence in the 1930s after it was found to cause serious problems when it formed in high-pressure natural-gas pipelines. During the most recent several decades there has been increasing interest in methane hydrates found on land under permafrost, and in the ocean within marine sediments mostly at depths of several hundred meters below sea level.

Gas hydrates (clathrates) are observed to form in one or another of several different cage-like crystalline structures; Figure 2.17 shows a common form, typical of methane clathrate. Formation of methane clathrate requires high pressure of the gas (Miller 1974); at 0 °C the required pressure is 26.4 bar (~ 26 atm). The pressure required increases with temperature, so, for example, it is about 73 bar at 10 °C. Figure 2.18 is a diagram showing how the pressure–temperature relationships constrain the occurrence of methane clathrates in marine sediments.

There are two sources of methane. As we will discover in later chapters, significant amounts of methane can be formed in sediments, enough to reach the necessary gas pressure. This requires a source of organic matter within the sediment, so commonly the circumstances are favorable under productive regions of the ocean, over continental and island slopes from depths of typically greater than 350 m down to considerable depths, depending on the availability of organic matter. The clathrate formed in sediments from methane produced there is often found as lenses or nodules dispersed within the sediment and, as shown in Figure 2.18, at some distance below the sediment surface. Most methane clathrate occurs in this situation.

A second (globally very much smaller) source of methane is leakage from much deeper natural-gas reservoirs. In this situation the gas can come up in a more concentrated flow; when it encounters the stability zone it can form massive deposits (up to at least several tons) and can even be exposed at the sediment surface, though it is unstable there.

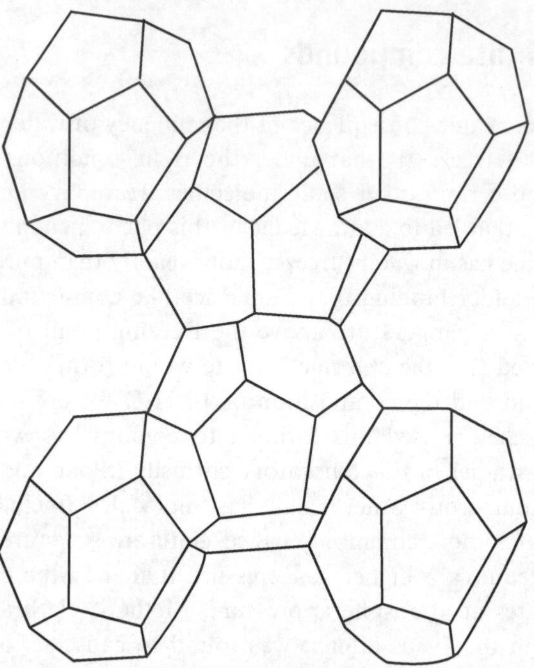

Figure 2.17 The most common arrangement of water molecules in a gas hydrate (redrawn from Sloan 2003). This arrangement, the smallest crystallographic unit cell, is called Structure 1 and contains 46 water molecules (there are at least two other less-common structures). A water molecule lies at each intersection of the lines. A single gas molecule (not shown) is sequestered inside each of the six individual cages in this structure. When the all the cages in the bulk crystalline hydrate are occupied by gas molecules there is about one gas molecule for each 5.7 molecules of water, but sometimes not quite all are occupied. As with ordinary ice, the open cage-like structure results in methane hydrate having a low density, about 0.94 g mL^{-1}, slightly more than that of simple water ice, but less than that of water. ■

The amount of methane clathrate present worldwide is enormous, though difficult to estimate precisely. Milkov (2004) suggested that the marine methane hydrate inventory must amount to between 500 and 2500 Gt of carbon. This is 5 to 25 times the known proved commercial reserves of natural gas. The hydrate is so widely dispersed, however, that I believe it is quite unlikely ever to be economically harvested. The amounts are so large, however, that if conditions should somehow change and the stored methane were released there could be a large effect on the Earth's carbon budget and consequently on the climate.

Methane hydrate can be a nuisance. The average composition of the ice-like material is about 5.7 water molecules for each methane molecule. Consider the situation when a sample core of sediment (suppose a 2-m-long core segment, with 10 L of sediment weighing 16 kg) is collected at sea and brought into the laboratory on the ship, and warms up to room temperature. Suppose this sediment contains just 100 g (about 106 mL) of methane hydrate dispersed within the whole core. When the core warms to 20 °C and the hydrate melts, the volume of gas released amounts to over 20 L. The resulting pressure can fracture the core tube, or cause the sediment to

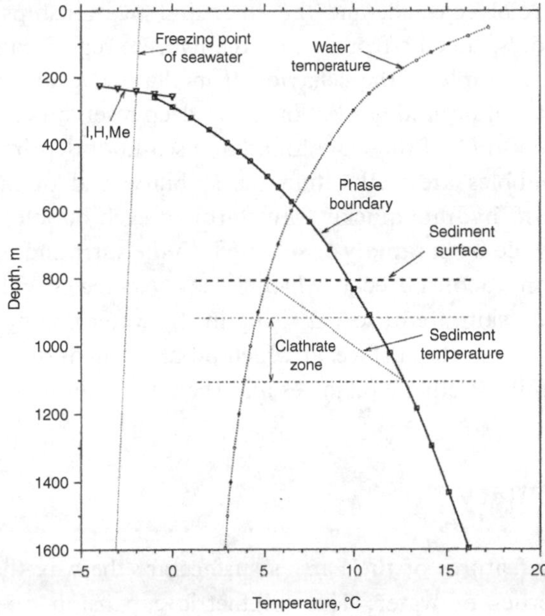

Figure 2.18 Diagram showing the conditions in the ocean that constrain where methane clathrate hydrate can occur. The water temperature is plotted as the global average for each depth, from Levitus and Boyer (1994b). The phase boundary marks the highest temperature at each depth where methane clathrate can be stable. Each point on the phase boundary is located according to the appropriate temperature and required partial pressure of methane gas dissolved in the water. The gas pressure required is plotted according to the hydrostatic pressure at depth in the ocean (taking the pressure as 1 dbar per meter). If the gas pressure were greater than this at each point, bubbles would form. The phase boundary is also sensitive to the salt content (the gas pressure required increases with salt content in the water); the values plotted are for a salinity of 35 ‰, using a relationship derived from data in Dholabhai *et al.* 1991). Clathrate will form and be stable at any point to the left of the phase boundary. There is no place in the open water column of the ocean where such gas pressures exist,and, if they did, the clathrate, having a density of about 0.94, nearly as low as ice, would float away. The required concentrations and pressures of methane are, however, often found in sediments. In the example here, the sediment surface is at a depth of about 800 m. Beginning at the sediment surface, the geothermal gradient of 25 °C per kilometer causes the temperature to intersect the phase boundary about 250 m below the sediment surface. Thus, no clathrate can be stable below this depth in this sediment. In principle, methane clathrate could be stable at any place between the sediment surface and the depth at which the geothermal temperature increase makes it impossible. It is rare for the necessary concentrations and partial pressures of methane to be found near the sediment surface. The methane near the surface can diffuse away too rapidly, and it is biologically oxidized by oxygen-using organisms near the surface and sulfate-using organisms somewhat deeper. In marine sediments conditions vary: depth, water temperature, types of sediment, strength of methane sources, geothermal gradients, and presence of different clathrate-forming gases that can affect the equilibrium. In the example shown here, there is a zone about 180 m thick where clathrates would be stable and could be found. The line labeled I,H,Me is part of the phase boundary for the equilibrium between pure-water ice, clathrate (hydrate), and methane (from data in Sloan and Koh 2008). This is an appropriate boundary for methane clathrates under permafrost in cold terrestrial regions. It is included here to note that this relationship has a different slope, and is somewhat offset from the relationship between methane clathrate and seawater. ■

be forcibly ejected, and the structural relationships within the sediment are destroyed. Sediments from within the hydrate region can be very difficult to study.

An example of the behavior of methane hydrates is provided by observations of bubbles of natural gas leaking from deep reservoirs under the sea bed. In the Gulf of Mexico some of these are found at a sea-floor depth of about 1500 m. At that depth the bubbles are well within the stability field of methane hydrate. A coating of methane hydrate quickly forms around each bubble, and this prevents the methane gas inside from rapidly dissolving into the surrounding seawater. The bubbles return a strong acoustic echo which is easily detected. The buoyant bubbles and their hydrate skin are observed rising in the water column until they are a few hundred meters from the surface, at which point the hydrate is unstable and decomposes; the remaining methane dissolves and the bubble disappears.

Summary

Many features of the Earth's surface are the way they are because of the physical properties of water. The volumetric expansion on freezing prevents fresh-water lakes, and the ocean, from freezing all the way to the bottom. The high heat capacity and the high heat of evaporation and condensation are important in buffering temperature changes and in transporting heat from low to high latitudes in both the ocean and in the atmosphere. The partly polar nature of the molecules is responsible for the high dielectric constant and the fact that water is a good solvent for many substances.

The isotopic composition of water molecules, and the fractionation effects associated with evaporation, condensation, and freezing provide important tracers of water movement and of various processes affecting the transport of water throughout the world, and its exchanges with various reservoirs. The association of isotopic composition with temperature that results from progressive temperature decreases and condensation processes provides important evidence on ancient climates.

The ability of water molecules to form ice-like structures in the form of a cage around small gas molecules results in the sequestering of very large amounts of methane in sediments where the temperature and pressure conditions are suitable and a sufficient source of methane is present. The amounts are so large that they can interfere with studies of the sediments, and potentially could influence the climate.

SUGGESTIONS FOR FURTHER READING

The following Wikipedia article "*Properties of water*" appears generally accurate, and covers some of the same material as this chapter: http://en.wikipedia.org/wiki/Properties_of_water.

Denny, M. W. 1995. *Air and Water*. Princeton University Press, Princeton, NJ.

A readable book with much discussion of the properties of water was written by Mark Denny some years go, and is still in print

Sloan, E. D. and C. A. Koh. 2008. *Clathrate Hydrates of Natural Gases*, 3rd edn. CRC Press, Boca Raton, FL.

The clathrate compounds, and especially the important methane clathrates, are very thoroughly reviewed in this book by Dendy Sloan and Carolyn Koh.

Franks, Felix, ed. (1972–1982) *Water, A Comprehensive Treatise*. Plenum Press, New York. 7 volumes.

For the masochist, determined to learn a *lot* more about water.

3 Salinity, chlorinity, conductivity, and density

No convenient, inexpensive, and accurate techniques exist for determining the total salt content of seawater.

FOFONOFF 1985

When dealing with any sample of seawater, the primary information required is usually a measurement of the salinity. Many properties, such as the concentrations of the conservative elements, can be calculated directly from the salinity, and others are related to it. The density is obtained from the salinity and the temperature, as are the solubilities of gases and other substances. The special requirements for exceedingly high accuracy and precision of this measurement, and the technical difficulties associated with the chemical composition of seawater, have influenced the early history of oceanography and the degree of international cooperation in marine science. An understanding of present practice and even some of the terms used requires an appreciation of the historical developments.

3.1 Need for accurate determination of salinity and density

In the latter decades of the nineteenth century, the question of variability in fisheries began to be addressed by several naturalists. The stocks of herring in the North Sea and along the coasts of Scandinavia were extremely variable, with great abundance in some years, and great scarcity in others. The Swedish chemist, Gustav Ekman, investigating the distribution of temperature and salinity in the Skagerack (between Sweden and Denmark), found that there were layers of water which differed in these respects, and he observed that herring seemed to prefer one layer over the others. The idea took hold that tracking the layers of water would help discover what happened to the fish, but one person alone could not do it. In 1890, Ekman and another Swedish chemist, Otto Pettersson, organized the first in a series of surveys of the Skagerack, the Kattegat, and adjacent regions of the Baltic and North seas, that involved coordinated sampling by vessels from Sweden and Denmark, and later Norway, Germany, and Scotland. These were the first examples of multi-ship surveys and of international cooperation in marine science, and were supported by the governments concerned in the interests of their respective fisheries. These cruises made possible for the first time the development of synoptic pictures of the distribution of temperature and salinity and some other properties of the water (concentrations of gases were measured, and it was observed that plankton were also characteristic of the different water masses).

Within the Kattegat and Skagerack the differences between the layers of water were rather easily detected by the titration methods available at the time, but, in the North Sea, and waters of the open Atlantic to the west into which the observations were being extended, the differences were less, and it became increasingly difficult to gain a sufficiently precise measure of the salinity. The difficulty was compounded by the fact that a number of investigators from several laboratories were involved, and it must have become apparent that they did not all get the same results from the same water samples. It became clear that satisfactory results could only be obtained by the most extreme care in standardization of the techniques. Enough had been learned, however, to show that the movements of water masses in the North Sea were quite complex and varied seasonally but did not necessarily repeat annually. The increased detail necessary to make sense out of these observations clearly required extensive international cooperation.

In 1899, upon the urging of Otto Pettersson and the Swedish Hydrographic Commission, King Oscar II of Sweden invited the governments of Germany, Denmark, Great Britain, Norway, The Netherlands, and Russia to send representatives to attend a meeting convened in Stockholm to develop international cooperation in marine science "in the interest of fisheries." An important item of business was the question of salinity determinations. At this meeting, Martin Knudsen, a Danish physicist, noted that to get useful results "it will be necessary to procure homogeneity in the measurement of salinity." Observing that the several countries involved would not likely allow all the samples to be measured in one laboratory, he suggested that all laboratories could obtain comparable results if they could test their techniques with samples of standard water from a single source, maintained the same from year to year. Knudsen proposed that an institution be established to prepare the standard water, and that the institution be supported by contributions from the interested nations. Knudsen was appointed chairman of a committee to evaluate, develop, and recommend the necessary techniques for the determination of salinity. A full report (Knudsen 1902) of the committee's deliberations was given by Carl Forch, Knudsen, and Søren Pedr Lauritz Sørensen, including details of the methods developed by Sørensen and Knudsen. (Sørensen later became even more famous for inventing the concept of pH.)

At a second meeting of the group (1901, in Oslo), it was recommended that there be established the *International Council for the Exploration of the Sea* (ICES), which indeed came into existence on June 22, 1902. This was the first international organization in oceanography, and possibly the first in science generally. In 1903, under the authority of ICES, a laboratory called the Standard Sea Water Service was set up in Oslo, under the direction of Fridtjof Nansen. Thus it was that the chemical complexity of the salts in seawater and the technical problems in their determination led to the establishment of the first internationally supported science laboratory, and perhaps the formation of ICES itself. For many years this organization played the strongest role in nurturing the growth of oceanography, and it still plays a significant role internationally.

In 1908, the laboratory was moved to Copenhagen, under the direction of Martin Knudsen (who had an active other-life carrying out research into the physics of gases, and teaching physics at the University of Copenhagen). For more than 40 years, Professor Knudsen was responsible for the preparation of standard seawater, then called Normal Water or, colloquially, "Copenhagen Water." Distributed throughout the world, it ensured that the investigators of all nations produced data that could be directly compared from place to place and from time to time. Copenhagen Water was prepared in large batches very carefully standardized against samples preserved from previous batches and also against silver nitrate solutions prepared from pure silver. Since about 1975, however, Copenhagen Water has come from England. The IAPSO Standard Sea Water Service became responsible for the management of standard seawater, which is now prepared and sold by Ocean Science International Ltd. This water is now sold in glass bottles containing about 200 mL and is adjusted to have a precisely specified salinity of about 35‰, and a precisely specified conductivity ratio (see Section 3.5 for a discussion of the current basis for standardization). In addition to the IAPSO Standard Seawater with salinity = 35, standards with $S = 10$, 30, and 38‰ are also now available, as is a special low-nutrient surface seawater.

As described earlier, the first impetus to obtain highly precise measurements of salinity was the desire to trace water masses in the interest of fisheries, and the movements of these water masses were of great concern. Until about 1900, all measurements of ocean currents were made by estimates of the drift of ships and other objects, and by crude current meters. In 1898, a Norwegian physicist, Vilhelm Bjerknes, published a paper on the physics of the circulation of fluids of different densities, and by 1900 the zoologist and explorer Fridtjof Nansen, and Bjorn Helland-Hansen (who had obtained some of the first samples of standard water prepared by Knudsen), were attempting to obtain highly accurate measurements of the density of seawater along the coast of Norway. These data were used by Helland-Hansen and Walfrid Ekman to make some of the first calculations of the movement of ocean currents from observations of the distribution of density. Later, convenient equations for the purpose were published by Johan Sandström and Helland-Hansen, and since then it has been a common practice to use the distribution of density and these geostrophic equations to calculate the motions of the major ocean currents. Variations in density within the ocean basins are, however, relatively small; making measurements accurate enough to be oceanographically useful requires that the determination of density should be accurate, or at least precise, to one part in 10^5 and equally reproducible from cruise to cruise. The provision of standard water made it possible to carry out the necessary chemical measurements with the required precision and accuracy.

3.2 Salinity

There is probably nothing more foundational to understanding the ocean than the study of salinity.
Richard Spinrad 2008

It was first discovered by Robert Boyle (1673), and confirmed by several investigators since, that drying and weighing the residue in order to determine the total salts in a

Table 3.1 Effects of heating the components of dry sea salt

Substance	With enough heat to eliminate all water
NaCl	Ok
KCl	Ok
K_2SO_4	Ok
$Na_2SO_4 \cdot xH_2O$	Ok (loses water at relatively low temperature)
$MgCl_2 \cdot xH_2O$	Some HCl may be lost
$CaCl_2 \cdot xH_2O$	Some HCl may be lost
$MgCO_3$	$\rightarrow MgO + CO_2\uparrow$
$Ca(HCO_3)_2$	$\rightarrow CaCO_3 + CO_2\uparrow \rightarrow CaO + CO_2\uparrow$
Organic matter	$\rightarrow H_2O\uparrow + CO_2\uparrow$

sample of seawater is not practical, because variable results are obtained. The reasons for this are: (1) the sea salts very tenaciously hold onto the water and (2) if the salts are heated enough to drive off all the water, the salts themselves start to decompose. The residue that appears on evaporating the water from seawater consists of mixtures of several substances; important examples are listed in Table 3.1. At temperatures required to completely dry sodium sulfate and magnesium chloride the magnesium carbonate and the calcium carbonate and bicarbonate will start to break down. There is also some loss of volatile salts of bromide and of iodide, and HCl can be lost from the chloride salts of magnesium and calcium. Thus, there is no real end point in the drying process, and the weight obtained varies with the conditions employed.

Nevertheless, it was felt that the total salt content of seawater should be defined in some way, and related in a precise manner to the density. The measurement of chloride or total halide, while known to be much more convenient and precise, was viewed as a surrogate for measurement of the total salt content.

At the international meeting in 1899, Martin Knudsen was asked to chair a committee to investigate and recommend the best procedures to follow in this regard, and he worked with Sørensen and Forch to evaluate the necessary methods and relationships. Sørensen developed a method for the determination of salinity that gave fairly reproducible results, although it was certainly not practical for routine use. He mixed the seawater samples with hydrochloric acid and water saturated with chlorine gas. This mixture was evaporated to dryness and heated at 480 °C for 72 hours. The remaining chloride in the dry salt was determined by titration. With this approach the bromide and iodide were replaced by chloride:

$$Cl_2 + 2\,Br^- \rightarrow Br_2 + 2\,Cl^-,$$
$$Cl_2 + 2\,I^- \rightarrow I_2 + 2\,Cl^-.$$

The bromine and iodine are volatile and so are driven off, along with excess Cl_2.

The salts of chloride are less volatile than bromide salts, so there was no loss of these salts by volatility. The carbon dioxide was all driven off, so no carbonates were formed. The $MgCl_2$ salts did lose HCl, leaving a residue of basic magnesium oxide.

Box 3.1 Definitions important to the concept of salinity

1. "*Salinity* is defined as the weight in grams of the dissolved inorganic matter in one kilogram of seawater after all the bromide and iodide have been replaced by the equivalent amount of chloride and all carbonate converted to oxide." Knudsen (1902).
2. "By *Chlorinity* is understood the mass of chlorine equivalent to the total mass of halogen contained in the mass of 1 kilogram of seawater." Sørensen (*in* Knudsen 1902, translated by Jacobsen and Knudsen 1940).
3. "*Salinity* ‰ = 1.805 Chlorinity ‰ + 0.030." Sørensen (*in* Knudsen 1902).
4. "The *Chlorinity* definition: $Cl = 0.3285234$ Ag ‰."
   Or: "The number giving the chlorinity in per mille of a sea-water sample is by definition identical with the number giving the mass with unit gram of Atomgewichtssilber just necessary to precipitate the halogens in 0.3285234 kilogram of the sea-water sample." Jacobsen and Knudsen (1940).
5. "S ‰ = 1.80655 Cl ‰." UNESCO (1962).
6. "The *practical salinity* [1978], symbol S, of a sample of seawater, is defined in terms of the ratio K_{15} of the electrical conductivity of the seawater sample at the temperature of 15 °C and the pressure of one standard atmosphere, to that of a potassium chloride (KCl) solution, in which the mass fraction of KCl is 32.4356×10^{-3}, at the same temperature and pressure. The K_{15} value exactly equal to 1 corresponds, by definition, to a practical salinity exactly equal to 35. The practical salinity is defined in terms of the ratio K_{15} by the following equation:
   $S = 0.0080 - 0.1692K_{15}^{1/2} + 25.3851K_{15} + 14.0941K_{15}^{3/2} - 7.0261K_{15}^{2} + 2.7081K_{15}^{5/2}$, formulated and adopted by the UNESCO/ICES/SCOR/IAPSO Joint Panel on Oceanographic Tables and Standards, Sidney, BC, Canada. 1 to 5 September 1980 ... This equation is valid for a practical salinity S from 2 to 42." UNESCO (1981a).

This loss was evaluated by titrating the dried salts and then accounted for with a small correction. All the organic matter was volatilized or oxidized. Measurements of the total salt content of seawater carried out by this technique yielded reproducible data even though the numerical value reported was not exactly equivalent to the weight of dissolved solid substances in the original seawater (the dissolved gases were ignored).

In 1902, Knudsen presented a famous definition (number 1 in Box 3.1) that was the beginning of many increasingly rigorous attempts to define the salinity of seawater. It should be noted that this definition does not include certain additional technical details, such as the necessity that the weights of the salts and of the seawater should both be referred to the weight *in vacuo*, which, as with all precise work, was taken for granted. It is also important to remember that the salinity thus defined is not as large a quantity as the total of the dissolved solids in ordinary seawater. In fact, it is probably about 160 mg

Table 3.2 Calculated loss of substance during a Sørensen/Knudsen determination of the salinity of a sample of seawater with $S = 35.00‰$

Substance	Calculation	Change, mg kg^{-1}
Bromide to chloride	$67.3\left(\frac{35.45 - 79.91}{79.91}\right)$	−37.5
$Ca(HCO_3)_2$ to CaO	$123\left(\frac{16 - 122}{122}\right)$	−106.9
Boric acid[1]	About 13 mg kg^{-1} lost	−13
Organic matter	1 to 2 mg kg^{-1}, all lost	−1 to −2
Total		About −160

[1]These are uncertain estimates because the carbon dioxide concentration in the seawater is only roughly estimated. Sørensen and Knudsen were probably unaware that some borate could be lost. In experiments replicating relevant parts of the Sørensen/Knudsen procedure, about half the borate was missing from the residue (R. Schaffenberg, personal communication, 1999).

less than the sum of the total dissolved solids in a kilogram of North Atlantic surface seawater (Table 3.2; later we will see a better estimate). It was presumed, however, that the salinity, so defined, was closely and linearly related to the total dissolved solids in ordinary seawater, a reasonably satisfactory assumption for the time.

3.3 Chlorinity

The Knudsen salinity was thought to be highly reproducible but it was tedious to measure. At that time, evidence from the work of Marcet, Forchhammer, Dittmar, and others, had suggested that the ratios of the major solid substances of seawater to each other were essentially constant. For example, the ratio of chloride to sodium, of sulfate to magnesium, of chloride to sulfate would all be constant, within the precision that such measurements could be made at that time, or nearly so. Therefore, if any one such substance could be determined, and if the ratio of the concentration of that substance to the total salinity were known, then the total salinity of the sample could be calculated accurately. About 55% of the total dissolved solids in seawater is chloride, and chloride can be determined with great exactness. Forchhammer (1865) had introduced the approach of measuring the chloride concentration and of multiplying by a coefficient to obtain the total salt concentration. His average coefficient was about 1.812, remarkably close to the value (1.817) of a modern estimate.

In order that all workers could produce comparable data, it was necessary to define the terms and procedures used, and to settle on the best estimate of the coefficient. A new measure of the dissolved substances was settled on: the "chlorinity," defined by Sørensen in 1902 (number 2, Box 3.1). This definition is a practical one. If a sample of seawater is titrated with silver nitrate to the end point, there has been a reaction such that all the chloride, bromide, and iodide have been precipitated.

It is difficult to measure separately the bromide and iodide, but, considering that they probably bear a nearly constant proportion to chloride, they can be treated as chemically equivalent. Rather than call the resulting measure the "halinity," or something like that, it was decided to call it simply the chlorinity.

The method for measuring the chlorinity introduced by Knudsen in 1902 is one of the most precise and accurate chemical measurement techniques known. It is based on the Mohr titration of silver nitrate with chloride using dichromate as an indicator. The details of this technique may be found in Oxner (1962). I know of no other chemical technique, in any field of investigation, with such precision (capable of measurements routinely to one part in one thousand or perhaps one part in three thousand, even at sea) that has been used for so many millions of samples, over so many years.

Unfortunately, advances in chemical knowledge required a redefinition of chlorinity. During the early decades of the twentieth century more precise determinations of the atomic weights of silver and chloride changed the accepted values. The change was great enough that the amounts of silver and chloride weighed out in the preparation of reagents would differ from earlier values, thus affecting the chlorinity determination. By 1937, it was recognized that this effect was sufficient to result in small but detectable differences in the chlorinity of carefully measured samples. To overcome this and make it possible to retain continuity between the chlorinity of samples measured at that time and samples measured between 1900 and 1937, it was decided to redefine chlorinity in units that would not change with adjustments in the accepted values of the atomic weights. In 1940, Jacobsen and Knudsen published a description of how this was done, and redefined the chlorinity (number 4, Box 3.1). This new definition preserved all the old relationships so that identical samples of seawater would yield identical measured chlorinities. The new definition, however, meant that the chlorinity no longer reported exactly the halide content of seawater; in fact, the halide content of a sample (expressed as chloride) is equivalent to 1.00043 times the chlorinity.

3.4 Relationships between chlorinity and salinity

Chlorinity . . . is still a corner-stone in the study of dissolved material in seawater.

Dan G. Wright et al. *2011*

Sørensen, between 1899 and 1902, measured very carefully the salinity of nine samples of seawater. He also measured their chlorinity by the Knudsen method. His nine water samples came from various parts of the ocean, all from the surface and varying in salinity from 2.6 to 40.2‰. Two were low-salinity waters from the Baltic, four were intermediate-salinity waters from the North Sea, one was a high-salinity water from the Red Sea and two were typical oceanic Atlantic waters. The nine determinations yielded a relationship between salinity and chlorinity, which was adopted as definition 3 (Box 3.1). This determination of the relationship between salinity and chlorinity has never been repeated, although it was used for more than 60 years to calculate all the millions of salinity values reported.

The Sørensen relationship does not extrapolate to zero. That is, at zero chlorinity there is still a finite amount of salt. This anomaly came about because of the inclusion of samples of Baltic Sea water much out of proportion to their volumetric importance in the world ocean. The chemistry of the Baltic Sea is slightly abnormal, having a salt composition differing from that of most seawater, due to dilution with river water. In practice, however, the failure of the S/Cl relationship to extrapolate to zero didn't matter very much, because salinity was only a dummy variable, anyway. The actual measurement was the chlorinity, and to calculate the density it was the usual practice to calculate the salinity and then enter the tables to find the density of seawater. These tables, however, were originally set up by measuring the density and chlorinity of a series of samples, calculating the salinity, and making up the tables accordingly. The real relationship was between density and chlorinity. It was still thought convenient to keep calculating salinity, because the salinity is a number that is relatively close to the total salt content of seawater, and it felt better to use it.

During the 1950s and 1960s it became an increasingly nagging problem that a relationship between salinity and chlorinity that did not reach zero at zero chlorinity, because it was biased by the inclusion of Baltic Sea samples, was unsatisfactory for the most precise general oceanographic use. An international panel of experts sponsored by UNESCO discussed the matter, and in 1962 it was recommended (UNESCO 1962) that Sørensen's expression be replaced by definition 5 (Box 3.1). During the subsequent 20 years this relationship was generally used to calculate the salinities of samples of seawater. At the chlorinity of standard seawater (Cl = 19.374‰, S = 35.00‰) this equation gives the same salinity as did the earlier equation. Of course, using this relationship the calculated salinity becomes slightly lower as the chlorinity decreases, relative to salinities calculated in the old way.

3.5 Conductivity and salinity

While even Knudsen had attempted to make use of the electrical conductivity of seawater to estimate the salt content, it was not until the late 1950s that electrical and electronic equipment became sufficiently sensitive and stable for practical use in marine investigations. Since about 1960 the use of so-called "salinometers," based on comparative measurements of conductivity, has almost totally supplanted the routine titrimetric determination of chlorinity. The latter procedure is now used only for research purposes, or sometimes in a generally simplified and less accurate form for inexpensive use in laboratories where the precision of the modern salinometer is not needed. Chlorinity is still, however, the fundamental measurement that underlies the salinity value reported for IAPSO Standard Seawater, and ties the present standards to the original measurements of Sørensen and Knudsen.

With the newer instrumentation it is possible to measure conductivity ratios to a precision of about one part in 40 000, so the salinity can be calculated to about $\pm 0.001S$. This is more precise than the best titration procedure by about a factor of five. Salinometers for use *in situ*, down to a depth of at least 6000 m, also report temperature and pressure, and highly accurate and detailed profiles of the vertical structure of temperature and salinity have been routinely taken throughout the world ocean for more than 40 years.

For some years, the standard against which the conductivity was determined was always standard seawater prepared by the Standard Sea Water Service, generally adjusted to a chlorinity of 19.374 (salinity, $S = 35.000$), or to other standards traceable to such standard seawater. The conductivity could not then (and still cannot) be measured absolutely with the requisite accuracy and precision, but the ratio of the two conductivities could be.

Around 1960, after good conductivity-type laboratory salinometers became available, several investigations were instituted to compare the relationships previously existing in Knudsen's hydrographic tables and similar compilations with the newer data obtained from conductivity measurements. Initially, the major work producing new and more precise measurements of the relationships between electrical conductivity, salinity, chlorinity, and density was carried out at the National Institute of Oceanography in England. This work resulted in a series of papers on these relationships, and in new data relating conductivity to salinity and density; the latter were published as the International Oceanographic Tables (UNESCO 1966). The descriptions of the methods used are in Cox *et al.* (1967, 1970).

The UNESCO tables were made up using only data from surface seawater, from above 200 m. Deep water has a slightly greater conductivity for a given chlorinity than surface seawater, and was excluded from the calculations for this reason. Therefore, seawater making up the bulk of the whole ocean was excluded from the data leading to the formulation of equations often used to deal with that water. In 1962, Cox *et al.* found that the relationship between density and salinity determined by conductivity showed much less scatter than the relationship between density and chlorinity determined by titration. In fact, the precision was about 10 times better. The precision of the conductivity measurement was about five times better than that of the chlorinity determination, and the rest of the improvement was due to the sensitivity of the conductivity method to small changes in the ratios of salts. From what we know of seawater chemistry this is reasonable, because small differences between samples in the ratios of the various salts to the chloride concentration might affect the density, would not be seen by the chlorinity determination, but could be partially taken account of in the determination of conductivity. For example, in a sample having slightly more calcium than normal the chlorinity may be the same but the density may be slightly greater; in this case the effect on conductivity should be in the same direction. Therefore, for measuring density it was evident that conductivity – more specifically, the conductivity ratio – was perhaps the better measurement to use.

In the early 1970s, the chemists involved in all this became dissatisfied with depending on what came to be seen as an uncertain preparation of standard seawater. By this time it was becoming appreciated that, at the precision needed for modern investigations, it was likely that variations in the ratios of some major ions (especially calcium and carbonate) to chlorinity in seawater, and even variations in the ratios of some of the minor constituents, were great enough that there could be measurable differences in the conductivity-to-chlorinity relationship as well as in the density-to-chlorinity relationship in samples of seawater from different locations, and even possibly in different batches of IAPSO Standard Seawater.

Earlier, it had been hoped that it would be possible to characterize a sample of seawater by an absolute measure of conductivity. For various technical reasons, this has so far proved a vain hope. The conductivity ratio between two samples, however, can be measured with extraordinary precision, so that the definition of some constant or primary standard seemed possible and desirable.

Lewis and Perkin (1978) concluded that any good standard should be defined in such a way that it could be made in any competent laboratory in the world, that it should not depend on the ionic composition of local water, that it should be based on a conservative property, and that it should be useful for the ultimate calculation of density to within acceptable limits. They recommended that a precisely defined solution of KCl in pure water could be a suitable standard. This is a standard that can always be re-synthesized when needed, and is not dependent upon the possible vagaries of the mixing of oceanic water masses and the consequent possible changes in ionic ratios over time. They recommended that the concentration be chosen so that its conductivity would be precisely equal to that of standard seawater at a chlorinity of 19.374‰, which by definition has a salinity of $S = 35.000‰$.

They further recommended that a "practical salinity scale" (PSS) be introduced, such that the salinity is defined in terms of the ratio of the conductivity of samples of seawater to that of the standard KCl solution. The scale would be calibrated by measuring the ratio of the conductivity of the standard KCl solution to that of standard seawater diluted or concentrated by the addition or subtraction of pure water.

Investigators in Canada, France, England, and the United States carried out several series of careful measurements; it was found (Lewis 1980) that 32.4356 g of KCl in 1 kilogram of solution would have exactly the same conductivity at 15 °C as seawater with a salinity of exactly 35.000‰ as determined by the classical chlorinity titration.

In 1980, an international panel considered and adopted, for use by the international oceanographic community, definition 6 in Box 3.1. This is termed the "Practical Salinity Scale 1978" or PSS 1978 (UNESCO 1981a), and is now the universal standard. For practical measurement at other temperatures, additional factors are required in the algorithm (Table 3.3). If the conductivity is measured at other than atmospheric pressure, as by an *in-situ* conductivity probe, an additional equation is required in order to take account of the effect of pressure on conductivity.

It should be noted that this definition of PSS 1978 says nothing about the mass of dissolved solids (or any chemical entity) in a sample of seawater. While it is based on a standard solution of 32.4356 g of KCl in one kilogram of solution, it is defined as the ratio of the conductivities. It has been argued that S, so defined, is a dimensionless number, with no units.

While Perkin and Lewis (1980; and other papers by the same authors) reported salinities in units of ‰, the SUN committee (UNESCO 1985) recommended that salinity be considered a dimensionless number (since it is calculated from the ratio of two conductivities) and that it be reported only in the following way,

Table 3.3 Measurement and reporting of the practical salinity, PSS 1978

$$S = a_0 + a_1 R_t^{1/2} + a_2 R_t + a_3 R_t^{3/2} + a_4 R_t^2 + a_5 R_t^{5/2} + \frac{t-15}{1+k(t-15)}(b_0 + b_1 R_t^{1/2} + b_2 R_t + b_3 R_t^{3/2} + b_4 R_t^2 + b_5 R_t^{5/2})$$

where: S is the salinity determined according to PSS 1978
t is the temperature in °C
R is the ratio of conductivities: $C(S,t,0)/C(35,t,0)$.
C is the conductivity at salinity S, temperature t and at 0 (= 1 bar) pressure.
a, b, k are empirically determined constants, as follows.
Note that the sum of the a_i terms = 35.0000.

$a_0 = +\ 0.0080$	$b_0 = +\ 0.0005$	$k = +\ 0.0162$
$a_1 = -0.1692$	$b_1 = -0.0056$	
$a_2 = +25.3851$	$b_2 = -0.0066$	
$a_3 = +\ 14.0941$	$b_3 = -0.0375$	
$a_4 = -7.0261$	$b_4 = +\ 0.0636$	
$a_5 = +\ 2.7081$	$b_5 = -0.0144$	

Definition number 6 in Box 3.1 is the formal definition of PSS 1978, and the ratio of the conductivities is expressed by the symbol K_{15}, to indicate that it is the ratio of a seawater sample against a specified solution of potassium chloride at 15 °C. In practice, one uses a standardized seawater of $S \approx 35$ and with an exactly known relation ($K_{15} \approx 1$) to the KCl standard during the measurement of actual samples. The measured ratios are designated by the symbol R. Because measurements are often made at temperatures other than 15 °C, the additional empirical terms (b and k) have been added to account for the effect of temperature.

The algorithm was introduced by Perkin and Lewis 1980; additional discussion is given in Lewis 1980. Other sources are UNESCO 1981a, Fofonoff and Millard 1983, and Fofonoff 1985).

e.g. "$S = 35.068$", with the specification somewhere in the paper that it is determined according to PSS 1978. According to SI rules, no unit should be attached to such a numerical value.

In order to call attention to the fact that salinity as now determined and defined is the result of a long and tortuous history of definition and measurement, to distinguish salinities on the PSS 1978 from earlier salinities, and to satisfy the common human need to designate and use a unit for a technically unitless numerical value, someone[1] suggested that salinities measured and calculated on the PSS 1978 could be designated in *practical salinity units* or *psu* (e.g. $S = 35.068$ psu). This usage spread rapidly, and is now almost universal. However, a joint committee of IOC, SCOR, and IAPSO has stated that psu should not be used.

It would seem that since the conductivity ratio (exactly 1 at $S = 35$ and $t = 15$ °C) is multiplied by the factor 35 to bring the numerical values into accord with the original definition and usage, and this factor of 35 may be considered as 35 g of salt (as specially

[1] Despite quite some hours of searching, I have so far not discovered who suggested this.

Box 3.2 International Thermodynamic Equation of Seawater 2010, TEOS-10

During the last 30 years a number of shortcomings associated with EOS-1980 have become apparent, and possibilities for improvement have been suggested. Feistel (2003) noted that EOS-1980 deviates detectably from the most recent and precise data for pure water in regard to density, compressibility, thermal expansion, and other properties, and departs detectably from direct measurements of saline samples in other properties. Valladares *et al.* (2011) conveyed the recommendation by committees of IOC, SCOR, and IAPSO that EOS-80 be replaced with TEOS-10, a new International Thermodynamic Equation of Seawater 2010. This formulation is thermodynamically more rigorous and consistent, and more accurate for calculating many thermodynamic properties.

In the proposed use of TEOS-10, some continuity with earlier work is maintained in that the salinity is still to be measured according to PSS 1978, and these primary measured values are to remain the archival record of salinity. However, for use with TEOS-10, the value measured according to PSS 1978, termed Practical Salinity, S_P, is converted sequentially into at least two other defined salinities. The term S_P is to be multiplied by the factor 35.16504/35 to yield S_R, the Reference Salinity. (We will learn in Chapter 4 that in 1978 the mass of total dissolved inorganic solids in water similar to IAPSO Standard Seawater with $S_P = 35‰$ was roughly 165 mg greater than the nominal salinity of 35 g kg^{-1}.) The S_R is then corrected by the addition of a small factor δS_A, to yield S_A, the Absolute Salinity. The factor δS_A is introduced to account for inorganic substances that are biologically transported and have some effect on the density (we will learn something about them in Chapters 7 and 8), and possibly any secular changes in the composition. The factor δS_A may be zero, or (for example in the extreme situation of intermediate water in the northern North Pacific) may amount to as much as 20 mg kg^{-1}. The factor δS_A is to be obtained from extensive look-up tables according to each location and depth. The S_A is now (once again) to be expressed in SI units as g kg^{-1}.

The value of S_A will then used to enter equations to calculate all other physical properties that are functions of the salinity.

A brief introduction to TEOS-10, as well as a full description, are available at the TEOS-10 website, www.teos-10.org, and these contain many references to the supporting literature. An extensive discussion of Absolute Salinity was presented by Wright *et al.* (2011).

defined) in one kilogram of seawater, it might still be appropriate to use the unit designation as ‰, as I generally do in this book, except when using or quoting material from some source. As we will see later (Box 3.2) further changes are in prospect.

The definition of PSS 1978 was selected so that the ratio of the conductivity of the standard to that of the then currently used IAPSO Standard Seawater having a

Table 3.4 Rough estimates of the precision easily attainable with several of the various ways to measure salinity

Method	Precision
Hydrometer	1 part in 50
Refractive index:	
hand-held refractometer	1 part in 70
bench-type refractometer	1 part in 700
Ordinary laboratory titration	1 part in 350
Knudsen titration:	
at sea	1 part in 3000
best laboratory practice	1 part in 8000
Modern (1990s) salinometer	1 part in 40 000
Calculation from direct measurement of density	1 part in 4000

For easy use in the laboratory or in the field, when only approximate values are needed, the hand-held refractometer is most useful; it only requires a drop of liquid, and the instrument costs only a few hundred dollars. Either of the two titration techniques require considerable care, but are easy to set up. A modern salinometer is a specialized piece of electronic and mechanical equipment, and costs several thousand dollars; once set up and calibrated, however, sample processing time is much shorter than by titration. A density meter at this level of precision is also expensive, but is even simpler to operate than a salinometer.

Grosso *et al.* (2010) claim that modern instrumental developments may have made it possible to measure salinity by refractometer to a precision of perhaps better than 1 part in 40 000. This would have the advantage that the refractive index relates to the total solids in solution rather that the total ions, and thus ought to scale with density more closely than does conductivity. As yet there is no commercial instrument employing this technique to the necessary precision.

chlorinity of 19.374‰ is exactly 1.0000, and the salinity is defined as $S = 35.000$ for this water. However, samples of seawater from other parts of the ocean than the surface North Atlantic may have slightly different ratios of the major constituents, so that in general neither the chlorinity nor the concentration of any individual salt may necessarily bear an exact proportionality to the salinity as here defined.

It must be kept in mind that, except for the nine samples measured by Sørensen in 1902, every reported value of salinity since 1900 is the result of a direct titration of chlorinity or, if measured in some other way, is ultimately traceable to a standard measured by a chlorinity titration. Since 1982, however, the traceability is through a standard KCl solution made up to match exactly the conductivity at 15 °C of IAPSO Standard Seawater at $Cl = 19.375$‰ that was prepared in 1978 or shortly before. IAPSO Standard Seawater is currently calibrated only by conductivity against the KCl solution.

Table 3.4 gives a comparison of several ways to measure salinity.

Box 3.3 The current version of the equation relating the density of seawater to the salinity and the temperature, under a total pressure of one atmosphere

The One Atmosphere International Equation of State of Seawater, 1980, Definition

The density (ρ, kg m^{-3}) of seawater at one standard atmosphere ($p = 0$) is to be computed from the practical salinity (S) and the temperature (t, °C) with the following equation:

$$\rho\ (S,t,0) = \rho_w + (8.244\,93 \times 10^{-1} - 4.0899 \times 10^{-3}t + 7.6438 \times 10^{-5}t^2 - 8.2467 \times 10^{-7}t^3 + 5.3875 \times 10^{-9}t^4)S + (-5.724\,66 \times 10^{-3} + 1.0227 \times 10^{-4}t - 1.6546 \times 10^{-6}t^2)S^{3/2} + 4.8314 \times 10^{-4}S^2$$

where ρ_w, the density of the Standard Mean Ocean Water (SMOW) taken as pure-water reference, is given by:

$$\rho_w = 999.842\,594 + 6.793\,952 \times 10^{-2}t - 9.095\,290 \times 10^{-3}t^2 + 1.001\,685 \times 10^{-4}t^3 - 1.120\,083 \times 10^{-6}t^4 + 6.536\,332 \times 10^{-9}t^5$$

The One Atmosphere International Equation of State of Seawater, 1980, is valid for practical salinity from 0 to 42 and temperature from –2 to 40 °C.

Alain Poisson in France and Frank Millero in the USA independently determined the density of seawater over ranges of both salinity and temperature. Their two data sets were combined and fitted to the equation given here (Millero and Poisson 1981, UNESCO 1981b). The salinity is based on the practical salinity scale (PSS 1978).

3.6 Salinity and density

The evidence that the conductivity ratio and the salinity (PSS 1978) calculated from it provide a better estimator of the density of seawater than any other available parameter led to a considerable effort to re-measure and redefine the density–salinity relationship to attain a new level of accuracy and precision.

This work was carried out largely in France and the United States (Millero and Poisson 1981, UNESCO 1981b). As summarized in UNESCO (1981b), an equation was developed to fit the data from some 467 measurements at atmospheric pressure, so that the density of a sample of seawater may be calculated at any temperature from –2 °C to 40 °C, once the salinity (PSS 1978) has been obtained from the measurements of the conductivity ratio. This equation is called "The One Atmosphere International Equation of State of Seawater, 1980" (Box 3.3).

Millero *et al.* (1980) measured the compressibility of water at many temperatures and salinities. Combining data measured at Miami and Woods Hole they developed an equation relating the density under pressure as a function of temperature and salinity, based on a total of 2023 data points. This equation incorporates the One Atmosphere

Box 3.4 The present version of the equation relating the density of seawater to salinity, temperature, and pressure

The High Pressure International Equation of State of Seawater, 1980, Definition
The density (ρ, kg m^{-3}) of seawater at high pressure is to be computed from the practical salinity (S), the temperature (t, °C) and the applied pressure (p, bars) with the following equation:

$$\rho(S,t,p) = \frac{\rho(S,t,0)}{1 - p/K(S,t,p)}$$

where $\rho(S,t,0)$ is the One Atmosphere International Equation of State of Seawater, 1980, and $K(S,t,p)$ is the secant bulk modulus given by:

$$K(S,t,p) = K(S,t,0) + Ap + Bp^2$$

where

$$K(S,t,0) = K_w + (54.6746 - 0.603\,459t + 1.099\,87 \times 10^{-2}t^2 - 6.1670 \times 10^{-5}t^3)S + (7.944 \times 10^{-2} + 1.6483 \times 10^{-2}t - 5.3009 \times 10^{-4}t^2)S^{3/2},$$

$$A = A_w + (2.2838 \times 10^{-3} - 1.0981 \times 10^{-5}t - 1.6078 \times 10^{-6}t^2)S + 1.910\,75 \times 10^{-4}\,S^{3/2}$$

$$B = B_w + (-9.9348 \times 10^{-7} + 2.0816 \times 10^{-8}t + 9.1697 \times 10^{-10}t^2)S$$

and the pure-water terms K_w, A_w and B_w of the secant bulk modulus are given by:

$$K_w = 19\,652.21 + 148.4206t - 2.327\,105t^2 + 1.360\,477 \times 10^{-2}t^3 - 5.155\,288 \times 10^{-5}t^4$$

$$A_w = 3.239\,908 + 1.437\,13 \times 10^{-3}t + 1.160\,92 \times 10^{-4}t^2 - 5.779\,05 \times 10^{-7}t^3$$

$$B_w = 8.509\,35 \times 10^{-5} - 6.122\,93 \times 10^{-6}t + 5.2787 \times 10^{-8}t^2$$

The High Pressure International Equation of State of Seawater, 1980, is valid for practical salinity from 0 to 42, temperature from –2 to 40 °C and applied pressure from 0 to 1000 bar.

This equation was the result of combining and fitting several sets of data (Millero *et al.* 1980, UNESCO 1981b).

equation, and reduces to the latter at zero pressure.[2] This equation was adopted as "The High Pressure International Equation of State of Seawater, 1980" (Box 3.4). The two equations are often referred to as "EOS 1980". The UNESCO/ICES/SCOR/IAPSO Joint Panel on Oceanographic Tables and Standards directed that after January 1, 1982, all reports of the density of seawater are to be calculated according to this equation (UNESCO 1981b). (But see discussion of TEOS-10, noted later, on p. 63).

Various algorithms for the computation of some additional physical properties of seawater were published by Fofonoff and Millard (1983), who provide tabulated values for checking the calculations.

It is evident from examination of EOS 1980 that the density calculated for any sample depends on ρ_w, the density calculated at each temperature for SMOW (that is, for salt-free pure water with the isotopic composition of SMOW). It is interesting to

[2] Conventionally, physical oceanographers consider zero pressure to be one atmosphere (101 325 Pa).

inquire into the source of the relationship giving the density of pure water, something that most people assume must be exceedingly well established.

The basic data on the density of water available when EOS 1980 was established were the measurements on redistilled and air-free water by Thiesen (1900) and Chappuis (1907), both widely used at the time, but differing from each other, the differences increasing with temperature. Tilton and Taylor (1937) re-evaluated Chappuis' data, and fitted to them an equation in the form used by Thiesen, then published a revised table of Chappuis' recalculated and smoothed data. Bigg (1967), attempting to improve our knowledge of the density, and also undertaking to convert the published data sets to SI units and to correct for the revised volume of the liter, made the latter adjustments, then fitted separate fifth-power polynomials to the adjusted data of Chappuis (1907) and Thiesen (1900) as well as to Tilton and Taylor's (1937) recalculation of Chappuis' data, weighted the three sets equally, and averaged the results at each interval. Bigg did not publish the equations involved in the calculations. His table gives the calculated density of pure air-free water, from 0 °C to 40 °C in 0.1 °C increments, in units of kg m^{-3}. The value at the maximum density of about 4 °C was 999.9719 kg m^{-3}. Characteristic differences between the original data sets (after polynomial fitting and smoothing) were commonly 1 to 10 g m^{-3} (g m^{-3} ≈ parts per million). It is therefore apparent that the last two significant figures in the values quoted above are uncertain, and the last one serves only to prevent the accumulation of rounding errors. The table by Bigg has been the standard of reference for most work since then.

By the early 1960s it was thought desirable to report the absolute density of seawater, whereas all previous seawater data were, in fact, based on measurements of the specific gravity relative to "pure water" at 4 °C. By this time it was known that, due to variations in isotopic composition, pure water could vary by up to 30 g m^{-3} depending upon its source (UNESCO 1976; this publication contains the 1962 and 1963 reports of the Joint Panel on the Equation of State of Seawater). In the 1976 report it was recommended that a provisional table of density be employed, based on a calculated density of SMOW. Since SMOW was becoming the universal standard for isotopic measurements and represented fairly closely the largest mass of water on Earth, it was the logical choice. But what is the density of SMOW?

The effect of variations in isotopic composition on the density of pure water has been investigated several times. Menaché and Girard (1970) and Girard and Menaché (1971) measured the isotopic effect on density and gave the relationship:

$$(\rho_{\text{sample}} - \rho_{\text{SMOW}}) \times 1000 = (0.211\delta^{18}\text{O} + 0.0150\delta\text{D}), \tag{3.1}$$

where ρ is in kg m^{-3}. Through the use of this equation Girard and Menaché (1972) made an estimate of what was most likely the density of SMOW at 4 °C, by comparison with the isotopic composition that, plausibly, might have characterized the water that yielded the density data published by Thiesen in 1900 and Chappuis in 1907 (and actually measured between 1891 and 1899). Their provisional estimate of the absolute density of SMOW at 4 °C, a value of 999.975 kg m^{-3}, was only about 3 g m^{-3} greater than the value in Bigg's table. Menaché (in UNESCO 1976) then converted all values in Bigg's table by the factor 999.975/999.972 and fitted the new values with a fifth-power polynomial. That is the polynomial that forms the basis for the assignment of ρ_w, in the One Atmosphere International Equation of State of Seawater, 1980 (Box 3.3).[3]

The absolute density of pure air-free water with a defined isotopic composition has since been measured again by several investigators, and the results are summarized by Tanaka *et al.* (2001), who provide an evaluation combining the various data sets and values for the absolute density adjusted to the isotopic composition of SMOW. The influence of isotopic composition on the density of water was calculated by a revised relationship, also due to Girard and Menaché:

$$(\rho_{\text{sample}} - \rho_{\text{SMOW}}) \times 1000 = 0.233\delta^{18}\text{O} + 0.0166\delta\text{D} \tag{3.2}$$

where ρ is in kg m^{-3}. The resulting estimates of the absolute density of SMOW differ by less than one part per million (less than one gram per cubic meter) from those utilized in EOS 1980 at all temperatures below 25 °C (Appendix C, Table C.7). Above 25 °C, the divergence increases so that at 40 °C the densities of SMOW measured in the 1990s are some 5 g m^{-3} less than estimated from the measurements made in the 1890s.

Water saturated with air has a lower density than air-free water, and the magnitude of the effect varies with temperature because the amount of gas in solution at equilibrium with the atmosphere varies with temperature. The measurements of Bignell (1983) provide the following relationship:

$$\Delta\rho = -4.612 + 0.106t \tag{3.3}$$

where t is in °C and $\Delta\rho$ is in g m^{-3}. It is evident that air-saturated water at 0 °C is about 4.6 g m^{-3} less dense than air-free water, and at 20 °C it is 2.1 g m^{-3} less dense.

The definition of the One Atmosphere International Equation of State of Seawater, 1980 appears to be slightly anomalous in one respect. The estimates of the absolute density of pure water (SMOW), which enter the equation as values of ρ_w, are derived by revisions of Bigg's tabulated values referring to air-free water. It seems anomalous, therefore, to state without qualification that this is a one-atmosphere equation of state, suitable for seawater under the pressure of a standard atmosphere (101 325 Pa). In the apparatus used to measure the density of seawater, both the pure water used for calibration and the seawater measured were in contact with air, and presumably air-saturated. Pure water saturated with air will, however, be roughly 2 to 4 g m^{-3} less dense than is calculated by the equation given. The effect will be a little less for seawater, which typically has only about 75% as much air dissolved in it as does fresh water.

[3] The history of the adjustments to the primary density data has been somewhat confusing. Poisson *et al.* (1980), in publishing the results of several determinations of the density of seawater, used an equation for ρ_w identical to that of Menaché (in UNESCO 1976) but stated only that it was based on the table given by Bigg (although their values were different) without citing Menaché as the source. Later, the EOS 1980, including the equation for ρ_w, was published by Millero and Poisson (1981) but they stated that the equation for ρ_w was based on the calculations of Bigg, citing neither Poisson *et al.* nor Menaché. In a summary of the data treatment for their equation, Millero and Poisson (in UNESCO 1981b) state "At present the density of SMOW is based on the equation of Bigg (1967)," although Bigg gave no such equation.

The extraordinary developments in accuracy and precision described earlier make it possible to specify that oceanographic salinity data should be taken with a precision of ± 0.001 and an accuracy of 0.002, or better. This precision corresponds to values for density of approximately ± 0.001 kg m^{-3}, or ± 1 g m^{-3}. The conductivity-to-salinity relationship depends on the nearly conservative relationships of the major chemical components of seawater. At the level of precision now achieved, the extremely small variations in the calcium-to-salinity ratio, and the non-conservative behavior of several minor components (especially silica) begin to affect the salinity-to-density relationship by more than 10 g m^{-3}. The highest precision is important in calculation of water motions within and between adjacent bodies of water, but if anyone needs to examine absolute-density differences between different major water masses or ocean basins to better than ± 10 g m^{-3}, much additional chemical information must be evaluated.

During the nearly 110 years since the measurements by Knudsen, Sørensen, and Forch, every change in the calibration of the chlorinity-to-salinity relationship was carried out in such a way that the reported salinity for similar water would be numerically the same as it was in 1902. Even the introduction of the PSS 1978 scale kept this relationship, albeit through a calibration of the KCl conductivity standard to the chlorinity of the IAPSO Standard Seawater that was available in 1978.

During the last decade, however, some people increasingly thought that EOS 1980 should be revised to take account of better data on the density of pure water, and of a revised temperature scale (it is now ITS 90; PSS 1978 was created using IPTS 68; differences vary, e.g. 15 °C in IPTS 68 is now 14.997 °C in ITS 90). Other changes could be made so that the value of salinity would more closely approximate the true mass of dissolved inorganic substance in seawater, and also make various thermodynamic relationships more rigorous. Numerous papers have been published on these issues, and committees of IOC, SCOR, and IAPSO were convened to discuss the matter. These agencies have recommended (Valladares *et al.* 2011) that EOS-80 be replaced with TEOS-10, a new International Thermodynamic Equation of Seawater (Box 3.2).

The proposed new measure of Absolute Salinity (S_A), expressed in g kg^{-1}, now departs from the 110-year tradition of maintaining continuity in salinity based on measurements of chlorinity. All equations relating chemical properties to salinity will eventually to be revised to account for (S_A) values that will be at least 1.00471 times greater for the same water than was the case earlier.

The relationships used later throughout this book continue to be based on PSS 1978 and EOS 1980. For our purposes, the differences are sufficiently small that it would make no substantive difference in the results if new S_A values were used rather than values based on PSS 1978.

An important consideration is that the actual measurements of salinity are still to be based on PSS 1978, and these are the data that should be archived. The measured data will then be transformed by calculation to S_A values, which will be used in new algorithms to calculate density and various thermodynamic properties of the water.

Summary

This chapter can best be summed up in a series of points as follows.

1. Salinity by the original definition was not exactly equal to the concentration of total salts.
2. In the original definition the chlorinity was equal to the total halide reduced to chloride.
3. The original relationship of *S*‰ to *Cl*‰ was non-conservative: *S*‰ = 0.03 + (1.805 × *Cl*‰).
4. The chlorinity definition was changed to account for changes in atomic weights, so that: total halide reduced to chloride = 1.00043 *Cl*‰.
5. The *S*‰ to *Cl*‰ relationship was changed to: *S*‰ = 1.80655 *Cl*‰. The new relationship crossed the old line at $S = 35.000$‰.
6. In 1978, *S* was defined in terms of a ratio of conductivity to a KCl standard. It was formally specified, therefore, to be a dimensionless number. However, since the proposed algorithm contains the factor 35 it could still be considered to represent approximately the weight in grams of the inorganic matter dissolved in one kilogram of seawater, according to the original definition. To avoid confusion with earlier definitions, the salinity may be written without units (e.g. $S = 35.03$), the PSS 1978 being specified, or in Practical Salinity units (e.g. $S = 35.03$ psu), though the latter is discouraged by relevant authorities. As noted earlier, in this book I have generally used the traditional unit: ‰.
7. Chlorinity no longer bears a unique relationship to salinity.
8. The ultimate standard for *S* is the conductivity of a solution made by dissolving a mass of KCl (32.4356 g) in pure water to make a total mass of exactly 1 kilogram.
9. Density has been (since 1980) calculated from *S*, *t*, and *p*, according to the One Atmosphere and High Pressure International Equations of State of Seawater, 1980.
10. The basis for the calculation of density is the density of SMOW.
11. Since 2011 there is a new International Thermodynamic Equation of Seawater, TEOS-10. Here the salinity is entered as the Absolute Salinity, S_A, derived by calculation from the Practical Salinity, (PSS 1978), to approximate the total inorganic salt content of the sample, now expressed in SI units as g kg^{-1}.

SUGGESTIONS FOR FURTHER READING

I don't know of any general reference that covers or reviews the material in this chapter. Most of the useful material is only in the original journal literature. The following are recommended as important contributions with some general discussion. In addition, the official source of information about the Thermodynamic Equation Of Seawater - 2010 (TEOS-10) is to be found at www.teos-10.org.

Lewis, E. L. 1980. The practical salinity scale 1978 and its antecedents. *IEEE J. Oceanic Eng.* **OE-5**: 3–21.

Lewis, E. L. and R. G. Perkin. 1978. Salinity: its definition and calculation. *J. Geophys. Res.* **83**: 466–478.

Millero, F. J., R. Feistel, D. G. Wright, and T. J. McDougall. 2008. The composition of Standard Seawater and the definition of the Reference-Composition Salinity scale. *Deep-Sea Res. I* **55**: 50–72.

Wright, D. G., R. Pawlowicz, T. J. McDougall, R. Feistel, and G. M. Marion. 2011. Absolute salinity, "Density Salinity" and the Reference-Composition Salinity Scale: present and future use in the seawater standard TEOS-10. *Ocean Sci.* **7**: 1–26.

4 Major constituents of seawater

The solid matter in seawater, though strictly speaking, and we may add necessarily, of a very complex composition, consists substantially of the muriates and sulfates of soda, magnesium, lime and potash. WILLIAM DITTMAR 1884

All the chemical elements in the periodic table that can be found on Earth (Appendix A) must be present to some extent in seawater, although not quite all have been detected there yet. For a variety of reasons it is convenient to consider separately the major and minor constituents. In this chapter only the major constituents will be considered. In a continuum of concentrations, any separation is arbitrary, but it is convenient to pick a value of one part in one million (=1 ppm or 1 mg/kg) as a lower limit for the concentrations of *major constituents*. Substances present above this concentration may have a detectable influence on the density, for example, while those present in lesser concentrations will, individually at least, generally not. Most of the substances that occur in concentrations greater than 1 mg/kg are *conservative* – that is, they are found in nearly constant proportions to each other and to the salinity. Most of the substances present in concentrations of less than 1 mg/kg are not conservative.

As defined above, there are 11 substances included among the major constituents, and there are a couple more that could be included but conventionally have not been.

4.1 Concentrations

The concentrations of the major constituents have been the subject of numerous investigations, as noted in Chapter 1. Currently accepted values for the concentrations of the 11 major constituents are given in Table 4.1 (important ionized forms of carbonate and borate are included). The concentrations are listed as a ratio to the salinity and, for $S = 35‰$, in grams per kilogram (g kg^{-1}), millimoles per kilogram (mmol kg^{-1}), and millimolar (mmol L^{-1} = mM). In each case the concentrations in "kg^{-1}" refer to units per kilogram of the whole solution. The one important

Table 4.1 Concentrations of the major constituents in surface seawater

	At salinity (PSS 1978): $S = 35.00‰$, $t = 20$ °C			
	mg kg^{-1} S^{-1}	g kg^{-1}	mmol kg^{-1}	mM
Na^+	308.1	10.782	469.00	480.61
K^+	11.40	0.399	10.21	10.46
Mg^{2+}	36.68	1.284	52.82	54.13
Ca^{2+}	11.76	0.4115	10.27	10.52
Sr^{2+}	0.223	0.0078	0.090	0.092
Cl^-	552.93	19.353	545.87	559.39
SO_4^{2-}	77.49	2.712	28.235	28.935
HCO_3^-	3.217	0.1126	1.845	1.891
CO_3^{2-}	0.317	0.0111	0.184	0.189
CO_2	0.017	0.0006	0.013	0.0133
Br^-	1.923	0.0673	0.842	0.863
$B(OH)_3$	0.591	0.0207	0.334	0.342
$B(OH)_4^-$	0.186	0.0065	0.082	0.0839
F^-	0.037	0.00130	0.068	0.070
Totals	1004.87	35.170	1119.87	1147.60
Alkalinity	–	–	2.300	2.37
Everything else	–	~0.03	–	–
Water	–	~964.80	~53, 554	~54, 880

All authorities recommend the mole (abbreviation: mol) as the unit for amount of substance (including the series: mmol, µmol, nmol, pmol, etc.). In marine studies, the most common convention for expressing concentration is the mol per kg of solution, sometimes called the "molinity" scale, by analogy with "salinity." (The term 'molinity' was originally suggested to me in 1969 by Kent Fanning.) This scale is conservative with changes in temperature and pressure, and with mixing. Many analysts, whose laboratory measurements are based on volume, use the "molarity" or "molar" scale (mol L^{-1} = M), but for maximum accuracy this requires that the density of the solution be specified. Here the millimolar values are calculated for $t = 20$ °C, density = 1.024763 kg L^{-1}.

The item "Everything else" includes about 19 mg of dissolved oxygen, nitrogen, and argon. The rest consists of additional dissolved solids such as micronutrients and all the other dissolved elements (about 0.6 mg in surface water) as well as organic matter, commonly about 1 to 2 mg kg^{-1}.

The data for the major constituents were obtained from papers by Culkin and Cox 1966, Riley and Tongudai 1967, Morris and Riley 1966, Culkin 1965, Wilson 1975, Carpenter and Manella 1973, and Uppström 1974. The CO_2 species were calculated for surface seawater in equilibrium with an atmospheric partial pressure of 400 µatm, appropriate for the year 2015. Values for Ca, Sr, and alkalinity are known to increase slightly with depth (relative to S) due to biologically mediated downward transport, many trace elements increase greatly, and total CO_2 increases with depth due to respiration and carbonate dissolution.

concentration scale not given in this table is the *molal* scale, or moles per kilogram of water. The concentration of water in seawater has never been independently measured. It is always calculated by difference, and the result might have to be adjusted if some substance were missed in the calculation (any uncertainty is, of course, entirely trivial). This table also contains some information that we are not yet ready to discuss, such as the alkalinity, which is not constant in seawater and is also a somewhat complicated matter to define. (The term "alkalinity" will be addressed in Chapter 7.) The concentrations of boric acid and total CO_2 are given in their dissociated and undissociated forms in order to closely approximate the masses involved. Their acid/base behavior will be discussed in Chapter 7. Of the major constituents listed in Table 4.1, total CO_2 is by far the most variable because CO_2 exchanges with the atmosphere, its complex chemistry is influenced by temperature and salinity, and it is both produced and consumed biologically.

The number of significant figures given for each of the elements varies. In most cases, the concentrations are known to approximately the number of significant figures given; in each case there is some uncertainty in the last significant figure. The values selected appear to be the best available estimates, or are the means of the two best estimates. The concentration of sodium in seawater is less well known than those of the other major elements because sodium is not determined directly. There is no method known by which it is possible to determine sodium in seawater with an accuracy much better than 1%. At the present time, the concentration of sodium is determined by adding up the chemical equivalent concentrations of all the anions and subtracting from this total the sum of the chemical equivalent concentrations of each of the cations (except sodium) as individually determined. The remainder is reported as the concentration of sodium. If it should be found, for example, that new determinations lead to a change in the accepted value for the concentration of magnesium, then the concentration of sodium would therefore be adjusted accordingly.

Values in Table 4.1 differ in minor details from a tabulation in Millero *et al.* (2008) because they included deep-water values for calcium and strontium, and they used a different temperature and a partial pressure of 320 μatm for CO_2.

It is convenient to make a bar graph which shows the concentrations of the various salts in seawater pictorially. This provides a helpful mnemonic by which we can more easily recall the relative concentrations of the major substances in seawater. In Figure 4.1 the substances are shown in mmol kg^{-1} of seawater, and also in milliequivalents per kilogram of seawater. In the latter case, the total charges of the anions and the cations are equal; this requirement is necessary to maintain electroneutrality. In a graph of this size only six or seven of the major constituents can be shown in blocks that are thicker than the width of a line. The sum total of all other substances present in seawater is approximately the width of one of these lines.

For comparison, the concentrations of substances in mammalian blood serum are also shown. This may be helpful to those whose previous experience with saline solutions was in studies of human or other mammalian physiology and who therefore have a feeling for the relative proportions and amounts of such salts in these blood fluids. Not only is seawater much more saline than mammalian blood serum, but the

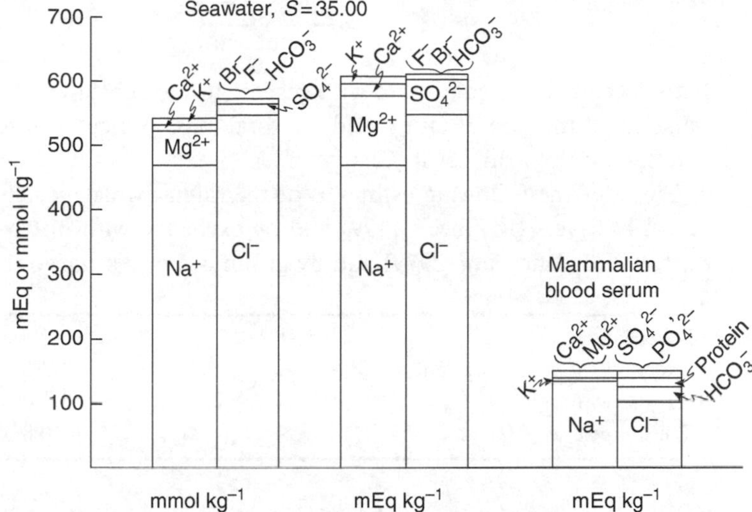

Figure 4.1 Concentrations of salts in seawater at salinity $S = 35.00‰$, expressed in mmol kg^{-1} and meq kg^{-1} (milliequivalents per kilogram of solution). Typical values for mammalian blood serum shown for comparison.

relative proportions of the various ions are different. In blood serum, sodium makes up a larger proportion of the total and the sodium-to-chloride ratio is much greater than it is in seawater. Calcium and magnesium make up a very small part of the total, but potassium is slightly more concentrated. The concentration of bicarbonate ion is much higher in mammalian fluids both absolutely and relatively. The concentration of sulfate is lower, and so on.

It is helpful to remember that in this graph most of the substances shown are essentially conservative in seawater; that is they occur in a constant ratio to the salinity. Much of marine chemistry is concerned with the study of those substances which are very important biologically or geochemically, but which occur in such small concentrations that even the total of their concentrations cannot be shown in a bar graph of this sort.

4.2 Residence times

The degree to which the concentration of any substance in the sea varies independently of the salinity (and thus departs from conservative behavior) depends on the balance between processes that tend to increase or decrease the local concentration (relative to the salinity) and the rate at which the ocean mixes and reduces such local differences. One approach to evaluating the possibilities for non-conservative behavior is to calculate the residence time of the substance. In its simplest form the residence time, or replacement time, is the length of time (using the symbol τ) it would take for the input of a substance into a well-mixed "box," in this case the ocean, to replace the amount present.

$$\tau = \frac{\text{mass present}}{\text{input rate}}$$

If the ocean is in steady state with respect to the substance, then the rate of output must equal the rate of input, and the total mass present will remain constant. (This will be discussed further in Chapter 13.)

Consider the following estimates of the rate at which calcium is carried into the ocean by rivers (Meybeck 1979) and by exchange with hot basalt at the spreading centers (Edmond *et al.* 1979) and by groundwater discharge (Milliman 1993):

River input	12.5×10^{12} mol yr^{-1}
Exchange at spreading centers	3.3×10^{12} mol yr^{-1}
Groundwater discharge	5.0×10^{12} mol yr^{-1}
Total input	20.8×10^{12} mol yr^{-1}

The total amount in the ocean at the present time (calculated from data in Tables 1.1 and 4.1) is about 14.2×10^{18} mol. The simple calculation:

$$\frac{14.2 \times 10^{18} \quad \text{mol}}{20.8 \times 10^{12} \quad \text{mol yr}^{-1}} = 0.68 \times 10^{6} \text{ yr}$$

gives the "residence time," in this case 680 000 years. This is the time required for the input of calcium to equal the amount present. It is usual to assume that most substances in the ocean are approximately in steady state, so that the total amounts in the ocean are nearly constant over time. The losses of calcium must therefore nearly match the inputs and are mostly due to the depositing of calcium carbonate in coral reefs and as a rain of shells from planktonic marine organisms onto the sea floor. Evidently the outputs are separated in space from the inputs, and we can ask if the ocean mixes quickly enough to erase the possible differences.

It is not easy to find a measure of the homogenization rate of the ocean. It is, however, known approximately how fast the deep water in the ocean is replaced. It takes about 1000 years for water sinking at high latitudes to replace the volume of water in the deep ocean. (Here, the somewhat arbitrary division between the deep ocean and the more rapidly replaced surface layers is set at 1000 m.) The oceans, of course, cannot be mixed homogeneously in only one such 1000-year cycle, but certainly a great deal of mixing must take place during this length of time. After a number of such replacement cycles, the ocean should be thoroughly mixed. Even if inputs of calcium to the ocean are localized near the edges where rivers enter and along the ridge-crest spreading centers, and the outputs are localized in other areas mentioned, the mixing of the ocean waters would not allow concentration differences from place to place that are great enough to be measurable, except perhaps in estuaries and shallow tropical bays. Simple comparison of the magnitudes of the timescale of mixing and the timescale over which the calcium in the ocean is replaced suggests that no gradients are likely to be observed in any water body not immediately adjacent to a source (such as rivers and ridge crests), and that calcium should appear to be a conservative element in seawater. Nevertheless, there has long been

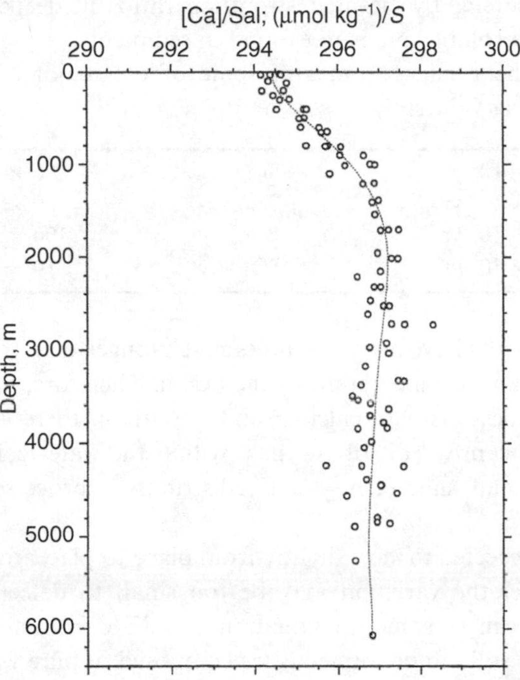

Figure 4.2 Concentrations of calcium in seawater, normalized to salinity (calcium in μmol kg^{-1} divided by salinity in ‰), at four stations in the South Pacific, all within a few hundred miles of each other in the vicinity of the Tonga Trench (15–20° S, 170–173° W). Data from Horibe *et al.* (1974).

evidence for the existence of vertical gradients in concentration. This was first shown by Dittmar in his analyses of seawater collected by the *Challenger* expedition. A more recent set of measurements of water collected in the Pacific shows that the surface water there (at least at those fairly typical locations) is depleted in calcium by about 1% relative to the deep water (Figure 4.2).

This vertical gradient is not large, and to observe it requires exquisitely precise analytical techniques, but it suggests that biological processes transporting calcium within the sea are likely to be acting on a rather large scale. These processes are carried out by calcareous plants and animals whose skeletons of calcium carbonate sink from the surface and, in part, dissolve at depth. If these data are characteristic of the differences between the surface and deep ocean generally, they imply a downward transport of calcium carbonate and dissolution at depth of probably more than 40×10^{12} mol yr^{-1}, a rate of internal transport and recycling significantly greater than the net inputs and outputs. This internal redistribution is just large enough to be determined by direct measurement of the calcium. Later, we will see that there are more sensitive ways to estimate this transport than by measuring the calcium concentration.

Surface water is also depleted in strontium, by about 1 to 3% (depending on location) relative to deep water (Bernstein *et al.* 1987). This decrease is even more difficult to observe, since the strontium concentration in seawater is less than 1% that of calcium. The decrease is also controlled biologically and, in this case, is thought to be due mostly to marine protozoa belonging to the group Acantharea, which live in surface waters and manufacture tests made of $SrSO_4$. The tests sink to deeper water and dissolve, causing the slightly elevated strontium ratios in deeper water. The compound $SrSO_4$ is quite soluble and is not found in sediments.

Rough estimates of the residence times of some of the major constituents in seawater give the following values:

Na^+	70×10^6 yr	Sr^{2+}	6×10^6 yr
K^+	7×10^6 yr	Cl^-	100×10^6 yr
Mg^{2+}	14×10^6 yr	Br^-	100×10^6 yr
Ca^{2+}	0.7×10^6 yr	SO_4^{2-}	10×10^6 yr

These additional constituents have residence times much longer than that of calcium and very much longer than the mixing time of the ocean. Therefore, they should all behave conservatively unless, as with calcium and strontium, there is some active process that might significantly redistribute them within the water column. Except for calcium, strontium, and sulfate, no such redistribution processes have been discovered.

Sulfate ion might be expected to vary slightly from place to place, for at least two reasons, although usually the variation may be too small to detect. During the freezing of sea ice there can be some fractionation of sulfate from chloride. When seawater freezes, the first solid to be formed is ice composed of pure water only; the growing crystals reject the more concentrated brine. Under usual conditions in nature, however, some of the brine is trapped among the crystals of ice. As the mixture of ice and brine cools, the brine is further concentrated by removal of water. At a temperature of –8.2 °C, crystals of $Na_2SO_4 \cdot 12H_2O$ begin to form and continue forming until –22.9 °C, where $NaCl \cdot 2H_2O$ begins to precipitate (Nelson and Thompson 1954). As the ice ages the residual brine can slowly melt its way down and escape from the bottom of the ice, leaving the $Na_2SO_4 \cdot 12H_2O$. If one takes sea ice and melts it, one finds that this ice has retained some salt and in this salt the sulfate-to-chloride ratio is elevated. Therefore, in seawater which is left behind by the freezing of sea ice, the sulfate-to-chloride ratio must be somewhat lowered. It was once speculated that it might be possible to use the sulfate-to-chloride ratio to trace water masses that have been formed through mixing with the residual brine from the freezing or melting of sea ice, and to differentiate these from water masses that have formed purely by cooling and evaporation. It is clear that this process is not very extensive in the ocean, and the prospect of tracing any major water mass this way is remote. Nevertheless, one paper has reported that it was possible to track for some short distance the water that was formed by the freezing of sea ice in the Antarctic.

The other process whereby the sulfate-to-chloride ratio can be altered occurs in anoxic basins. In Chapter 12 we will discuss the anoxic basins in detail, but at this point we note that in such basins some of the sulfate is reduced to sulfide. Because of its very different chemistry, the sulfide may, in one way or another, be removed from the water so that both the sulfate-to-chloride ratio and the total sulfur-to-chloride ratio may be altered. This commonly changes the sulfate-to-chloride ratio in the pore waters of marine sediments, but in the present ocean this process is not extensive enough to cause detectable variations within the major ocean basins.

At the present time there is no convincing evidence that, relative to the salinity or to the chloride concentration, the concentrations of sodium, magnesium, potassium, sulfate, fluoride, or bromide vary detectably from place to place within the major water masses in the ocean.

Summary

The list of 11 major constituents of seawater includes those substances present in normal average seawater at concentrations greater than 1 mg kg^{-1}. It is conventional that nitrogen and oxygen are not included in this list, which includes only substances that are present in the solid matter remaining when the water is evaporated off. Of the 11 substances listed, carbon dioxide is the most variable, because it exchanges as a gas with the atmosphere and is biologically both consumed and produced.

The remaining 10 substances are conservative or nearly so. A conservative substance is one that is found in constant proportion to the salinity or to the concentration of chloride. Slight departures from conservative behavior are observed for calcium and for strontium, because these two elements are transported vertically downwards at significant rates in the form of biologically produced calcium carbonate and strontium sulfate, respectively.

The reason that the 10 substances are found to be conservative (or nearly so) is that they have long residence times. The ocean is mixed to a considerable extent in about a thousand years, and must be pretty well homogenized in several thousand years. These substances all have residence times exceeding several hundred thousand years, so their input or output rates are very small compared to the amounts present, and spatial variations of these rates will have little measurable effect on the concentrations.

SUGGESTIONS FOR FURTHER READING

I have not found a recent discussion that provides additional information on the major constituents of seawater. A useful survey of the subject, while somewhat dated, is provided in this chapter by T.R.S. Wilson.

Wilson, T. R. S. 1975. Salinity and the major elements of sea water. In *Chemical Oceanography*, 2nd edn, vol. **1**, ed. J. P. Riley and G. Skirrow. Academic Press, New York, ch. 7, pp. 365–413.

5 Simple gases

... under equal circumstances of temperature, water takes up, in all cases, the same volume of condensed gas as of gas under ordinary pressure. W. HENRY 1803

In this chapter we consider those gases present at constant concentrations in air; these are nitrogen, oxygen, and the noble gases (except radon). These gases occur in essentially constant relative proportions everywhere in the atmosphere up to an altitude of about 95 km. The relative proportions of these gases dissolved in seawater are different than in air and vary according to conditions. Oxygen varies because it is biologically produced and consumed, and several physical processes affect the relative concentrations of all of them. Investigation of these variations leads to important insights into several oceanic processes.

Air always contains water vapor. At high altitudes and at such cold high regions as the South Pole the water vapor content may be less than 1 part in 10^5 (mole fraction < 0.00001), while in the humid tropics it may exceed 1 part in 20 (mole fraction > 0.05). Water vapor often must be accounted for in various calculations dealing with components of air.

Other gases of some interest in the marine environment, such as radon, carbon dioxide, carbon monoxide, methane, ammonia, nitrous oxide, dimethyl sulfide, hydrogen sulfide, etc. are all more or less variable in concentration and some will be discussed in later chapters.

5.1 General considerations

The concentrations in air of the six gases considered here (Table 5.1) are listed as the mole fraction for each gas, which is very nearly the fractional composition by volume.

It is sometimes useful to know the mean molecular weight of air. Taking the respective molecular weights, and data from Table 5.1, this is 28.95 g mol^{-1} for dry air. When CO_2 is included, the value becomes 28.97 g mol^{-1}. Since the molecular weight of water is about 18, humid air is always slightly less dense than dry air at the same temperature.

Gas pressures are reported in a great variety of units. The recommended SI unit of pressure is the pascal, Pa, which is one newton per square meter. Some authorities discourage the use of the bar ($= 10^5$ Pa), but it is in common use, and has the

Table 5.1 Concentrations of conservative gases in dry air, the solubility of each pure gas in pure water, and the concentrations in pure water in equilibrium with the atmosphere

Gas	Symbol	Mole fraction in air	Solubility, μmol kg^{-1} bar^{-1}	Concentration, mol kg^{-1}
Nitrogen	N_2	0.7808	695.0	537.4×10^{-6}
Oxygen	O_2	0.2095	1372.2	284.7×10^{-6}
Argon	Ar	0.00934	1508	13.95×10^{-6}
Neon	Ne	18.2×10^{-6}	462.0	8.328×10^{-9}
Helium	He	5.24×10^{-6}	385.4	2.00×10^{-9}
Krypton	Kr	1.14×10^{-6}	2764	3.12×10^{-9}
Xenon	Xe	0.087×10^{-6}	4996	0.434×10^{-9}

Atmospheric data from Warneck (1988). "Mole fraction in air" means moles of each gas per total moles of gas, commonly expressed as, for example: $X_{N2} = 0.7808$. The solubility of each gas is given as the concentration at 20 °C in pure water relative to the pure gas at a gas partial pressure of 1 bar (= 100 kPa). This is one form of the true Henry's law constant. The last column gives the concentration in pure water at 20 °C in equilibrium with wet air at a total pressure of 101.325 kPa (standard atmospheric pressure). Solubility and concentration values for He, Kr, Xe carry probably one more significant figure than is currently justified, as the solubilities of these gases have not yet been re-examined to the best current standards. Concentrations and solubilities calculated from relations in Appendix D.

advantage that 1 bar is close to the standard atmospheric pressure of 101 325 Pa, while the decibar (dbar) is close to the pressure exerted by one meter of water. Meteorologists use the millibar (= 100 Pa). The standard atmospheric pressure widely used for more than 100 years by physical chemists to define thermodynamic standard states was 1.01325 bar = 101 325 Pa. In recent decades, however, 100 000 Pa is generally used as the standard state. The actual pressure of the atmosphere varies, of course, with altitude, but even at sea level it may vary in the extremes by more than 7% around the global average of about 1011 millibar, due to variations in temperature and the dynamic motions of the atmosphere.

5.2 Simple gas laws

In any sample of mixed gases, the total pressure (p_T) is the sum of the partial pressures (pp) of the individual components. For air:

$$p_T = pp(N_2) + pp(O_2) + pp(Ar) + pp(H_2O) + \ldots. \quad (5.1)$$

In each case, the partial pressure of the gas is in very nearly the same ratio to the total pressure as the number of molecules of that gas to the total number of gas molecules present.

For any gas, the pressure, volume, and temperature are related by the ideal gas law:

$$pV = nRT, \quad (5.2)$$

where p is the pressure of the gas; V is the volume of the gas; n is the number of moles of gas present (each mole contains Avogadro's number of units: 6.022×10^{23} molecules, or atoms for monatomic gases); T is the absolute temperature, K (= °C + 273.15); R is the gas constant, and has units consistent with the other parameters ($R = 0.082057$ L atm K^{-1} mol^{-1}, when the pressure is in atmospheres; if the pressure is in bars, the constant has the value: $R = 0.083145$ L bar K^{-1} mol^{-1}).

Thus, at STP (standard temperature and pressure: 0 °C or 273.15 K, and 1 atmosphere pressure = 101 325 Pa) 1 mole of an ideal gas occupies a volume of 22.414 L, while if the standard pressure is 1 bar, the molar volume is 22.711 L.

All real gases deviate slightly from the "ideal" behavior expressed by Eq. (5.2). For precise work (Weiss 1974), it is sometimes necessary to account for these deviations by using one of several possible equations of state with additional empirically measured terms, such as the so-called "virial equation":

$$PV_{\mathrm{m}} = RT\left(1 + \frac{B_{\mathrm{T}}}{V_{\mathrm{m}}} + \ldots\right), \tag{5.3}$$

where V_{m} is the molar volume, and B_{T} is the second of several virial coefficients (RT conventionally being the first; subsequent terms after the second are only of interest in extreme circumstances). The coefficient B_{T} is a function of temperature and is different for each gas; tabulated values may be found in the *CRC Handbook of Chemistry and Physics* (Haynes 2011). For all gases considered here, departures from ideality under atmospheric conditions (total pressure near 1 atmosphere and oceanic temperatures) are less than 1% (e.g. less than –0.10% for N_2 and O_2, but –0.69% for Xe); so, for the light gases, Eq. (5.2) is usually satisfactory. However, for accurate calculation of the situations that might be found under high pressures in the ocean, such as, for example, in the swim bladders of fish at great depths or in the gas vesicles of undersea basalts, the virial equation or some other formulation is necessary.

5.3 Solubility in water

The amount of a gas dissolved in a quantity of water, when the gas and water phases are in equilibrium, is directly proportional to the pressure of the gas. This fact was discovered by William Henry in 1803 and is expressed by the simple relationship now known as Henry's law:

$$[\mathrm{G}] = H_{\mathrm{G}} \times pp(\mathrm{G}), \tag{5.4}$$

where [G] is the concentration of gas G in solution; pp(G) is the partial pressure of gas G; H_{G} is the Henry's law constant for gas G, and is a function of T and S. H_{G}, in addition to being a function of temperature and salinity, is characteristic for each gas.[1]

While a confusing plethora of units and conventions has been used for expressing the concentrations and solubilities of gases in seawater (see Appendix D), preferred

[1] Some authors use a Henry's law constant expressed in the inverse way; i.e., the constant goes down as the solubility goes up, as in $[\mathrm{G}] = pp(\mathrm{G})/k_{\mathrm{G}}$. This is something to beware of.

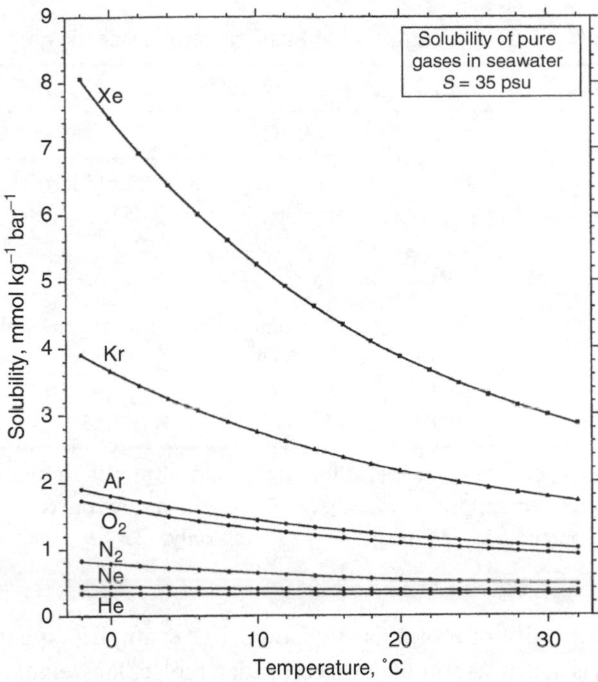

Figure 5.1 Solubilities of the noble gases plus nitrogen and oxygen in seawater ($S = 35‰$) as a function of temperature. Each case shows the concentration in a solution in equilibrium with a gas phase where the partial pressure of the gas is exactly one bar (100 000 Pa). (Data derived from tables in Appendix D.) Note both the very large differences in the intrinsic solubilities of the different gases and the range in sensitivity to temperature. Note also that in the series of noble gases the solubility increases greatly with atomic weight, though not in a linear fashion. The diatomic molecules (N_2 and O_2) are more soluble than monoatomic molecules would be at similar weights. Evidently the sizes, shapes, and electronic structures of gas molecules must strongly influence their interactions with water molecules and thus the observed solubility.

concentration scales should give values that reflect the stoichiometry of chemical relationships and do not change with pressure or temperature. Therefore, following Kester (1975), the concentrations will usually be expressed here in moles of gas per kilogram of solution (generally $\mu mol\ kg^{-1}$ or $nmol\ kg^{-1}$).

Since the major conservative gases in the atmosphere are present in very nearly constant proportions, it has been the usual oceanographic practice not to calculate from the Henry's law constant for each gas, or to give solubilities as concentrations in equilibrium with the pure gas. Instead, their solubilities are provided by tabulating their concentrations in seawater in equilibrium with the atmosphere, at 1 atmosphere total pressure, under specified conditions of temperature and salinity. The conventions of (a) having the atmosphere dry, or (b) saturated with water vapor at the temperature of the measurement, have both been used (Appendix D), and the latter convention is used for the solubility tables in the Appendix.

As shown in Table 5.1 and in Figure 5.1, the solubilities of the various pure gases are quite different, and increase with increasing molecular weight, and with increasing

Table 5.2 Concentrations of three major atmospheric gases in water and seawater

Gas	Air, %	Pure water		Seawater, $S = 35‰$	
		0 °C	30 °C	0 °C	30 °C
		Concentrations (μmol kg^{-1})			
N_2	—	830.5	457.9	622.0	362.5
O_2	—	457.0	237.3	347.9	190.7
Ar	—	22.30	11.65	17.02	9.38
		Composition of total gas (mol%)			
N_2	78.08	63.4	64.8	63.02	64.4
O_2	20.95	34.9	33.6	35.3	33.9
Ar	0.934	1.70	1.65	1.72	1.67

Concentrations calculated for equilibrium with wet air at a pressure of 1 atmosphere. Percent composition of the gases in air and in solution, calculated in each case for simplicity as percentage of $\Sigma([O_2] + [N_2] + [Ar])$ only.

complexity of molecular structure. For example, oxygen, with a molecular weight of 32, is nearly as soluble as argon with a molecular weight of 40. The triatomic molecules N_2O (nitrous oxide) and CO_2 (carbon dioxide), each with a molecular weight of 44, are roughly six and eight times more soluble than xenon with an atomic weight of 131, so they cannot be shown on Figure 5.1. As a consequence of the differing solubilities, the mixture of gases dissolved in water in equilibrium with air does not have the same percentage composition of each gas as the air from which it came. This is shown in Table 5.2. An examination of this table reveals a number of the important properties of atmospheric gases relative to their solubilities in water.

First, the gases are much less soluble at high temperatures than they are at low temperatures. This fact is quite general for all gases.

Second, the gases are less soluble in seawater than they are in fresh water. The effect of salt is always to reduce the solubility of gases in water.

Third, as already noted, the gas extracted from a sample of water which has been in equilibrium with the air does not have the same percentage composition as does the air from which it came.

Fourth, the change of solubility with changing temperature is different for each gas (Figure 5.2).

The solubilities of these gases in either pure water or seawater do not change in a linear fashion with temperature, but show a curved relationship. The solubilities are not linear with salt content either, but the curvature is less than for temperature, over the salinity range of natural waters (Figure 5.3).

The observation that the solubilities of these gases show non-linear relationships with temperature and with salinity, and that the relationships between the solubilities of the various gases are not proportional under various conditions, might seem to complicate the use of these gases in oceanographic studies but in fact provide many opportunities to gain additional useful information.

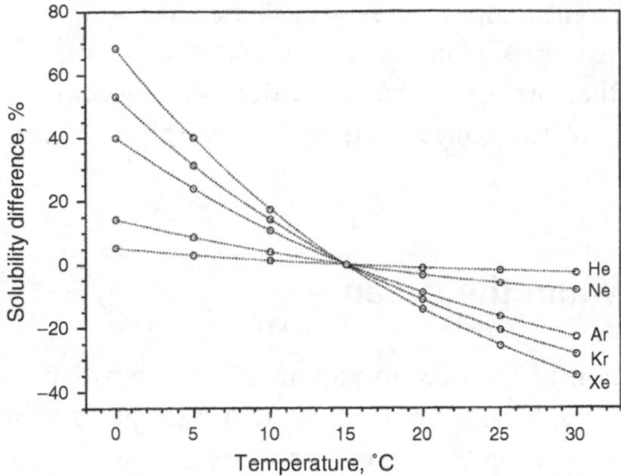

Figure 5.2 Effect of temperature on the solubilities of the noble gases in seawater at a salinity of 35‰, expressed as the percentage change from the solubility for each gas at 15 °C. (Data from Appendix D.)

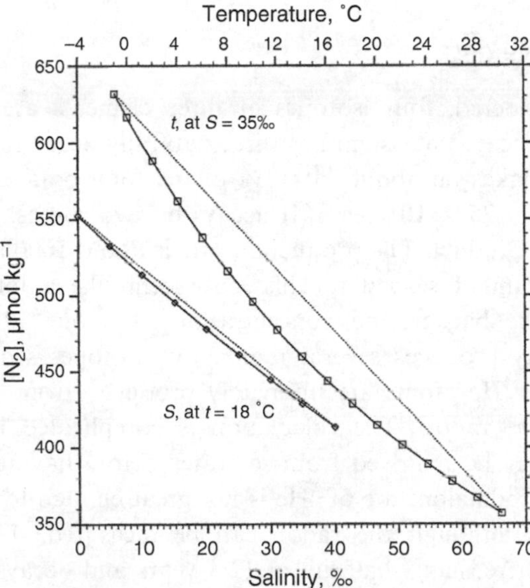

Figure 5.3 Solubility of N_2 in seawater as a function of temperature at $S =$ 35‰ (upper curve) and of salinity at 18 °C (lower curve), relative to wet air. The dotted lines show the concentrations that would be found if two water masses, with the characteristics of the end members and saturated against air, mixed in various proportions without any opportunity to re-equilibrate with air. (Data from Appendix D.)

Consider, for example, the curved relationship for the solubility of nitrogen with temperature. If one mixes two water masses at different temperatures, and in each water mass the nitrogen was in equilibrium with the atmosphere at the temperature of the water, then the resulting mixture will have a temperature which is nearly the average of the two temperatures and a gas concentration which is the average of the two gas concentrations, but the gas concentration will be greater than the saturation value and the nitrogen will appear to be supersaturated. For example, suppose that in investigating conditions in some water mass a careful measurement of the concentration of nitrogen in this water showed that it is supersaturated. This might be taken as evidence that nitrogen was being formed there, or introduced by some process, whereas in fact it might be only the consequence of the mixing of two water masses that were both exactly saturated with nitrogen but at quite different temperatures. In practice, it is difficult to determine the nitrogen with sufficient precision for this kind of analysis. As we will see later, there are several other processes that can affect the concentrations of gases, and careful measurement of the concentrations of several gases can help to sort things out.

5.4 Sources and sinks within the ocean

Before further consideration of the oceanographic information that can be derived from a study of various gases, we must consider where these gases might come from and where they may go. In addition to gain or loss by exchange with the atmosphere, there are two possibilities. First, gases may be produced or consumed *in situ*. Second, they may exchange with the sediments.

5.4.1 Production by radioactive decay

Of the noble gases considered, four isotopes of three elements are produced in seawater by radioactive decay. Potassium has three naturally occurring isotopes; of these, ^{40}K, which now makes up about 0.0117% of the total potassium, is radioactive, with a half-life of 1.25×10^9 years. It decays by two routes, one of which yields ^{40}Ar as a daughter product. The production rate is about 5000 atoms of ^{40}Ar per hour in each kilogram of seawater. This may seem like a lot, but it is a negligible addition to the background concentration of about 10^{19} atoms per kilogram. Uranium decay produces several radioactive isotopes sequentially (see Chapter 10), and several ^{4}He atoms are ultimately produced from each atom of uranium, as well as some radon. The calculation is complicated because some intermediate products may be removed from seawater before having a chance to decay. In any case, the production rate of ^{4}He is not great enough to be detectable against the background, although the radon can be measured. Tritium, which originates in the atmosphere, has a half-life of 12.3 years and decays to ^{3}He. The background concentration of ^{3}He is extremely low, so even the very low production rate from tritium is easily detectable. Therefore, the occurrence of ^{3}He is, in part, a tracer for the past presence of tritium. It can be a sensitive tracer for short-term processes affecting surface waters.

5.4.2 Biological fluxes

The production of oxygen by photosynthesis and its consumption by respiration will be discussed in much more detail later. Oxygen is produced during the daytime in all parts of the ocean where light is sufficient for photosynthesis. It is consumed everywhere in the ocean, all the time, except for those regions where the concentration has been reduced to zero. Nitrogen also is biologically produced and consumed through denitrification and nitrogen-fixation reactions, respectively. These latter processes operate only slowly in seawater relative to the large amounts of nitrogen present, their rates are not well understood, and effects on the concentration of nitrogen gas in normal seawater are generally too small to measure directly. In sediments, however, the process of denitrification, to be discussed later, can yield measurable excesses of nitrogen in the interstitial water. Nitrogen fixation in sediments has also been quantified in laboratory measurements by measuring the loss of nitrogen from the overlying water (Fulweiler *et al.* 2007).

5.4.3 Exchange with the sea floor

Radon and helium, and to some extent nitrogen, are produced in the sediments. Radon is a product of radioactive decay from radioactive minerals in the sea floor. It has a half-life of only a few days, so it is found in the largest concentrations near the sediment–water interface; measurements of the distribution of radon within the bottom several hundred meters of the sea can yield estimates of the rate of vertical mixing of the water there. The production of ^4He from the sediments, while greater than in the water column, cannot be seen against the background. A sizeable increase in the concentration of both ^4He and ^3He is, however, observed at mid-depths in the Pacific, and this has been shown to come from out-gassing from the Earth's mantle. A considerable volume of lava and included volatile substance is continually extruded at a depth of around 2000 meters in the region of the East Pacific Rise, one of the most active mid-ocean-ridge areas, so the increased helium at mid-depths is most easily seen there, but must also be introduced in many parts of the ocean along the mid-ocean ridges and other regions of undersea volcanism. Nitrogen is produced by denitrification reactions in the sediments, and oxygen is consumed by respiration of benthic organisms on and in the sediments.

5.5 Atmospheric exchange by diffusion

Exchange with the atmosphere is, of course, the process with which we are most familiar, and is the dominating influence on the concentrations of most gases in seawater. All atmospheric gases must exchange with surface waters; the net rate must be in the direction towards equilibrium under the ambient conditions. In considering the physical processes involved in this exchange at the air–water interface, it becomes evident that the situation is not entirely simple. Both the air and the water are always in motion to some extent. Turbulent eddies commonly must keep both air and water mixed and chemically homogeneous at some distance above and below the interface. However, such eddies cannot penetrate exactly to the surface, and indeed, at some distance, perhaps very close to the interface, eddy motions must nearly vanish. A common way to imagine the situation is to consider that right at the interface there are unstirred laminar boundary layers (Figure 5.4) that do not readily mix with the air above or the water below. For non-reactive gases, such as O_2, N_2, and the noble gases, it can be shown that they diffuse rapidly enough through the unstirred

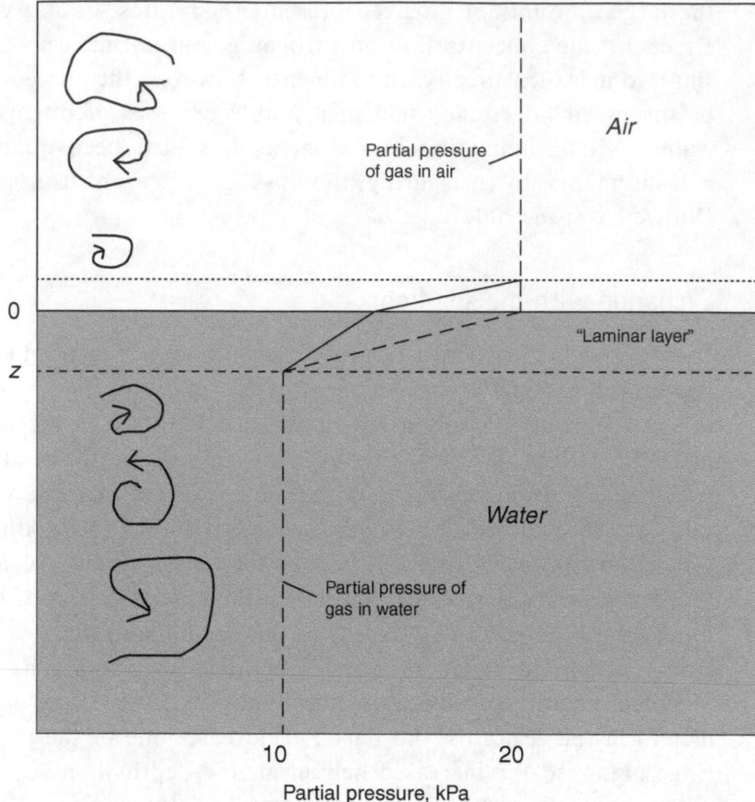

Figure 5.4 Conceptual model of air–sea gas exchange, according to the theory that unstirred laminar boundary layers form a barrier at the surface. The drawing is made for a situation where some gas has a partial pressure in the liquid phase only one-half that in the air, so that transport of gas is from the air to the liquid. Irregular turbulent eddies maintain a homogeneous partial pressure (and concentration) in the bulk of each phase, but rarely penetrate the boundary layers. Within the boundary layers the transport is by molecular diffusion alone, a much slower process (Broecker and Peng 1974, 1982).

laminar boundary layer in the air that this is not a significant barrier. However, there is certainly a barrier imposed by the layer of unstirred water. For very reactive gases, such as SO_2 or NH_3, the reaction rate at the water surface is so great that diffusion in the boundary layer of the air limits transport rates so in such a situation this must also be considered (Liss and Slater 1974).

This unstirred layer is too thin for its characteristics to be measured directly. It must be in part a transient feature, with water entering and leaving the layer according to turbulence below and the stress of the wind above. A great deal of effort has gone into examining the structure of this layer and devising various models of the transport of gases and other substances across it.

What can be measured, in favorable circumstances, is the actual flux of gas across the air–water surface or interface, and some of the conditions that affect this, whether in beakers or some other laboratory apparatus or in lakes or the ocean. The flux is usually reported as mol m^{-2} s^{-1} or some related unit. Any net flux across the surface requires a difference in partial pressure of gas between that in the water phase and in the gas phase, and will be driven by this difference. Since the partial pressure is proportional to the concentration, the flux of gas will be related to the difference between the concentration deep in the body of the liquid and the concentration that would exist exactly at the interface, assuming that the gas there is in equilibrium with the air above:

$$\text{flux} = e_G \times ([G]_I - [G]_W), \tag{5.5}$$

where the flux is the mass transported per unit area and per unit time (mol m^{-2} s^{-1}); $[G]_I$ is the calculated gas concentration at the interface (mol m^{-3}), from $[G]_I = H_G \times pp(G)$, $[G]_W$ is the measured gas concentration in the bulk liquid (mol m^{-3}); and e_G is the exchange coefficient for this gas (m s^{-1}), under existing conditions.

Because e_G has the units of m s^{-1}, which is a velocity, e_G has been variously called the "transfer velocity," "exchange velocity," "exit coefficient," "exchange coefficient," or "piston velocity." (I suppose that in thinking of the velocity upward or downward in a column of water, the image of a piston came to mind, and the term was thereby introduced.) Here we will use the term *exchange coefficient*.

As a conceptually simple introduction (Figure 5.4), imagine that in the boundary layer of unstirred water the transfer of gas is by molecular diffusion only. In this case, the flux of the gas in question is a function of its diffusion coefficient (which in turn depends on the type of gas and the temperature and viscosity of the water; a typical value of D_G for O_2, for example, might be 2×10^{-9} m^2 s^{-1}). A formulation for transport by molecular diffusion (Fick's law) is as follows:

$$\text{flux} = \frac{D_G \times ([G]_I - [G]_W)}{z} \tag{5.6}$$

and

$$e_G = \frac{D_G}{z}, \tag{5.7}$$

where z is the thickness of the unstirred laminar boundary layer and D_G is the molecular diffusion coefficient of the gas in water.

From these equations, it would be predicted that the fluxes of various gases should be directly proportional to their respective diffusion coefficients, other factors being equal. For some time it was thought that if the flux of one gas under known conditions of wind speed could be measured, and if its diffusion constant (D) in water was known, then the thickness of the unstirred boundary layer could be estimated. From this, the fluxes of other gases could be calculated from their diffusion coefficients and relative concentrations in air and water. Estimates of the thickness of the unstirred boundary layer in the open ocean typically range from 30 to 70×10^{-6} m, depending on the state of the wind, under conditions that allow the measurements to be made, and thicknesses several times greater can be calculated from flux measurements in tanks with low wind speeds.

The exchange coefficient varies quite strongly with conditions. To study the exchange of gases between the air and the sea, the challenge has been to measure the flux of gases under a variety of conditions, of which wind speed is generally considered to be the most important, and to develop a relationship between e_G and the conditions of interest. This has been done in laboratory experiments and by measuring the flux of gas into or out of the water column of lakes and even the whole ocean (Broecker and Peng 1974, 1982; Liss and Slater 1974; Ledwell 1984; Wanninkhof 1992). It turns out that the simple formulation above does not describe the data at all well.

Ledwell (1984) modeled the transfer of turbulent motions from the body of a liquid into the boundary layer near the air–water interface, and the corresponding effect on the diffusion of gases through the layer. His evaluation showed that the exchange coefficient for a variety of gases should not vary as D but as $D^{0.67}$ for a quite smooth surface and about $D^{0.5}$ for the much more common condition in nature where some waves are present. The viscosity of water in the "unstirred layer" just below the interface is important in affecting the transfer of momentum from wind into the water and from the bulk water towards the interface. Both viscosity and D vary with temperature and (very slightly) with salinity. In order to take account of these various effects, it was convenient to introduce the Schmidt number:

$$\mathrm{Sc} = \frac{\nu}{D}, \text{ the units of which are } \frac{\mathrm{m^2 s^{-1}}}{\mathrm{m^2 s^{-1}}},$$

where ν is the kinematic viscosity (dynamic viscosity/density) of the seawater; D is the diffusion coefficient for each gas in seawater; and Sc is the Schmidt number, a dimensionless number, characteristic of each gas; it varies strongly with temperature and weakly with salinity.

Values for the dynamic viscosity and the kinematic viscosity of water and seawater are tabulated in Appendix C, Table C.4, some recently measured values for the diffusion constants of several gases are tabulated in Table D.10, and some estimates of the Schmidt number for various gases in seawater are in Table D.11.

Since the Schmidt number is inversely proportional to the diffusion constant, it is evident that gas exchange rates must be proportional to the inverse of the Schmidt number, and the flux will be formulated as follows:

$$\text{flux} = \frac{f}{\mathrm{Sc}^{0.5}}([\mathrm{G}]_\mathrm{I} - [\mathrm{G}]_\mathrm{W}), \tag{5.8}$$

where $f/\mathrm{Sc}^{0.5}$ is the exchange coefficient e_G, f is some function of the wind speed that replaces the unstirred boundary layer thickness z in the earlier formulation, and the superscript 0.5 is for the most common conditions but may be as great as 0.67.

Studies of the rates at which these gases exchange across the air–water interface under varying wind conditions have shown that the transfer rates increase greatly as the wind speed goes up (Figure 5.5). The comparison of Wanninkhof's data with

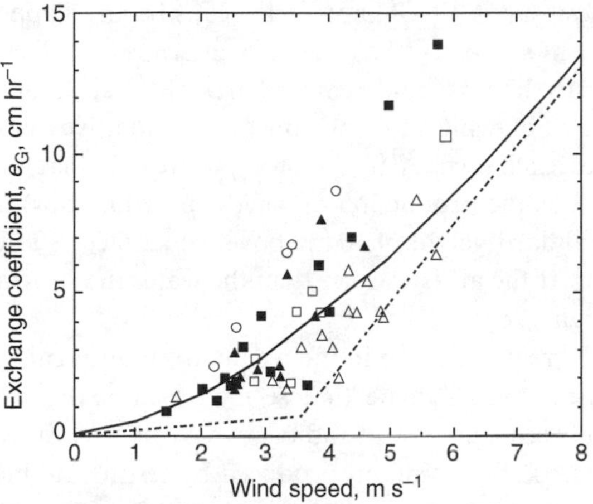

Figure 5.5 Gas transfer velocities as a function of wind speed; experimental data from lakes. These data were obtained by introducing the tracer gas SF_6 into lake water, and observing the rate at which the gas was lost from the lake under a variety of wind speeds. The wind speed is plotted as the speed at the standard height of 10 m above the water surface. The exchange coefficient, e_G, has been normalized from that for SF_6 at various temperatures to a standard Sc of 600, as this latter value is close to that (at 20 °C in fresh water) for CO_2, a gas of more general interest. The dashed line is from a fit by Liss and Merlivat (1986) to an extensive set of data from earlier studies in lakes and the ocean. The solid line is a fit to the data presented here; the equation for this line is $e_G = 0.45\ u^{1.64}$, where u is the wind speed (redrawn from Wanninkhof 1992).

earlier results, and the scatter of the data, show how difficult it is to obtain precise evidence from field data, and also provide evidence that instantaneous wind speed alone may not be a sufficient measure of the conditions which control air–water gas exchange. The effect of wind speed may be qualitatively understood as composed of the following factors: the increase in surface area by ripples and waves, a thinning of the boundary layer by an increase of turbulent motion in the water below, and the conversion of wave energy (originally from the wind) into turbulent energy. The effects increase dramatically as wind speed increases, but as winds exceed 10 m s^{-1} (~20 knots) it becomes increasingly difficult or impossible to make meaningful measurements. Average wind speeds in low and mid latitudes are less than 10 m s^{-1}, however, and many measurements have been made of the gas fluxes in various regions, both in lakes and in the ocean. The data from different investigations are quite scattered. It is argued by Jähne *et al.* (1987a) that the scatter is not necessarily due to errors in measurement, but to the fact that at any location in the ocean the wave field that contributes to turbulence in the water may often derive from somewhere else and not be in equilibrium with the instantaneous wind field at the time of measurement.

Additional factors that may affect the relationship of a measured exchange coefficient to the wind speed include the following. Small ripples form very quickly, but it takes some time for larger waves to adjust to a change in wind speed. Above a speed of about 4 m s^{-1} (~8 knots) whitecaps begin to appear, and become visibly common over 8 m s^{-1}. Also, the onset of whitecaps is earlier and coverage is much greater at tropical than at cold temperatures (Monahan and O'Muircheartaigh 1986). The presence of whitecaps introduces a qualitative change in the nature and extent of air–sea exchange. A further complication is that the wind stress on the sea surface (and thus the production of waves and whitecaps), relative to the measured wind at the standard height of 10 m above the surface, is affected by the thermal structure of the air. If the air is warmer than the water the stress is less (Monahan and O'Muircheartaigh 1986).

The great variation in the results from different studies led Wanninkhof (1992) to focus on one estimate that seemed secure. Broecker *et al.* (1985a) calculated the global ocean uptake of radioactive carbon-14, in the form of $^{14}CO_2$. They used as data the known amount produced by testing atomic bombs in the atmosphere, and measurements over time of the amount of $^{14}CO_2$ in the atmosphere and in the ocean. This estimate did not depend on any details of turbulence, meteorology, etc. The only major uncertainty was whether the ocean inventory was adequately captured by the incomplete surveys available. The estimated uptake rate could then be related to the known average wind speed over the whole ocean (Figure 5.6). Since a variety of measurements had suggested that the exchange velocity seems to depend on the square of the wind speed, Wanninkhof forced an equation containing this function of the wind speed through that point and another for the Red Sea only and adjusted the equation to take account of the statistical distribution of wind speeds, the non-linearity of the equation, and for the fact that the global average value is the result of summing air–sea exchanges over the whole range of wind speeds (Figure 5.6). Wanninkhof's suggested relationship for the exchange velocity became (slightly modified):

$$e_G = \frac{8.0u^2}{\mathrm{Sc}^{0.5}}; \tag{5.9}$$

so the rate of exchange would be calculated as:

$$\mathrm{flux} = \frac{8.0u^2}{\mathrm{Sc}^{0.5}}([\mathrm{G}]_\mathrm{I} - [\mathrm{G}]_\mathrm{W}), \tag{5.10}$$

where e_G is the exchange velocity in cm hr^{-1} (for m d^{-1}, use $e_G = 1.92\,u^2/\mathrm{Sc}^{0.5}$); u is the wind speed in m s^{-1}, normalized to the standard height of 10 m; and Sc is the Schmidt number, is calculated for each gas, temperature, and salinity.

The flux will be in mol m^{-2} hr^{-1} if the concentrations are given as mol m^{-3}.

A number of other formulations of exchange rate with wind speed have been published (e.g. Wanninkhof *et al.* 2004), but at the present time this one published by Wanninkhof (1992) seems satisfactory for illustrative purposes, and indeed is commonly used.

Typical magnitudes of the exchange coefficients calculated with this relationship are shown in Figure 5.7. Here, the effect of increasing wind speed is dramatically evident, as are the effects due to different molecular diffusivities of different gases, and the effect of temperature. The resulting exchange rates may be seen by

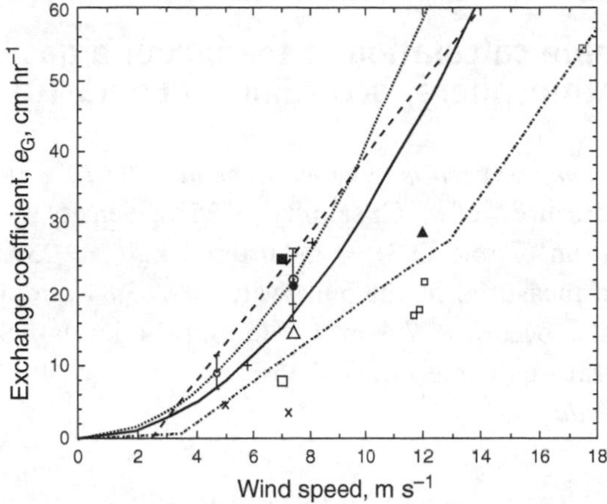

Figure 5.6 Some of the gas transfer coefficients measured in ocean regions, collected from the literature by Wanninkhof (1992). These include several measurements of the loss of radon from surface waters, and experiments obtained by injection of gas tracers. The solid circle is the global average uptake of bomb-produced $^{14}CO_2$, and the open circle is for $^{14}CO_2$ uptake into the Red Sea. The estimated uncertainties of the oceanic ^{14}C inventories are indicated. The exchange velocities are normalized to a Schmidt number of 600 to allow comparison between observations. The wind speeds are all normalized to the standard height of 10 m above the water surface. The dotted line is an equation with an assumed zero exchange at zero wind speed, and an assumed relation to the square of the wind speed, passed through the ocean $^{14}CO_2$ uptake rate. The solid line is the relationship developed by Wanninkhof to allow for the non-linearity of the relationship and the statistical distribution of wind speeds; the equation for this line is given in the text. Two other published formulations to predict the rate of gas exchange are also on the graph, identified by the two dashed lines. (Redrawn from Wanninkhof 1992).

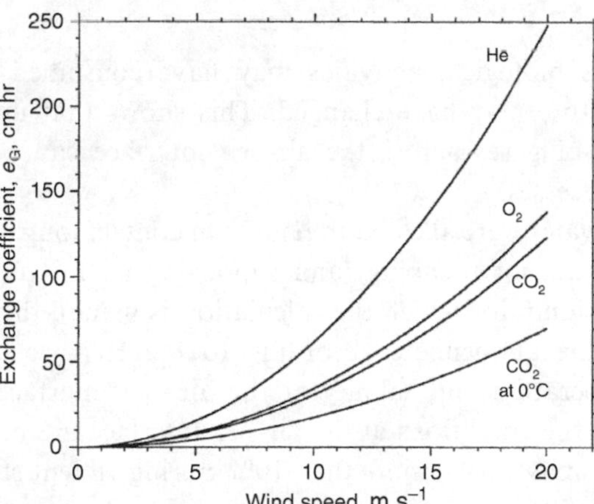

Figure 5.7 Calculated values of the exchange coefficient for several gases as a function of wind speed at the standard height of 10 m above the water surface, according to Eq. (5.9). The values for He, O_2, and CO_2 are calculated for a temperature of 20 °C, and values for CO_2 are also calculated for 0 °C. The values are plotted up to a wind speed of 20 m s^{-1}, but observational support for any relationships in the ocean does not extend above 15 to 18 m s^{-1}.

considering the example worked out in Box 5.1. It is evident that for common wind speeds, in this case a moderate breeze of about 14 knots, sufficient to cause small waves and some whitecaps, a 50-m water column supersaturated with oxygen by 10% will still have a concentration nearly 1% above equilibrium even a month later.

Box 5.1 Example calculation of the flux of a gas between the ocean and the atmosphere, according to Eq. (5.10)

Conditions; oxygen is observed to be initially 10% supersaturated:
Temperature = 18 °C; salinity = 35‰; density = 1025 kg m^{-3}; Oxygen conc. at saturation (Table D.3) = 234.0 μmol kg^{-1}, = 239.85 mmol m^{-3}; Oxygen concentration measured in the bulk water = 263.84 mmol m^{-3}; Wind speed, u, (at 10 m) over the ocean = 7.4 m s^{-1} (about 14 knots); Schmidt number for O_2 at this temperature (Table D.11) = 651

Calculation:

$$\text{From Eq. (5.10), flux} = \frac{1.92u^2}{\text{Sc}^{0.5}}([\text{G}]_\text{I} - [\text{G}]_\text{W})$$

$$= -98.86 \text{ mmol m}^{-2}\text{d}^{-1}.$$

In this case the (–) sign indicates that the oxygen is leaving the water, as it should.

Suppose the mixed layer depth is 50 m. The initial excess O_2 above the concentration at saturation amounts to 1200 mmol in a water column 1 square meter in surface area. The initial loss rate therefore amounts to about 0.0824 of the excess per day, but if O_2 is not being formed or consumed, the loss rate will decrease as the concentration gradient decreases. In fact, under the conditions specified it will take about 8.4 days for the water column to become 5% supersaturated, and 28 days to reach 1% above saturation.

During this time various biological activities may have consumed or produced oxygen, and the temperature may have changed. This shows that under ordinary conditions the transport of gases across the air–sea interface can be quite slow, relative to some other processes.

If the atmosphere and water were allowed to remain in contact long enough, under conditions where waves are not breaking (and supposing no mixture with water below the depth of the column for which the calculation is made), the water would reach equilibrium with the atmosphere according to the Henry's law solubility appropriate for the temperature and salinity at the air–sea interface. In oceanographic studies, however, the conditions at the air–sea interface are never constant. The atmospheric pressure may vary – more than 10% during violent storms, and up to several percent seasonally – especially in those cold, stormy regions where the water sinks. At high latitudes the variation between summer and winter conditions is considerable, and atmospheric pressure varies accordingly. Therefore, the concentration at equilibrium is equally variable. Indeed, given the natural variability of both air and water, the two are probably only at exact equilibrium during some chance time while shifting from one state to another. It seems that it is difficult or impossible to know precisely what were the conditions to which, at some time in the past, a given water mass may have been exposed or potentially equilibrated. Fortunately, these effects are usually small enough that it is common practice to assume equilibrium with standard atmospheric pressure. Furthermore, by taking account of the different solubilities and exchange coefficients of various gases some corrections are possible, at least in principle.

A word of caution should be noted here, however, with respect to the relationships expressed in Eqs. (5.9) and (5.10) and shown in Figures 5.6 and 5.7. These equations may represent the closest approach we yet have to relating wind speed to the exchange fluxes of gases between the air and the ocean, and they are commonly used. They derive from some statistical function of the wind in the world ocean with respect to both speed and variability, and the exchange coefficient explicitly goes to zero at zero wind speed. In the real ocean, things are more complicated.

First, the function according to the square of the wind speed appears plausible, but some data suggest a lower exponent at low speeds, a higher exponent at higher speeds, and (for rather obvious reasons) there are no satisfactory measurements under really strong wind conditions. One can imagine extreme exchange rates under extreme winds. We later discuss the effect of a qualitative change in the conditions at the sea surface when, at wind speeds over about 10 knots, white caps can appear and waves begin to break.

Second, it can be argued that it is not the wind speed itself that is important. Rather it is the underlying turbulence in the water and the consequent replacement of the thin surface layers that controls the rate of air–sea exchange. The wind speed itself is only a surrogate measure (and not necessarily a very accurate measure) of the consequent turbulence.

Third, a light wind may soon reach a steady state with respect to the waves it causes, but a strong wind may require a very long fetch and hours to days to reach steady state with respect to the wave action, and it may often change direction and speed before it does so. Gas exchange measurements at any given time may be done before a steady state has been reached, or after, and after the wind dies wave action may continue for a long time and far from the source. As Jähne *et al.* (1987a,b) point out, these effects may cause variability in gas exchange measurements when related to the instantaneous wind speed at the time of measurement.

Fourth, the assumption of zero exchange at zero wind speed refers only to the effect of the wind itself. At zero wind speed there must often be some gas exchange, although there is little experimental evidence under these conditions. Consider a situation where there is absolutely no wind, but the sky is clear. At night the surface layer of the sea loses heat to space by radiative cooling. This layer must sink and be replaced and as the night progresses the surface skin may be replaced many times. This replacement of the surface layers must be an effective way to encourage gas exchange, even in the total absence of wind. Furthermore, the temperature of the skin exposed to the atmosphere will be lower than that of the measured value in the bulk water. During the day, evaporation and the consequent increase in salinity and density leading to convection of the surface layers may also contribute to gas exchange. These processes have not yet been subjected to substantive evaluation, but it seems compelling that they must contribute to gas exchange even in the absence of wind and at very low wind speeds. The only question is their quantitative significance.

Fifth, even under what would generally be considered constant conditions with a light wind, the diel changes in temperature of the top few meters of the ocean can, over several days, result in a kind of "thermal pumping" which may cause a small (less than 1%) undersaturation relative to equilibrium conditions (McNeil 2006).

Sixth, the temperature of the exact surface of the water is generally different from that of the bulk water usually used in an evaluation. In a series of observations Donlon and Robinson (1997) found that typically the surface skin was about 0.5 °C colder during the day and about 0.3 °C colder at night. The difference increased with wind speeds up to about 10 m s^{-1} (~19 knots) but at higher wind speeds the increasing turbulence drove the difference to zero. The observed values, while quite variable, tell us that measurements in the bulk water will commonly overestimate the temperature of the surface skin actually exposed to the air by 0.3 to 0.5 °C, with an uncertainty of close to ± 1 °C. This could lead, all by itself, to a slight apparent supersaturation of all gases measured in the bulk water, at the temperature of the bulk water.

Seventh, even the measurement of wind speed is fraught with difficulty. The speed of the wind is slowed by friction with the surface, so that the measured speed increases with height. The speed at 10 m is about three times greater than at 10 cm. This is the reason that a standard height is specified for the reporting of wind speed. The shape of the vertical relationship of speed with height, however, also varies according to whether the water and the lowest layer of air are colder or warmer than the air higher up (Stull 1988). Accordingly, the stress on the water surface for a given 10-m wind speed will vary depending on the temperature structure in the air, a factor not usually taken into account.

5.6 Air injection

The air–sea interface is generally not smooth, and this greatly complicates any evaluation of the concentrations of various gases in seawater. Winds above a certain velocity cause breaking or plunging waves that can carry bubbles of trapped air to surprising depths (often greater than 10 m) below the surface. Langmuir circulation can entrain and carry small bubbles down close to the bottom of the mixed layer. Such bubbles are compressed by hydrostatic pressure, and the gases in the bubble are driven to dissolve in the surrounding water. The extent of solution will vary with the circumstances, up to complete solution. Calculation of the effect on the concentrations of gases in water, using the case of complete solution for ease of calculation, shows that not only does the concentration of gas in the water go up, but the relative proportions of the various gases change. This is because these gases are not dissolving from the bubble in proportion to their respective solubilities; instead, a complete little

Table 5.3 Mole fraction (X) of gases in dry air, concentrations in air-saturated seawater, and the percentage increase caused by the injection of 1 mL of air

Gas	X_{air}	[G] in seawater, mol kg^{-1}		Percent increase
		Air-saturated	Added with1 mL air	
N_2	7.808×10^{-1}	4.337×10^{-4}	3.27×10^{-5}	7.54
O_2	2.095×10^{-1}	2.340×10^{-4}	8.77×10^{-6}	3.75
Ar	9.34×10^{-3}	1.149×10^{-5}	3.91×10^{-7}	3.40
Ne	1.83×10^{-5}	6.915×10^{-9}	7.66×10^{-10}	11.06
He	5.24×10^{-6}	1.670×10^{-9}	2.19×10^{-10}	13.13
Kr	1.14×10^{-6}	2.56×10^{-9}	4.77×10^{-11}	1.86
Xe	8.7×10^{-8}	3.54×10^{-10}	3.64×10^{-12}	1.03

Conditions: $S = 35‰$, $t = 18\ °C$.

parcel of air is injected into the water. Since the ratios of gases in air are different than the ratios in seawater under equilibrium conditions (Table 5.2), the result is that those gases which are less soluble than the average will be an increased fraction of the total gas in the water, while those which are more soluble than the average will be a decreased fraction of the total gas in the water. Table 5.3 gives a set of numerical values calculated for a plausible magnitude of air injection. When the fractional increase for one gas is plotted against that for another gas, there must be a characteristic relationship for each pair of gases. Figure 5.8 shows such a relationship for the pair helium and argon. An evaluation of the situation in a sample of water requires considerable precision and accuracy. Provided accuracy is somewhat better than 1%, ideally around +/–0.1%, it is possible to sort out the various processes that may have affected the concentration of gases in a sample of water. For example, deep seawater is generally supersaturated with the least-soluble noble gases. Craig and Weiss (1971) and Bieri (1971), by an examination of the concentrations of several noble gases in deep water from the Pacific, showed the magnitude of air injection that must have influenced their concentrations. They also found, however, that there was more helium than could have been introduced by air injection (the values would have plotted above and to the left of the He/Ar air injection line on Figure 5.8). Since there was no apparent way the excess helium, above the quantity from air injection, could have come from the atmosphere, they concluded that the excess helium must enter the deep Pacific Ocean from the Earth's mantle. This was several years before the first discovery of the famous black smokers and associated exchange processes at the places where the great crustal plates are pulled apart and mantle material is upwelled and exposed to seawater.

The gases that are of most interest and importance in seawater are oxygen and carbon dioxide. They are studied in order to evaluate biological processes in the water and as tracers of ocean circulation. Carbon dioxide is of special interest because at present about a quarter of this gas produced by fossil-fuel combustion ends up in the ocean. Concentrations of both oxygen and carbon dioxide are affected

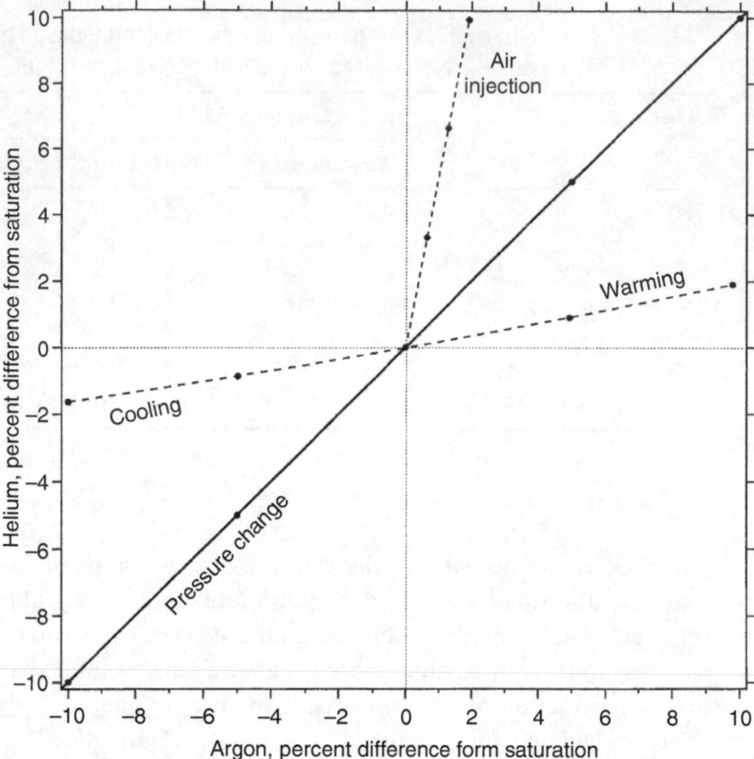

Figure 5.8 Effect of pressure change, air injection, and temperature change on the saturation of surface seawater with atmospheric helium and argon. Imagine that a body of seawater is exactly at saturation equilibrium with the atmosphere at the defined standard pressure of 1013.25 mbar; both gases will be exactly saturated, and their concentrations will plot at the intersections of the zero lines on this graph. Three situations are independently plotted here.

[1] If the atmospheric pressure when equilibrium is attained is greater or lesser than the assumed standard, the concentrations will depart from the standard zero point along the line labeled "pressure change"; the percent change for each gas will be the same because they both follow Henry's law. The points on the line identify atmospheric pressure differing by 5% and 10% from the assumed standard.

[2] If bubbles of air are driven into the water by breaking waves, and forced to dissolve by the additional (hydrostatic) pressure, the amounts added relative to the concentrations present at equilibrium will be different for each gas (Table 5.3). Thus, for He, the most insoluble gas, the proportional increase in concentration will be greater than for argon, a more soluble gas. A plot of this type was first shown by Bieri (1971). The points on the line correspond to 0.25, 0.5, and 0.75 mL of air injected into 1 kg of seawater at 2 °C and $S = 35‰$.

[3] As was evident from the discussion in Box 5.1, equilibration of gases in seawater with the atmosphere can be quite slow. It is easily possible for a significant change in temperature of the water to occur too quickly for the gas exchange to keep up. Suppose that the water has recently warmed; the measured temperature may be higher than the temperature to which the water is equilibrated. It will appear that all gases are supersaturated, since all are less soluble at higher temperatures (Figure 5.1). Conversely, if the water has recently cooled, the gases will appear to be undersaturated. Since the solubility of each gas has a unique response to temperature, the ratios of one gas to another will appear to differ from that expected for equilibrium at the temperature of measurement. The warming/cooling line shows the apparent difference from saturation with He and Ar for changes in temperature; the points on the line correspond to temperature changes of 2 and 4 °C, in each case from assumed equilibrium at 2 °C. ■

by biological processes and also by the physical processes described above. It should be possible to account for the physical processes by examining the concentrations of the non-reactive noble gases. Measurements show that the noble gases are often not in equilibrium at the temperature and salinity of the samples. Procedures worked out originally by Craig and Weiss (1971), Bieri (1971), and extended by Hamme and Emerson (2002), based on the relationships of the type illustrated in Figure 5.8, make it possible to show the extent of air injection and non-equilibrium temperature that must have affected the water when in contact with the atmosphere. From this, the processes that affected the reactive gases, oxygen and carbon dioxide, can be worked out.

The regions that supply the deep water of the world ocean, the high latitudes in the North Atlantic and in the Antarctic region, are characterized by stormy weather, so it seems likely that deep water will generally appear strongly affected by air injection. Tropical and temperate regions have storms, but also long periods of relative calm, so that surface water may tend to approach closer to equilibrium. With the limited data available, it appears that this is what is observed.

An example of the use of careful measurements of gas concentrations to sort out oceanographic processes is that of Craig and Hayward (1987), who were addressing the current controversies as to the magnitude of biological production in the open ocean. These authors showed that under favorable circumstances the concentrations of argon and nitrogen could be used to evaluate the magnitude of air injection. From this they could evaluate how much of the oxygen supersaturation in surface waters could be ascribed to biological production.

Summary

The physical exchanges of simple gases between the atmosphere and seawater initially appear deceptively straightforward to measure and investigate. The major gases depart from ideal behavior to such a small extent that this can often be ignored. The equilibrium solubility according to Henry's law is easy to understand, and for each gas is a simple matter of partial pressure, temperature, and salinity. The real ocean is, however, more challenging. It is not even easy to know whether the temperature measured in the upper mixed layer is exactly the temperature to use when considering the equilibrium solubility.

Rates of exchange are generally modeled as functions of wind speed, even though there must usually be some small exchange at zero wind speed, and waves may be the result of wind some distance away. In any case, turbulence below the surface must be of crucial importance in controlling air–sea exchange rates and the relation of this to the wind speed is poorly known. The relationships to wind speed are subject to many confounding factors, and results of different investigations are often inconsistent. The process of air injection further complicates the various relationships. The considerable range in solubilities and exchange rates of the different gases, however, provide the possibility of sorting out the various processes, provided sufficiently accurate measurements of several gases can be made.

SUGGESTIONS FOR FURTHER READING

Emerson, S. and J. Hedges. 2008. *Chemical Oceanography and the Marine Carbon Cycle*. Cambridge University Press, New York.

Chapter 10 in this book is recommended as it provides clear descriptions of additional material, examples, and perspective.

Nightingale, P. D. and P. S. Liss. 2006. Gases in seawater. In *The Oceans and Marine Geochemistry*, ed. H. Elderfield (*Treatise on Geochemistry*, vol. 6, ed. H. D. Holland and K. K. Turekian). Elsevier, New York, pp. 49–81.

This review includes an extensive survey of studies of surface exchange processes as well as discussions of the many additional gases of interest from biogeochemical and other perspectives. This is the most thorough survey available at the present time.

6 Salts in solution

It was found . . . that with sufficiently dilute solutions the osmotic pressure was the same as the gas pressure, i.e. the pressure which the dissolved substance would exert as a gas.

JACOBUS H. VAN'T HOFF, NOBEL LECTURE 1901

While seawater may be a very dilute solution of some substances, it is, from the point of view of the physical chemist, a quite concentrated solution, and difficult to model theoretically. Substances dissolved in seawater do not behave in a manner that can be easily predicted from their properties in dilute solutions. One of the more important examples of this concerns the solubilities of various salts in seawater.

The effects of salts on the physical properties of water are remarkably diverse. The lowering of the freezing point and raising the boiling point are well known; the associated changes in osmotic pressure are less familiar, and are startling in their magnitude. Studies of the latter effect were important during early attempts to comprehend the nature of salt solutions. The volume changes and associated effects on the absorption of sound are also remarkable. Lastly, the salt in seawater complicates the measurement of pH.

6.1 Solubility of salts

To describe the solubility of a single salt in pure water, it is necessary only to express the concentration of a saturated solution in an appropriate way; for example, in units of mass per unit volume (g L^{-1} or mol L^{-1}). However, in complex solutions with many ions present and where the ions of the salt being considered are often not present in the same ratio as they are in the solid pure salt, it is necessary to use the *solubility product constant*, which is formulated as follows:

$$K_{sp} = [M^+]^{\nu+}[A^-]^{\nu-}, \tag{6.1}$$

where K_{sp} is the solubility product constant; $[M^+]$ and $[A^-]$ are the concentrations of cations and anions, respectively, when the solution is just saturated and is therefore in equilibrium with the solid phase; and v^+ and v^- are the numbers of cations and anions, respectively, required for the simplest molecular formula.

As an example, if a solution of calcium fluoride is measured when the liquid is exactly saturated, in equilibrium with the solid salt, the K_{sp} of CaF_2 has the value of

$$K_{sp} = [Ca^{2+}] \times [F^-]^2 = 3.45 \times 10^{-11}.$$

The K_{sp} values of numerous salts are tabulated in handbooks; it should be noted that these vary with temperature. As formulated above, the K_{sp} is the so-called thermodynamic constant, suitable only for working with pure solutions of very insoluble salts, where the total concentration of ions in solution is extremely small. In practice, many solutions have appreciable concentrations of other ions, and additional information is needed.

In order to appreciate the difficulties of working in seawater, consider the solubility of calcium carbonate. This is an important example in marine science, one that historically played a role in early attempts to understand the chemistry of seawater. The value for the K_{sp} of this substance is reasonably well known from studies in distilled water and, at 20 °C, is reported to be 3.35×10^{-9} for calcite and 5.3×10^{-9} for aragonite, these being the two most common crystalline forms of $CaCO_3$.

In order to establish the degree to which some solution is undersaturated, or supersaturated, one calculates the product of the concentrations of the appropriate ions, to get the *ion product*, and compares this with the solubility product constant. If the ion product calculated as above is less than the solubility product constant, the solution is undersaturated, and if it is greater the solution is supersaturated. From Table 4.1, the concentration of calcium in seawater at $S = 35‰$ is 10.52×10^{-3} M. In surface seawater at a typical pH the concentration of carbonate ion, as we will see later, is about 200×10^{-6} M. Accordingly, the ion product is

$$IP = [Ca^{2+}] \times [CO_3^{2-}] = (10.52 \times 10^{-3})(\sim 200 \times 10^{-6}) \approx 2104 \times 10^{-9}.$$

The degree of saturation, often identified with the symbol Ω, may be given by the ratio of the ion product to the solubility product constant:

$$\Omega = \frac{IP}{K_{sp}} \approx \frac{2104 \times 10^{-9}}{3.35 \times 10^{-9}} \approx 628. \tag{6.2}$$

From this result we observe that, in typical surface seawater, a simple-minded calculation would show $CaCO_3$ to be supersaturated by a factor of about 630. Obviously, the oceans are not now instantaneously turning milky-white with a precipitate of calcium carbonate, so we may ask: is the ocean really this supersaturated, or is the above calculation too simplistic and somehow in error? In order to begin to address this question we must first learn something about the properties and physical chemistry of concentrated solutions.

There are three or four properties of solutions which are called "colligative properties," because they act together; that is, there is a generally constant proportionality between them. These are:

depression of the freezing point;
depression of the vapor pressure;
elevation of the boiling point; and
elevation of osmotic pressure.

The magnitudes of these effects appear to be proportional to the ratio of the number of moles of solute to the number of moles of solvent. Such concentrations are often expressed in molal units. A brief consideration of the colligative properties is a useful background to the later discussion of the activity coefficients of dissolved substances in Section 6.4. For the purpose of this discussion, water will be the only solvent considered.

6.2 Freezing point and boiling point

Consider the phase diagram for water (Figure 2.6). Any substance dissolved in water tends to increase the area occupied by liquid in this phase diagram; the ice–water boundary is shifted to the left, and the freezing point is lowered, while the liquid–vapor boundary is shifted to the right and the vapor pressure is lowered, thus the boiling point is raised. The effects of solutes in causing this are linearly proportional, at least at low concentrations, to the molal concentrations of the solutes. (Recall that a one molal solution is 1 mole of solute in 1 kilogram of solvent.) For water

the molal boiling-point elevation is +0.514 °C per molal unit;
the molal freezing-point depression is –1.855 °C per molal unit.

For example, a solution in which the concentration of some substance is 0.1 molal will have a boiling point at standard atmospheric pressure of 100.051 °C. The elevation of the boiling point is not convenient to measure routinely, in part due to the difficulty of controlling the pressure. The freezing-point depression can be measured precisely (to better than 0.001 °C), so this approach has been very popular.

6.3 Osmotic pressure

The French physicist Jean Antoine Nollet (1748) was the first person to observe the phenomenon known as osmotic pressure. For some reason,[1] he filled a tube of glass with spirits of wine, which we call alcohol, covered the top with a tightly fitting skin or membrane made out of the bladder of a pig, immersed the tube in water, and left it for some period of time (Figure 6.1). After some hours the bladder bowed out under internal pressure. This observation aroused great interest among the natural

[1] He was experimenting on the effects of gases on boiling liquids. Having removed dissolved gas from some alcohol he wished to protect it from air, and stored it in a tube under water.

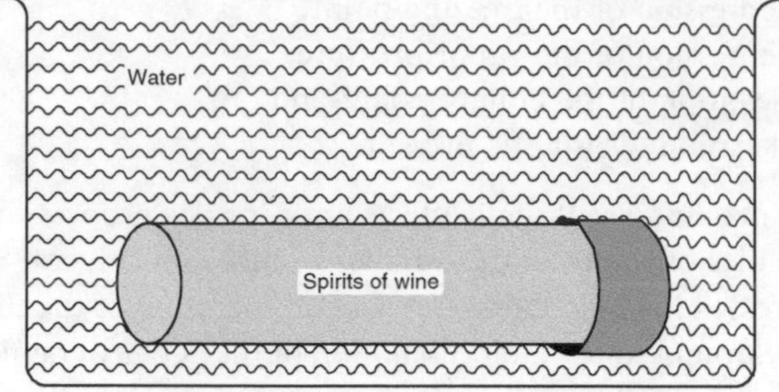

Figure 6.1 Nollet (1748) covered the end of a glass container full of spirits of wine (alcohol) with a membrane made out of the bladder of a pig and immersed the container in water. Gradually the membrane bowed out under internal pressure. When pricked with a pin, the liquid was ejected with considerable force. It was found that whenever a membrane, which was diffusively permeable to water, separated pure water from a solution containing some substance that could not diffuse through the membrane, water would pass into the solution, raising the pressure in that compartment. Thus began the study of osmotic pressure.

scientists of the day, and the experiment was repeated many times with many substances – salts and sugar, for example. Sugar was a particular favorite.

The phenomenon is quite general. For example, suppose a gas-tight chamber is divided into two sections with a very thin palladium foil and one side is filled with nitrogen gas and the other side with hydrogen gas at same total pressure (Figure 6.2). The hydrogen will gradually diffuse across the membrane, from the side that is all hydrogen to the side which is mostly nitrogen. The hydrogen will do this because the partial pressure of hydrogen is greater on the right hand side than on the left hand side, and the foil is permeable to the hydrogen molecules. Nitrogen is too large a molecule to diffuse through the palladium foil, so it all stays on the one side. The result of this situation is that the total pressure on the nitrogen side of the palladium foil gradually increases, and the foil bows out and eventually bursts. It is less obvious why an apparently similar series of events should occur in liquid solution.

The study of osmotic pressure became quantitative with the introduction of methods for measuring the actual pressure generated. One of the most common methods was to measure the hydrostatic head generated by a system such as is shown in Figure 6.3. The water enters the dialysis bag, the solution rises in the capillary, and the osmotic pressure is the pressure necessary to prevent further rise. The hydrostatic head just balances the osmotic forces.

In 1885, the Dutch scientist Jacobus Henricus van't Hoff discovered that in dilute solutions the osmotic pressure follows the relationship:

$$\Pi = cRT, \tag{6.3}$$

where: Π is the pressure; c is the concentration in moles per liter; R is the gas constant, as in the equation: $pV = nRT$; and T is the absolute temperature in kelvins.

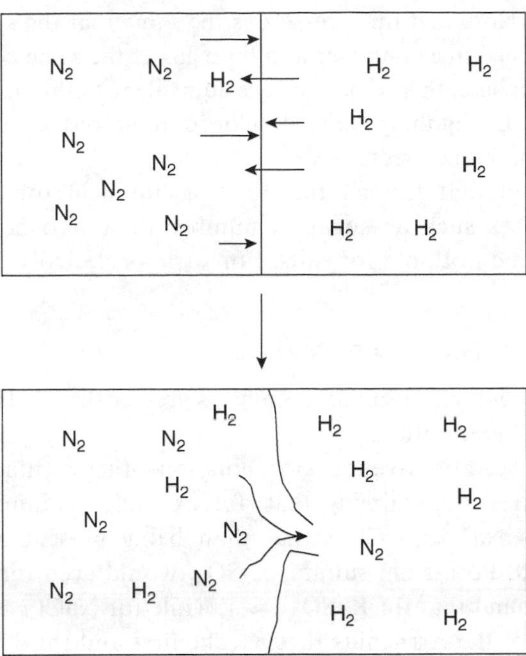

Figure 6.2 Experiment to show the phenomenon in gases analogous to the osmotic-pressure effect in liquids. Nitrogen and hydrogen are placed in opposite sides of a container separated by a thin palladium foil. At the beginning, the total pressure is precisely equal on both sides. The foil is permeable to hydrogen but not to nitrogen. The hydrogen diffuses into the nitrogen and the total pressure on the nitrogen side is thereby increased. Eventually the foil bursts.

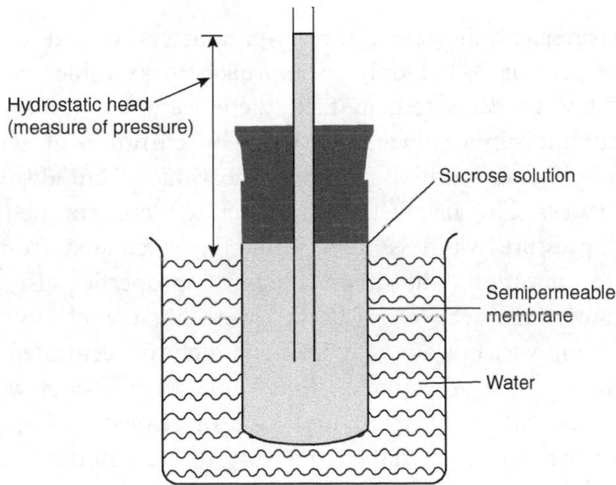

Figure 6.3 In the very simplest form of osmometer, the solution of interest is contained in a dialysis bag that is permeable to water but not to the substance of interest. Water enters the bag until the increasing hydrostatic pressure just balances the entering tendency of the water. This pressure of the water, conveniently visualized as the hydrostatic head, is the osmotic pressure.

For example, the osmotic pressure of a 1 molar solution of sucrose in water at 27 °C might be calculated as follows (ignoring non-linearities at such high concentrations):

$$1 \times 0.082 \times 300 = 24.6\,\text{atm}(= 2493\,\text{kPa}).$$

This is the hydrostatic pressure that must be applied to a solution 1 M in sucrose to prevent water from entering the solution through a semipermeable membrane immersed in pure water. Note that this pressure is the same that the sucrose would have if it were a gas confined in a container as a pure gas at the same concentration. At about 10 m per atmosphere, this is the pressure equivalent to that of a column of water nearly 250 m high. Living things have developed numerous ways to deal with or make use of these remarkable pressures.

Van't Hoff also found that though this relationship held for a variety of mainly organic substances such as sugar, a number of amino acids, alcohol, and so on, in general it did not hold for salts. For various electrolytes van't Hoff found that

$$\Pi = icRT, \tag{6.4}$$

where i is a characteristic of a given salt and is always greater than 1. He also found that i was a function of concentration.

In 1887, the Swedish scientist Svante Arrhenius, on other grounds, had proposed his theory of ionization, claiming that, for example, sodium chloride in solution forms the ions Na^+ and Cl^- rather than being present as the single dissolved molecule NaCl. Potassium sulfate, K_2SO_4, would accordingly produce three ions. Van't Hoff found that for K_2SO_4 $i \approx 3$, while for NaCl $i \approx 2$. This was a striking confirmation of the Arrhenius theory. The first and third Nobel prizes in chemistry were awarded to van't Hoff and Arrhenius for these observations and theories that are fundamental to our understanding of the chemistry of solutions.

On careful measurement, however, these i-parameters turned out to be not exactly unit values of 2 or 3, but only to approach these values as the solution becomes less and less concentrated; in fact, these were the values found upon extrapolation of the measured values back to a concentration of zero or *infinite dilution*. In moderately concentrated solutions, the values were always somewhat less than the unit values. This meant that in moderately concentrated solutions of salts, the osmotic pressure was less than would be calculated from Arrhenius' theory of complete ionization. The other colligative properties also gave similar evidence that salt solutions behaved as if they were not completely ionized. The freezing point lowering was not as great in moderately concentrated solutions as would have been predicted. For example, from Table 4.1, seawater with $S = 35‰$ has a total ion concentration of 1.120 mol kg^{-1} of seawater. The extrapolated freezing-point lowering (based on data from very dilute solutions, and ignoring non-linearities at concentrations this great) is:

$$(1.120\text{mol}) \times (1.855\ ^\circ\text{Cmol}^{-1}) \times 1000/965 = 2.153\ ^\circ\text{C}.$$

The measured freezing point of such seawater is –1.922 °C. The osmotic pressure of seawater has not been measured, but could be estimated from the freezing-point depression to be 24.5 atm at 25 °C, about 89% of the theoretical value of 27.5 atm predicted from relationships in dilute solution.

The explanation of these deviations from the simple model first imagined had to wait for the development of more detailed theories of ionic solutions.

6.4 Activity coefficients

In 1923, the Dutch scientist Peter Joseph William Debye and the German scientist Erich Armand Arthur Joseph Hückel (it was clearly the fashion then to have long names) developed detailed models of the behavior of charged ions in solution, taking account of the electrical interactions of individual ions with the surrounding cloud of ions of both opposite and like charges. They found that the *activity* (as they termed it) of each ion was reduced by the presence of the other ions, and that (at least in simple salt solutions) this effect was sufficient to account for the reduction in Van't Hoff's *i*-terms and also the other effects on colligative properties. This success applied only to relatively dilute solutions (much more dilute than seawater); over the successful range it showed that most salts were indeed completely dissociated, but that the activity and mobility of each ion were reduced in some proportion to the total concentration of all ions. This reduction could be allowed for by incorporating a factor called the *activity coefficient*.

According to current usage, the activity of a salt is expressed in concentration units and the ratio of the activity to the actual concentration is the activity coefficient, γ. Thus:

$$\{NaCl\} = [NaCl] \times \gamma_{NaCl}, \tag{6.5}$$

where [] denotes the chemical concentration of a substance, and {} denotes the "activity" of that substance. As the concentration of all ions, including the salt of concern, in the solution approaches zero, the freedom of action of each ion approaches a maximum, and the activity coefficient approaches 1.

In order to establish a common basis for data describing the chemistry of solutions, it is conventional to report the properties of ionic substances as the numerical values they would have at infinite dilution in pure water. These are the so-called "thermodynamic values" and include properties such as freezing-point lowering and the dissociation constants of acids. The departure of actual from thermodynamic values is taken care of by incorporation of the activity coefficients. In many cases, especially in very dilute solutions, numerical values for activity coefficients can be theoretically estimated, so that the effective activities of some substances can be calculated from the chemically determined concentrations.

Through use of the techniques listed above, there have been many thousands of measurements of the mean activities and the mean activity coefficients of many ionic substances under a variety of conditions. Table 6.1 and Figure 6.4 give some indication of the range of activity coefficients commonly found in such measurements. These data lead to a number of observations. It should be noted, for example, that the activity coefficient (that is, the mean ion activity coefficient) of HCl goes through a minimum at approximately 0.5 molal concentration and then begins to rise, so that at a concentration of about 2 molal the activity coefficient rises above 1. Not all substances behave in the same way, however, and sodium chloride shows only a small rise at higher concentrations. Calcium chloride and sodium sulfate, salts containing one divalent ion each, exhibit much lower activity coefficients. This phenomenon is

Table 6.1 Molal activity coefficients of some common solutes at 25 °C

Concentration, molal	HCl	NaCl	Na_2SO_4	$CaCl_2$
0.001	0.966	0.965	–	–
0.005	0.929	0.927	–	–
0.01	0.905	0.902	–	–
0.05	0.830	0.819	–	–
0.1	0.796	0.778	0.452	0.518
0.2	0.767	0.735	0.371	0.472
0.5	0.757	0.681	0.270	0.448
2.0	1.009	0.668	0.154	0.792
3.0	1.316	0.714	0.139	1.483
4.0	1.762	0.783	0.376	2.93
5.0	2.38	0.784	–	5.89

Data extracted from Robinson and Stokes (1970).

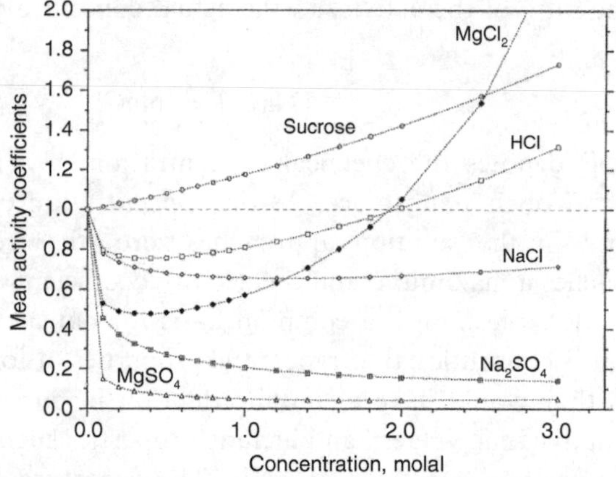

Figure 6.4 Activity coefficients at moderately high concentrations. Mean ion activity coefficients of a variety of individual ionic substances in pure solutions in water at 25 °C, and the activity coefficient of sucrose in pure solution at the same temperature. Data from concentrations below 0.1 m are not plotted, and the curvature of the relationships below this concentration is not properly represented here. The relationship shown by uncharged substances is different from that exhibited by ionized substances, and the various salts differ dramatically in their behavior as the concentration is changed. (Data from Robinson and Stokes 1970.)

quite general and, at least over the best-understood dilute range of concentrations, solutions of salts with divalent ions exhibit low ionic activities. A variety of salts show quite different behaviors at higher concentrations, however, as shown in Figure 6.4. It is evident that the mean activity coefficient, $\gamma\pm$, differs greatly between salts at high concentrations, and cannot be easy to calculate theoretically. However, at very low concentrations the activity coefficients can indeed be estimated from basic physical principles and the nature of the surrounding medium.

In order to examine the influence of the surrounding ions on the activities of an ion of interest, it is necessary to have a quantitative measure of the ionic *milieu*. The effects of the cloud of ions in a salt solution depend both on their concentrations and on the magnitudes of their electric charges. For example, the electrostatic force between two doubly charged ions is four times that between two singly charged ions. Therefore, the magnitude of the charge must be taken into account in attempting to derive a quantitative measure of its effect. The measure which was devised to take account of the total effects of all the charged particles or ions in a solution is called the *ionic strength*, introduced by Gilbert Newton Lewis and Merle Randall in 1921. The ionic strength, I, is calculated as follows:

$$I = \frac{1}{2}\sum m_i(z_i^2), \tag{6.6}$$

where $m_i =$ the molal concentration of the ith ion and $z_i =$ the charge on the ith ion. For example, the ionic strength of a 1 molal solution of sodium chloride, NaCl, is

$$I = \frac{1}{2}[(1 \times 1^2) + (1 \times 1^2)] = 1.0,$$

and the ionic strength of a 1 molal solution of potassium sulfate, K_2SO_4, is

$$I = \frac{1}{2}[(2 \times 1^2) + (1 \times 2^2)] = 3.0.$$

In dilute solutions, a number of properties are functions of the ionic strength. Among these are:

colligative properties;
rates of ionic reactions;
solubilities of slightly soluble salts; and
electromotive force of electrochemical cells.

The contribution of Debye and Hückel was to take account of the ionic strength and other properties of a salt solution in order to calculate the activity coefficients. In 1923, they introduced what is now called the Debye–Hückel equation:

$$\log\gamma\pm = -\frac{z^+z^- \times A \times \mathrm{I}^{0.5}}{1 + (B \times a \times \mathrm{I}^{0.5})}. \tag{6.7}$$

It will not be useful here to derive or develop this equation further, but it is presented in order to provide an opportunity to note the considerations involved in a satisfactory theoretical development. In this equation, the z parameters are the charges on the various ions and A and B are complex factors that include the dielectric constant of water, the absolute temperature, the Boltzman constant (gas constant per molecule), Avogadro's number, and the charge on the electron, while a is the mean distance of closest approach of the ions. The numerical values of A and B are about 0.5 and 0.33, respectively, and do not vary greatly. This equation may be used to calculate the mean activity coefficients of the ions of simple salts in dilute solutions and is experimentally shown to be accurate up to a concentration of

Table 6.2 Estimated free-ion activity coefficients for some seawater constituents

Cations	Activity coefficient	Anions	Activity coefficient
H^+	0.967	OH^-	0.863
Na^+	0.707	Cl^-	0.623
K^+	0.623	Br^-	0.642
NH_4^+	0.618	NO_3^-	0.490
Mg^{2+}	0.285	SO_4^{2-}	0.219
Ca^{2+}	0.256	CO_3^{2-}	0.205

Values above are calculated for $S = 35‰$, $t = 25$ °C, and atmospheric pressure. Selected from extensive tabulations in Millero and Schreiber 1982.

approximately 0.1 molal. For very dilute solutions, up to 0.001 molal, the Debye–Hückel equation simplifies to

$$\log\gamma = -[z^+z^- \times A \times I^{0.5}]. \tag{6.8}$$

Within this concentration range the error is less than 3%.

The equation above gives the mean γ for both the positive and negative ions. Just as it is impossible to have a solution with only cations or only anions, it is theoretically impossible to make rigorously defensible measurements or calculations of the activity coefficients of individual ions. For ordinary salt solutions, only the mean activity coefficient, $\gamma\pm$, is measured. Seawater is a mixture of many ions: it is about 0.5 molal in each of the two most concentrated ions, and it has an ionic strength, I, of about 0.7, so it is both complex and relatively concentrated. As such, it poses a considerable challenge to the physical chemist. A great deal of effort has gone into devising various theoretical and empirical approaches to the prediction and measurement of individual ion activity coefficients in such solutions (Pytkowicz 1979). By combining various theoretical assumptions and empirical relationships from a variety of salt solutions, it is possible to obtain working estimates of individual ion activity coefficients and to predict the behavior of individual ions in seawater, with results that appear to be realistic (Whitfield 1975a,b). The full derivation of these approaches is beyond what we can attempt here.

As shown in Table 6.2, the estimated activity coefficients of calcium and carbonate ions in seawater are quite low, and these can be accounted for when calculating the solubility constant of calcium carbonate, as follows.

The thermodynamic constant, $K_{sp} = \{Ca^{2+}\} \times \{CO_3^{2-}\}$ at saturation, by definition. Including the activity coefficients, the ion product, IP, is

$$\begin{aligned} IP &= [Ca^{2+}] \times \gamma_{Ca} \times [CO_3^{2-}] \times \gamma_{CO3} \\ &= (10.53 \times 10^{-3}) \times 0.256 \times (200 \times 10^{-6}) \times 0.205 \\ &= 1.10 \times 10^{-7}. \end{aligned}$$

At this point the degree of saturation (symbol Ω) is calculated to be

$$\Omega = \frac{1.10 \times 10^{-7}}{3.35 \times 10^{-9}} = 33.$$

From this evidence, we may conclude that, even taking the estimated activity coefficients into account, the ion product of calcium and carbonate ions in seawater is still 33 times greater than the thermodynamic solubility product constant. It still appears that calcium carbonate is greatly supersaturated in the typical surface seawater considered here.

Evidently, there is still more to learn, and we will have to revisit this matter after some consideration of the carbon dioxide system in seawater.

The example given above, of the use of activity coefficients to evaluate the solubility of a salt in seawater provides an opportunity to note two common conventions in the use of solubility-product constants and other analogous constants in studies of these complex solutions. For example, the solubility-product constant may be listed as the thermodynamic constant, in which case it is defined as follows:

$$\begin{aligned} K_{sp(CaCO3)} &= \{Ca^{2+}\} \times \{CO_3^{2-}\} \text{at saturation} \\ &= [Ca^{2+}] \times \gamma_{Ca} \times [CO_3^{2-}] \times \gamma_{CO_3^{2-}} \text{ at saturation.} \end{aligned}$$

In using the thermodynamic (or infinite dilution) constants when working with any solution that has appreciable amounts of salt present, it is necessary to have an estimate of the activity coefficients for each component investigated. Such estimates are fraught with difficulties, especially when a wide range of conditions must be considered.

In practice, the apparent constants are measured under a variety of conditions, and the data fitted to empirical equations so that values for the constants can be calculated throughout the appropriate ranges. These constants, designated K^*, are called apparent constants or empirical constants. They are also called *stoichiometric constants*, because they relate to the measured concentrations rather than to activities. For work in seawater, such constants have been evaluated by measurement over a range of controlled conditions of temperature, salinity, pressure, etc. These measurements are the source of the tables of numerical constants (and the algorithms for calculating them) used in evaluating the carbon dioxide system and other chemical properties of seawater. An important example will be seen in Chapter 7 when dealing with the measured solubility of calcium carbonate in seawater.

6.5 Electrostriction

An interesting way to note the rather dramatic effects of the presence of dissolved salts on the properties of solutions is to observe their effects on the density or, inversely, on the specific volume of a solution. If one places, for example, 30 g of large crystals of ordinary salt (NaCl) in the bottom of a 1-L volumetric flask, quickly adds water up to the mark, and then allows the crystals to dissolve, it is observed that as the salt dissolves the volume of the mixture decreases, and the water level falls.

Table 6.3 Effect of electrostriction on the volume of salt solutions

Substance	Density, g L^{-1}	Volume, mL	Calculated volume	Observed volume	Δ Volume Obs. – Calc.
H_2O	998.21	971.74			
NaCl	2165	13.86	985.60	980.78	–4.82
$MgCl_2$	2320	12.93	984.67	977.90	–6.77
$MgSO_4$	2660	11.28	983.02	971.91	–11.11

In each case the solution is made to a concentration of 3.0% w/w by adding 30 g of the dry anhydrous salt to 970 g of water at a solution temperature of 20 °C. The calculated volume is the simple sum of the volumes of the components. Calculated from data in Haynes (2011).

A more precise way to measure this effect is to determine the density of a salt solution, and relate this to the density that would be calculated from a simple addition of the two substances. The data in Table 6.3 and Figure 6.5 show that the effect can be surprisingly large. The volume constriction is greater for divalent than for monovalent salts. The double divalent salt, $MgSO_4$, in low concentrations causes a volume constriction apparently slightly greater than the total volume of added salt. A non-ionic substance such as glucose causes little effect. What is happening here?

The answer is that water molecules, having an asymmetric distribution of positive and negative charges, orient around the charged ions, with their positive sides attracted to the negative ions, and their negative sides attracted to the positive ions. This attraction is sufficiently powerful, and the structure of the adjacent liquid water sufficiently disrupted, that the total volume is considerably reduced. In a 3% NaCl solution (as in Table 6.3) there is about 1 mole of ions, and about 54 moles of water, with each mole of water normally occupying about 18 mL. It seems implausible that 1 mole of water could be so strongly affected as to reduce its effective volume by nearly one quarter, and in fact it is clear from other evidence that several water molecules must be associated with each ion. The effect for divalent ions is even greater, because in a 3% $MgSO_4$ solution there is only about 0.50 moles of ions. Per mole of ions, the effect in the $MgSO_4$ solution is more than four times greater. In this solution the electrostriction effect, as it is called, is sufficient to reduce the volume of the solution by more than the volume of 1 mole of water for each mole of ions. Several water molecules must be, to some extent, oriented and immobilized around each ion. Extensive efforts to measure or to estimate the number of water molecules associated with each ion have shown that the result depends on the method of measurement, and that the result also depends on the type of ion involved. Small ions appear to immobilize more water molecules than large ones. Estimates for different ions range from about one water molecule per ion to perhaps eight or ten per ion, with those water molecules nearest to the ion most strongly affected, and with the effect decreasing with distance from the ion.

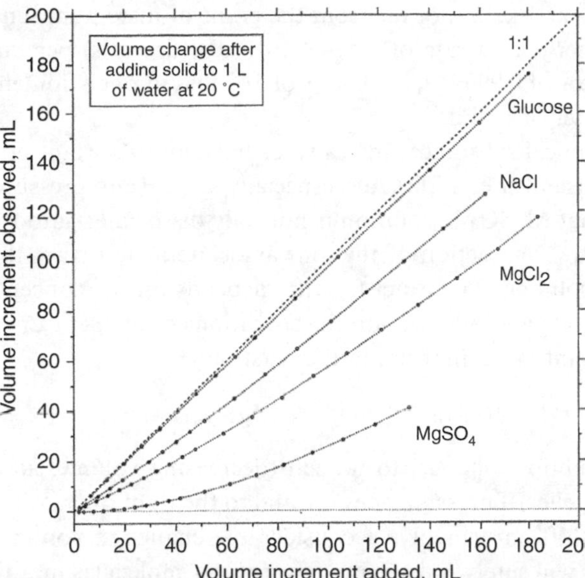

Figure 6.5. Effect of electrostriction. Calculated values of the volume of solution after addition of a known volume of salt to 1 liter of pure water. For each salt the volume of solution is less than the sum of the volumes of water and the volume of added solid salt. The dashed line is the increment of volume if the volumes are additive. Values for glucose, a non-ionic substance, are shown for comparison. Calculated from data in Haynes (2011).

In a kilogram of seawater at a salinity of 35‰, there are altogether about 1.12 moles of ions and about 53.5 moles of water molecules, so it is clear that a significant fraction of the water molecules is affected by the presence of salt, and the structural arrangements of the water molecules are disturbed in some fashion. This provides some qualitative insight into the reasons why seawater, unlike pure water, does not show a maximum in the density as it is cooled towards the freezing point.

These and other effects in ionic solutions have been extensively investigated in order to gain an understanding of the properties of such solutions, but such studies go beyond what we can usefully consider here, except to make a comment on the absorption of sound. This property is of some importance in marine acoustics, because sound is attenuated in seawater much more strongly than in fresh water.

6.6 Absorption of sound

Many marine organisms, and humans, use sound to search for food and other objects, and to transmit information. When a sound is emitted, the sound waves are weakened as they spread out in various geometric patterns, depending on conditions. The sound waves also lose energy as they pass through the water. One of the remarkable characteristics of seawater is that sound waves lose much more energy passing through the sea than they do in pure water.

The absorption of sound energy by pure water depends on the square of the frequency, and is also somewhat affected by the temperature and pressure. An example of the magnitude of the effect is given by the following: at 18 °C, in surface water, sound waves at 1 kHz (not far from the point of maximum sensitivity of the human ear) lose energy at a rate of 2.51×10^{-4} dB (decibels) per kilometer. This works out to a factor of 0.999942, or 0.0058 of 1% loss in one kilometer. Water is a very good transmitter of sound.

At least two chemical substances in seawater have an important effect on sound transmission, because their equilibria are especially sensitive to pressure. We will see later (Chapter 7) that $MgSO_4$ in solution is not fully dissociated into the constituent ions Mg^{2+} and $SO_4{}^{2-}$. A fraction of the ions associate as ion pairs that then exist, uncharged, in the solution. The exact fraction depends on the concentration of the magnesium and sulfate ions and also on the conditions of temperature and pressure. The quantitative point of equilibrium in the reaction:

$$Mg^{2+} + SO_4^{2-} \leftrightharpoons MgSO_4. \tag{6.9}$$

will shift if the conditions change, to the left (decreasing volume, according to the principle of Le Châtelier) if pressure goes up, and to the right (increasing volume) if it decreases. Such a shift must involve a considerable change in volume, because the uncharged ion pair will surely have many fewer water molecules in a tightly packed arrangement closely surrounding it, compared to the large number bound to or packed around the doubly charged individual ions. Compared to most ionic reactions, this shift of the equilibrium is relatively slow, possibly because so many tightly bound water molecules must be rearranged.

Now imagine what happens when a sound wave passes through water. The sound wave is a region of increasing pressure followed by decreasing pressure. As pressure increases some of the ion pairs will dissociate, and many water molecules will become packed in some tight arrangement around the charged ions, and the volume of the solution will slightly decrease. As the sound wave passes the pressure decreases and the reverse process will take place. As the water molecules move around and are rearranged, some frictional energy will be dissipated as heat. This is energy lost from the sound wave as it does work on these molecular rearrangements.

The effect is significant. Under the same conditions noted above, the loss due to the magnesium sulfate amounts to 6.082×10^{-3} dB km^{-1}, which works out to a loss of 0.14% of the energy in the sound wave in passing through 1 kilometer of water, 25 times more than is due to the water alone. The loss of energy in the sound wave due to $MgSO_4$ also increases as the square of the frequency, up to a frequency of about 100 kHz, after which the effect of $MgSO_4$ decreases, presumably because the reaction is too slow to be much affected above this frequency (Figure 6.6.)

Another substance in seawater has an even bigger effect. The borate ion causes, under the above conditions and with the further specification that the pH is such that the water is in equilibrium with current atmospheric CO_2, a further loss of 1.54% of the energy in a sound wave passing through 1 kilometer of seawater. It is not known what chemical reactions are involved in this case, only that the borate ion must be involved in some way. The chemical reaction involved must be slower than the magnesium sulfate reaction, because the borate effect begins to decrease above about

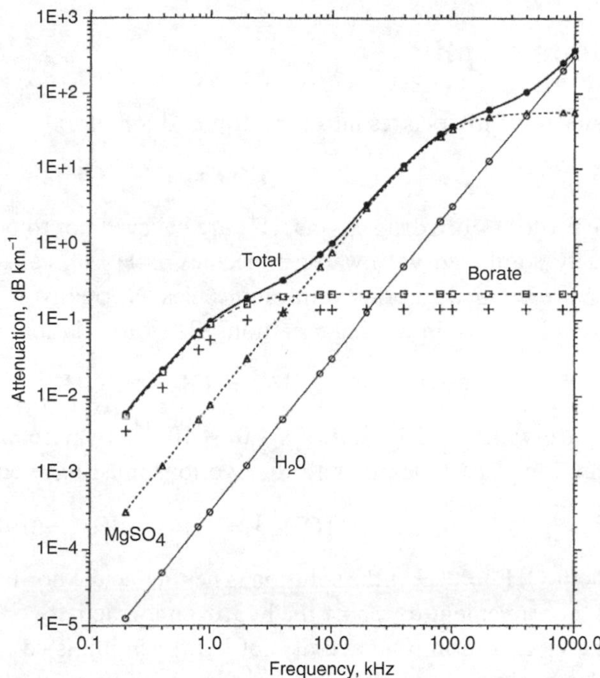

Figure 6.6 Attenuation of sound energy in seawater by water, by the shifting equilibrium between magnesium and sulfate ions and magnesium sulfate ion pairs, and by some unknown shifting equilibrium involving borate ion. The top line gives the total absorption in seawater. The algorithm used for making these calculations is in Appendix K. The data plotted here, and joined by lines, are for conditions as follows: $S = 35‰$, $t = 12$ °C, pH = 8.292, TA = 2.32 mmol kg^{-1}, $pp\mathrm{CO_2} = 280$ μatm, depth = 10 m. The data for borate that are plotted here as crosses are for the situation where the $pp\mathrm{CO_2}$ is increased to 560 μatm, double the pre-industrial concentration.

1 kHz (Figure 6.6). There is some evidence that additional uncharacterized chemical species have a small effect on sound absorption; these are included within the borate effect as shown in Figure 6.6.

In any sample of seawater (as we will see in Chapter 7) the fraction of the total boron present as borate ion depends on the pH; for this reason the energy lost from sound waves passing through the sea is influenced by the pH of the water. As the pH decreases the concentration of borate decreases and the portion of the sound absorption due to the borate also decreases, and this effect is large enough to have some consequence (Hester *et al.* 2008). The data plotted in Figure 6.6 were calculated for surface seawater in equilibrium with a partial pressure of carbon dioxide, $pp\mathrm{CO_2}$, of 280 μatm (pH = 8.292), appropriate for the latter half of the nineteenth century. If the atmospheric CO_2 concentration doubles to 560 μatm, with other conditions the same, the pH will be 8.032, and at 0.2 kHz the total absorbance of sound will be reduced by 2.09×10^{-3} dB km^{-1}, a reduction of 35%. As CO_2 increases, there will be more noise in the ocean.

6.7 A note on pH

Water itself dissociates into a proton and a hydroxyl ion:

$$H_2O \leftrightharpoons H^+ + OH^-.$$

The protons or hydrogen ions, H^+, are believed not to occur free in solution; they are largely combined with water molecules as H_3O^+, called hydronium ions. These in turn are bonded to other water molecules. Normally, it is satisfactory to use just H^+ for convenience in writing equations. The dissociation constant is given by:

$$K_w = \{H^+\} \times \{OH^-\}. \tag{6.10}$$

In pure water at 25 °C, K_w is 1.0×10^{-14}. In pure water, activity coefficients by definition must be unity, and the two ions must have equal concentrations, so that:

$$\{OH^-\} = \{H^+\} = [H^+] = 10^{-7}.$$

When $\{OH^-\} = \{H^+\}$ the solution is neither acidic nor basic, and is termed "neutral."

It is convenient to report the hydrogen ion activity on a logarithmic scale, because the concentration (and activity) of hydrogen ion is commonly a very small number, and the glass electrode commonly used to measure the activity of hydrogen ions responds linearly to the logarithm of the activity of the ion. The convenient convention is to use the pH scale, proposed by S. P. L. Sørensen in 1909 as p_{H^+}, later changed to pH. In this scale:[2]

$$pH = -\log\{H^+\}. \tag{6.11}$$

Accordingly, at 25 °C, pure water has a pH of 7.0.

The dissociation constant of water depends on the temperature, so that the pH of (neutral) pure water ranges from 7.47 at 0 °C to 6.76 at 40 °C.

The empirical or apparent dissociation constant, K_w^*, is, of course, also a function of salinity, so that, for example, at neutrality the pH of seawater with $t = 25$ °C may range from 7.0 at $S = 0$‰ to 6.7 at $S = 40$‰. Table F.1 in Appendix F presents a convenient tabulation of values for the constant K_w^*, defined in such a way as to make it possible to calculate the concentration of $[OH^-]$ in seawater after measurements of pH, provided certain definitions and corrections are taken into account.

The activity of H^+ strongly determines a great deal of seawater chemistry, controlling as it does the relative concentrations of many species of weak acids, such as those derived from carbon dioxide. However, neither the concentration nor the activity of H^+ can be measured directly.

A close approximation to the $\{H^+\}$ can be obtained by utilizing the effect of H^+ on the electrical potential of a glass electrode. This potential is quite closely proportional

[2] The small "p" in pH has variously been stated to refer to power, potential, puissance (French: strength), potenz (German: power, potential), and other similar words. In fact, originally it was used by Sørensen as an arbitrary label of a sample solution, the reference solution being "q" (Nørby 2000). Today, however, we may still think of pH evocatively as meaning "puissance d'Hydrogen."

to $\{H^+\}$, if the composition of the solution is not too different from that of the standards used in calibration. The potentials involved are small, and the glass electrode has a high impedance, so voltmeters used for this purpose must be properly designed and quite sensitive. The usual commercial instruments used to make these measurements also calculate the voltage results in terms of pH and are called *pH-meters*. The procedures are very well described in Grasshoff *et al.* (1999).

Glass electrodes are commonly calibrated against a series of standard buffers that have been evaluated and recommended by the US National Bureau of Standards, and the results reported in the so-called NBS pH scale, widely used in all fields of investigation.

It has long been recognized that measurements of $\{H^+\}$ with the available glass-electrode/reference-electrode systems are theoretically unsound, because there are electrical potentials in the systems that respond to properties of the ionic medium other than just the $\{H^+\}$. The biggest problem occurs with the *liquid junction potential*, an electrical potential that develops across the liquid junction that forms a salt bridge between the reference electrode and the test solution. These electrical potentials cannot be entirely accounted for either theoretically or empirically, and they can vary somewhat according to the design of the electrode, and with the medium to be measured. Nevertheless, the great convenience and general reproducibility of measurement with pH-meters and glass electrodes has led to almost universal use. Changes in electrode potentials associated with transfer from low-ionic-strength buffers to high-ionic-strength seawater, however, must ultimately be included within the empirical dissociation constants and provide some small degree of uncertainty in their values. Some amelioration of these problems can be achieved by the use of buffers made up in high-ionic-strength synthetic seawater (e.g. Millero *et al.* 1993).

As measurements have become more and more precise, small, uncontrolled variations from electrode to electrode become more important, and the theoretical unsoundness of the technique has led to a number of proposals for different scales to represent the concentration or activity of hydrogen ion (see evaluations by Dickson 1984, 1993, Dickson *et al.* 2007). More detail is given in Appendix F. In their simplest form, these newer scales assume the constant composition of seawater with respect to some of the major ions, so are restricted in use to saline waters close to seawater in composition. The recent extensive work with the total hydrogen scale (pH_T), the development of buffers for use with the glass electrode, the calibration of carbonic acid dissociation constants to these (e.g. Roy *et al.* 1993), and the development of a remarkably precise and stable spectrophotometric technique for measuring pH using indicator dyes (Byrne and Breland 1989, Clayton and Byrne 1993), have led to the use of the pH_T scale for the highest-precision work with the carbon dioxide system (Dickson *et al.* 2007). Present practice is to use the pH_T scale for work on the carbonate system in the open ocean.

For measurements in solutions whose salinity varies widely or whose composition departs from the usual constancy of relative proportions of the major ions, there may not be an advantage in changing from the usual glass-electrode measurements calibrated on the pH_{NBS} scale against NBS buffers, or possibly buffers in synthetic seawater. This also applies to physiological work on marine organisms. In the derivations and calculations carried out in the next chapter, the constants used were based on measurements with glass electrodes and NBS buffers or were adjusted to the pH_{NBS} scale.

Summary

The saltiness of seawater causes many properties of the solution to differ in remarkable ways from those observed in more dilute solutions.

The freezing point of most seawater is close to 2 °C below zero. The osmotic pressure is about 24 atmospheres; this is the pressure that must be overcome to produce fresh water from seawater by forcing the fresh water out through a semipermeable membrane.

The dissolved ions in solution create an environment or milieu, quantitatively expressed by the term "ionic strength," such that the activities of all ions in the solution are significantly reduced; the effect is much greater for divalent than for monovalent ions. One important consequence is that the solubilities of substances composed of one or more divalent ions (such as calcium carbonate) are much greater in seawater than in fresh water, although additional factors are also involved.

Ions in solution attract water molecules to orient around them in some closely packed fashion; the volume occupied by these water molecules is less than in the more open arrangement where there are no ions. Thus, the volume of the solution is less than the sum of the volumes of the constituents. This effect, called electrostriction, is much stronger for divalent ions than for singly charged ions. In seawater a significant proportion of the water molecules is affected.

The reduction in volume by electrostriction leads to much greater attenuation of sound in the ocean, compared to the situation in fresh water. Some substances, such as $MgSO_4$, exist in solution both as ions and as uncharged ion pairs in equilibrium with them. The volume change associated with a shift in the equilibrium caused by changes in pressure causes energy to be lost from sound waves passing through the ocean. Borate ions in seawater also participate in some pressure-sensitive reaction; since the concentration of borate is a function of the pH of the seawater, the attenuation of sound is also sensitive to pH.

The high ionic strength of seawater causes additional uncertainty in the measurement of pH using electrodes calibrated against standard NBS buffers with low ionic strength. A different pH scale is currently in use for the highest precision in open ocean work. In this book, however, the pH_{NBS} scale is used throughout.

SUGGESTIONS FOR FURTHER READING

Stumm, W. and J. J. Morgan. 1996. *Aquatic Chemistry; Chemical Equilibria and Rates in Natural Waters*, 3rd edn. John Wiley & Sons, New York.

The book by Werner Stumm and James Morgan has long been a classic text, and is still perhaps the best place to go for a wide range of information covered by its title. It is a useful reference for many chapters in the present volume, not just this one.

Riebesell. U., V. J. Fabry, L. Hansson, and J.-P. Gattuso, eds. 2010. *Guide to Best Practices for Ocean Acidification Research and Data Reporting*. European Commission, Luxembourg.

This document contains a chapter by Andrew Dickson on the various pH scales used for ocean water and problems with measurement. It is currently the most up-to-date discussion of this subject.

7 Carbon dioxide

. . . we are carrying out a tremendous geophysical experiment of a kind that could not have happened in the past or be repeated in the future — ROGER REVELLE 1956

The study of the carbon dioxide system, the other buffers in seawater, and the so-called alkalinity of seawater has historically been a major focus of research by chemical oceanographers. The importance of this system is evident in the following considerations:

- The carbon dioxide system is responsible for about 95% of the acid–base buffering over the normal range of pH in ordinary seawater. On short timescales (up to at least several thousand years), the pH of seawater is controlled mostly by this system, and changes in pH are caused mostly by changes in the various components of the CO_2 system.
- Short-term changes in the concentration of total CO_2 in seawater are due largely to the photosynthetic and respiratory activities of organisms, so a great deal can be learned about biological activity by monitoring this system.
- The vexed questions associated with the problems of precipitation and solution of calcium carbonate in the oceans can only be approached on the basis of a thorough understanding of the CO_2 system itself.
- The climate of the Earth is strongly affected by the concentration of CO_2 in the atmosphere, although the exact quantitative relationships are still subject to investigation.
- The reservoir of CO_2 in the oceans is much greater than that in the atmosphere, so small changes in processes affecting the oceanic reservoir could have comparatively large effects on the concentration in the atmosphere.
- Humans are significantly increasing the atmospheric concentration of CO_2, and the oceans play an important role in modulating this increase.

7.1 Reservoirs of carbon dioxide

The largest reservoirs of carbon accessible on the surface of the Earth are not in the form of gaseous or dissolved carbon dioxide; most of the carbon is in the rocks, and most of this is in the form of limestone. This rock type is composed largely of calcium carbonate and calcium–magnesium carbonate (dolomite). These and other sedimentary rocks also generally contain some organic carbon; the latter is usually in low concentration, but the large mass of rock involved results in this sedimentary organic matter being a very large reservoir of carbon, second only to that of the limestone. In addition, some carbon is found in beds of coal. In some locations liquid and gaseous hydrocarbons, formed by the heating of the more or less diffusely distributed organic carbon, have migrated to places where they have become trapped in various sedimentary formations. Of the total organic carbon in the sedimentary rocks, only about 0.03% is considered to be recoverable fossil fuel. Table 7.1 gives estimates of the amounts of carbon in these and other reservoirs.

It is a matter of considerable interest to discover how the distribution of carbon amongst the various reservoirs may have changed through geological time, and may be changing today. This would help to set in perspective the changes that humans are

Table 7.1 Total carbon in the atmosphere and crust of the Earth

	Total mass of C in units of:	
	Gt (10^9 t)	Pmol (10^{15} mol)
Precipitated carbonates[a]	82 370 000	6 860 000
Organic carbon in sediments[b]	13 500 000	1 120 000
Oceanic dissolved CO_2[c]	38 000	3164
Atmospheric CO_2[d]	830	69
Recoverable fossil fuel[e]	3710	309
Humus, = non-living organic carbon:		
Terrestrial soils[f]	1790	149
In seawater, as dissolved organic matter[g]	800	67
Biota[h]	560	47

[a] From Hay (1985). Includes only Phanerozoic carbonates.
[b] Includes both marine sediments and Phanerozoic sedimentary rocks. From Hay (1985), assuming that pelagic marine sediments have 0.4% organic carbon.
[c] From Denman *et al.* (2007).
[d] Calculated for the year 2010, when the atmosphere had about 390 ppmv of CO_2.
[e] From Table 7.8.
[f] From Prentice *et al.* (2001). This is the average of two estimates differing from the average by ±12%.
[g] Derived from estimates of concentration reported by Sharp (1993).
[h] From Prentice *et al.* (2001). This is the average of two estimates differing from the average by ±18%. Most of the mass of biota is trees.

causing, and help to predict the ultimate effect that humans will have. At the present time we have only a very limited understanding of the causes and magnitude of past changes. It is obvious that geological processes that would tend to change the amount of carbon in the rock reservoirs, even if by only a tiny fraction, could have a profound effect on the concentrations of CO_2 in the atmosphere and in the oceans. The huge masses of both carbonates and organic carbon in the rocks are largely the result of biological activities in seawater, and it is conditions in the sea which dictate whether the overall rate at which carbon is stored in the sediments is less or greater than the long-term average. Similarly, the physical and biological processes that affect the exchange between the sea and the atmosphere must, to a considerable extent, control the concentration in the atmosphere. An understanding of the carbon dioxide system in seawater is therefore central to a serious consideration of both natural and human-induced changes in the climate of Earth.

7.2 Relationships in solution

7.2.1 Hydration rates

When CO_2 gas dissolves in acidified water most of the molecules remain as free, unassociated CO_2 (Butler 1982). Some of the molecules combine with water:

$$CO_2(f) + H_2O \rightarrow H_2CO_3. \tag{7.1}$$

The rate of reaction (7.1) is pseudo first order, which means that the rate of the reaction may be calculated using the concentration of only one reacting substance. (In formulating reactions with water molecules in aqueous solution, it is conventional to treat the water as having unit concentration and thereby eliminate it from the rate expression.):

$$\frac{d[CO_2(f)]}{dt} = -k_{CO_2}[CO_2(f)].$$

It is observed (Soli and Byrne 2002) that, at 25 °C, $k_{CO2} \approx 0.03\ s^{-1}$. This shows that the direct hydration reaction is remarkably slow. In one second, only about 3% of the molecules of free CO_2 in the solution combine with a water molecule. The back reaction:

$$H_2CO_3 \rightarrow CO_2(f) + H_2O \tag{7.2}$$

is first order:

$$\frac{d[H_2CO_3]}{dt} = -k_{H_2CO_3}[H_2CO_3],$$

where $k_{H_2CO_3} \approx 26 s^{-1}$.

Reactions (7.1) and (7.2) are perfectly reversible, and are commonly represented this way:

$$CO_2(f) + H_2O \leftrightarrows H_2CO_3. \tag{7.3}$$

When the rates of the forward and back reactions are the same, the reactants are at equilibrium, and

$$-k_{CO_2}[CO_2(f)] = -k_{H_2CO_3}[H_2CO_3].$$

The equilibrium constant (designated K_0) for reaction (7.3) is thus the ratio of the two rate constants:

$$K_0 = \frac{[H_2CO_3]}{[CO_2(f)]} = \frac{-k_{CO_2}}{-k_{H_2CO_3}} \approx \frac{0.03}{26} \approx \frac{1}{870}.$$

This value is only approximate, different workers having obtained slightly different values, but it serves to show that, of the two forms considered here, by far the greatest proportion is present in the form of $CO_2(f)$, rather than H_2CO_3.

If the solution is alkaline, the following reaction also becomes significant:

$$CO_2(f) + OH^- \rightarrow HCO_3^-. \tag{7.4}$$

The forward reaction given in Eq. (7.4) is second order, which means that the rate of the reaction must be calculated by multiplying together the concentrations of the two reacting substances, along with the appropriate constant:

$$\frac{d[CO_2(f)]}{dt} = -k_{OH^-}[OH^-][CO_2(f)].$$

The rate constant, k_{OH}, $\approx 8500\ s^{-1}\ (mol\ L)^{-1}$.

The rate constant of the reverse reaction, $k_{HCO_3^-}$, $\approx 2 \times 10^{-4}\ s^{-1}$.

The various rates of reactions are either directly dependent on the pH (because the k_{OH} term is multiplied by the OH^- concentration) or indirectly dependent on the pH in that the relative proportions of the various other carbonate species present are pH-dependent, as we will see later. Since the overall rates of reactions are pH-dependent, so is the rate of approach to equilibrium, and it happens that there is a minimum in this rate at intermediate values of pH. Figure 7.1 shows the half-time for the carbonate system to attain equilibrium in seawater at four different temperatures, after a small perturbation. Seawater and biological fluids are very close to the pH at which the minimum rates are found. It can take a very long time to reach equilibrium; the $t_{1/2}$ at 0 °C and at pH 8 is about 5 minutes, and even at 20 °C it can take nearly 30 seconds to reach halfway towards equilibrium.

The extreme slowness of these hydration and dehydration reactions causes problems for organisms that must expel CO_2 through their gills or lungs. The residence time of blood in these respiratory structures is not long enough to allow the slow equilibration to occur. Organisms have dealt with this difficulty by developing the enzyme *carbonic anhydrase*, which greatly speeds up the hydration–dehydration reactions. This enzyme is nearly ubiquitous in the respiratory organs of all types of animals. Plants also utilize this enzyme to increase reaction rates during the absorption of CO_2 used in photosynthesis.

It has been suggested that the slowness of the hydration–dehydration reactions of CO_2 might impede air–sea gas exchange rates, but it seems that this is likely to be significant only under conditions of extreme calm. Bolin (1960) developed a kinetic argument to show that the slowness of the hydration reactions ought not to influence

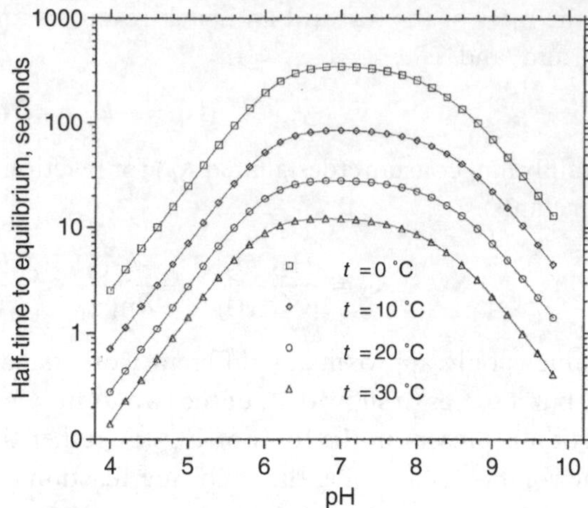

Figure 7.1 Half-time to attain equilibrium after some small perturbation of the carbon dioxide system, in seawater with a salinity of 35‰ and over the temperature range from 0 to 30 °C, as a function of pH. The formulation used here derives from Kern (1960) and the data for the rate constants are from Johnson (1982).

$$t_{1/2} = \frac{\ln 2}{k_{CO2} + k_{OH}[OH^-] + \alpha k_d\{H^+\} + \beta k_{HCO3}}$$

where:
k_{CO2} is the rate constant for the reaction: $CO_2 + H_2O \rightarrow H_2CO_3$,
k_{OH} is the rate constant for the reaction: $CO_2 + OH^- \rightarrow HCO_3^-$,
k_d is the rate constant for the reaction: $H_2CO_3 \rightarrow CO_2 + H_2O$,
k_{HCO3} is the rate constant for the reaction: $HCO_3^- \rightarrow CO_2 + OH^-$,
α is the value of the fraction $[CO_2aq]/[TCO_2]$,
β is the value of the fraction $[HCO_3^-]/[TCO_2]$.
Some of these terms will be defined later. At this point it is sufficient to show that the rates of hydration and dehydration of CO_2 are strongly dependent on temperature and pH. The rates of these reactions are difficult to measure precisely and individually might differ by at least several percent from the reported values. Those who wish to learn more about the kinetics in this system should consult the detailed evaluation, including other reactions, presented by Zeebe and Wolf-Gladrow (2001). ■

the rate of exchange under normal surface conditions on the sea. Kanwisher (1963) carried out experiments with carbonic anhydrase in beakers of seawater. He found that under quiescent conditions this enzyme could accelerate the rate of exchange, but that in stirred beakers the effect was not measurable. The rate constant for the reaction of OH^- ion with CO_2 (Eq. (7.4)) is fast, but the hydroxyl ion is present in low concentration. Under especially alkaline conditions the rate of this reaction is high enough that accelerated uptake might be expected. Emerson (1995) and Wanninkhof and Knox (1996) investigated this possibility in acid and alkaline lake waters, and in seawater, and showed that in alkaline lakes under conditions of low wind speed the rate of CO_2 uptake was considerably greater than that of non-reactive gases. In seawater, under conditions of low wind speeds over the ocean, they found that the

CO_2 exchange rate was likely to be enhanced by a few percent, while under normal or high wind speeds the effect was likely to be quite negligible.

7.2.2 Equilibria in solution

The carbon dioxide that has combined with water is called *carbonic acid.* This dissociates as an acid; the various equilibrium constants are conventionally expressed and defined as follows:

$$CO_2(f) + H_2O \overset{K_0}{\rightleftharpoons} H_2CO_3 \overset{K_1}{\rightleftharpoons} HCO_3^- + H^+$$
$$\uparrow\downarrow K_2$$
$$CO_3^{2-} + H^+$$

Here $CO_2(f)$ represents un-hydrated free CO_2 molecules in solution. The dissociation constants are formulated as follows:

$$K_0 = \frac{\{H_2CO_3\}}{\{H_2O\} \times \{CO_2(f)\}} \tag{7.5}$$

$$K_1 = \frac{\{H^+\} \times \{HCO_3^-\}}{\{H_2CO_3\}} \tag{7.6}$$

$$K_2 = \frac{\{H^+\} \times \{CO_3^{2-}\}}{\{HCO_3\}} \tag{7.7}$$

These formulations are the basic equations from which most subsequent manipulations are derived. Note that these equilibrium constants are the thermodynamic constants, and are expressed in terms of the activities of the ions involved. These thermodynamic constants vary with temperature, of course, but not with changes in the ionic milieu, which exerts its effects by changing the activities of the ions involved, and thus the ratios of their concentrations one to another.

If we wish to know the concentrations and activities of the various CO_2 species in a solution, we must make either the appropriate direct measurement, if possible, or else measurements of other parameters and calculate the value of interest. It is not technically feasible at this time to measure directly either the concentrations or the activities of HCO_3^-, CO_3^{2-}, or H_2CO_3 (Table 7.2). While the concentration of the dissolved $CO_2(f)$ cannot be measured, its activity can, as this is defined as being equal to the partial pressure of CO_2 gas in equilibrium with the solution.[1] From this, and the Henry's law solubility, the concentration of $CO_2(f) + H_2CO_3$ can be calculated. The concentration of H_2CO_3 is always very small relative to that of $CO_2(f)$, and under constant conditions they should always be in a constant ratio to each other, so

[1] Because CO_2, at normal ambient temperatures and pressures, exhibits detectably non-ideal relationships, a small correction is needed for precise work to translate the concentration in the gas phase to the *fugacity* or thermodynamically accurate activity (Weiss 1974).

Table 7.2 Experimental accessibility of the parameters important in an evaluation of the CO_2 system

Empirical constants from prior experimental measurement		
Parameter	**Method**	
K_1^*	By titration	
K_2^*	By titration	
H_{CO_2}	By gas analysis or chemical analysis	
Parameters that can be measured or easily calculated		
Parameter	**Method**	
Temperature, salinity, and pressure		
$\{H^+\}$	By glass electrode	
$[TCO_2]$[(a)]	By chemical analysis	
$[HCO_3^-] + 2[CO_3^{2-}]$	By titration	
$ppCO_2$	By measurement of CO_2 in gas phase	
$[CO_2\,(f)] + [H_2CO_3]$	Calculation from H_{CO_2} and $ppCO_2$	
Parameters that cannot be directly measured or calculated		
$\{H_2CO_3\}$	$[H_2CO_3]$	γH_2CO_3
$\{HCO_3^-\}$	$[HCO_3^-]$	γHCO3–
$\{CO_3^{2-}\}$	$[CO_3^{2-}]$	γCO_3^{2-}

(a) $[TCO_2]$ means total CO_2, or $[CO_2(f)] + [H_2CO_3] + [HCO_3^-] + [CO_3^{2-}]$

it is convenient to lump them together. Conventionally, the $CO_2(f) + H_2CO_3$ species are treated together as if they were one substance, the concentration of which is given as $[CO_2aq]$; the usual values of the constants reflect this assignment. The sum of all the CO_2 species, including the ionized forms, is called the total CO_2 or TCO_2, and can be measured by chemical analysis. The sum of the concentrations of the bicarbonate ion and twice the carbonate ion, $[HCO_3^-] + 2[CO_3^{2-}]$, can be measured by titration. The activity of H_2O ($\{H_2O\}$) can be measured, but is always close to the standard value of 1.00 (See Appendix C, Table C.3.). Since $\{H_2O\}$ does not change much with variations in the salt content of ordinary solutions not much more concentrated than ordinary seawater, this parameter is normally lumped in with other constants.

In order to obtain relationships that are expressed in terms of parameters that can be measured, we must rearrange Eqs. (7.5), (7.6) and (7.7). The first step is to set up the equations in terms of those parameters that can be measured. From Eqn. (7.5):

$$K_0 = \frac{[H_2CO_3] \times \gamma_{H_2CO_3}}{\{H_2O\} \times [CO_2(f)] \times \gamma_{CO_2(f)}}. \tag{7.5a}$$

We cannot easily know the various activity coefficients in the CO_2 system; as a practical matter we may not want to know them. Instead, they are subsumed into the empirical dissociation constants. It is helpful to remember that they contribute to the numerical values of those constants. From Eq. (7.5a) a new hydration equilibrium constant is defined:

$$K_0^* = \frac{[H_2CO_3]}{[CO_2(f)]}, \tag{7.8}$$

where $K_0^* = K_0 \dfrac{\{H_2O\} \times \gamma_{CO_2(f)}}{\gamma_{H_2CO_3}}$

and K_0^* is called an *apparent dissociation constant* or *empirical dissociation constant* or *concentration dissociation constant.*

As noted before, it is not possible to easily measure either the concentration of free CO_2 or the concentration of H_2CO_3, but in the right circumstances their sum can be measured. The solubility of CO_2 in seawater is measured after making the seawater acid enough that no dissociation of H_2CO_3 takes place. Under these circumstances, the measured total CO_2 in the solution is the sum of $[H_2CO_3] + [CO_2(f)]$, designated $[CO_2aq]$ to distinguish it from CO_2 in the gas phase with which it is in equilibrium, so:

$$[CO_2aq] = [H_2CO_3] + [CO_2(f)] \tag{7.9}$$

and

$$[CO_2aq] = H_{CO_2} \times ppCO_2, \tag{7.10}$$

where H_{CO2} is the Henry's law constant and $ppCO_2$ is the partial pressure of CO_2.

From Eq. (7.8)

$$[CO_2(f)] = \frac{[H_2CO_3]}{K_0^*},$$

and from Eq. (7.9)

$$[CO_2aq] = [H_2CO_3]\frac{K_0^* + 1}{K_0^*},$$

and

$$[H_2CO_3] = [CO_2aq]\frac{K_0^*}{K_0^* + 1}. \tag{7.11}$$

Combining Eqs. (7.6) and (7.11), the following expression is derived,

$$K_1 = \frac{\{H^+\} \times [HCO_3^-] \times \gamma_{HCO_3^-}}{[CO_2aq] \times \dfrac{K_0^*}{(K_0^* - 1)} \times \gamma_{H_2CO_3}}$$

and a new empirical constant is defined as follows:

$$K_1{}^* = \frac{\{H^+\} \times [HCO_3^-]}{[CO_2aq]}, \tag{7.12}$$

where K_1^* is the first *apparent dissociation constant* of carbonic acid. This constant is related to the *thermodynamic constant* as follows.

$$K_1^* = K_1 \frac{K_0 \times \{H_2O\} \times \gamma_{CO_2(f)} \times \gamma_{H_2CO_3}}{\gamma_{HCO_3^-} \times (K_0 \times \{H_2O\} \times \gamma_{CO_2(f)} + \gamma_{H_2CO_3})}.$$

Similarly, the expression for the second dissociation constant, from Eq. (7.7), is treated as follows:

$$K_2 = \frac{\{H^+\} \times [CO_3^{2-}] \times \gamma_{CO_3^{2-}}}{[HCO_3^-] \times \gamma_{HCO_3^-}}$$

and

$$K_2^* = \frac{\{H^+\} \times [CO_3^{2-}]}{[HCO_3^-]}, \tag{7.13}$$

where $K_2^* = K_2 \dfrac{\gamma_{HCO_3^-}}{\gamma_{CO_3^{2-}}}$.

The values of $\{H_2O\}$ and the various activity coefficients absorbed within the first and second apparent dissociation constants are each dependent more or less strongly on the ionic strength and on the ionic composition of the solution. Accordingly, the apparent dissociation constants change quite considerably with changes in salinity. Moreover, the effects of salinity also vary with the temperature. In the case of seawater, the ratios of the major ions one to another are nearly constant, so the variation with salinity is normally uncomplicated by variations in composition. It must be remembered, however, that in unusual cases, where such variations occur, associated variations of the apparent dissociation constants are also likely.

While $\{H_2O\}$, the activity of water, may be measured accurately and the other activity coefficients estimated roughly in one way or another, the combined uncertainty is still large, and the apparent dissociation constants cannot be predicted from basic principles to an acceptable level of accuracy. In practice, the most accurate approach has been to measure the empirical apparent dissociation constants, K_1^* and K_2^*, at various temperatures and salinities, and to develop and tabulate a set of values for use under the appropriate conditions.

The first acceptable modern measurements of the dissociation constants of carbonic acid in seawater were made by Kurt Buch in the 1930s. The development of Buch's constants is historically important since they were in use for more than 20 years. Subsequently, the formal definition of the constants was slightly changed, and improved accuracy has been obtained. The first improvement was by John Lyman, a graduate student who enrolled at Scripps Institution of Oceanography in 1937. Lyman defined the constants in a fashion similar to that given here. From January to March, 1941, he carried out a total of about 15 titrations of seawater samples at three temperatures and various salinities. Later, he calculated the results, interpolated graphically to obtain estimates at even intervals, and then submitted his thesis (Lyman 1956) with tables of the constants for the carbonate system and the borate system. While he never published these tables himself, Lyman's constants were the general standard in use for nearly 20 years. Even now, they are the only ones for which there are supporting data up to salinities as high as $S = 90‰$.

The apparent constants were again re-determined by Mehrbach *et al.* (1973), and their observed values refitted to improved algorithms (Box 7.1) by Plath *et al.* (1980).

Box 7.1 Algorithms for calculating the constants associated with the major components of the buffering system or alkalinity of seawater

In the following formulations the dissociation constants are based on measurements made on the NBS pH scale.

T	is the temperature in kelvin: $K = t\,°C + 273.15$
S	is the salinity in ‰
$ppCO_2$	is the partial pressure of CO_2, in atmospheres (1 atm = 101 325 Pa)
$[CO_2aq]$	is the sum of $[CO_2\,(f)] + [H_2CO_3]$, in mol kg^{-1} of seawater
H_{CO_2}	is the Henry's law constant, as in $[CO_2aq] = H_{CO2} \times ppCO_2$
pK_1^*	is negative log of the first apparent dissociation constant of carbonic acid
pK_2^*	is negative log of the second apparent dissociation constant of carbonic acid
$\ln K_B^*$	is natural log of the apparent dissociation constant of boric acid

(1) Solubility of CO_2 (modified from Weiss 1974); see Appendix E, Table E.2:

$$\ln H_{CO_2} = 9345.17/T - 167.8108 + 23.3585 \ln T + [0.023517 - 2.3656 \times 10^{-4}T + 4.7036 \times 10^{-7}T^2] \times S$$

(2) Dissociation constants of carbonic acid (from Mehrbach *et al.* 1973 as modified by Plath *et al.* 1980); see Appendix E and Tables E.3a, E.3b, and E.4:

$$pK_1^* = 17.788 - 0.073104T - 0.0051087S + 1.1463 \times 10^{-4}T^2$$
$$pK_2^* = 20.919 - 0.064209T - 0.011887S + 8.7313 \times 10–5T^2$$

(3) Dissociation constant of boric acid (from Dickson 1990b); see Table F.3:

$$\ln K_B^* = (-8966.90 - 2890.53\,S^{0.5} - 77.942S + 1.728S^{1.5} - 0.0996S^2)/T + (148.0248 + 137.1942S^{0.5} + 1.62142S) + (-24.4344 - 25.085S^{0.5} - 0.2474S) \ln T + (0.053105S^{0.5})T$$

The Mehrbach constants have been extensively tested (Millero *et al.* 2002) and appear to fit overdetermined seawater data better than several other sets of constants subsequently published. The Mehrbach constants were determined using the NBS pH scale and in this book are used with the NBS scale for all the calculations presented. Readers should be aware that for open-ocean work with water that has only small variations in salinity current practice is to use a different pH scale with the constants appropriately adjusted (see Chapter 6 and Appendix F).

7.2.3 Calculation of CO_2 species from pH, TCO_2, and $ppCO_2$

Once the dissociation constants K_1^* and K_2^* and the solubility constant H_{CO_2} have been established by experiment over the necessary ranges of salinity and temperature, it requires (after the salinity and temperature are known) only two measurements on an actual sample of seawater to establish the state of the CO_2 system in that sample

(Park 1969). Deferring until later the determination of carbonate species by titration, the accessible pairs of measurements are:

$\{H^+\}$ and $ppCO_2$,
$\{H^+\}$ and $[TCO_2]$,
$ppCO_2$ and $[TCO_2]$,

where $\{H^+\}$ is the activity of hydrogen ion, obtained from a measurement of pH; $ppCO_2$ is the partial pressure of CO_2, obtained from measuring the CO_2 in a small volume of gas phase equilibrated with a large volume of aqueous phase; and $[TCO_2]$ is the total CO_2 in all forms in the solution, obtained by acidification and complete extraction of the CO_2 followed by chemical measurement of the extracted gas.

$\{H^+\}$ and $ppCO_2$

Combining Eqs. (7.10) and (7.12), which relate the activity of hydrogen ion and the partial pressure of CO_2 to the other forms of CO_2 in solution, gives the following:

$$[HCO_3^-] = \frac{ppCO_2}{\{H^+\}}(H_{CO_2} \times K_1^*). \tag{7.14}$$

Then, combining Eqs. (7.13) and (7.14):

$$[CO_3^{2-}] = \frac{ppCO_2}{\{H^+\}^2}(H_{CO_2} \times K_1^* \times K_2^*). \tag{7.15}$$

From Eqs. (7.10), (7.14), and (7.15):

$$[TCO_2] = \frac{ppCO_2}{\{H^+\}^2}(H_{CO_2})(\{H^+\}^2 + \{H^+\}K_1^* + K_1^*K_2^*) \tag{7.16}$$

$\{H^+\}$ and $[TCO_2]$

Equations relating measurements of pH and $[TCO_2]$ to the other species in solution are derived as follows.

From Eq. (7.16)

$$ppCO_2 = \left(\frac{[TCO_2]}{H_{CO_2}}\right)\frac{\{H^+\}^2}{\{H^+\}^2 + \{H^+\}K_1^* + K_1^*K_2^*}. \tag{7.17}$$

From Eqs. (7.10) and (7.17):

$$[CO_2aq] = [TCO_2]\frac{\{H^+\}^2}{\{H^+\}^2 + \{H^+\}K_1^* + K_1^*K_2^*}. \tag{7.18}$$

From Eqs. (7.14) and (7.17):

$$[HCO_3^-] = [TCO_2]\frac{\{H^+\}K_1^*}{\{H^+\}^2 + \{H^+\}K_1^* + K_1^*K_2^*}. \tag{7.19}$$

From Eqs. (7.15) and (7.17):

$$[CO_3^{2-}] = [TCO_2]\frac{K_1^* K_2^*}{\{H^+\}^2 + \{H^+\}K_1^* + K_1^* K_2^*}. \tag{7.20}$$

ppCO_2 and [TCO_2]

With measurements of *pp*CO_2 and [TCO_2], it is possible to determine the components of the CO_2 system without measuring the pH. If all the constants form a self-consistent set of parameters, it is therefore possible to calculate the activity of hydrogen ion. The equations are developed as follows.

From Eqs. (7.10), (7.13), and (7.14):

$$\frac{ppCO_2 \times H_{CO_2} \times K_1^*}{[HCO_3^-]} = \frac{K_2^* \times [HCO_3^-]}{[CO_3^{2-}]},$$

and the relationship that

$$[TCO_2] = [CO_2aq] + [HCO_3^-] + [CO_3^{2-}], \tag{7.21}$$

the quadratic equation may be solved to arrive at

$$[HCO_3^-] = \frac{ppCO_2 H_{CO_2} K_1^*}{2K_2^*}\left(-1 + \sqrt{4K_2^*\left(\frac{[TCO_2] - ppCO_2 H_{CO_2}}{ppCO_2 \times H_{CO_2} \times K_1^*}\right) + 1}\right). \tag{7.22}$$

From Eqs. (7.10), (7.12), and (7.22):

$$\{H^+\} = \frac{2K_2^*}{-1 + \sqrt{4K_2^*\left(\frac{[TCO_2] - ppCO_2 H_{CO_2}}{ppCO_2 H_{CO_2} K_1^*}\right) + 1}}. \tag{7.23}$$

Having values for [HCO_3^-] and $\{H^+\}$,[CO_3^{2-}] is obtained from Eq. (7.13):

$$[CO_3^{2-}] = \frac{[HCO_3^-] \times K_2^-}{\{H^+\}}.$$

There is some advantage in determining the components of the carbonate system by measuring the pH and either the *pp*CO_2 or the [TCO_2]. These measurements are free of many of the complexities associated with interpreting the results of titration measurements. Measurement of pH is relatively simple and cheap to carry out, especially if using the NBS scale (though for the ultimate in precision the seawater scale is currently used, see Appendix F, Table F.1). Measurement of either the *pp*CO_2 or the total CO_2 are technically somewhat more difficult, complex, and expensive to carry out with the necessary precision and accuracy. Nevertheless, these measurements are now frequently made, even on shipboard. The choice of methods depends on the accuracy and precision of the measurement techniques available, and their convenience.

Most observations on the carbonate system in seawater have, however, been made by measurement of pH followed by titration with acid to establish the total concentration of proton-reacting components, and calculation of the contribution of carbonate and bicarbonate ions to this total. In order to understand the reactions involved and the general method of calculation we will have to consider the other buffers in seawater and the concept of the *alkalinity* of seawater.

7.2.4 Boric acid equilibria

The second most important buffer in seawater after the CO_2 system is composed of the couple, boric acid and borate ion. Since our main interest is usually in the carbonate system, a means must be developed to subtract the effects of borate ion from the total titration result.

The boric acid dissociation is given by:

$$B(OH)_3 + H_2O \leftrightharpoons B(OH)_4^- + H^+. \tag{7.24}$$

(The alternate expression, which is less correct,

$$H_3BO_3 \leftrightharpoons H_2BO_3^- + H^+. \tag{7.24a}$$

is not experimentally distinguishable from Eq. (7.24) during an acid–base titration.)

The thermodynamic dissociation constant, K_B, is converted to an apparent dissociation constant, in the same manner as for the first and second dissociation constant of carbonic acid:

$$K_B = \frac{\{H^+\} \times \{B(OH)_4^-\}}{\{B(OH)_3\}}$$

and

$$K_B^* = \frac{\{H^+\} \times [B(OH)_4^-]}{[B(OH)_3]}. \tag{7.25}$$

Values of K_B^* were also determined by Lyman and tabulated in his thesis. The values used here, however, (Box 7.1) are those of Andrew Dickson (1990b). See also Appendix F, Table F.3.

In order to subtract the contribution made by boric acid to the buffering by seawater, we need to know the concentration of $B(OH)_4^-$, the borate ion. Since boron in seawater is thought to be conservative, i.e. to have a constant ratio to the salinity, the concentration of total boron (TB, in mol kg^{-1}), may be calculated by the relationship:

$$[TB] = 11.88 \times 10^{-6} \times S,$$

where

$$[TB] = [B(OH)_4^-] + [B(OH)_3]. \tag{7.26}$$

From Eq. (7.25):

$$K_B^* \times [B(OH)_3] = \{H+\} \times [B(OH)_4^-],$$

and from Eq. (7.26):

$$K_B^* \times ([TB] - [B(OH)_4^-]) = \{H^+\} \times [B(OH)_4^-]$$
$$K_B^* \times [TB] = \{H^+\} \times [B(OH)_4^-] + K_B^* \times [B(OH)_4^-]$$
$$[B(OH)_4^-] = \frac{K_B^* \times [TB]}{\{H^+\} + K_B^*} \tag{7.27}$$

At this point we now have expressions for the most important constituents of the alkalinity of seawater.

7.2.5 Alkalinity of seawater

Seawater normally has a pH that ranges between 7.9 and 8.4. For surface water, the average value is near 8.2; it is, therefore, alkaline, because it is on the alkaline side of the pH of a neutral solution.

The term *alkalinity*, as used in referring to seawater, has little to do with the fact that seawater is slightly basic, however. Instead, the term refers to the ability of substances in seawater to combine with hydrogen ions during the titration of seawater with strong acid to the point where essentially all the carbonate species are protonated. The substances in seawater that can react in this way are listed in Table 7.3. Under good conditions, the precision with which the total alkalinity of seawater can be determined is now about ± 1 μmol kg^{-1} (e.g. Bradshaw *et al.* 1981). For some purposes this precision is not required, and the calculation procedures commonly adopted for ordinary work ignore the contributions of all substances except bicarbonate, carbonate, borate, and, sometimes, hydroxyl ion. For ordinary work with open-ocean surface water, this is a relatively safe procedure, but for deep water, near-coastal or estuarine water, or in experimental situations, it may not be. In water from the deep Pacific, for example, phosphate and silicate may contribute significant additional alkalinity (Table 7.3).

When seawater is titrated with acid the pH drops, at first slowly and then more rapidly, as the major species that react with hydrogen ions are protonated:

$$CO_3^{2-} + H^+ \rightarrow HCO_3^-$$
$$HCO_3^- + H^+ \rightarrow H_2CO_3$$
$$B(OH)_4^- + H^+ \rightarrow B(OH)_3 + H_2O$$

The maximum change in pH per unit of added acid occurs at about pH 5, but the titration is usually carried to an end point a little below pH 4, and the contribution of other species reacting below pH 5 subtracted by calculation. Beyond this point other species in seawater, primarily SO_4^- and F^-, dominate the reactions with hydrogen ions.

In most samples of surface seawater the end point is reached at about 2.2 ($\pm$ 0.3) mmol of acid protons added per kilogram of seawater, normalized to a salinity of 35‰.

The value determined in this way is the *total alkalinity* or *the titration alkalinity* of the water, and (ignoring the constituents after the fourth in Table 7.3) it is constituted as follows:

Table 7.3 Substances contributing to the alkalinity of seawater

Substance	μmol kg$^{-1(a)}$
HCO_3^-	1861
CO_3^{2-}	182
$B(OH)_4^-$	82
OH^-	4
Organic matter	(3 to 8)[b]
$HPO_4^{2-} + PO_4^{3-}$	(3)[c]
$MgOH^+$	(2)
$H_3SiO_4^-$	( )[c]
NH_3	( )[c]

[a] None of these substances is conservative with salinity. The values given are approximate concentrations, calculated for $S = 35$ ‰, pH = 8.2, and $t = 18$ °C, that can combine with measurable proportions of hydrogen ions upon the addition of small amounts of acid, and thus can make a detectable contribution to the alkalinity. Values in parentheses are quite variable or are uncertain.

[b] Huizenga and Kester (1979) showed that in samples of organic matter extracted from seawater there were about 13 mol of titratable groups per 100 mol of carbon. (Humic matter from rivers may also contain 12 to 14 mol of such groups per 100 mol of carbon.) Here I assume that the concentration of organic carbon in seawater is 40 to 100 μmol kg^{-1}, and that for each 100 atoms of carbon in the organic matter there are 13 free carboxylic acid groups or other groups able to combine with a proton. The average pK of these groups was about 3.5, so an alkalinity titration to this end point would result in only about one half of the groups becoming protonated. Concentrations of organic carbon could be greater in some surface waters.

[c] Concentrations of phosphate, silicate, and ammonia are very low in surface waters. Phosphate concentrations can be as much as 3 μmol kg^{-1} in Pacific deep water. Silicate is abundant in pore waters (up to several hundred micromol per kilogram) and in deep water of the Pacific may reach 180 μmol kg^{-1}, but the dissociation constant is poorly known. If the dissociation constants of silicic acid increase in seawater as much as do those for phosphate, relative to fresh water or solutions of NaCl, then silicate might contribute 10 μmol of alkalinity per kilogram, or more. Ammonia is commonly abundant only in pore waters or in other anoxic waters.

$$[TA] = 2[CO_3^{2-}] + [HCO_3^-] + [B(OH)_4^-] + [OH^-] - \{H^+\}. \quad (7.28)$$

The $\{H^+\}$ term above is not important at pH 8.2, being a value of less than 10^{-8}, but it is included to be formally complete (at least with respect to these components) and because we can know its value.

The carbonate alkalinity, CA, is that portion of the total alkalinity that is contributed by the carbon dioxide species:

$$[CA] = [TA] - [B(OH)_4^-] - [OH^-] + \{H^+\},$$

and from Eq. (7.27)

$$[\mathrm{CA}] = [\mathrm{TA}] - \frac{K_\mathrm{B}^* \times [\mathrm{TB}]}{\{\mathrm{H}^+\} + K_\mathrm{B}^*} - \frac{K_\mathrm{W}^*}{\{\mathrm{H}^+\}} + \{\mathrm{H}^+\}. \tag{7.29}$$

After the measurement of total alkalinity is made, the contributions from borate and hydroxyl are subtracted, according to Eq. (7.29), and the remainder is the *carbonate alkalinity*.

$$[\mathrm{CA}] = 2[\mathrm{CO_3^{2-}}] + [\mathrm{HCO_3^-}] \tag{7.30}$$

Having an estimate of the carbonate alkalinity, it is then possible to calculate other important parameters of the carbonate system.

7.2.6 Calculation of the carbonate species from CA and pH

After obtaining the carbonate alkalinity, the next step is to calculate the concentrations of the individual carbonate species. The appropriate expressions are derived as follows.

From Eq. (7.13):

$$[\mathrm{CO_3^{2-}}] = \frac{K_2^* \times [\mathrm{HCO_3^-}]}{\{\mathrm{H}^+\}}. \tag{7.31}$$

From Eqs. (7.30) and (7.31):

$$[\mathrm{CA}] = [\mathrm{HCO_3^-}] + 2\left(\frac{K_2^* \times [\mathrm{HCO_3^-}]}{\{\mathrm{H}^+\}}\right),$$

$$[\mathrm{CA}] = [\mathrm{HCO_3^-}]\left(\frac{\{\mathrm{H}^+\} + 2K_2^*}{\{\mathrm{H}^+\}}\right),$$

and

$$[\mathrm{HCO_3^-}] = [\mathrm{CA}]\left(\frac{\{\mathrm{H}^+\}}{\{\mathrm{H}^+\} + 2K_2^*}\right). \tag{7.32}$$

From Eqs. (7.30) and (7.32):

$$[\mathrm{CA}] - 2[\mathrm{CO_3^{2-}}] = [\mathrm{CA}]\left(\frac{\{\mathrm{H}^+\}}{\{\mathrm{H}^+\} + 2K_2^*}\right),$$

$$2[\mathrm{CO_3^{2-}}] = [\mathrm{CA}] - [\mathrm{CA}]\left(\frac{\{\mathrm{H}^+\}}{\{\mathrm{H}^+\} + 2K_2^*}\right),$$

$$[\mathrm{CO_3^{2-}}] = \frac{[\mathrm{CA}]}{2}\left(1 - \frac{\{\mathrm{H}^+\}}{\{\mathrm{H}^+\} + 2K_2^*}\right),$$

$$[\mathrm{CO_3^{2-}}] = [\mathrm{CA}]\left(\frac{K_2^*}{\{\mathrm{H}^+\} + 2K_2^*}\right) \tag{7.33}$$

An expression for the sum of $[H_2CO_3] + [CO_2(f)]$ is developed as follows. From Eqs. (7.12) and (7.32):

$$[CO_2aq] = [CA]\left(\frac{\{H^+\}}{K_1^*}\right)\left(\frac{\{H^+\}}{\{H^+\} + 2K_2^*}\right),$$

$$[CO_2aq] = [CA]\frac{\{H^+\}^2}{K_1^*(\{H^+\} + 2K_2^*)}. \tag{7.34}$$

We now have expressions for each of the carbonate species; their sum is [TCO_2], the concentration of all forms of carbon dioxide in a sample of seawater:

$$[TCO_2] = [CO_2aq] + [HCO_3^-] + [CO_3^{2-}]. \tag{7.35}$$

From Eqs. (7.32) to (7.35):

$$[TCO_2] = \frac{[CA] \times \{H^+\}^2}{K_1^*(\{H^+\} + 2K_2^*)} + \frac{[CA] \times \{H^+\}}{\{H^+\} + 2K_2^*} + \frac{[CA] \times K_2^*}{\{H^+\} + 2K_2^*},$$

and

$$[TCO_2] = [CA]\frac{\{H^+\}^2 + \{H^+\}K_1^* + K_1^*K_2^*}{K_1^*(\{H^+\} + 2K_2^*)}. \tag{7.36}$$

The partial pressure of CO_2 gas in equilibrium with a sample of seawater can also be calculated from the carbonate alkalinity and the pH. Combining Eqs. (7.10) and (7.34),

$$ppCO_2 = [CA]\frac{\{H^+\}^2}{H_{CO_2} \times K_1^*(\{H^+\} + 2K_2^*)}. \tag{7.37}$$

The partial pressure derived from equations relating the various parameters in the water phase is in fact the *fugacity*; this expresses the tendency of the CO_2 to escape from the water. It is necessary to introduce this term because CO_2 is a slightly non-ideal gas under usual conditions in the surface ocean. The partial pressure differs from fugacity that would be calculated from the concentration by a factor of about 0.996, varying somewhat with temperature. The difference is small, and for simplicity I mostly use the term partial pressure, where the proper term in thermodynamics is fugacity. Appendix E provides further details and an algorithm for calculation.

A typical set of data describing the CO_2 system in surface seawater is given in Table 7.4, where the results of calculating the individual species, according to the relationships developed above, are presented. For the purpose of this table, the total alkalinity was held constant, and the pH was varied. This corresponds to the situation in real seawater where there is no precipitation or solution of calcium carbonate, but the pH is caused to vary by introduction or removal of CO_2. Data such as are given in Table 7.4 may be calculated for any sample of seawater if the salinity, temperature, total alkalinity, and pH are known.

An important conclusion to be drawn from the table is that the partial pressure of CO_2 is extremely sensitive to variation in pH. When the pH changes from 8.00 to 8.30 the $\{H^+\}$ decreases by a factor of 2.00, and bicarbonate and carbonate ions change by less than a factor of 2.0. The partial pressure of CO_2, however, changes by the

Table 7.4 Carbon dioxide system in surface seawater: typical data

Pressure = 1 atm (101 325 Pa) Salinity = 35.00‰
Temperature = 18 °C Density = 1025.273 kg m^{-3}
$K_1^* = 9.079 \times 10^{-7}$ $K_2^* = 6.168 \times 10^{-10}$ $K_W^* = 2.38 \times 10^{-14}$
$H_{CO2} = 3.429 \times 10^{-2}$ mol kg^{-1} atm^{-1} (= 338.4 nmol kg^{-1} Pa^{-1})
Total borate = 416 μmol kg^{-1} $K_B^* = 1.540 \times 10^{-9}$
Total alkalinity = 2320 μmol kg^{-1} ppH_2O = 2032 Pa

pH	8.000	8.100	8.200	8.300
$\{H^+\}$	0.01000	0.00794	0.00631	0.005012
BA	55.5	67.5	81.6	97.7
OH^- alk	2.4	3.0	3.8	4.7
CA	2262	2249	2235	2218
$[CO_2]$ (ppmv)	662.4	508.8	388.0	293.4
$ppCO_2$ (μatm)	646.9	496.9	378.9	286.5
$ppCO_2$ (Pa)	65.55	50.35	38.39	29.03
$[CO_2aq]$	22.2	17.0	13.0	9.82
$[HCO_3^-]$	2014	1947	1869	1780
$[CO_3^{2-}]$	124.2	151.2	182.7	219.0
$[TCO_2]$	2160	2115	2065	2008

All values in the table are in μmol kg^{-1}, except for $ppCO_2$ (properly, the fugacity) which is in micro-atmospheres and pascals, while the equilibrium concentration in a dry atmosphere (the usual convention) is in ppmv. The difference between the $ppCO_2$ and the ppmv at 101 325 Pa is due to a small correction for the fugacity and a much larger, in this case, correction for the presence of water vapor.

In this table the changes in pH (and thus $\{H^+\}$, the hydrogen ion activity) are due only to changes in the concentration of total CO_2. The values were calculated, using data and algorithms from Chapter 7 and Appendices E and F for t = 18 °C and TA = 2320 μmol kg^{-1}, close to average values for surface water in the global ocean (Levitus 1982, Millero *et al.* 1998). In all cases above (except for density), uncertainties in the accuracy of the constants lead to uncertainties in the absolute values of the fourth and possibly of the third significant figures.

factor 2.26, and thus is the most sensitive of the variables that could be measured to describe the system. This is one reason that, as mentioned before, it has been suggested that measurements of partial pressure of CO_2 should be used to characterize the CO_2 system in samples of seawater, as a substitute for or supplement to the other measurements usually made. An additional reason for making direct measurements of the partial pressure of CO_2 is that this is a crucial parameter used in calculating the air–sea fluxes of CO_2.

7.2.7 What happens when we add acid?

The titration of a sample of seawater with acid can be carried out in two ways (Figure 7.2). If the CO_2 is allowed to escape to the atmosphere after each addition

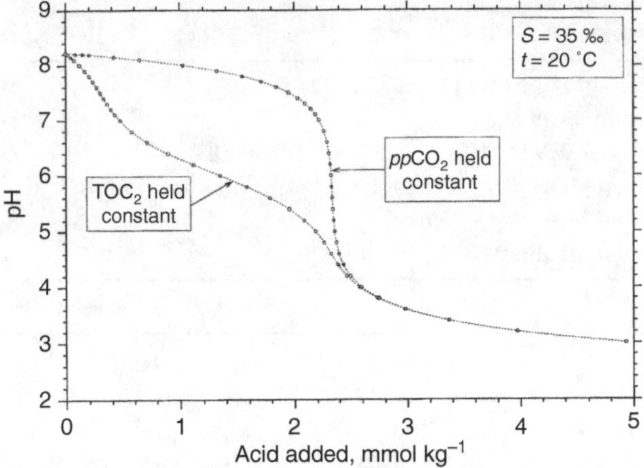

Figure 7.2 Titration curves of seawater calculated for two conditions. Top curve: the partial pressure of CO_2 is held constant at the present atmospheric value by carrying out the titration with stirring in an open vessel exposed to the atmosphere and allowing the CO_2 to escape as the $ppCO_2$ rises after each incremental addition of acid. Bottom curve: the titration is carried out in a closed container, retaining the CO_2; in this situation the higher concentrations of the various CO_2 species cause the pH to be considerably lower during the early part of the titration. In order to evaluate the concentrations and the constants involved in the CO_2 system itself, the titration must be carried out in such a way that the CO_2 cannot escape.
These titration curves were calculated using combinations of Eqs. (7.16), (7.19), (7.20), (7.27), (7.28), and equations for HSO_4^- and HF in the same form as Eq. (7.27), using constants as given in Appendix F. An accurate titration curve below pH ~5 requires the inclusion of the latter two components, as they are the main buffers in seawater in the pH range below this. The individual components that contribute to the titration curve (TCO_2 constant) are shown in Figure 7.3.

of acid the pH does not change very much in the early stages of the titration. If the container is closed to the atmosphere the CO_2 must stay in solution, even though the $ppCO_2$ rises considerably. This CO_2 affects the equilibria so that at each stage there is a higher concentration of protons than if the CO_2 had been allowed to escape; the pH falls more rapidly at first, until all the CO_3^{2-} and the HCO_3^- are eventually protonated. The two curves come together at about pH 4.0. Below this pH other species become more important, especially SO_4^{2-}, which becomes protonated to HSO_4^-, and also F^- which becomes protonated to HF. These last two species are therefore responsible for most of the buffering below pH 4.

Some appreciation for the concentrations of most of the substances important to the buffer system in seawater can be gained by examination of Figure 7.3. Here the calculations were made for a situation where the total CO_2 is maintained constant and the pH is caused to vary by addition of strong acid or base (for simplicity, calculated without change in volume). Note that the pH of typical surface seawater is generally between 7.9 and 8.4, only a narrow part of the range on this graph. This figure also includes sulfate and fluoride, and shows how these become important in the buffering of seawater under acid conditions. From this graph one can see the

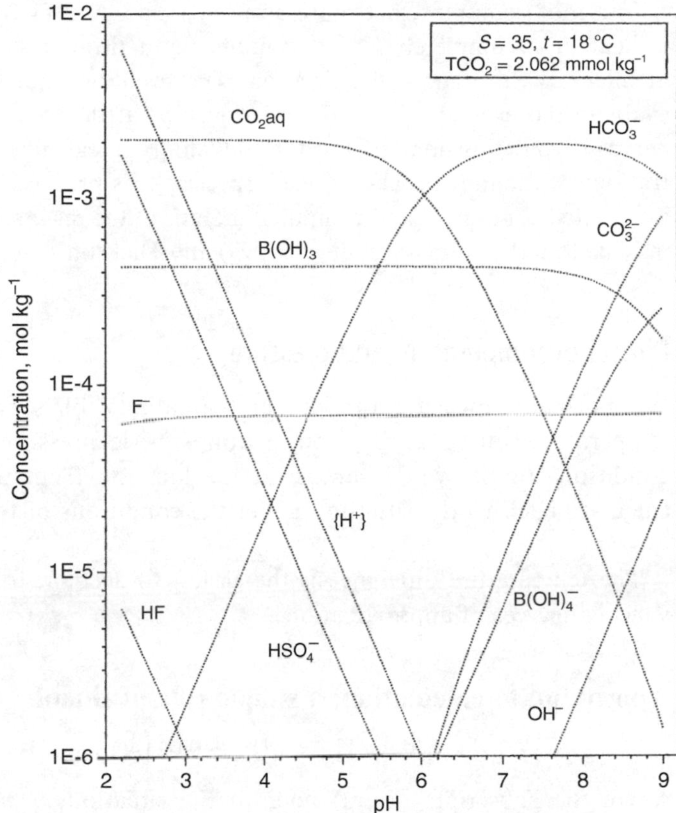

Figure 7.3 Buffer systems in seawater calculated for a situation where total CO_2 is held constant and the change in pH is caused only by the addition of strong acid or base. The chosen values are characteristic of surface seawater. Various other constituents that might appear near the bottom of the graph, between 10^{-6} and 10^{-5} mol kg^{-1}, are not shown. These include silicate and phosphate in deep water, and organic matter everywhere. The titration behavior of organic matter in seawater is poorly known and cannot be represented with certainty. Also not shown is the sulfate ion, SO_4^{2-}; at 28 mmol kg^{-1} this is above the top of the graph.

relative importance of carbonate, fluoride, sulfate, hydroxyl, borate, etc., in different parts of the pH range. With uniform additions of acid the pH change is greatest just above pH 4 (Figure 7.2).

The two curves in Figure 7.2 provide insight into the two most frequently used procedures for determining the titration alkalinity and the carbonate species in seawater (Grasshoff *et al.* 1999). The simplest procedure is a single point titration. A known quantity of acid, the quantity established by experience as that which brings the final pH close to 3.5, is added to a sample in an open container. The solution is stirred until a consistent pH reading is obtained. The pH is read; and the total alkalinity is calculated from the final pH, the amount of acid added, and a salinity-dependent correction factor (*f*). The correction factor, *f*, allows for the final concentration of excess protons, and also for the uptake of protons by the sulfate and to a lesser extent by the fluoride ion at the final pH value.

In a more complex, but more accurate, procedure (a *Gran titration)* the sample is held in a completely closed container and titrated stepwise with pH readings at intervals throughout the titration. The resulting data are analyzed in detail to evaluate the exact shape of the relationships at different portions of the titration curves. From this analysis the total alkalinity is calculated as well as the concentrations of the individual carbonate species. This procedure is best carried out with automated equipment and computer analysis of the results. Details of the procedures may be found in Grasshoff *et al.* (1999) and Dickson *et al.* (2007).

7.2.8 Effects of temperature and pressure

We normally measure the pH and alkalinity of seawater samples at room temperature, 20 to 25 °C, and at atmospheric pressure. In order to know the conditions in the water sample at the location from which it came, we must calculate back to the situation under the conditions of temperature and pressure *in situ*.

The temperature difference is the easiest to account for, which can be done to varying degrees of approximation.

Approximate calculation, a simple rule of thumb

$$\mathrm{pH_{is}} = \mathrm{pH_m} + 0.0114(t_\mathrm{m} - t_\mathrm{is}), \tag{7.38}$$

where the subscripts (is, m) refer to the situation *in situ* and when measured, respectively. Note that in the usual situation, where the *in-situ* temperature is colder than the temperature of measurement, the *in-situ* pH is higher than the measured pH. The reason for this is that the dissociation constants of the various substances reacting with protons decrease at lower temperatures, so there is a decreased concentration of hydrogen ions. The effect amounts to a little more than 1/100 of a pH unit per degree celsius.

More exact calculation

The pH change with temperature is a weak function of both the salinity of the samples and of the initial pH, and is not quite linear with temperature, so for careful work additional factors must be taken into account. This can be partly accomplished by the approach of deriving an equation containing hydrogen ion, the necessary constants, the total alkalinity, and the total borate (ignoring the lesser constituents of the buffer system in seawater), and then solving for the activity of hydrogen ion. Such a relationship may be obtained by combining Eqs. (7.29) (omitting the small correction for $[\mathrm{OH^-}]$) and (7.36):

$$[\mathrm{TCO_2}] = \left([\mathrm{TA}] - \frac{K_\mathrm{B}^*[\mathrm{TB}]}{\{\mathrm{H^+}\} + K_\mathrm{B}^*}\right)\left(\frac{\{\mathrm{H^+}\}^2 + \{\mathrm{H^+}\}K_1^* + K_1^*K_2^*}{K_1^*(\{\mathrm{H^+}\} + 2K_2^*)}\right) \tag{7.39}$$

This equation is a cubic in $\{\mathrm{H^+}\}$, and can be rearranged as follows.

Table 7.5 Effect of pressure on carbonic and boric acid dissociation constants

Pressure, dbar	$\frac{K^*_{1(P)}}{K^*_1}$	$\frac{K^*_{2(P)}}{K^*_2}$	$\frac{K^*_{B(P)}}{K^*_B}$
1000	1.11	1.07	1.13
2000	1.24	1.15	1.28
4000	1.52	1.32	1.63
6000	1.87	1.52	2.06
8000	2.28	1.75	2.60
10 000	2.77	2.01	3.26

The pressure effect is expressed as the ratio of the value of each constant at high pressure to the value at atmospheric pressure, calculated for 5 °C. Pressure increases approximately one decibar for each one meter in depth. From Appendix H.

$$A\{H^+\}^3 + B\{H^+\}^2 + C\{H^+\} + D = 0 \tag{7.40}$$

It may be evaluated by substitution of the appropriate numerical values for the dissociation constants and concentrations, and extraction of the appropriate root.

The easiest and most accurate approach to correcting to *in-situ* temperature is to set up (for example) an Excel spreadsheet with the appropriate constants and measured values (and including the hydroxyl ion, and any other appropriate components), and solve for the TCO_2 at the temperature of measurement. Then by iteration find the pH that results in the same TCO_2 at the *in-situ* temperature.

The effect of increasing pressure is opposite to that of decreasing temperature, so that even though the deeper parts of the ocean are nearly always colder than the temperature of measurement, the pH may be lower than its measured value in the laboratory. The pressure effects on the various dissociation constants have been determined by Culberson and Pytkowicz (1968). All the dissociation constants increase with increase in pressure, and the increases are quite significant – in the deepest parts of the ocean the constants exceed twice their values at atmospheric pressure (Table 7.5).

The way the correction is applied is to calculate or look up the dissociation constant for the *in-situ* temperature, then correct this using a factor appropriate for the pressure. The appropriate values are then substituted into the equations, which are then solved for $\{H^+\}$ or pH. Selected typical values are given in Table 7.6 to show the magnitude of the effect.

More extensive tabulations may be found in Culberson and Pytkowicz (1968). Millero (1979) developed several algorithms for calculating temperature and pressure dependencies of the constants.

When surface seawater is warmed or cooled, the partial pressure of CO_2 is strongly affected (Figure 7.4). This effect is very important in the exchange of CO_2 between the atmosphere and the sea. The sensitivity of the relationship is great enough that small errors in the measurement of surface temperature can significantly affect the calculation of fluxes to or from the ocean. It should be appreciated that the

Table 7.6 Examples of the effect of pressure on the pH of seawater

Pressure, dbar	2 °C	10 °C	15 °C
0	8.132	8.163	8.180
1000	8.093	8.126	8.144
2000	8.053	8.089	8.108
5000	7.934	7.979	8.002
10 000	7.735	7.798	7.831

At each temperature, the pH was established for seawater with $S = 35‰$, TA = 2.320 mmol kg^{-1}, in equilibrium with an atmosphere having 390 µatm of CO_2. The equations were then solved at each pressure, holding the TCO_2 constant. The resulting pH is that which a parcel of water would have if it were to sink from the surface to each specified depth. The calculation was done for an isothermal compression. In the real world, the compression would be essentially adiabatic. The temperature increase to a depth of about 9700 m, corresponding to a pressure increase of 10 000 dbar, amounts to 1.28, 1.724, and 1.987 °C, respectively. The corresponding pH values would be 7.722, 7.780, and 7.811. The pressure effect also varies a little with changes in the starting conditions; for example, it would have been slightly different when the partial pressure in the atmosphere was 280 µatm.

temperature of the top meter, or even the top few millimeters, may be slightly different than that of the bulk liquid below. During the direct measurement of the partial pressure of CO_2 in surface seawater it is important that the temperature be carefully controlled.

The temperature relationship results in significant oceanic transports of CO_2. When examining Figure 7.4, imagine that water in the Caribbean Sea at a temperature close to 30 °C is carried by the Gulf Stream north into the Nordic seas east of Greenland; during this trip it cools and may have time to take up enough CO_2 to approach equilibrium with the cooler atmosphere. Then, carrying the increased concentration of CO_2, it sinks, forms North Atlantic Deep Water, and flows south towards the Antarctic, where it contributes to the water that fills much of the deep Pacific. If this water eventually upwells to the surface, as for example in the great upwelling region off Peru or in the equatorial current systems, it will warm and release CO_2 back to the atmosphere. This uptake by the ocean in high northern latitudes and release mostly in low latitudes is balanced by atmospheric transport in the opposite direction. This transport by the so-called *solubility pump* occurs by entirely physical processes. The transport by biological processes, the *biological pump*, will come up in later discussion.

7.3 Calcium carbonate

As was shown in Chapter 4, the residence time of calcium in the oceans is about 700 000 years. The total calcium in the oceans has therefore been replaced probably several thousand times since the earth began, and some 800 to 900 times since the

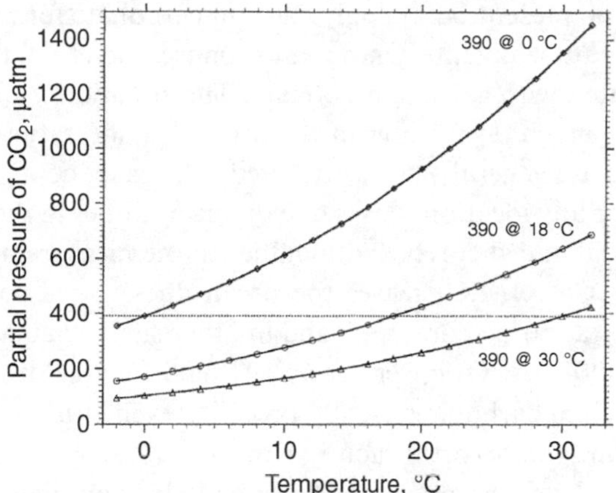

Figure 7.4 Effect of temperature changes on the partial pressure of CO_2. Warming or cooling seawater has a large effect on the $ppCO_2$, amounting to very approximately 4% per 1 °C when close to a partial pressure of 390 μatm of CO_2, but the effect varies to some extent with the salinity, the alkalinity, and the starting concentration of TCO_2. For the curves in this figure, the salinity was always 35‰ and the total alkalinity was 2.320 mmol kg^{-1}. Three cases are presented:

(a) 390 μatm at 0 °C, where the seawater starts at a temperature of 0 °C, $ppCO_2 = 390$ μatm and $TCO_2 = 2.204$ mmol kg^{-1};

(b) 390 μatm at 18 °C, where the seawater starts at a temperature of 18 °C, $ppCO_2 = 390$ μatm and $TCO_2 = 2.071$ mmol kg^{-1};

(c) 390 μatm at 30 °C, where the seawater starts at a temperature of 30 °C, $ppCO_2 = 390$ μatm and $TCO_2 = 1.967$ mmol kg^{-1}.

The 18 °C points, for example, can be fit by the equation: $ppCO_2 = 172.634 + 8.592t + 0.1473t^2 + 0.0028t^3$.

It should be pointed out that the concentration of CO_2 in a dry atmosphere differs at each of the points of equilibrium above; it requires concentrations of 394.1 ppmv at 0 °C, 399.4 ppmv at 18 ° C and 408.0 ppmv at 30 °C to achieve the partial pressures (actually the fugacities) as plotted above. The largest cause of this effect is the presence of water vapor. In the real world, the concentration of CO_2 in the dry atmosphere would be nearly the same everywhere, and the partial pressures at equilibrium for each temperature would be correspondingly a little bit different. It seemed that the presentation as above made for a clearer view of the relationships.

beginning of the Cambrian. The amount entering the sea today is (roughly) estimated at about 21×10^{12} moles per year or about 840 million tons of calcium per year.

If the ocean is in steady state, the calcium entering the ocean each year must be removed at a similar rate, and most of this must be in some mineral phase. Examination of the sedimentary record shows that the only solid sedimentary phases where calcium is abundant are calcium carbonate ($CaCO_3$) and dolomite, which has the approximate composition $CaMg(CO_3)_2$. Dolomite is found extensively in ancient deposits, and appears to be an alteration product of calcite and aragonite. Exactly how this alteration comes about is poorly known.

In the present ocean, only calcium carbonate forms a significant part of the recent carbonate sediments. One might suppose, therefore, that the ocean is, on the average, saturated with calcium carbonate. The situation is quite complex, however, because in fact much deep water in the ocean is undersaturated with $CaCO_3$, while surface waters are generally supersaturated. The exact depth of changeover is slightly uncertain for any location; this is due to uncertainties in the physical constants and various practical and theoretical difficulties in the measurements.

The $CaCO_3$ that leaves the ocean does so via some deep-sea sediments and by precipitation on coral reefs and in other shallow areas. Sediments composed of more than 30% $CaCO_3$ cover nearly one-half the area of the deep-sea floor.

Calcium carbonate usually precipitates in either of two major crystal forms:

aragonite, orthorhombic; or

calcite, rhombohedral (also contains varying concentrations of magnesium).

Aragonite is less stable and more soluble under normal conditions than calcite, so in sediments aragonite is much less abundant than calcite.

Evidence from the sediment shows that the calcium carbonate there is all, or nearly all, the product of biological activity. Many kinds of organisms make calcareous shells that eventually sink towards the bottom of the sea. Among the major agents of calcium carbonate removal are the following.

Foraminifera. These single-celled protozoa are abundant in the upper layers of the ocean, although some species live on the bottom. Most *forams*, as they are called, have a calcareous skeleton or *test* made of $CaCO_3$, crystallized as calcite. Compared to most single-celled organisms the forams are large – they can be seen with the naked eye – and therefore their tests can sink rapidly to the bottom. They contribute the largest part of the $CaCO_3$ falling from the plankton to the sediments each year, and many parts of the ocean floor are covered with a *foram ooze*, in which the $CaCO_3$ content may sometimes even exceed 98% of the total mass of dried sediment.

Coccoliths. The second most important planktonic precipitation agent for $CaCO_3$ is the group of single-celled planktonic algae known as the coccolithophorids. These carry on the surfaces of their cells numerous coccoliths – tiny calcitic plates of complex and varied design. The sinking of the coccoliths, either individually or incorporated in the fecal pellets of plankton-feeding herbivores, is an important pathway for the transport of $CaCO_3$ to the deep water and to the sediments.

Mollusc shells. The clams and snails are well known. The small planktonic molluscs, the pteropods, are perhaps more important in terms of total $CaCO_3$ precipitated, and in some parts of the ocean the bottom is covered with a *pteropod ooze*. The crystal form of pteropod $CaCO_3$ is aragonite. Since aragonite is more soluble than calcite, pteropod shells dissolve more readily than foram shells or coccoliths, and their distribution in the sediments is much more limited.

Coral reefs. These were probably even more important in earlier geological epochs than at the present time. While coral animal skeletons are the most spectacular and well-known components of coral reefs, the coralline red algae (such as *Lithothamnion* and *Porolithon*) may be as important, both as main structural components and as binding agents. Many ancient reefs have a higher proportion of mollusc shells than do the modern reefs. Kinsey and Hopley (1991) estimated that coral reefs precipitate about 900 × 10^6 tons of $CaCO_3$ each year. This rate translates to a loss of calcium from the ocean amounting to 9 × 10^{12} moles per year, or about 43% of the total input. Some of this calcium carbonate may be transferred as detritus carried by wave action and currents into deeper water, where it may accumulate on deeper slopes, instead of on the top of the reef.

Other algae. The green calcareous alga *Halimeda*, a common form in all warm shallow areas, is often responsible for much of the mass of smaller particles of aragonite that fill back-reef and lagoon areas (Hillis-Colinvaux 1980). *Halimeda* is also found outside reef environments, and produces significant accumulations of $CaCO_3$ in sea-grass regions and along the deeper edges of shelves (Jensen *et al.* 1985). Other calcareous algae may also be locally significant.

Inorganic precipitation. The question of whether $CaCO_3$ may precipitate inorganically to a significant extent in the ocean has been a matter of controversy. Leaving aside the precipitation of carbonate cements in some shallow sediments and other localized occurrences, one focus of debate has been on the phenomenon of *whitings*. In warm, shallow, tropical water such as is found in the Persian Gulf or over the Bahama Banks there is often observed a whitening of the water, caused by a fine suspension of $CaCO_3$. The solubility of $CaCO_3$ decreases as temperature rises and, in addition, the temperature sensitivity of the various constants of the CO_2 system result in the following reaction being driven to the right as the temperature rises:

$$2\,HCO_3^- \rightleftharpoons H_2O + CO_3^{2-} + CO_2 \uparrow . \tag{7.41}$$

The CO_2 is lost to the atmosphere, the concentration of carbonate ion increases and therefore so does the tendency for $CaCO_3$ to precipitate. It has been thought that this could cause the appearance of whitings. Other evidence suggests, however, that the crystals formed are all associated with organic matter or with the remnants of old phytoplankton cell membranes. It appears plausible that the organic particles form a nucleus upon which crystals of $CaCO_3$ can precipitate (Robbins and Blackwelder 1992). Whether this is considered biological or inorganic is perhaps a matter of semantics. On the Bahama Banks, there are hundreds of square kilometers of *oolitic sands*, small rounded particles of $CaCO_3$ thought to be at least partially produced by inorganic precipitation, but for which a biological process has also been invoked.

Taking all these outputs together, Milliman (1993) and Milliman and Droxler (1995) estimated that the accumulation rate of calcium carbonate in the modern ocean amounts to about 3.2 Gt per year, or 32 × 10^{12} moles per year. This comprised an estimated 11 × 10^{12} moles per year in the deep-sea sediments, 14.5 × 10^{12} moles per year in coral reefs and other shallow-water ecosystems, and the remainder mostly on the slopes. This rate is so much greater than the estimated input of calcium (from Chapter 4: 21 × 10^{12} moles per year, already a high-end estimate because of the inclusion of what looks like a rather high value for groundwater input) that it seems likely that, at the present time, the ocean is not in steady state, but instead is precipitating and accumulating calcium carbonate in the sediments at a faster rate than calcium is being supplied. A plausible explanation is that sea level has risen since the end of the last glaciation by about 125 m, flooding large shallow areas of the shelves and providing substrate for greatly increased coral reef growth and other carbonate deposition. According to this scenario, the ocean is now losing calcium as $CaCO_3$ and has not caught up and reached a new steady state since sea level rose. It is difficult to be sure of this conclusion, because values for both the inputs and outputs (especially the latter) are estimates with considerable quantitative uncertainty, which can only be improved by acquiring considerably more data on the various fluxes.

7.3.1 Solubility of calcium carbonate

As mentioned earlier, the difficulties in accounting for the solubility behavior of calcium carbonate in seawater have exercised the minds of marine chemists for several decades. In this section we will consider the solubility of $CaCO_3$ in seawater, in the somewhat artificial (but partly historical) sequence beginning from estimates in distilled water, then sequentially adding correction factors, until no further corrections are known.

The thermodynamic (infinite dilution) solubility product constant for calcium carbonate is expressed as follows:

$$CaCO_3 \leftrightharpoons Ca^{2+} + CO_3^{2-},$$

$$K_{sp} = \{Ca^{2+}\}\{CO_3{}^{2-}\}.$$

For the two common mineral phases of calcium carbonate the constants are:

$$\text{aragonite } K_{sp} \approx 5 \times 10^{-9},$$

$$\text{calcite } K_{sp} \approx 3.35 \times 10^{-9}.$$

Calcite has a lower solubility and accordingly is the more stable form. Under most conditions, aragonite tends to dissolve and to reprecipitate as calcite. This recrystallization is usually very slow, so aragonite may persist in sediments for a considerable time.

As shown earlier (Chapter 6), a measure of the degree of saturation of a substance is given by the ratio of the ion product to the solubility product constant. This ratio is 1.0 in an exactly saturated solution. For normal subtropical surface seawater (Table 4.1; Table 7.4, pH 8.20, $S = 35‰$, but with $t = 25\ °C$) the calculation could be made as follows:

$$[Ca^{2+}][CO_3^{2-}] = (10.28 \times 10^{-3})(2.15 \times 10^{-4})$$

$$IP = 22.2 \times 10^{-7}.$$

For calcite, the ratio of the ion product over the solubility product constant is:

$$\frac{IP}{K_{sp}} = \frac{22.2 \times 10^{-7}}{3.35 \times 10^{-9}} = 663.$$

In this case the concentrations are compared to the thermodynamic constant (from Jacobson and Langmuir 1974), which we know is not suitable for seawater solutions.

If we did not know about activity coefficients we would say that the solution is 663 times supersaturated, or that the actual concentrations of both ions are about 25 times greater than in a saturated solution; or that the concentration of one ion is 663 times too high.

Next, we take the activity coefficients into account. For a salinity of about 35‰ these (Table 6.2) are estimated to be:

$$\gamma_{Ca^{2+}} = 0.256, \gamma_{CO_3^{2-}} = 0.205.$$

Including these γ values in the calculation of the ion product gives (as in Chapter 6):

$$\frac{IP}{K_{sp}} = \frac{[Ca^{2+}]\gamma_{Ca^{2+}}[CO_3^{2-}]\gamma_{CO_3^{2-}}}{3.35 \times 10^{-9}} = 35.$$

With this correction the calculated degree of supersaturation is much smaller. However, the seawater is still supersaturated, and by the large factor of 35-fold.

This difficulty was recognized from at least the 1930s through the 1950s, but no satisfactory explanation was advanced; there was sort of an idea around that seawater possessed certain almost mystical properties that made it an exceptional solvent for calcium carbonate.

The difficulty was first resolved in the famous papers of Garrels *et al.* (1961) and Garrels and Thompson (1962). These investigators drew upon earlier work by Greenwald (1941), a biochemist who had improved our understanding of the solubility of $CaCO_3$ in biological fluids by showing that much of the analytically determined CO_3^{2-} ion is not present as a free ion, but rather is complexed to form ion pairs with various cations. For example

$$Mg^{2+} + CO_3^{2-} \leftrightarrows MgCO_3^0, \tag{7.42}$$

$$K^*_{(MgCO_3)} = \frac{[MgCO_3^0]}{[Mg^{2+}][CO_3^{2-}]}. \tag{7.43}$$

Garrels and Thompson measured the apparent association constants of various ion pairs and showed that, in solutions similar to seawater, up to 90% of the $CO_3{}^{2-}$ might be paired with cations, and thus made unavailable to react with and precipitate calcium. (Also, the presence of these ion pairs changes slightly the ionic strength of the solution.) Garrels and Thompson (1962) developed a model for the associations of each of the various ionic species in seawater and included the constants for each of the ion pairs that were then known to exist. They solved the many simultaneous equations that result from this approach, and obtained estimates of the concentration of each species in seawater. Table 7.7 is an extract from their results. They concluded that some of the Ca^{2+} and most of the CO_3^{2-} is tied up as ion pairs with other ions, so that when calculating the solubility it is necessary to take this phenomenon into account. With this, the IP should now be calculated as follows (using the values of 0.91 and 0.09 as the fractions of the Ca^{2+} and CO_3^{2-} ions, respectively, that are in a free state):

Table 7.7 Effect of ion pairing on the carbonate system in seawater

Ion pair	Concentration	Ion pair	Concentration
$CaSO_4^0$	8% of Ca^{2+}	$MgSO_4^0$	11% of Mg^{2+}
$CaHCO_3^+$	1% of Ca^{2+}	$MgHCO_3^+$	1% of Mg^{2+}
$CaCO_3^0$	0.2% of Ca^{2+}	$MgCO_3^0$	0.3% of Mg^{2+}
TOTALS	9.2% of Ca^{2+}		12.3% of Mg^{2+}
$CaCO_3^0$	7% of CO_3^{2-}	$CaHCO_3^+$	4% of HCO_3^-
$MgCO_3^0$	67% of CO_3^{2-}	$MgHCO_3^+$	19% of HCO_3^-
$NaCO_3^-$	17% of CO_3^{2-}	$NaHCO_3^0$	8% of HCO_3^-
TOTALS	91% of CO_3^{2-}		31% of HCO_3^-

The major species thought to be important are listed, along with the estimated degree of association. Derived from Garrels and Thompson (1962).

$$\mathrm{IP} = \left([\mathrm{Ca}^{2+}](\gamma_{\mathrm{Ca}^{2+}})\frac{\text{free Ca}}{\text{total Ca}}\right)\left([\mathrm{CO}_3^{2-}](\gamma_{\mathrm{CO}_3^{2-}})\frac{\text{free CO}_3^{2-}}{\text{total CO}_3^{2-}}\right) = 9.5 \times 10^{-9}.$$

This revised value for the ion product can now be compared to the value of the solubility product constant:

$$\frac{\mathrm{IP}}{K_{sp}} = \frac{9.5 \times 10^{-9}}{3.35 \times 10^{-9}} = 2.8.$$

No further general corrections are known, and this evaluation shows that in typical warm surface waters $CaCO_3$ is indeed supersaturated, in this case by about three-fold.

Although there are no additional correction factors, there has been some improvement in the values of certain constants. As with dissociation constants, it is a convenient practice to define and measure empirical apparent solubility-product constants, defined according to stoichiometric concentrations, and measured as functions of salinity and temperature:

$$K_{sp}^* = [\mathrm{Ca}^{2+}][\mathrm{CO}_3^{2-}].$$

Directly measured values (Appendix G, Table G.2) of the apparent constant tend to show even greater supersaturation than calculated above. For example, at $S = 35‰$ and 25 °C, the K_{sp}^* of calcite is 4.27×10^{-7}, suggesting that this typical surface seawater used for an example may actually be supersaturated with calcite by a factor of about five.

Direct measurements of the apparent solubility product constants, K_{sp}^*, for calcite and aragonite in seawater have been a technical challenge, and results have not been entirely consistent, but all show that the K_{sp}^* values are dramatically sensitive to changes in salinity (Figure 7.5). Recently measured values for K_{sp}^* of calcite at 25 °C

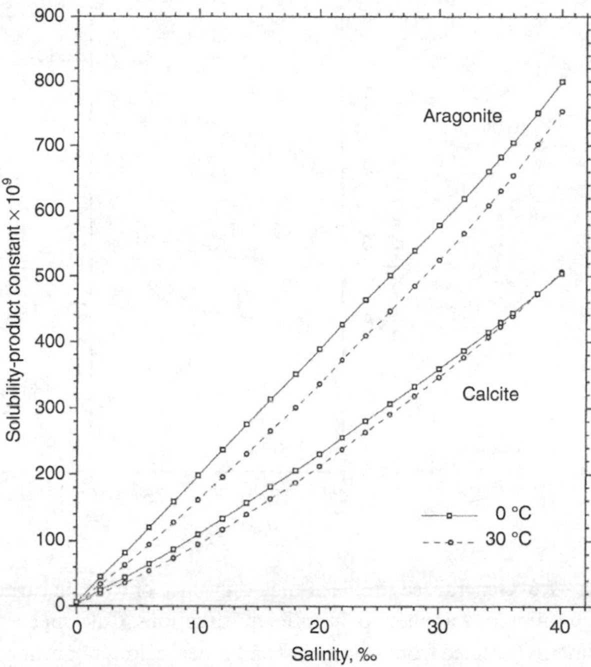

Figure 7.5 Solubility of calcium carbonate in seawater. The stoichiometric solubility product constants ($K_{sp}* = [Ca^{2+}]\,[CO_3^{2-}]$) for calcite and for aragonite in seawater, in relation to the salinity of the water and at temperatures of 0 °C and 30 °C, were calculated using the algorithms presented by Mucci (1983). Tabulated values are given in Appendix G.

and $S = 35‰$ range from 4.27×10^{-7} (Mucci 1983) to 4.70×10^{-7} (Plath *et al.* 1980), while the K^*_{sp} of aragonite under the same conditions has been reported as 6.48×10^{-7} (Mucci 1983) to 9.46×10^{-7} (Plath 1979). The situation is discussed in UNESCO (1983a). The general conclusion that surface waters are quite supersaturated still holds, and this appears to be universal in ordinary warm surface waters.

Why does surface seawater not yield a precipitate of calcium carbonate? This matter is not completely understood, but some information is available to help provide an explanation. Pytkowicz (1973) found that Mg^{2+} ions effectively delayed or prevented the nucleation of $CaCO_3$ crystals so that precipitation was not easily initiated in solutions with a Mg^{2+} concentration similar to that of seawater. Chave and Suess (1967) observed that fresh calcite crystals, placed in contact with seawater, would react at first fairly rapidly (either to precipitate more $CaCO_3$ or to dissolve, depending on conditions) while crystals that had been exposed to seawater for some time would at first react slowly if at all. This led to the suggestion that something (presumably organic matter) in seawater may coat the crystals so that their surfaces become less reactive with the surrounding medium.

It may be that in nature all exposed surfaces of $CaCO_3$ are coated with organic matter or other materials and are thus rendered so passive that only under special conditions is $CaCO_3$ able to precipitate inorganically in surface seawater.

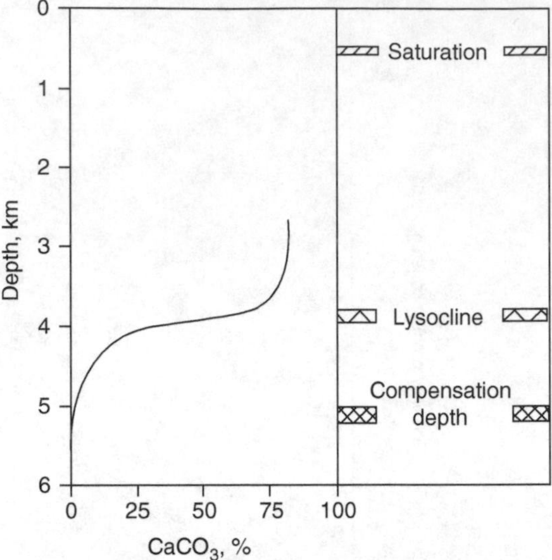

Figure 7.6 Generalized profile of the occurrence of calcium carbonate in Pacific Ocean sediments, and terminology applied to the different situations. Calcium carbonate is abundant in shallow sediments (selected from sea mounts and other regions not near coastal areas where the carbonate is usually diluted by clastic sediments); below about 4000 m the concentration decreases rapidly and little or none is found in sediments lying deeper than 5000 m. The depth at which the decrease begins is termed the *lysocline*, and the depth below which there is essentially no calcium carbonate is termed the *compensation depth*. Water layers above about 500 m are generally supersaturated, while water below this depth is slightly undersaturated. The degree of undersaturation increases with depth. Strictly speaking, the curve relates to the calcite compensation depth (CCD); at some distance above this there is an aragonite compensation depth (ACD), because aragonite is more soluble than calcite. The exact depth of the lysocline varies regionally within major ocean basins and between oceans. For example, both the lysocline and the CCD occur deeper in the Atlantic than in the Pacific.

7.3.2 Solution of calcium carbonate

Carbonate sediments (defined as those with more than 30% $CaCO_3$) cover about one-half of the deep-ocean floor, and the coverage of foram ooze alone is about 35%. These calcareous sediments are restricted, however, to depths shallower than about 4000 m in the Pacific and 4500 m in the Atlantic. Below these depths the concentration of $CaCO_3$ in the sediments rapidly decreases and below about 5000 m in the Pacific there is no $CaCO_3$ at all (Figure 7.6). The depth at which a rapid decrease in concentration begins is called the *lysocline*, and the depth below which no $CaCO_3$ is found is called the *compensation depth*. This latter term, however, does not seem especially evocative of the situation that all the calcium carbonate that must have fallen onto the sea floor at that depth has completely dissolved.

Why is calcium carbonate more soluble at depth in the ocean than at the surface? Several factors contribute. First, calcium carbonate is an unusual salt, in that it appears to become slightly more soluble as the temperature drops (Figure 7.5), but the effect is small. Second, increasing pressure increases the K^*_{sp} (Ingle 1975), and this effect is more pronounced at lower temperatures. At 2 °C and a pressure corresponding to a depth

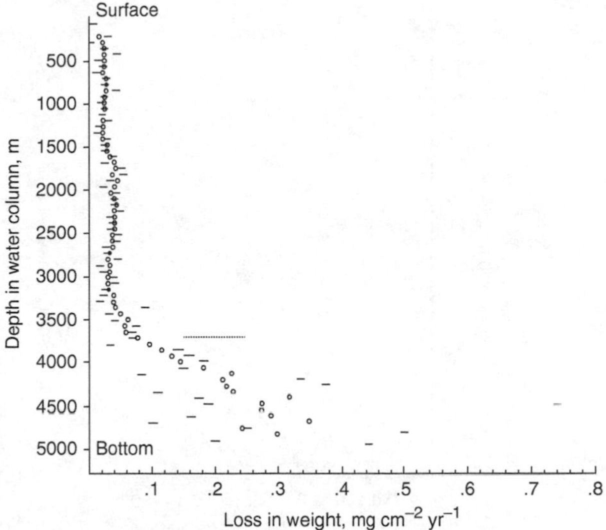

Figure 7.7 Data from the experimental dissolution rate observations of Peterson (1966). Calcite spheres were suspended for several months at various depths in the eastern Pacific Ocean. Bars show rates of dissolution for individual calcite spheres (length of bars represents uncertainty due to assigned weighing errors). Circles show rates of dissolution averaged over five adjacent spheres. Dashed line shows the level of abrupt increase in rate.

of about 6500 m, Ingle found that the K^*_{sp} was about 3.7 times greater than at atmospheric pressure. In addition, the effect of pressure on the dissociation constants of boric and carbonic acids causes the pH to decrease (Table 7.6) and in consequence the concentration of CO_3^{2-} ion is also decreased. The combined effects cause $CaCO_3$ to become considerably more soluble at depth. Third, respiration of organisms at depth, living on organic carbon particles falling from the surface, releases carbon dioxide into the water, further decreasing the pH, decreasing the carbonate ion concentration, and thus further reducing the ion product of calcium and carbonate ions. Measurement of the various coefficients involved in these effects is difficult and subject to some uncertainty, so it is not easy to predict the exact degree of saturation in seawater at depth, but none of the relationships noted above lead to a prediction that there should be a sudden increase in the rate of solution of $CaCO_3$, or such a sharply decreased amount in the sediment, with increased depth.

A classical experiment by Mel Peterson (1966) provided some insight into the processes responsible for the distribution of sedimentary carbonate. He suspended calcite spheres, contained in plastic containers open to the water, from a buoyed wire at a location in the North Pacific. After some months the calcite spheres were recovered and the change in weight measured. Peterson found that at all depths below approximately the top 200 m there was some loss of $CaCO_3$. This loss was small and relatively constant at all depths below 200 m down to 3600 m, about the depth of the lysocline. Below this depth the rate of dissolution increased sharply (Figure 7.6). Peterson's evidence suggests that sedimentary particles of carbonate may dissolve so slowly above the lysocline that the rate of supply exceeds the rate of dissolution, and carbonate builds up in the

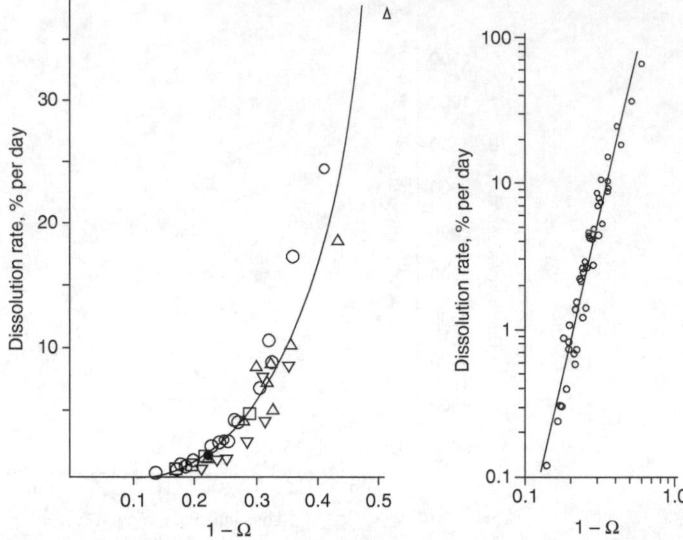

Figure 7.8 Dissolution rate versus degree of undersaturation for reagent-grade calcite, measured in artificial seawater at 20 °C (Keir 1980). Undersaturation is expressed by the relationship: degree of undersaturation = 1 – Ω,

where $\Omega = \frac{\mathrm{IP}}{K^*_{sp}} = \frac{[Ca^{2+}][CO_3^{2-}]}{K^*_{sp}}$.

On a linear scale (left side) it looks like no dissolution occurs unless the solution departs some distance from a saturated state. On a log–log scale (right side), however, it becomes apparent that the rate simply drops off so fast at low degrees of undersaturation that it is very difficult to measure. The equation of the straight line above is:

$$\log R = n \log(1 - \Omega) + \log k,$$

where R is dissolution rate in % per day, $n = 4.54$, $k = 1310$.

Most of the other carbonate materials examined (foram tests, sediment samples, etc.) exhibited dissolution rates that fell along the same slope as above ($n \approx 4.54$), but were offset by varying and generally much lower values of k.

sediment. Below the compensation depth the rate at which the particles dissolve is fast enough that no significant accumulation is found on the bottom.

Experiments on the kinetics of dissolving calcium carbonate have demonstrated that the rate of dissolution is a remarkably strong function of the degree of undersaturation. Keir (1980) exposed powdered reagent-grade calcite to artificial seawater with varying degrees of undersaturation and monitored the rate of solution by changes in pH and total alkalinity. He found that the data (Figure 7.8) could be described by a relationship in which the rate of dissolution is related to the degree of undersaturation raised to some exponent:

$$R = k(1 - \Omega)^n. \tag{7.44}$$

(for definitions see Figure 7.8). The high value of the exponent, 4.5, means that the rate at which $CaCO_3$ dissolves in seawater responds very strongly to small changes in

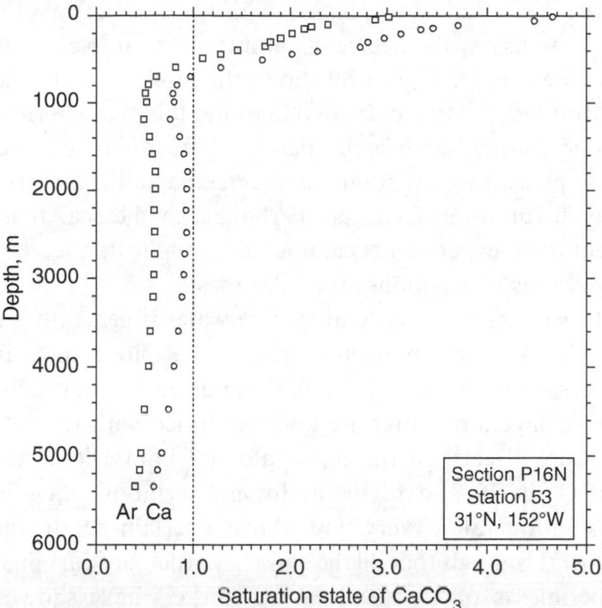

Figure 7.9 Saturation state of calcite and aragonite in the water column at a station in the central North Pacific. The basic data come from Feely *et al.* (2006). The data from their cruise report include for each sample the pressure and temperature measured *in situ*, and the salinity, total alkalinity, total CO_2, and pH, measured on shipboard at atmospheric pressure and at 25 °C. The dissociation constants for borate and carbonate species and the solubility product constants for calcite and aragonite come from Appendices E, F, and G. The pressure effects on these constants come from Appendix H. The equations for the carbonate system were set up, the dissociation constants adjusted to the temperature and pressure for each sample *in situ*, and the equations solved by iteration to obtain the same total CO_2 as calculated or measured on shipboard. From this the carbonate ion was calculated for each sample, and the ion product of calcium and carbonate ions calculated. The latter value was divided into the K^*_{sp} appropriate for the temperature, salinity, and pressure of each sample. The results in the graph above are expressed as Ω (defined in Figure 7.8). They are plotted against a depth axis, as the depth for each sample can also be calculated from the pressure.

the degree of undersaturation. Most calcium carbonate minerals examined behaved similarly, with exponents near 4.5, but most were offset to the right on a graph such as in Figure 7.8, so that the *k* value was smaller. In general, the larger the grain size, the lesser the *k* value, but, even at comparable grain sizes, natural samples dissolved more slowly than reagent-grade calcite.

Earlier work by Morse and Berner (1972) had also shown that rates of dissolution are very strongly and non-linearly influenced by the degree of undersaturation, and Berner and Morse (1974) showed that the presence of phosphate in the water greatly retards the rate of dissolution. Possibly, dissolution is slowed down by the phosphate in seawater, although Broecker and Takahashi (1978) could find no evidence for this in an examination of field data.

It takes a lot of calculation with a variety of relationships and estimated coefficients to assess the degree of saturation in samples of water from deep in the ocean. As an example, Figure 7.9 shows the result of such calculations for a station in the central North Pacific. It is evident that this characteristic part of the North Pacific is undersaturated with both calcite and aragonite at all depths below about 600 m.

At present we do not quite understand all the factors involved in the precipitation and dissolution of calcium carbonate in the sea. It is a difficult subject, because laboratory experiments cannot easily duplicate the very long time course and exact conditions of reactions in the deep sea.

It is clear, however, that surface water is generally supersaturated with respect to $CaCO_3$, but this mineral does not precipitate inorganically unless the degree of supersaturation becomes much greater than is normally present. It is also clear that deeper layers of water are undersaturated with respect to $CaCO_3$ (due to both the pressure effects on the carbonate and borate ionic equilibria in seawater and on $CaCO_3$ itself, and to the additional metabolic CO_2 in deep water), but that the dissolution rate is very slow above a certain depth (the lysocline) and fast enough below this depth that all the sinking particles of carbonate mineral are dissolved. The experiments of Keir and earlier workers have shown that the very rapid rise in dissolution rate with small changes in the degree of undersaturation might well account for the observed distribution. The matter is also complicated by another factor. Most particles sink rapidly, relative to their dissolution rates, so some may dissolve while sitting on or just beneath the surface of the sediment. Under the sediment surface they are also affected by somewhat higher concentrations of metabolic CO_2 from the organisms in the sediment (Emerson and Bender 1981). This effect is great enough that $CaCO_3$ should not be preserved, or be poorly preserved, in sediments that lie considerably above what might otherwise be considered a level of saturation equilibrium.

7.3.3 Effects of solution and precipitation

One of the apparently counter-intuitive properties of the carbonate system is the effect of precipitation or dissolution of calcium carbonate on the partial pressure of CO_2. When the equilibrium with calcium carbonate is added to the relationships in solution (Figure 7.10), the effect can be described qualitatively. If calcium reacts with carbonate ion to form calcium carbonate, excess hydrogen ion is left behind. This drives some portion of the bicarbonate ion to react to form undissociated carbonic acid (a component of CO_2aq), driving the equilibrium described by the constant K^*_{sp} to the left. This increases the partial pressure of CO_2. Figures 7.11 and 7.12 show the results of quantitative calculations of this effect. Imagine a coral reef, growing healthily,[2] precipitating calcium carbonate at a rate of 1 kg m^{-2} yr^{-1} (= 10 mol m^{-2} yr^{-1}). This would result in the release to the atmosphere of 6.6 moles of CO_2 per year

[2] The calculations resulting in Figures 7.11 and 7.12 were done for the global average surface temperature of 18 °C, and apply to any calcium carbonate-precipitating organisms at this temperature. In fact, however, most corals require a temperature 5 to 10 °C warmer for healthy growth.

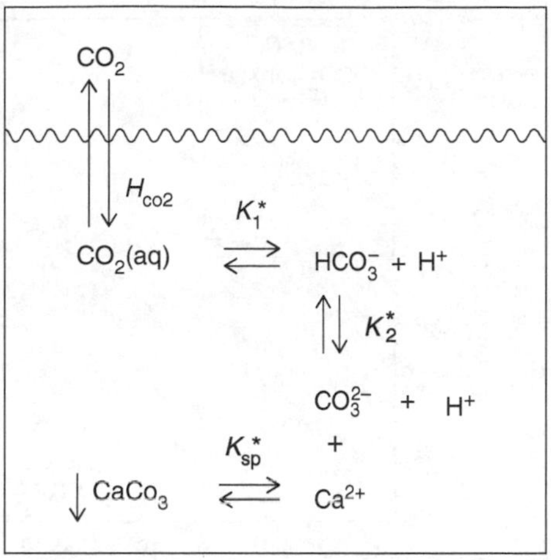

CO_2

H_{co2}

$CO_2(aq) \overset{K_1^*}{\rightleftarrows} HCO_3^- + H^+$

K_2^*

$CO_3^{2-} + H^+$

$+$

$CaCo_3 \overset{K_{sp}^*}{\rightleftarrows} Ca^{2+}$

Figure 7.10 Equilibria between calcium carbonate and the dissolved and gaseous species of carbon dioxide.

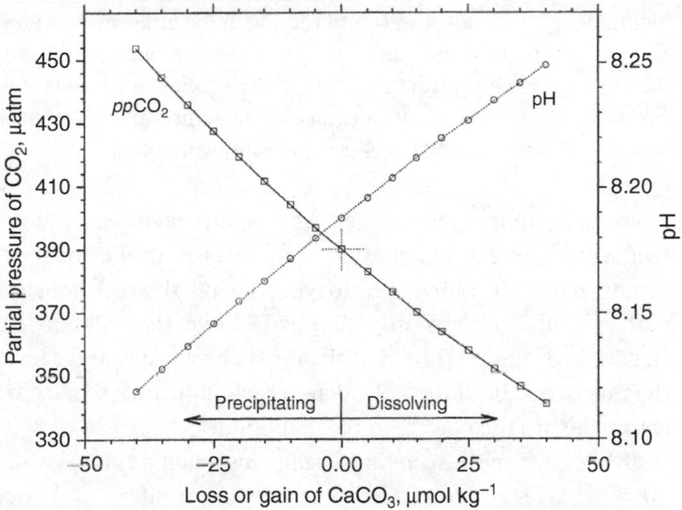

Figure 7.11 Effect of precipitating calcium carbonate from seawater, or dissolving calcium carbonate into seawater, on the partial pressure of CO_2 and on the pH of the solution. The starting solution was the same as in Table 7.4 ($S = 35$ ‰, $t = 18$ °C, TA = 2320 µmol kg^{-1}), and the water was at equilibrium with an atmosphere containing CO_2 at a partial pressure of 390 µatm. At each point on the graph the calculated amount of calcium carbonate was subtracted or added, and the partial pressure of CO_2 and the pH of the resulting solution were calculated. No loss of CO_2 was allowed in this calculation. Surface seawater is supersaturated with calcium carbonate so the inorganic or biological precipitation of calcium carbonate is in principle straightforward. Calcium carbonate will not dissolve into this water, however, so that half of the graph is included primarily to show the theoretical magnitude of the effect. It is possible for some biota to dissolve calcium carbonate in their food and excrete the dissolved products into the water, achieving the same effect. Calcium carbonate can of course dissolve in deep water, where the starting conditions are more acid.

Note that the precipitation of $CaCO_3$ from seawater involves a loss of one unit of CO_3^{2-} for each unit of $CaCO_3$ precipitated; according to Eq. (7.28) this means the loss of two units of total alkalinity.

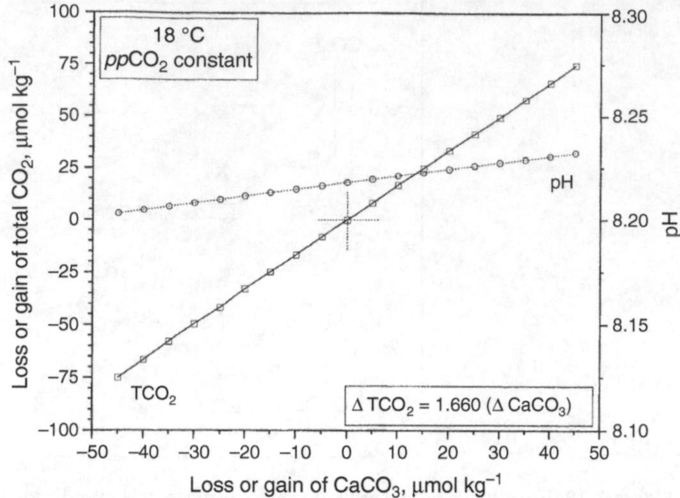

Figure 7.12 Effect of adding or removing calcium carbonate from seawater under the same conditions as in Figure 7.11, but here the CO_2 is allowed to enter or leave the water, maintaining the solution in equilibrium with the atmosphere at a partial pressure of 390 µatm. Here are plotted the amount of CO_2 lost to maintain equilibrium with the atmosphere, and the amount taken up from the atmosphere if calcium carbonate is dissolved into the solution. For example, if 40 µmol of calcium carbonate precipitates from the solution, an additional loss to the atmosphere of 26.4 µmol of CO_2 will occur.

from each square meter of reef. At this rate, each year one square kilometer of growing reef would release about 80 tons of carbon, in the form of carbon dioxide, into the atmosphere. Of course, photosynthesis on the reef consumes CO_2 to make organic matter, but this is mostly recycled, while the calcium carbonate is stored in the structure of the reef. In fact, the most convenient and sensitive method of measuring the rate of accumulation of calcium carbonate on a reef (or in a laboratory experiment) is to measure the change in total alkalinity.

Other carbonate-secreting organisms, such as clams and snails and, more importantly, pelagic organisms such as the foraminifera and coccolithophorids, also contribute to the change in alkalinity and the release of CO_2 to the atmosphere, to the extent that their calcareous remains are permanently buried.

7.4 Anthropogenic carbon dioxide

Our interest in the effects of carbon dioxide introduced by humans into the atmosphere stems largely from the fact that this causes an increase in the so-called greenhouse effect. Water vapor, CO_2, and some other less-abundant gases absorb some of the infrared heat rays that otherwise the Earth would radiate to outer space, and thus these gases help to keep the temperature of Earth considerably warmer than it would be in their absence. Only recently has it become generally appreciated that acidification of the ocean caused by increased CO_2 is also a matter of considerable concern.

The first measurements of the infrared absorbing capacity of CO_2 and water vapor were made in the nineteenth century by John Tyndall, who suggested that glacial episodes might have been caused by a lowered atmospheric concentration of CO_2 (Tyndall 1861). Several decades later, motivated largely by a concern about the causes of the ice age, Svante Arrhenius set out to calculate the magnitude of the effect (Crawford 1996, Pilson 2006). On the basis of poorly known physical constants and atmospheric models of the time, Arrhenius calculated that reducing the concentration of CO_2 in the atmosphere by one third would cool the Earth an average of about 3 °C, while doubling the CO_2 concentration in the atmosphere would lead to a world average temperature increase of 5 to 6 °C (Arrhenius 1896). He later (1908, p. 53) lowered this estimate to 4 °C. Modern estimates have revised the latter prediction downwards, to about 3 ± 1 °C, but this is quite enough to cause large changes in the global ecosystem. In addition, Hansen *et al.* (2008) suggested that longer-term changes (at least several hundred years) due to slow feedbacks may raise the value for a doubling of CO_2 up to as much as 6 °C. Both the Arrhenius and modern estimates predict that the temperature increase should be greater at high than at low latitudes. In addition, a number of other gases that absorb infrared radiation, manufactured or discovered in recent decades, are also increasing in the atmosphere. Additional factors involved in a full assessment of the various causes of climate change are discussed by Hansen and Lacis (1990) and by Solomon *et al.* (2007).

7.4.1 Atmospheric increase

Atmospheric CO_2 concentrations are variable both seasonally and geographically, and the variations are often large in comparison with longer-term secular trends that can be seen over several years. Such variations initially led to disparity between investigators and regions, because most sampling locations were influenced by local effects such as uptake by vegetation during daytime in the summer, respiration at all times, industrial production, and variable winds, all of which result in very erratic values especially in the lower layers of the atmosphere.

Appreciating the urgent need to understand the impact of anthropogenic CO_2 in the atmosphere, Roger Revelle, in 1957, urged the establishment of a monitoring station at a location that would produce the best representative data. Such a station was established near the top of Mauna Loa in Hawaii, at an altitude of 3400 m; the analytical program was managed by David Keeling, beginning in 1958 and continuing for more than 45 years, and now also continued by the National Oceanic and Atmospheric Administarion (NOAA). The results of this work (Figure 7.13) show the value of sampling far removed from continental effects. On the top of Mauna Loa, air has not been affected by Hawaiian vegetation or other near-surface activities. Samples from this station provide, as nearly as can be obtained from one location, mixed air typical of this latitude averaged around the world. Seasonal effects are dramatically evident. The concentration of CO_2 reaches a maximum in late spring and a minimum in late fall. The annual increase was compellingly evident after only three or four years.

These data are the primary record of the concentration of CO_2 in the atmosphere, and their publication, year after year since soon after the work began, has convinced the

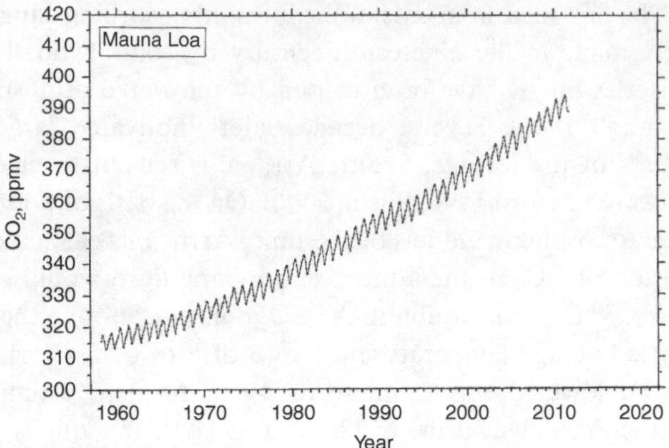

Figure 7.13 Atmospheric CO_2 concentrations recorded at Mauna Loa Observatory, Hawaii. Data from Keeling and Whorf (1994), with updates to 2011 from NOAA (Dr. Pieter Tans, NOAA/ESRL, www.esrl.noaa.gov/gmd/ccgg/trends/). The concentrations are expressed in parts per million by volume (ppmv) in dry air. The plotted points are monthly averages of measurements made several times per day. The January values for each year are plotted at the year marks. Some useful annual averages are *1960:* 316.91; *1970:* 325.68; *1980:* 338.69; *1990:* 354.19; *2000:* 369.48; *2010:* 389.78.

world of the (so far) inexorable increase in the concentration of CO_2. The upward slope of the line varies somewhat inter-annually, but, until now, also continues to increase.

At the same time as the Mauna Loa sampling started, samples were collected from several other stations in both the northern and southern hemisphere, including a series at the South Pole. From these it became evident that the seasonal variations in concentrations are very much smaller in the southern than in the northern hemisphere, and, as with the seasons, the two regions are out of phase with each other (Bolin and Keeling 1963). Samples from a 10-year period of collection at the South Pole, and at Point Barrow, Alaska (Figure 7.14) clearly show the dramatic effect of location on the seasonality of the concentrations. Seasonality is greatest at the most northerly station, and least at the South Pole. The obvious latitudinal differences would not be seen if it only took a few weeks for the atmosphere to be well mixed globally. Indeed, the inter-hemisphere mixing time is somewhat longer than one year, so seasonal effects can remain evident in each hemisphere.

The seasonal changes must be due largely to the influence of plant photosynthesis drawing down the concentration during the summer, while respiratory processes continue to some extent all year, returning the CO_2 back to the atmosphere. The use of fossil fuel for heat during the winter may also contribute to the seasonality. The smaller effect in the southern hemisphere is due to the smaller land mass for terrestrial plant photosynthesis there, combined with some effect from a generally lower level of fossil-fuel consumption. The larger proportion of ocean in the southern hemisphere may also act to smooth out the seasonal changes.

A detailed examination of latitudinal variations in concentration shows an increased concentration of atmospheric CO_2 in the tropics, centered over the

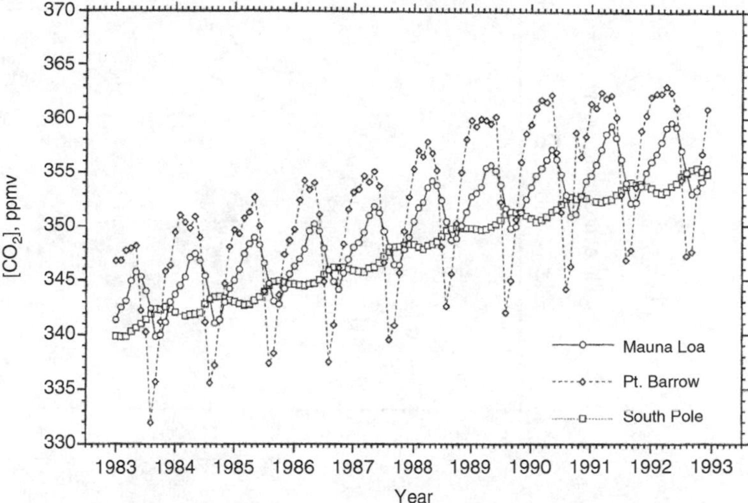

Figure 7.14 Monthly mean concentrations of CO_2 in the atmosphere at three stations selected to cover a broad latitudinal range: Point Barrow, Alaska (71° N); Mauna Loa, Hawaii (20° N); and the South Pole (90° S). Data cover the span from January, 1983 to December, 1992. (Data are from the Carbon Cycle Group, Climate Monitoring and Diagnostics Laboratory, NOAA, Boulder, Colorado.) These records show the large seasonality at Point Barrow, which is near sea level but is influenced very little by the ocean and very much by air from the large continental land masses in the northern hemisphere. The Point Barrow record also appears noisier than the others, due to the effects of relatively unmixed swirls and eddies of air from different near-surface regions. The record at Mauna Loa is smoother, because it samples air from a height of 3400 m and is much farther from local sources and sinks. The Mauna Loa station probably samples air masses that are a better representation of the northern-hemisphere average than is obtainable at any other single station. The seasonal maxima and minima lag behind those at Point Barrow because of the time taken to mix the air to that altitude and distance from the continents. The small seasonality evident at the South Pole is about six months out of phase with that in the northern hemisphere. The small magnitude of the southern hemisphere seasonal changes is due to the relatively small area of land in that hemisphere, compared with the area of the ocean, and also the height and remoteness of the station. The small southern seasonal signal may also be damped by some mixing of air carrying the larger northern signal across the Equator. ■

Equator. This increase is ascribed to release from the ocean in the equatorial and near-equatorial upwelling regions. A gradient from north to south is ascribed to a combination of anthropogenic input in the mid latitudes of the northern hemisphere and to the oceanic uptake of CO_2 by cold water sinking in the circum-Antarctic regions (Bolin and Keeling 1963). Quantification of such global-scale gradients in the atmosphere is important in attempting to understand the role of the oceans in controlling the CO_2 concentration in the atmosphere, and in modulating the increases caused by humans.

If one attempts to assess the nineteenth-century and pre-1958 information on CO_2 in the atmosphere, it is clear that the rather limited sampling at a few mostly

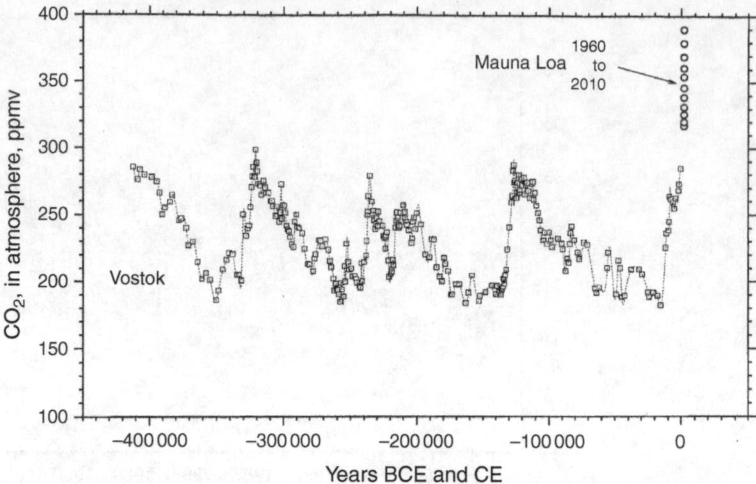

Figure 7.15 Concentrations of CO_2 in the atmosphere during the last 420 000 years. Data from an ice core drilled by Russian engineers at their Vostock base in Antarctica, and analyzed by Russian, French, and American scientists (Petit *et al.* 1999). Recent data from the atmosphere at Mauna Loa (Figure 7.13); each fifth-year annual average is plotted from 1960 to 2010. Data from other ice cores that were analyzed in detail over times in the nineteenth and early-twentieth centuries tie the data from this core directly up to the modern data from the atmosphere.

continental locations, largely in the northern hemisphere, must have produced data that are unlikely to be reliable. Beginning in the 1980s, however, analysis of gases extracted from ice cores drilled in Greenland and Antarctica (summarized by Raynaud *et al.* 1993 and Petit *et al.* 1999) has provided a remarkably detailed record of CO_2 and other greenhouse gases in the atmosphere during the last 420 000 years (Figure 7.15), and the concentrations in the ice tie directly into the last 50 years of accurate measurements of the atmosphere. At the present time the concentrations of CO_2, CH_4, and N_2O are all higher than at any earlier time within this long record, as well as other records going back more than 650 thousand years.

7.4.2 Total quantities involved

In order to set the anthropogenic release of CO_2 (Figures 7.16, 7.17) in perspective we should compare it to the sizes of the various reservoirs of carbon on the surface of Earth, and to the natural fluxes of carbon into and out of these reservoirs. Table 7.1 shows that by far the greatest reservoir of carbon in the Earth's crust is that present as biologically precipitated carbonates, mostly calcium carbonate, but also as dolomite. Atmospheric CO_2 is a tiny fraction of the total, and even amounts to less than 2% of the CO_2 dissolved in the oceans. While the carbonate reservoir is mined for various purposes, and some CO_2 is released in the manufacture of cement and other products, the major anthropogenic additions to atmospheric CO_2 are from the burning of fossil fuel and from the reduction in forest biomass, mostly in tropical regions.

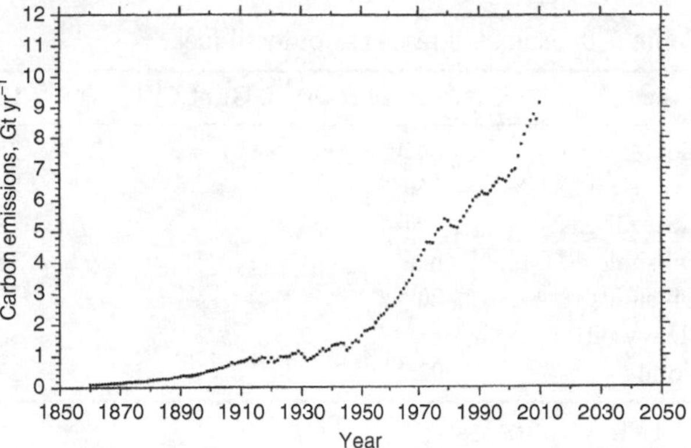

Figure 7.16 Global annual total release of carbon dioxide from the burning of fossil fuel and the calcining of limestone to make cement, expressed as Gt (= 10^9 t) of carbon per year. Data prior to 1950 from Keeling (1973) and later data same as for Figure 7.17.

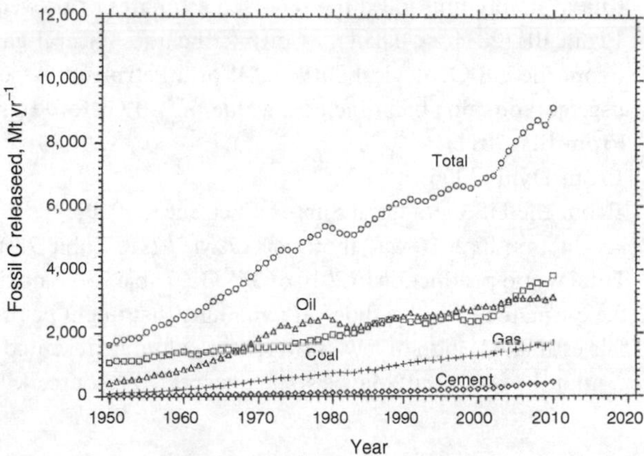

Figure 7.17 Global annual release of carbon dioxide from the burning of the several major classes of fossil fuel and the calcining of limestone to make cement, expressed as Mt (= 10^6 t) of carbon per year. Data up to 2008 from Boden *et al.* (2011). Data for 2009 and 2010 calculated from BP (2011).

Future effects from the addition of CO_2 to the atmosphere will ultimately be limited by the mass of fossil fuel yet remaining in the ground, which is not well known. Various estimates (Table 7.8) differ; this is perhaps expected, considering the extremely localized distribution of many deposits, frequently at great depths underground. Despite decades of intensive geological investigations in the most promising regions, there may be significant resources yet undiscovered; it is also likely that some of the projected ultimate resource will not be found. The size of the estimated resource is also dependent on the assumed cost of getting the fuel out. A useful deposit must be exploitable with a yield of energy considerably greater than that spent to do the exploiting. Using the estimate given, the quantity of CO_2 produced by burning all the potentially accessible fossil fuel would be some four times greater than

Table 7.8 Estimated resources of fossil fuel

Type	Proved reserves, Gt of C[a]	Ultimate resource, Gt of C
Coal	642[b]	2568[c]
Crude oil	160[d]	283[e]
Natural gas	104[f]	380[e]
Oil shale	n.a.	352[g]
Oil sands	20[f]	75[h]
Heavy oil	n.a.	50[h]
Totals	926[i]	3708[j]

[a] 1 Gt = 10^9 t.
[b] From BP (2011) × 0.746 t C (t coal)$^{-1}$. Average factor from the The Carbon Dioxide Information Analysis Center (CDIAC).
[c] I have not found satisfactory speculations on ultimate resources. The matter needs thorough review (Energy Watch Group, in 2007, suggests approximately the value given here. "Coal: Resources and Future Production". See www.energywatchgroup.org). Here I have simply multiplied the reported estimates of proved reserves by 4.
[d] From BP (2011) × 0.85 t C (t oil)$^{-1}$. Includes natural-gas liquids.
[e] From the US Geological Survey, "World Petroleum Assessment 2000". See http://pubs.usgs.gov/dds/dds-060/. Includes value of 210 Gt for "tight gas"; See: www.iea.org.
[f] From BP (2011).
[g] From Dyni (2006).
[h] From the US Geological Survey Fact Sheet 70–03.
[i] Production in 2010 was about 9.1 Gt yr^{-1} (see Table 7.9).
[j] Total world production to 2010 of 365 Gt (Table 7.9) is not included in this total; the total is the estimate of fossil carbon yet remaining that might be produced. I cannot guess whether this estimate is high or low. Surely, some ultimate resource does not exist, or will not, or cannot, be produced. Maybe some unexpected resource will be found somewhere.

the amount now present in the atmosphere, but only about 9% of the amount now dissolved in the ocean.

The rate at which CO_2 has been released by humans during the last several decades through the burning of fossil fuels is given in Table 7.9, while Table 7.10 and Figure 7.18 provide additional detail on global carbon fluxes. Comparison with Table 7.8 shows that by 2010 possibly about 10% of the total ultimate resource may have been burned. The fraction used is greater for oil resources alone, however. Much of the remaining resource is in the form of coal, but the ultimate resource of coal is poorly documented. It seems likely that in the near future the use of petroleum will decrease, and the use of coal will increase. For a unit of energy produced, however, the release of CO_2 from coal is considerably greater than from oil or natural gas, so any substitution of coal for oil will additionally increase the production of CO_2.

Examination of Table 7.10 shows that the amount of CO_2 released by humans to the atmosphere is now appreciable in the overall CO_2 fluxes and balance of Earth. For example, the annual discharge of CO_2 from the burning of fossil fuel is now about 14%

Table 7.9 Anthropogenic emissions of fossil CO_2, according to source

Year	Coal	Petroleum	Natural gas	Gas flared	Cement	Total
1950	1070	423	97	23	18	1630
	65.6	*26.0*	*6.0*	*1.4*	*1.1*	
1960	1410	849	227	39	43	2569
	54.9	*33.1*	*8.8*	*1.5*	*1.7*	
1970	1556	1839	493	87	78	4053
	38.4	*45.4*	*12.2*	*2.15*	*1.9*	
1980	1947	2422	740	86	120	5316
	36.6	*45.6*	*13.9*	*1.6*	*2.3*	
1990	2419	2515	1020	40	157	6151
	39.3	*40.9*	*16.6*	*0.7*	*2.6*	
2000	2370	2818	1288	48	226	6750
	35.1	*41.8*	*19.1*	*0.7*	*3.4*	
2010	3807	3120	1690	73	449	9138
	41.7	*34.1*	*18.5*	*0.8*	*4.9*	

Amounts are given in units of Mt (= 10^6 t) of C released for the year listed, and underneath (in italics) is the percentage of the total carbon for that year. The column for coal includes lignite and peat.

Values to 2008 from Boden *et al.* (2011). Later values estimated from data published by BP (2011). The estimate of total emissions from the eighteenth century until 2010 is 365 Gt. of carbon.

It is perhaps worth noting that if the present world population had achieved the USA standard of living by burning fossil fuel at the present USA rate, the release to the atmosphere would be about 34 Gt yr^{-1}, rather than the 2010 rate of 9.1 Gt yr^{-1}.

of the total worldwide terrestrial fixation of carbon by photosynthesis. In addition to the release from fossil fuel, there has been a possibly large but poorly quantified reduction in the standing crop of biota (mostly trees). When forest land is cleared or grassland is converted to tilled agriculture a considerable quantity of soil carbon (humus) is also gradually lost, apparently by oxidation under the changed conditions. Some estimates suggest that the total mass of biota and soil humus has been reduced by 50% during the last several centuries. Uncertainty in the present rates of deforestation, combined with the evidence that perhaps one and a half times as much carbon may be stored again on land due to regrowth of some forests and other biota, leads to uncertainty in the estimates of the total anthropogenic release of CO_2. It is evident from the data in Table 7.10, however, that during the 1990s the atmospheric increase amounted to about one half of the amount discharged into the atmosphere from the burning of fossil fuels, while about one third entered the ocean. These proportions may be changing with time, but so far is unclear.

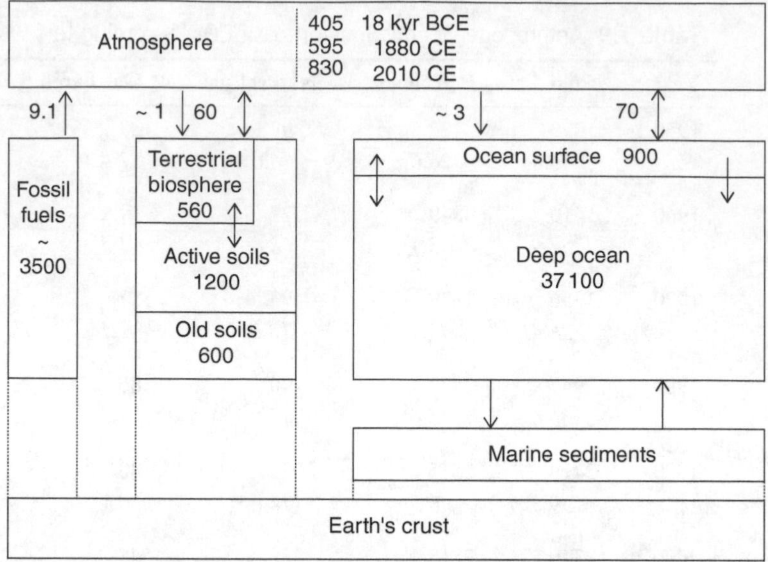

Figure 7.18 Principal reservoirs and fluxes of carbon (arranged as in Sundquist 1993). All values are in Gt (= 10^9 t) of carbon or in Gt yr^{-1}. The only well-constrained estimates are the anthropogenic flux of fossil-fuel carbon, and the masses of carbon in the atmosphere and in the ocean. The one-way vertical arrows to and from the atmosphere are estimates of the anthropogenically driven fluxes. Two-way arrows indicate approximate magnitudes of the exchanges between reservoirs; these must vary to some small extent from year to year. The 60 Gt yr^{-1} exchange between the atmosphere and the terrestrial biosphere is intended to represent the net primary production; the total gross photosynthetic fixation of carbon might be twice as large. The upper two values for the atmospheric reservoir represent measured values from the time of the last glacial maximum and at the beginning of large-scale fossil use, while the bottom value is the measured concentration in the atmosphere, selected for the year 2010. ■

7.4.3 Sinks for CO_2

In order to estimate where the CO_2 that has not remained in the atmosphere has gone, it is instructive to attempt a quantitative assessment of each of the various known possibilities, if only to discard some of them. It must be emphasized that at present it is not known for sure where quite all the CO_2 has gone, however.

Rock weathering

The weathering of rocks often consumes CO_2. The simplest reaction involves the dissolution of limestone through the action of water made slightly acid with dissolved CO_2. This is the source for much of the calcium found in rivers.

$$\begin{gathered} CO_2 + H_2O \rightarrow H_2CO_3 \rightarrow H^+ + HCO_3^- \\ CaCO_3 + H^+ + HCO_3^- \rightarrow Ca^{2+} + 2HCO_3^-. \end{gathered} \tag{7.45}$$

Other rock types, however, also dissolve with a concomitant uptake of CO_2. Siever (1968) presented the generalized weathering reaction of a potassium feldspar (a representative rock mineral) to yield kaolinite, a common clay mineral.

Table 7.10 Estimates of labile reservoirs of CO_2 and some current fluxes

	Gt of C
Inventories	
Total ocean water column (TCO_2) [(a)]	38 000
Atmosphere: estimate of 288 ppmv in 1850[(b)]	613
Atmosphere: 390 ppmv in 2010	830
Terrestrial biota (1850)	716
Terrestrial biota (1990)[(c)]	560
Changes	
Atmospheric increase, 1850 to 2010	217
Terrestrial biota decrease, 1850 to 2000[(c)]	–156
Fluxes, annual	
Net primary production (mostly recycled) – ocean[(e)]	50
Net primary production (mostly recycled) – terrestrial[(d)]	60
From fossil fuels and cement (estimated for 2010)[(f)]	9.14
From fossil fuels and cement (average for 2000s)[(g)]	8.05
Atmospheric increase (average for 2000s)	4.32
From tropical deforestation (estimated for 1990s)[(a)]	1.6
Uptake on land (estimated for 1990s)[(a)]	–2.6
Uptake into ocean (estimated for 1990s)[(a)]	–2.2
From volcanoes and related sources[(h)]	0.07
Fluxes to atmosphere, cumulative	
From fossil fuels (cumulative 1750 to end of 2010)[(f)]	365
From plants and soils (cumulative 1850 to 2000)[(d)]	156
Total anthropogenic (cumulative to end of 2010)[(i)]	521
Rate of atmospheric increase: 2000 to 2010, Gt yr^{-1}	4.32

(a) Denman *et al.* (2007)
(b) From data in Table J-1 and the mean molecular weight of dry air (28.97) it works out that each 1 ppmv of CO_2 in the atmosphere = 2.128 Gt of carbon.
(c) Houghton (2003).
(d) Schlesinger (1997); also see Field *et al.* (1998): ocean 49, land 56 Gt yr^{-1}.
(e) Berger *et al.* 1988, Martin *et al.* 1987; also, see Field *et al.* (1998), as above.
(f) From Table 7.9.
(g) From Boden *et al.* (2011).
(h) Gerlach (2011).
(i) Assuming 1990s' rate for plants and soils continued through 2010.

Many values are provisional, and subject to varying amounts of revision. (For estimates of uncertainties consult the original sources.) For several values I have used those accepted in the IPCC fourth report (Denman *et al.* 2007). Values for primary production are uncertain by perhaps 10% or more, as are estimates of the inventory of carbon in terrestrial biota. Net fluxes from plants and soils (decreases and increases of biomass and humus) are decidedly uncertain. Primary production is of course largely balanced by respiration in each ecosystem.

$$10CO_2 + 10H_2O \rightarrow 10H_2CO_3 \rightarrow 10H^+ + 10HCO_3^-;$$
$$10KAlSi_3O_8 + 5H_2O + 10H^+ + 10HCO_3^- \rightarrow 5Al_2Si_2O_5(OH)_4 + 20SiO_2 + 10K^+ + 10HCO_3^- \quad (7.46)$$

This reaction is similar to those of many other rock minerals, releasing, in addition, Na^+, Ca^{2+} and other cations; in effect these reactions convert free CO_2 into fixed CO_2. The clay mineral is relatively insoluble, but during this reaction the silica (SiO_2) is released in a soluble form ($Si(OH)_4$); reactions of this type are the source of most of the dissolved silica in river water.

The average concentration of HCO_3^- in the world's rivers (Berner and Berner 1987) is 0.87 mol m^{-3}. This quantity is an overestimate of the amount of CO_2 "fixed" by the above weathering reactions, if only because some of the HCO_3^- must derive from solution of limestone. Nevertheless, taking this value as a considerable overestimate, and the total river flow (3.74×10^{13} m^3 yr^{-1}) we can calculate:

$$0.87\,\text{mol}\,m^{-3} \times 3.74 \times 10^{13} m^3 yr^{-1} = 3.25 \times 10^{13} \text{mol yr}^{-1}.$$

Suppose that weathering reaction rates are exactly proportional to the total CO_2 concentration in the atmosphere (which is very unlikely; indeed, there is good reason to believe that such rates are less sensitive than this), then the increase in fixation would be proportional to the increased CO_2 concentration; that is, it would increase at about 0.51% per year. This would consume an additional 1.66×10^{11} moles or 0.002 Gt of C, much less than 0.1% of the CO_2 introduced by humans from fossil fuels alone. Increased weathering of rocks, therefore, seems quite an insignificant sink for the CO_2 produced, at least on the short term. To the slight extent that it occurs, it will increase the alkalinity of river water and of seawater. (Over millions of years, however, changes in the rate of rock weathering could have a profound effect on the concentration of CO_2 in the atmosphere.)

Biosphere

In the ocean, CO_2 is generally not a limiting nutrient, but on land it may be. In terrestrial systems, increased concentrations of CO_2 can increase the rate of photosynthesis. If this increase were exactly proportional to the atmospheric increase, the effect would be to raise terrestrial photosynthesis by about:

$$0.0051 \times 60 \text{ Gt yr}^{-1} = 0.31 \text{ Gt yr}^{-1},$$

which is less than 4% of present fossil-fuel CO_2 release. The physiological effect is not generally so large, however, and many other factors may limit production. Therefore, in the world as a whole, any increase in photosynthetic rate caused by increased CO_2 will likely be much less than proportional to the increase in atmospheric CO_2. Furthermore, such an increase in photosynthetic rate will not necessarily lead to an increased standing stock of biomass or humus. Respiration must also be increased. Most of the CO_2 fixed by photosynthesis appears again in the respiration of plants, animals, and bacteria during the next few hours, days, weeks, or months depending on the types of organisms involved. If trees grow faster, they may fall down faster (Phillips and Gentry 1994). Possibly, in some whole ecosystems, only the turnover rate would increase, not the standing stock.

Only that carbon which ends up in tree trunks, or humus, or other storage reservoirs that are not oxidized for at least several years is removed from the atmosphere during the timescale of interest here. Values for the change in both these large reservoirs are decidedly uncertain, due to the difficulty in making global inventories of forest biomass, and have been subject to considerable debate. It seems clear that the biomass of tropical forests must still be decreasing, but there is also considerable re-growth in mid- and high-latitude forests, and with a warming climate the tree line is moving north. Additional factors that might result in greater growth rates and, possibly, storage of carbon in terrestrial biomass, are fertilization with air-borne anthropogenic nitrogen compounds and, in sulfur-deficient regions, air-borne sulfur compounds.

In 1988, Ralph Keeling developed a remarkably precise method to measure the concentration of atmospheric oxygen with sufficient precision that it is possible to observe how the concentration varies inversely with the seasonal changes in carbon dioxide, and also to see that there is a decrease (of a few parts per million) each year caused by the burning of fossil fuels (Keeling and Shertz 1992, Keeling *et al.* 1996, Manning and Keeling 2006). Since these observations began, however, the concentration of oxygen has not decreased as much as it should have, considering the amount of fossil fuel that has been burned during this time. That is, there is a little more oxygen in the atmosphere than there should be. The only explanation of this that anyone has thought of is that some of the CO_2 must have been reformed into organic matter that was sequestered somewhere, almost certainly on land, and thus causing some oxygen to be released. This geochemical evidence seems more compelling than results from attempts to measure, directly in the field, global changes in the standing stock of biomass.

By measuring both the increasing concentration of carbon dioxide and the decreasing concentration of oxygen, it is possible to apportion the anthropogenic carbon dioxide that is not in the atmosphere into the two other apparent sinks; dissolving into the ocean and being sequestered as increased biomass on land. An example of a graphical approach to show the relationships among the various processes is in Figure 7.19.

The calculation is, of course, dependent on the accuracy of each of the factors involved. The increased concentration of CO_2 in the atmosphere is the best-known value, thanks initially to the analytical skill and dedication of David Keeling, and the change in concentration of O_2 is nearly as well known, thanks to the analytical skill and persistence of Ralph Keeling. The accuracy of the estimate of land uptake depends on the accuracy of the value for total CO_2 released from burning fossil fuel, which we can't discuss here, but I guess might have an uncertainty of perhaps 5%. The estimate of O_2 out-gassed from the warming ocean may be problematical. There may be some uncertainty in the estimates of the ratios of O_2/CO_2 associated with burning fuel and growing trees. Nevertheless, overall, the approach summarized in Figure 7.19 seems to yield the best available estimate of the uptake of carbon by increasing biomass on land. Most of this carbon is likely to be sequestered in the

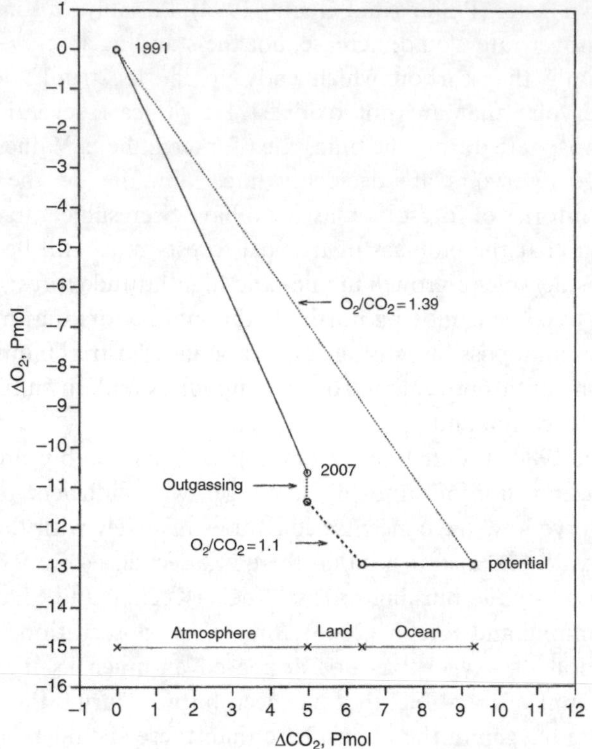

Figure 7.19 Apportioning the anthropogenic carbon dioxide.

The concentration of CO_2 in the atmosphere at Mauna Loa in 1991 (355.59 ppmv, Figure 7.13) and an estimate of the concentration of oxygen that year (http://bluemoon.ucsd.edu/data.html), are plotted at zero on the graph. The *increase* in the concentration of atmospheric CO_2 up until 2007 (28.13 ppmv, Figure 7.13) is expressed as an *increased* total amount of carbon in the atmosphere (4.984×10^{15} mol of C = 4.984 Pmol), and the *decrease* in the concentration of oxygen between 1991 and 2007 ($\delta[O_2/N_2] = -290$ per meg, or parts per million) is expressed as a *decrease* in the total amount of oxygen (–10.66 Pmol of O_2), and the values plotted and identified for the year 2007. If all the CO_2 from fossil fuel and cement for those years (111.8 Gt of C; Boden *et al.* 2011) had stayed in the atmosphere the increase would have been 9.31 Pmol, and if nothing else had perturbed the concentration of oxygen the total amount in the atmosphere would have decreased by 12.94 Pmol (using the value from Manning and Keeling (2006) of 1.39 mol O_2 per mol of CO_2), and these two values are plotted and labeled "potential."

A correction is necessary because the ocean is warming and causing oxygen to leave and escape to the atmosphere (some nitrogen also escapes into the atmosphere, and this has already been taken into account while calculating the final oxygen concentration). Manning and Keeling (2006) adopted an escape flux of 0.45×10^{14} mol of O_2 per year during the 1990s; if that applies to the 16-year interval plotted here it would amount to 0.72 Pmol, identified here as "out-gassing."

The increased biomass on land and the associated release of oxygen are estimated by following the line back from the (corrected) 2007 point down to the level where it should be when all the "excess" O_2 is accounted for by uptake of CO_2 in the production of biomass. This line has a slope of 1.1 O_2 for each CO_2, as this ratio is believed to typical of terrestrial photosynthetic production of biomass (Manning and Keeling 2006). The resulting estimate for the partitioning of the distribution of released CO_2 during these 16 years is: 53.5% in the atmosphere, 31.3% in the ocean, and 15.2% on land. ■

regrowth of north temperate forests, and their expansion as the treeline moves north and the taiga grows over the tundra.

The increased biomass on land has to account for the CO_2 released during the ongoing tropical deforestation, as well as some of the CO_2 released by burning of fossil fuels, so that the total amount sequestered on land is greater than shown in Figure 7.19 by the amount of CO_2 released by deforestation.

Ocean

During the period 1991 to 2007 the atmosphere experienced an annual increase in the total amount of CO_2 amounting to 3.74 Gt of carbon (Figure 7.19). If the calculation shown in Fig 7.19 is correct, some 1.06 Gt per year was stored in terrestrial biota. Since the average fossil-fuel consumption during that time was 6.99 Gt per year, the remainder, some 2.19 Gt per year, must be in the ocean. Much effort has been devoted to trying to find this anthropogenic carbon in the ocean, a difficult task because of the large quantities already present, the complex nature of ocean circulation and mixing, and the lack of long-term records of sufficient quality.

One important technique involves the use of changes in the isotopic composition of the carbon. The ^{13}C content of both fossil-fuel carbon and wood is less than that of atmospheric CO_2, so that, as fuel is burned, the $\delta^{13}C$ characteristic of the CO_2 in the atmosphere is measurably changing. So, too, is the $\delta^{13}C$ in the upper layers of the ocean. Quay *et al.* (1992) utilized changes in $\delta^{13}C$ values measured in surface waters of the Pacific Ocean between 1970 and 1990 to estimate that the oceans during this time had been taking up CO_2 at a rate amounting to about 2.1 Gt of carbon per year. This is not inconsistent with the estimates shown in Figure 7.19. These atmospheric changes in the isotopic composition of the carbon in carbon dioxide due to the burning of fossil fuel, and involving both ^{13}C and ^{14}C, are called the *Suess effect*.

It is in principle possible to estimate the amount of anthropogenic CO_2 taken up by the ocean by sampling throughout the water column in each ocean basin, measuring the total CO_2 in the water and, for each sample, subtracting the amount that should be present (assuming a certain degree of equilibration with the atmosphere), the amount that is contributed by respiration (estimated by the oxygen concentration) and the amount contributed by dissolution of calcium carbonate (estimated from the total alkalinity corrected for certain biological effects). This approach was carried out on results from two large international sampling programs (WOCE and JGOFS) in the 1990s during which over 9600 hydrographic stations were occupied throughout the world oceans (except for the Arctic). The results (Sabine *et al.* 2004) provide a detailed picture showing that most of the surface ocean is affected, and there is detectable anthropogenic CO_2 down in many places to 1000 m or more, and down to the bottom in the North Atlantic carried by the recently formed deep water. Sabine *et al.* (2004) estimated that for the time interval of 1880 to 1994 the ocean had taken up 118 Gt of carbon, while the atmosphere had increased by 165 Gt. Given that the fossil-fuel plus cement production was 244 Gt during this time, that left a net loss of biomass on land of 39 Gt (plus or minus a considerable uncertainty, estimated at 28 Gt). They suggest that in more recent decades the land may have become a net sink for CO_2.

Several complex global carbon-cycle models also converge on estimates that the ocean must be taking up CO_2 at rates of around 2 Gt per year (Prentice *et al.* 2001); they also predict that the rates will increase with increasing concentrations in the atmosphere.

We cannot here develop such global models in detail, but we can examine in a simple-minded fashion the consequences of what we do know about the capacity of seawater to take up CO_2 and how this will change with changing conditions.

In order to gain some quantitative appreciation of the question of uptake by the sea, we will first make an estimate of the amount of CO_2 that would have been taken up by the surface layers of the sea if they remain today as close to equilibrium with the atmosphere as they were in 1850. Of course, in any location an exact air–sea equilibrium seldom exists; the $ppCO_2$ in surface waters is sometimes greater than that in the atmosphere, sometimes less, depending on vagaries of biological activities, temperature changes, and the degree of local upwelling. Nevertheless, one can imagine that the surface ocean was, on a global average, close to equilibrium with the atmosphere then, and is now. On this basis the calculation is relatively straightforward. We can calculate the total change over time, and can also estimate the rate of change at any given time.

Most of the CO_2 in the ocean does not reside in the pool of CO_2aq, which is proportional to the $ppCO_2$; rather, it is in the form of HCO_3^- and CO_3^{2-}, and all these forms are in equilibrium with each other. In addition, the equilibrium is affected by the other components of the total alkalinity (mainly borate). Therefore, all these species must be taken into account when calculating the change of $[TCO_2]$ in surface seawater corresponding to a change in the partial pressure of CO_2 in the atmosphere. As CO_2 is added to a sample of seawater, the ability of the water to take up more CO_2 becomes less and less. The change varies with conditions, and can be expressed as an *uptake factor* (Figure 7.20). From this figure it is evident that in 1850 1 kilogram of average 18 °C ocean water would take up about 0.78 μmol of CO_2 for each increase of 1 ppmv of CO_2 in the atmosphere, while today such water in equilibrium with the atmosphere could only take up about 0.50 μmol of CO_2 after an increase of 1 ppmv in the atmosphere. If surface seawater were not replaced, its capacity to take up additional CO_2 would be continually decreasing, but it is replaced by water from below; unfortunately we do not know very well the rate at which this replacement takes place.

The change in partial pressure of CO_2 in seawater corresponding to a change in the concentration of TCO_2 can be expressed as a proportional change through the use of the Revelle factor. This is shown in Figure 7.21, from which it can be seen that, at the present time in water at ~18 °C and a $[TCO_2]$ of about 2.07 mmol kg^{-1}, an increase of about 1% in $[TCO_2]$ will result in about an 11% increase of the $ppCO_2$ in a gas phase in equilibrium with the water.

The surface layer of the ocean is mixed on various timescales from hours to days to seasons; the thickness of this mixed and vertically nearly uniform layer varies greatly from place to place, but the worldwide average depth to the thermocline is about 70 m. Through the course of one or two years, however, much of it is probably mixed occasionally to a depth of 200 m or deeper, so, for the purpose of calculation, I take this depth as including water that is exposed to atmospheric exchange during a very few years. Suppose a 200-m surface mixed layer has not been replaced since 1850.

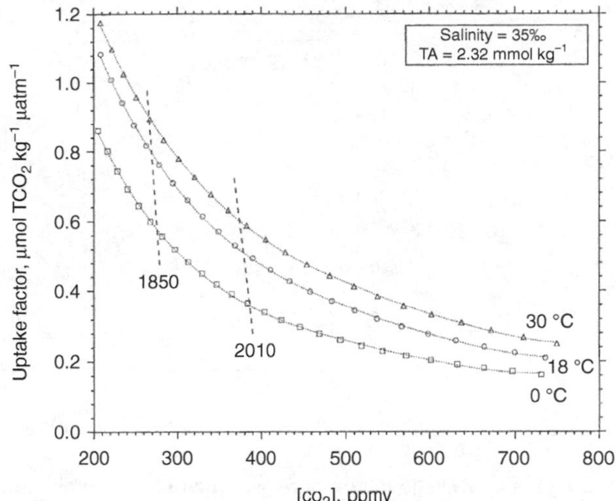

Figure 7.20 The uptake factor. This factor expresses the increase in the concentration of total CO_2 in seawater corresponding to an increase in the partial pressure of CO_2. It is defined (Pilson 1998) as:

$$UF = \frac{\Delta[TCO_2]}{\Delta ppCO_2}$$

This is the incremental increase in the concentration (μmol kg^{-1}) of total CO_2 in seawater corresponding to an incremental increase in the equilibrium partial pressure (μatm) of CO_2. The factor was calculated over short increments using a combination of Eqs. (7.29), (7.36), and (7.37). The uptake factor varies with temperature, salinity, total alkalinity, and with the partial pressure of CO_2. These values were calculated for a constant salinity of 35‰ and a constant total alkalinity of 2.320 mmol kg^{-1}. Calculations were done at 18 °C because that is the world average ocean surface temperature, and at 0 °C and 30 °C to cover most of the observed range. The range of partial pressures of CO_2 corresponds to those present during the last glaciation, up to values that might be expected some time in the next century.

The sloped lines indicate the values in 1850 with a concentration in the atmosphere of 280 ppmv, and values for the year 2010 with an atmospheric concentration of 390 ppmv. The effective partial pressure is offset from the concentration value because of a correction to account for the dilution of the dry atmosphere by water vapor and a smaller correction to account for the non-ideal behavior of CO_2. There is more water vapor at 30 °C than at 0 °C, accounting for the slope.

Then we can calculate the amount of CO_2 that should have entered the surface ocean between then (when the $ppCO_2$ was ≈ 288 μatm) and 2010 (Box 7.2). The increase of CO_2 in this layer is 53.2 Gt, about 15% of the amount released from fossil fuels.

Also, as shown in Box 7.2, with an annual increase in the CO_2 partial pressure of about 2.0 μatm, the top 200 m of the ocean should currently be taking up CO_2 at a rate of about 0.84 Gt of carbon each year. It is evident (Figure 7.20) that in future years, as the $ppCO_2$ continues to rise, and the $[TCO_2]$ in surface water increases, the capacity of such surface seawater to continue taking up CO_2 must slowly decrease, if the water is not renewed.

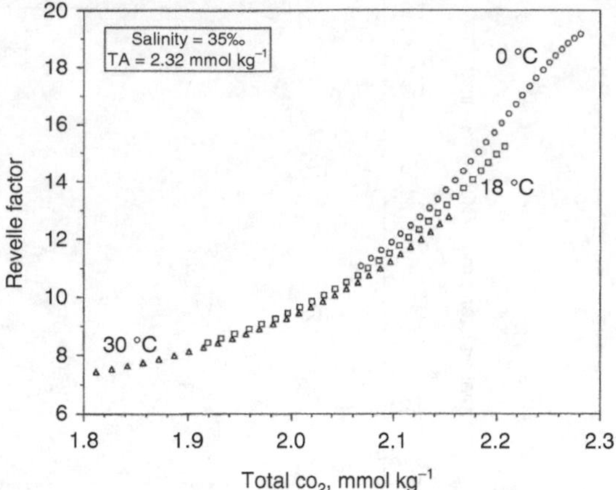

Figure 7.21 The Revelle factor expresses the proportional change in $ppCO_2$ corresponding to a change in the concentration of TCO_2 in seawater. Assuming the total alkalinity does not change, it is defined as follows (Sundquist *et al.* 1979):

$$RF = \frac{d(ppCO_2)/(ppCO_2)}{d[TCO_2]/[TCO_2]}$$

This factor was introduced (as γ) by Revelle and Suess (1957), in their classic paper on the increase of atmospheric CO_2, in order to account for the effects of the complex chemistry of carbon dioxide and the alkalinity of seawater on the capacity of the ocean to absorb or release CO_2. It was termed the homogeneous buffer factor by Sundquist *et al.* (1979), and christened the Revelle factor by Wally Broecker (Broecker and Peng 1982) to honor Roger Revelle for his influence on the modern study of carbon dioxide. The Revelle factor is related to the previously defined uptake factor approximately as follows:

$$RF = \left(\frac{1}{UF}\right)\left(\frac{[TCO_2]}{ppCO_2}\right)$$

As shown by Butler (1982) the RF is difficult to define and calculate rigorously, due to the complexity of the relationships and the resulting differentials. It is, however, easy to approximate numerically; the calculations leading to the graph above and to the previous uptake factor graph can be done on a spreadsheet, solving for the components of the carbonate system [Eqs. (7.29), (7.36), and (7.37)] at constant total alkalinity, temperature, and salinity, for a series of situations corresponding to increasing concentrations of TCO_2.

In the polar oceans in the wintertime there are regions where surface water becomes dense enough to sink to the bottom, thus ultimately forming the water that fills the ocean basins below ~1000 m. On a long-term basis, this deep water is formed at a rate commonly estimated to be about 30 Sv. If we assume (1) that this water comes from the surface 200 m, (2) that the water is now as close to equilibrium with atmospheric CO_2 as it was in 1850, and (3) that it is replaced by upwelling elsewhere with water that has not previously been exposed to anthropogenically elevated CO_2, then, as shown in Box 7.2, it should now be carrying down a burden of CO_2 greater than it was in 1850 by about 45 μmol kg^{-1}, leading to an increased transport of 0.52 Gt per year. The calculated sum of

Box 7.2 Calculated uptake of anthropogenic CO_2 by the ocean based on very simple assumptions

This estimate of the plausible uptake from the atmosphere into the surface water of the ocean, and carried down by the sinking of surface water to depth in the polar regions, is based only on surface-water equilibrium with the atmosphere and assumed replenishment rate of deep water.

For purpose of calculation take the following values:
$ppCO_2$ (1850) ≈ 288 μatm
$ppCO_2$ (2010) ≈ 390 μatm
Present rate of increase ≈ 2.0 μatm yr^{-1}
S = 35‰; TA = 2.32 mmol kg^{-1}
Assume: Surface water is on average close to equilibrium with atmospheric CO_2

1. Surface water (without replacement):
   a. Take the temperature as 18 °C, about average for world ocean
   b. Under these conditions [TCO_2] (1850) = 2.009 mmol kg^{-1}, [TCO_2] (2010) = 2.071 mmol kg^{-1}
   c. Increase = 62 μmol kg^{-1}
   d. Use 200 m as depth, then volume (from Table J-6) ≈ 69.65 × 10^6 km^3
   e. Then total absorbed to reach equilibrium with present atmosphere ≈ 53.2 Gt
   f. Present value of uptake factor ≈ 0.49 μmol kg^{-1} μatm^{-1}
   g. Present rate of uptake if equilibrium is maintained ≈ 0.84 Gt yr^{-1}
2. Sinking in polar regions:
   a. Take the temperature as 0 °C
   b. Under these conditions [TCO_2] (1850) = 2.159 mmol kg^{-1}, [TCO_2] (2010) = 2.204 mmol kg^{-1}
   c. Increase ≈ 45 μmol kg^{-1}
   d. Use rate of sinking as 30 Sv (= 9.47 × 10^{14} m^3 yr^{-1})
   e. Then increased rate* of CO_2 transport (2010 – 1850) ≈ 0.52 Gt C yr^{-1}
   f. Present value of uptake factor ≈ 0.365 μmol kg^{-1} μatm^{-1}
   g. Present (yr 2010) increase in rate per year ≈ 0.010 Gt yr^{-1}
3. Sum of these calculated rates: oceanic uptake of anthropogenic CO_2 = 1.36 Gt C yr^{-1}
4. The most recent release rate of CO_2 (yr 2010, Table 7.9) is 9.138 Gt of C.

   Suppose the fraction entering ocean can be estimated from Figure 7.19 as 31.3%
   Then amount entering the ocean should be ~2.86 Gt in 2010.
   The rate calculated above falls short by 1.50 Gt.

---

* For example, this calculation is as follows:
$(45 \times 10^{-6}\ \text{mol kg}^{-1})(1028\ \text{kg m}^{-3})(30\ \text{Sv})(10^6\ \text{m}^3\ \text{s}^{-1}\ \text{Sv}^{-1})$
$(3.16 \times 10^7\ \text{s yr}^{-1})\ (12.011\ \text{g mol}^{-1})\ (10^{-15}\ \text{Gt g}^{-1}) = 0.53\ \text{Gt C yr}^{-1}$.

the CO_2 taken up by the surface ocean and carried to depth by polar sinking amounts to 1.36 Gt per year, far less than the 2.86 Gt of carbon in CO_2 that was estimated (Figure 7.19 and Box 7.2) to be likely taken up by the ocean.

In addition to polar sinking, however, there are other routes by which surface water sinks below the upper mixed layer. Surface water is continually mixed downwards into the main thermocline. Of perhaps greater importance, a number of intermediate water masses are formed in all oceans and at various latitudes, and these are found generally at depths shallower than 1500 m. Examples are Antarctic Intermediate Water found in the south Atlantic, south Indian, and south Pacific oceans; North Pacific Intermediate Water, and several so-called Mode Waters such as the "18-degree water" in the North Atlantic; and intermediate water formed in the northern Indian ocean by salt-rich water discharged from the Red Sea and the Persian Gulf. The total rates at which these water masses are renewed is not well known, but they account for substantial flows.

Using the data from Box 7.2 we can calculate the rate at which surface seawater would have to be subducted into these intermediate waters to account for the 1.50 Gt per year of additional carbon entering the ocean. The calculation shown in Box 7.3 shows that this quantity of CO_2 could be accounted for by subduction of intermediate waters at a rate of about 66 Sv. Since large areas of the ocean are characterized by substantial volumes of water between 200 and 1000 m that contain measurable anthropogenic CO_2 as well as other anthropogenic substances, such as various chlorofluorocarbons, a subduction rate of ~66 Sv into these intermediate waters seems entirely plausible.

Another way to think about the flow rates just discussed is to consider the effect on the residence time of water in the upper mixed layer of the ocean. As shown in Box 7.3, the residence time with respect to polar deep-water formation may be about 74 years. Including a plausible value for subduction of water into intermediate layers within the main thermocline lowers the residence time in the upper 200 m to 23 years or less.

The calculations presented above are straightforward but very simplistic. Much more complex models of oceanic transport and uptake of CO_2 have been examined (see summaries in Prentice *et al.* 2001 and Sarmiento and Gruber 2006); these along with the data summarized by Sabine *et al.* (2004) show in greater detail the uptake and distribution of anthropogenic CO_2 in the world ocean.

It is helpful, in contemplating these matters, to know how the increase in the atmosphere corresponds to the release rates. Figure 7.22 shows that the cumulative increase in the atmosphere amounts to about 56% of the amount released from the burning of fossil fuels and making cement. Allowing for significant year-to-year variations, probably due largely to biological effects, the proportion has not changed much, but there is some indication that it might be decreasing. Nothing related to the balance between release of CO_2 from land-use changes and uptake through biomass increases is considered here. In any case, if surface water were not replaced at some significant rate the decreasing uptake factor would lead to an increasing fraction of the released CO_2 staying in the atmosphere. Surface water is replaced, however, by upward mixing everywhere, and by direct upwelling in certain regions. This upwelled

Box 7.3 Calculation of the residence time of surface seawater required if all the anthropogenic CO_2 is taken up from the atmosphere into surface seawater and subducted by the sinking of surface water to depth in the polar regions and also into intermediate water masses

Assume: surface water is on average in equilibrium with atmospheric CO_2. For purpose of calculation take the same values as in Table Box 7.2, with the addition that non-polar subducted water is at a temperature of 15 °C.

1. From Box 7.2: surface water (no replacement) is absorbing: 0.84 Gt C yr^{-1}
   Sinking polar water carries down: 0.52 Gt C yr^{-1}
   Total removed from atmosphere: 1.36 Gt C yr^{-1}
   Additional sink needed: 1.50 Gt C yr^{-1}

2. Additional sinking required to carry down fossil-fuel carbon at rate of 1.50 Gt yr^{-1}:
   a. 1.50 Gt C = 1.248×10^{14} mol
   b. Increased concentration = 59 μmol kg^{-1}
   c. Flow required is 2.063×10^{15} m^3 yr^{-1} = ~66 Sv

3. Residence time of surface (0 to 200 m) seawater:
   a. With respect to polar sinking only $= \dfrac{69.7 \times 10^{15} m^3}{0.947 \times 10^{15} m^3 yr^{-1}} = \sim 74 yr$
   b. With respect to polar +2.c (above)$= \dfrac{69.7 \times 10^{15} m^3}{3.01 \times 10^{15} m^3 yr^{-1}} = \sim 23 \text{ yr}$

It should be cautioned that these illustrative calculations in Boxes 7.2 and 7.3 incorporate a simplifying caricature of the mixing and transport processes in the complex real ocean. They do provide a certain plausibility and a sense of the scale and magnitude of the processes discussed. Much more can be learned by examining and experimenting with whole ocean numerical models, but up to the present time even these complex models do not capture or predict from basic physics all the observed behavior of the actual ocean.

water comes from some (generally not very great) depth, and has therefore been out of contact with the atmosphere for some length of time. The source water for this upwelling water will eventually become equilibrated with higher concentrations of atmospheric CO_2, which could lead to an increasing fraction remaining in the atmosphere. Whether this happens or not depends on details of ocean circulation and on the extent to which the atmospheric concentration continues to increase.

7.4.4 Ocean acidification

Carbon dioxide reacts with water to form carbonic acid, so adding CO_2 to a sample of water will make it more acid (Table 7.4). As the partial pressure of CO_2 in the atmosphere increases, and the concentration of CO_2 in surface water correspondingly

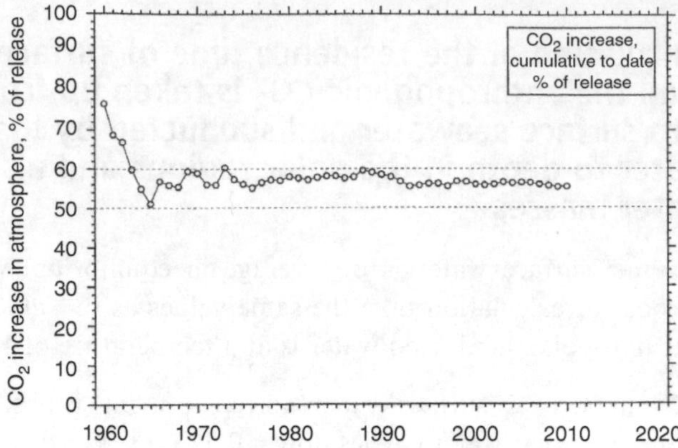

Figure 7.22 Increase of CO_2 in the atmosphere as a fraction of the quantity discharged into the atmosphere from the burning of fossil fuels and making cement, plotted as the cumulative average. The effect of changes in storage in or release from biota on land do not enter into this calculation. The amount of CO_2 in the atmosphere undergoes "natural" variations that are an appreciable fraction of the annual increase. For this reason, the data are "noisy" when plotted year by year; the first few years of data in the plot above show this effect. There is some indication of a decrease in the values over time.

increases, the relative proportions of all the components of the carbonate system, and borate, are changed (Figure 7.23). The pH of surface seawater has declined markedly since the peak of the last glaciation (Figure 7.24). It had decreased about 0.1 pH units by 1850, and has decreased a further 0.1 pH units since then, all calculated for comparable temperatures.

A decrease of 0.1 pH unit corresponds to an increase in the hydrogen ion activity of about 26%. There is little information available on which to base a guess on whether this much of a change is likely to have a significant effect on phytoplankton. Hinga (2002) reviewed the pH effects on the growth of phytoplankton and showed that phytoplankton appear to vary greatly in sensitivity to pH, and that some appear to respond physiologically to changes as small as 0.1 pH unit. It must be remembered, however, that growth of a plankton bloom can by itself change the pH by several tenths of a pH unit. It remains an open question whether shifting the average equilibrium concentration of hydrogen ion by 0.1 pH units will result in a change in the species composition of the phytoplankton. It is quite likely that larger changes can have a noticeable effect, especially on those that secrete calcium carbonate.

Most of the attention devoted to biological effects of increased CO_2 in the ocean has focused on organisms that construct shells or skeletons of calcium carbonate. Since calcium is nearly conservative with salinity in the ocean, it is changes in the carbonate ion that are important. As the partial pressure of CO_2 increases in the atmosphere, and correspondingly increases in the ocean, the concentration of carbonate ion decreases (Figure 7.25). Surface seawater in equilibrium with the atmosphere is everywhere considerably supersaturated with calcium carbonate, but the degree of supersaturation is least in cold water. Not until the partial pressure reaches about 600 μatm (a concentration of about 608 ppmv in the atmosphere) will such

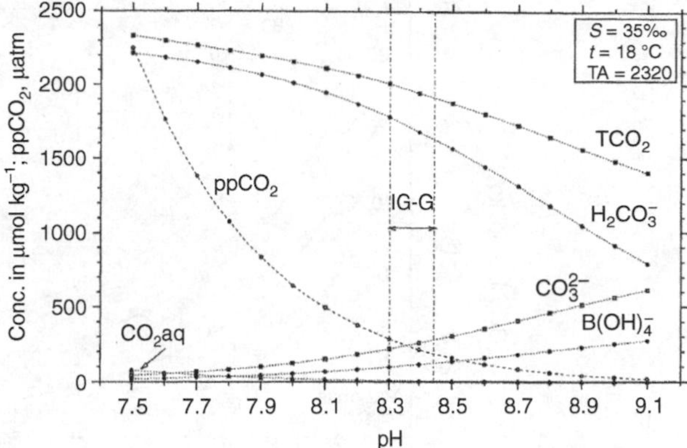

Figure 7.23 Effect of changing the concentration of total CO_2 in seawater on the distribution of the various carbon species and borate. In this case the variation of pH is due only to the addition or subtraction of CO_2. At each point the seawater is in equilibrium with a gas phase having the partial pressure (fugacity) as shown. The data were calculated for salinity $S = 35‰$, a constant total alkalinity of 2320 μmol kg^{-1} and a temperature of 18 °C, typical of surface ocean water. The boundaries labeled IG-G indicate the range of equilibrium pH values in surface seawater at 18 °C that would characterize a peak during glacial times, 20 000 years ago, and the present interglacial before the introduction of anthropogenic CO_2. In 2010, the pH was about 8.2, based on the concentration of 390 ppmv in the atmosphere at that time.

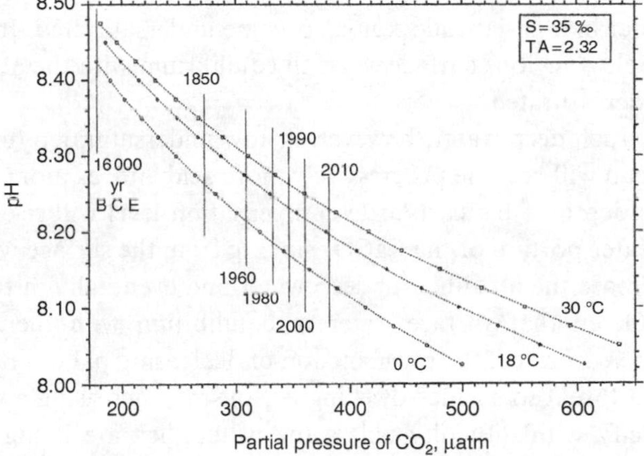

Figure 7.24 Calculated pH of surface seawater as a function of the partial pressure of CO_2 with which the water is in equilibrium, at three different temperatures. The total alkalinity and salinity were held constant; the calculation was done for temperatures of 0 °C, 18 °C and 30 °C, using Eqs. (7.29) and (7.37). Average $ppCO_2$ noted for several individual years: for the year 1850 from the ice-core data (Raynaud *et al.* 1993), and for subsequent years from the Mauna Loa data as in Figure 7.13. The lowest values of $ppCO_2$, less than 200 μatm, correspond to the situation near the last glacial maximum, about 18 000 years ago.
In each case above, the line corresponding to the year is set at the $ppCO_2$ at 18 °C corresponding to the concentration in a dry atmosphere as generally reported.
For example, a concentration of 390 ppmv in 2010 corresponds to a partial pressure (fugacity) of 381 μatm.

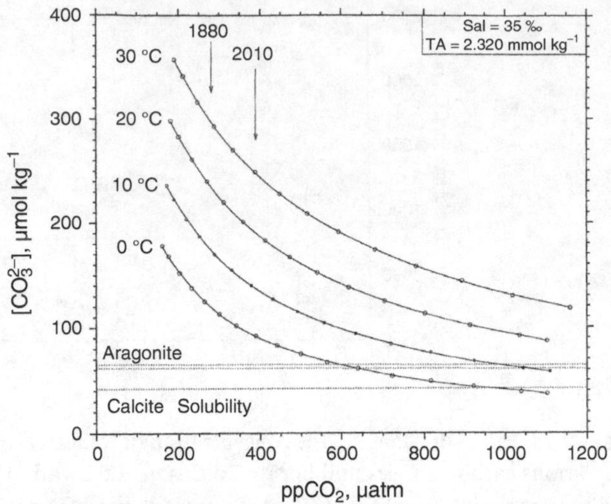

Figure 7.25 Concentration of carbonate ions in seawater as related to the partial pressure of CO_2 in the water. The values are calculated for a salinity of 35‰ and a total alkalinity of 2.32 mmol kg^{-1}, and at four temperatures. The dashed lines identify the concentrations of carbonate ion such that calcium carbonate is exactly saturated, according to the relationships published by Al Mucci (see Figure 7.5). The saturation state of calcite is shown as a single line, because its K_{sp} is not very temperature sensitive; aragonite is shown as two lines, the upper one for 0 °C and the lower one for 30 °C.

water in high-latitude regions become undersaturated. It appears that in the warmest tropical regions surface water in equilibrium with the atmosphere will never become undersaturated.

Much deep water, however, is now undersaturated (e.g. see Figure 7.9). The deep ocean will become progressively more acid and, as more CO_2-rich water mixes down, the depth of the carbonate compensation level will rise closer to the surface, and a greater portion of the $CaCO_3$ sinking from the surface will dissolve. This process will increase the alkalinity in deep water and eventually in the whole ocean.

Given that surface water in equilibrium with the atmosphere is considerably supersaturated, the phenomenon of decreasing pH did not at first seem to pose much of a threat to surface-dwelling organisms. Increasing evidence that many organisms calcify with difficulty unless the water they are living in is supersaturated led to experiments with enclosed ecosystems (mesocosms) containing corals and other calcifying organisms. These experiments showed that corals appear to calcify less quickly at $ppCO_2$ levels where the water is still considerably supersaturated (e.g. Langdon *et al.* 2000, Leclercq *et al.* 2000, and many more since). It appears that corals and a number of other types of organisms require their environment to be highly supersaturated. Since corals make their skeletons of aragonite (which is more soluble than calcite) they are more at risk than calcite-secreting organisms. Recent evidence (De'ath *et al.* 2009) suggests that corals on the Great Barrier Reef have grown more slowly during the last several years; the speculation is that they may be already experiencing the effects of higher $ppCO_2$.

The calculations and comments above are based on the situation where the surface water is in equilibrium with atmospheric CO_2. Most of the time, however, surface water is probably not in equilibrium with the atmosphere, because exchange with the atmosphere is quite slow. Either recent net photosynthetic activity has reduced the TCO_2 and raised the pH, or there has been net respiration, increasing TCO_2 and lowering the pH. In productive waters, and especially in shallow coastal and estuarine waters, these swings can be quite considerable. It is these natural extremes that must have the greatest effect on organisms, and as the concentration of CO_2 increases in the atmosphere, the low pH extremes will be still lower.

There are also deep-water corals that live at cold temperatures in some places a few hundred meters below the surface. These have not yet been experimented with, but they must operate under lesser degrees of supersaturation. As water in their environment gains increased CO_2 they will also presumably be at risk.

Many organisms other than corals are likely to experience difficulties in calcification as the degree of saturation decreases. Recent reviews of the subject and speculations about the future are in Andersson *et al.* (2005), The Royal Society (2005), and Doney *et al.* (2009).

7.5 Longer-term issues

When considering the possible effects caused by anthropogenic increases in atmospheric CO_2, it is important to examine the record of changes that have taken place in the past over a much longer span of time. Dated samples of ice collected by coring in the Antarctic and Greenland ice caps provide a remarkable record of past changes in climate and of the concentrations of many substances, including the concentration of CO_2 in the atmosphere (Figure 7.15). It is evident that relatively warm interglacial times (such as we are currently living in) are characterized by concentrations of CO_2 in the atmosphere of 280 to perhaps 300 ppmv, while cold glacial times are characterized by concentrations around 180 ppmv.

We cannot discuss here the difficult questions of what causes the natural rise and fall of CO_2 concentrations in the atmosphere. It seems that some slight warming caused by changes on the Earth's orbit causes release of CO_2 from the ocean, which causes further warming. The atmospheric changes are rapid enough, on a geological timescale, that the ocean seems the only likely source for the large amounts of CO_2 required. A knowledge of processes which contribute to net uptake or release of CO_2 from the oceans is crucial to gaining an understanding of past and future changes in the atmosphere.

On a geological timescale, the atmospheric increase in this century is dramatic, appearing almost instantaneous, and the present concentration of CO_2 in the atmosphere is greater than any seen on Earth in at least the last 650 000 years. Extrapolation suggests that the concentration will pass the 400 ppmv level well before the year 2020.

We have already seen (Section 7.2) that the ocean may have been, for several thousand years, precipitating more calcium carbonate than is supplied by the rivers, and thereby decreasing the total alkalinity of the ocean and releasing CO_2 into the atmosphere. As the ocean becomes more acid, due to the discharge of anthropogenic CO_2, the loss of calcium carbonate from the ocean will likely decrease, and might reverse, increasing the total alkalinity of the ocean. In addition, eventually increasing CO_2 deep in the ocean will increase the dissolution of calcium in the sediments there, and cause the CCD to rise, and will also increase the alkalinity of the ocean.

Increasing alkalinity increases the ability of the ocean to take up CO_2 (in effect it increases the uptake factor), but since the turnover time of the deep water is close to 1000 years, it will take a long time for the dissolution of calcium carbonate at depth to have much effect on atmospheric CO_2. In fact, it appears likely that humans will have burned up most of the available fossil fuel long before this effect becomes significant. Indeed, if the total fossil-fuel resources of the world are near 3700 Gt (Table 7.8), and humans manage to limit their consumption and release rate to not more than 10 Gt per year, the fossil-fuel reserves of the world will be effectively exhausted in 370 years, much less than the time needed for one recycling of the deep water of the world ocean. Eventually, over several thousand years, the dissolution of $CaCO_3$ on the sea floor will remove some of the excess CO_2 in the atmosphere. Over several hundred thousand years, increased weathering of silicate rocks will neutralize much of the remaining excess CO_2 (Archer 2005, Archer *et al.* 2009).

For further discussion of the future of climate, see Solomon *et al.* (2007, 2009).

Summary

The subject of carbon dioxide in the ocean is important because:

a. The CO_2 system in the dissolved state is the major short-term acid/base buffer, and sedimentary carbonates provide the next most important buffer over a timescale of many thousands of years.
b. The CO_2 system is intimately involved with and responds to photosynthesis and respiration.
c. The ocean is the largest readily available reservoir of CO_2, and is a major sink for anthropogenically discharged CO_2. As such, it has important influences on the climate of the Earth.

Carbon dioxide reacts very slowly with water molecules, a fact of great importance to biota, because without the enzyme carbonic anhydrase both respiration and photosynthesis would be difficult or impossible to carry forward at rates that are usual at the present time.

The state of the CO_2 system in a sample can generally be assessed by measuring one of several possible pairs of variables (along with the temperature and salinity), and then calculating the results according to the appropriate empirical constants. The partial pressure of CO_2 is in some ways the most sensitive and important variable.

The borate system is the next most important buffer, and for many purposes it is essential to know the state of the borate system in order to evaluate the carbonate system, and to calculate the total alkalinity.

Large parts of the sea floor are covered with carbonate sediments; these reveal much about the history of the Earth, and their formation and dissolution influences Earth's climate. Carbonate precipitation and solution are difficult to evaluate quantitatively because of the complexity of the carbonate system and possibly because of still unknown interactions with seawater.

Much anthropogenic CO_2 is taken up by the ocean, but the capacity for uptake decreases as the uptake proceeds. The pH drops as CO_2 is absorbed, and this will certainly have profound biological consequences.

The ultimate fate of most anthropogenic CO_2 is to end up in the ocean, and it will be increasingly important to evaluate the rates and mechanisms by which this occurs.

SUGGESTIONS FOR FURTHER READING

Zeebe, R. E. and D. Wolf-Gladrow. 2001. *CO_2 in Seawater: Equilibrium, Kinetics, Isotopes*. Elsevier, New York.

This book is a classic. It is the most complete presentation available of the intricacies of the CO_2 system in seawater. Importantly, it includes sections on the isotopic differences between the various species in solution and between the solution and the solid phases.

Dickson, A. G., C. L. Sabine, and J. R. Christian, eds. 2007. *Guide to the Best Practices for Ocean CO_2 Measurements*. PICES Special Publication 3. Available from CDIAC; see http://cdiac.ornl.gov/oceans/Handbook_2007.html.

This book provides an authoritative review of the CO_2 system, especially focused on the issues involved with making high precision measurements in the open ocean. It includes detailed information on the best available techniques for making the necessary measurements.

Doney, S. C., V. J. Fabry, R. A. Feely, and J. Kleypas. 2009. Ocean acidification: the other CO_2 problem. *Annu. Rev. Mar. Sci.* 1: 169–192.

This excellent review summarizes a wealth of mostly quite recent measurements and speculation on the many possible biological effects of increased CO_2 and consequent lower pH of ocean water. Highly recommended for anyone interested in this potentially important aspect of global change.

Zeebe, R. E. 2012. History of seawater carbonate chemistry, atmospheric CO_2 and ocean acidification. *Ann. Rev. Earth Planet. Sci.* 40: 141–165.

Those who wish to explore further the effect of increased atmospheric CO_2 on the chemistry of seawater, in the light of what we know about high CO_2 excursions in the past, should examine this paper by Richard Zeebe.

8 Nutrients

... the biochemistry of the ocean is curiously complex ... and its processes are conducted upon an enormous scale.

FRANK WIGGLESWORTH CLARK 1908

As used in the marine-science literature, the word *nutrient* does not mean the food we eat; instead, it usually refers to the important and commonly measured elements needed for the growth of plants: phosphorus, nitrogen, and silicon. Of course, many other elements are necessary for life. Hydrogen, carbon, oxygen, and sulfur are essential, and such elements as iron, copper, cobalt, zinc, and boron are required in traces. In seawater, carbon dioxide, sulfate ions, and borate are abundant and never limiting, so, conventionally, they have not been included among the nutrients. The status of several other elements is uncertain; because of the very low concentrations in which they are needed it is only in recent decades that much experimental work could be accomplished. This chapter deals primarily with the so-called major nutrients: phosphorus, nitrogen, and silicon.

8.1 Phosphorus

Because of its crucial role in biological activity, phosphorus has been extensively studied. Phosphorus has certain unique properties that appear to qualify it for its role as both an essential constituent of the genetic material (RNA and DNA) of all organisms, and an essential participant in many energy-transforming mechanisms (via ATP, etc.) of all organisms (Westheimer 1987). Since all living things require phosphorus, and the element is present in seawater in very low concentrations, the amount present in a body of water must often set an upper limit to the biomass of living organisms that can grow there. (There are, however, many localized situations where one or another substance may take over the role of *the* limiting nutrient. There is further discussion of this in Sections 8.4 and 8.5.)

8.1.1 Crustal abundance and geochemical transport

The abundance of phosphorus varies greatly both within and between rock types, but is typically low in granites, with concentrations of about 0.13 to 0.27% expressed as

Table 8.1 Various estimates of important fluxes of phosphorus

Contemporary continental erosion:		
River-borne sediment, flux to ocean[(a)]	$\approx$	20×10^9 t yr^{-1}
River-borne dissolved substance, flux to ocean[(b)]	$=$	4.1×10^9 t yr^{-1}
Total present denudation rate of continents	$\approx$	24.1×10^9 t yr^{-1}
Phosphorus content of Earth's continental crust[(c)]	$\approx$	24.5 mol t^{-1}
Total "natural" phosphorus transport to oceans (24.5 mol t^{-1} $\times 24.1 \times 10^9$ t yr^{-1})	$\approx$	590×10^9 mol yr^{-1}
Pre-human, possibly	$\approx$	300×10^9 mol yr^{-1}
Flux of total dissolved phosphorus by rivers:		
River flow to ocean[(d)]	$=$	37.4×10^{12} m^3 yr^{-1}
Estimated pre-human average concentration[(e)]	$=$	0.81 mmol m^{-3}
Pre-human flux of dissolved phosphorus	$\approx$	30×10^9 mol yr^{-1}
Present flux of dissolved phosphorus[(e)]	$=$	62×10^9 mol yr^{-1}
Geochemical transport by humans:		
Total P mined[(f)], average 2009–2011[(g)]	$\approx$	780×10^9 mol yr^{-1}
Increase of total geochemical flux, due to humans:		
$\left(\frac{590+780}{590}\right)$ or $\left(\frac{590+780}{300}\right)$	$\approx$	Factor of 2.3 or 4.6

Except for the estimate of phosphorus mined, most values above are perhaps not known to within plus or minus 20 to 30%.

(a) Poorly known; see Chapter 13. A current guess is that the present delivery rate is twice the pre-human rate.

(b) Meybeck 1979.

(c) Appendix Table J.2.

(d) Baumgartner and Reichel 1975 (Appendix Table J.2).

(e) Meybeck 1982: includes both organic P and inorganic P, where $P_t \approx 2.5\ P_i = P_i + P_{org}$.

(f) About 99.9% from phosphate rock, 0.1% guano.

(g) USGS (2012). Use: 87% fertilizer, 6% detergents, 7% industry and food additives (Llewellyn 1994).

PO_4, while basalts may average 0.40 to 0.8% (Kornitnig 1978). Average values for shales range from 0.15 to 0.4%, expressed as PO_4. An uncertain estimate for the whole crust is 25 moles per ton (Appendix J, Table J.2). The world's rivers have been inadequately analyzed for phosphorus, and estimates of the average river concentration vary by a factor of two or more, so the geochemical flux is poorly known. The natural geochemical transport can be approximated very roughly, as in Table 8.1, by using estimates of the crustal abundance of phosphorus and of continental denudation rates. Some authors (e.g. Froelich *et al.* 1982) have used estimates of pre-agricultural and present rates of continental denudation to estimate the effect of humans on the transport of phosphorus and other substances. Humans have greatly increased the rate of erosion over large areas of land, but by building dams in many places have also greatly reduced the transport of sediment to the sea. It is not at all clear that the net transport of sediment to the ocean is much greater now than in pre-agricultural times (Milliman and Meade 1983). The overall uncertainties are, however, not great enough to weaken the general conclusion. From Table 8.1 it is evident that humans have more than doubled the total geochemical transport of phosphorus.

Humans also appear to have more than doubled the average concentration of dissolved phosphorus in the world's rivers (Table 8.1), but the dissolved inorganic phosphorus is only a small part of the total transport. Some is in a dissolved organic form, but most is included with the particulate matter. There are not enough analyses of particulate phosphorus to allow a satisfactory independent estimate of this, however, on a global basis. Some of the phosphorus carried on the surfaces of particles is released upon mixing with seawater (Froelich 1988), but the quantitative relationships are poorly known and have been examined in very few rivers. Some of the particulate phosphorus is trapped in estuaries and near-shore sediments, but in general it is not known what proportion this amounts to. Therefore, the flux of river-borne phosphorus to the ocean is poorly known. For rough calculation, it seems reasonable to take the value of 30×10^9 moles per year as the pre-human transport of dissolved phosphorus to the oceans, and to take the present rate at about 62×10^9 moles per year.

While some polluted rivers now carry dissolved phosphorus loads more than 10 times the natural levels (Meybeck and Helmer 1989), and global averages are still uncertain, it nevertheless seems evident that much of the phosphorus mined by humans remains in agricultural soil and in lake sediments, only slowly making its way to the sea. The lack of sufficient measurements on enough rivers, especially measurements of the important particulate forms, results in the situation that, at present, estimates of the fraction of the anthropogenically mobilized phosphorus that is entering the ocean have only limited accuracy.

8.1.2 Forms of occurrence in seawater

Inorganic phosphate

In deep water, nearly all dissolved phosphorus is present as simple inorganic phosphate (often called *orthophosphate*) in its various ionized forms, while in open-ocean surface waters some other form, apparently phosphorus combined with dissolved organic matter, may often exceed 50% of the generally much smaller total concentrations there.

Phosphoric acid, H_3PO_4, has three hydrogens (Figure 8.1); the first of these dissociates as a strong acid, and essentially no undissociated H_3PO_4 is present in normal seawater. This phosphate in seawater exists only as the three phosphate anions, and as ion pairs of these with various cations. The ionization constants of phosphoric acid are very different in seawater from those determined in fresh water or in NaCl solution (Table 8.2). The effects of ionic strength alone (absent specific ionic interactions, but neglecting a small effect of sodium ion) on the activity coefficients of the various species may be seen by the differences between the dissociation constants in distilled water and NaCl solution. The very much larger second and third apparent dissociation constants in seawater compared to those in NaCl solution are due to the very strong ion-pairing observed for the phosphate species with

Table 8.2 Dissociation constants of phosphoric acid under several conditions at 20 °C

	Distilled water	NaCl, 0.68 M	Seawater, $S = 35‰$
K_1^*	7.52×10^{-3}	1.91×10^{-2}	1.83×10^{-2}
K_2^*	6.23×10^{-8}	4.03×10^{-7}	7.42×10^{-7}
K_3^*	2.20×10^{-13}	6.41×10^{-12}	9.27×10^{-10}
Speciation in seawater at 20 °C and at pH 8.0:			
	% of total P	*% as ion pairs*	
H_3PO_4	–[(a)]	–	
$H_2PO_4^-$	1.2%	21%	
HPO_4^{2-}	90.3%	68%	
PO_4^{3-}	8.4%	99.9%	

Values in distilled water from the *CRC Handbook of Chemistry and Physics* (Haynes 2011); values in 0.68 M NaCl solution from Atlas *et al.* (1976); values in seawater at $S = 35‰$ from Appendix F, Table F.7. Calculation of the fraction present in each form at pH 8 from the seawater dissociation constants; in each case the species listed includes the concentration of the free ion as well as the ion pairs of that ion with Na^+, Ca^{2+} and Mg^{2+}. The percentage of each phosphate species present as ion pairs is from Atlas *et al.* 1976.

(a) In seawater at this pH, the undissociated form comprises less than one millionth of 1% of the total inorganic phosphate present.

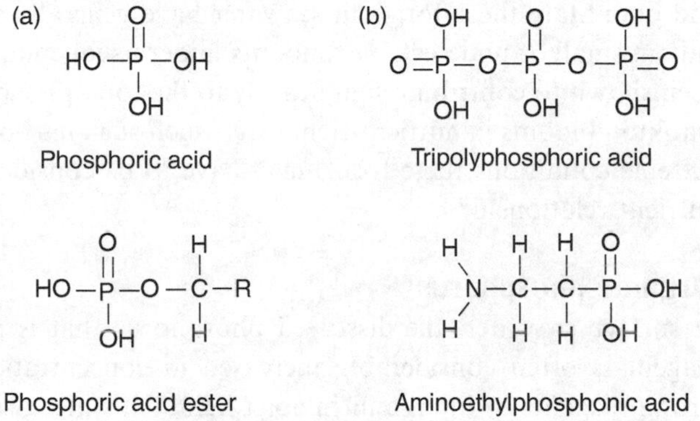

(a)

HO–P(=O)(OH)–OH

Phosphoric acid

(b)

O=P(OH)(OH)–O–P(=O)(OH)–O–P(OH)(OH)=O

Tripolyphosphoric acid

HO–P(=O)(OH)–O–CH₂–R

Phosphoric acid ester

H₂N–CH₂–CH₂–P(=O)(OH)–OH

Aminoethylphosphonic acid

Figure 8.1 Examples of inorganic and organic phosphorus compounds. Most phosphorus in seawater is simple inorganic phosphate, represented above as phosphoric acid. At normal seawater pH, essentially all the hydrogens are dissociated, and only the ionized forms (phosphate, PO_4^{3-} hydrogen phosphate, HPO_4^{2-}; and dihydrogen phosphate, $H_2PO_4^-$) are present. Similarly, the other forms shown are also ionized to a varying extent. Tripolyphosphoric acid is an example of polyphosphates, which range from pyrophosphate with two units up to very long chains. Most organic phosphorus is combined through ester linkages with an oxygen between the phosphorus and the carbon; the "R" represents one of many organic carbon groups, including lipids, sugars, proteins, and nucleic acids. Phosphonic acids, which have a direct phosphorus-to-carbon bond, are found in significant amounts in some marine organisms. The carbon-to-phosphorus bond is not hydrolyzed by boiling with concentrated acids or bases. The example here, aminoethylphosphonic acid, was the first example of such compounds found in nature.

calcium and magnesium. The situation is similar to that seen with carbonic acid in seawater (Chapter 7), but the effect is greater because of the very strong ion-pairing observed for the PO_4^{3-} ion (Table 8.2).

Polyphosphates

These polymeric substances are technically inorganic, but in nature they are made biologically. Phosphorus can occur in an extraordinary variety of straight-chain, branched, and cyclic polymeric forms (Toy 1973). The simplest of these is pyrophosphoric acid, with two phosphate groups. The tripolyphosphoric acid illustrated in Figure 8.1 is familiar as the phosphorus part of ATP (adenosine triphosphate). As with phosphate itself, it is largely ionized in aqueous solution at neutral to basic pH.

The longer-chain inorganic polyphosphates may extend to contain at least dozens of units. Under ordinary conditions in solution the thermodynamically stable form is the ionized monomer phosphoric acid, but the rates of hydrolysis of the long chains to the monomers are slow at normal seawater temperatures. If only inorganic processes are involved, the half-lives of the simple polyphosphates in dilute aqueous solutions may range from hours to decades, depending on the temperature and pH.

Polyphosphates, often highly polymerized, are made in large amounts by certain algae and bacteria, where they may act as storage products. They can be released into the surrounding environment when cells lyse or are otherwise disrupted. Under normal conditions, polyphosphates are not reactive to the usual molybdate reagent (see Section 8.1.3) employed for measuring inorganic phosphate. Efforts to identify and quantitate these forms in seawater have generally shown that they are present only in small or undetectable amounts under usual conditions. They have sometimes been shown to contribute significantly to the total phosphorus during intense phytoplankton blooms in nutrient-rich water, such as in harbors, and under such relatively extreme conditions these forms may have to be considered during an evaluation of nutrient relationships.

Organic phosphorus

In surface seawater, the dissolved phosphorus that is measured by the molybdate reagent is often considerably increased in concentration by prior oxidation with strong reagents or intense ultraviolet light. This increase is ascribed to the presence of organic compounds of phosphorus, *mineralized* (this term means *broken down and the components released in non-organic forms*) during the oxidation process. It is not known what types of organic molecules comprise most of the organic phosphorus; many kinds are possible. For example:

phospholipids,
phosphoproteins,
nucleic acids and nucleotides,
phosphocarbohydrates, and
phosphonic acid derivatives.

In this list the first four classes are familiar from organic chemistry or elementary biochemistry. In these four types of compounds the phosphate unit is attached through an ester bond to the carbon skeleton, as illustrated in Figure 8.1. Most such ester bonds are relatively easily *hydrolyzed* (meaning a bond is broken with the addition of water) by enzymes, by acid, or even just by hot water. For example:

$$RCH_2-O-PO_3^{2-} + H_2O \rightarrow RCH_2OH + HPO_4^{2-} \quad (8.1)$$

The derivatives of phosphonic acid, however, are extremely stable. These compounds have a direct phosphorus–carbon bond, as for example in aminoethylphosphonic acid (Figure 8.1). The direct carbon–phosphorus bond is so difficult to break that the only feasible way is to oxidize the carbon chain attached to it. This compound was first identified as a biological product by Horiguchi and Kandatsu (1959) in ciliates from sheep rumen, and independently discovered by Jim Kittredge in 1962 in extracts from sea anemones (Kittredge and Roberts 1969). It is synthesized by many organisms, including some marine phytoplankton. The stability of this class of compounds suggests that they should be good candidates for being part of the organic phosphorus dissolved in seawater, though some bacteria, at least, can metabolize them. They do comprise about 25% of the phosphorus in high-molecular-weight dissolved organic matter, and this fraction appears surprisingly constant in the deep water of all ocean basins (Kolowith *et al.* 2001). Exactly what compounds are involved is still a mystery.

Several investigators have shown that all the classes of phosphorus-containing compounds mentioned above may be found within the pool of organic compounds dissolved in seawater (Paytan and McLaughlin 2007), but detailed information on these structures is lacking.

8.1.3 Oceanic occurrence

The presence of phosphorus in seawater was suspected at least as early as 1865, because it is mentioned by Forchhammer. Accurate knowledge of its concentration and variation in seawater was, however, only obtained after the development in 1924 of sensitive colorimetric techniques appropriate for seawater. These methods all depend upon the reaction of phosphate with molybdic acid followed by chemical reduction of the complex. Quantitation is by measurement of the absorbance of red light by solutions of the blue-colored reduced phosphomolybdic acid. The most convenient modern technique is that of Murphy and Riley (1962). The method is very sensitive; of all the nutrient substances routinely measured in seawater, phosphorus is present in the smallest concentration.

Because of its importance as a primary nutrient for all photosynthetic and other autotrophic organisms, phosphorus has been studied extensively throughout the oceans and its distribution is known in great detail. The range of concentration is large, from zero (below detection limit, conventionally about 0.02 μmol kg^{-1}) to about 3 μmol kg^{-1}. Because of biological uptake at the surface and downward transport by organisms and their detritus, the concentration is nearly always less at the surface than at depth. Figure 8.2 shows three vertical profiles from typical regions in the central North Atlantic, central North Pacific, and the Mediterranean. The dramatic differences between these major oceanic regions are due to the interaction

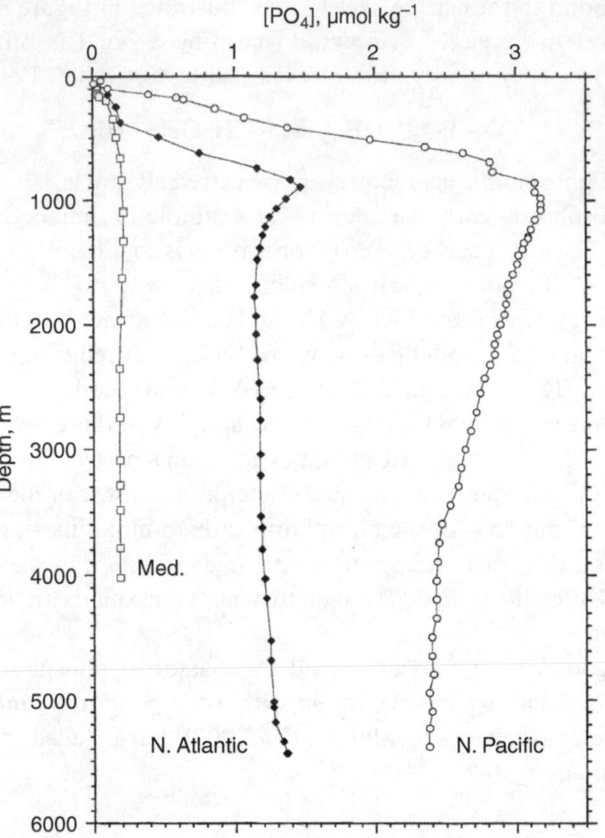

Figure 8.2 Vertical distribution of inorganic phosphate at three stations characteristic of major oceanic regions. Pacific data from GEOSECS Sta. 204 at 31° N, 150° W (Broecker *et al.* 1982); Atlantic data from TTO Sta. 241 at 36° N, 56° W (PACODF 1986). Mediterranean data from GEOSECS Sta. 404 at 35° N, 17° E (Weiss *et al.* 1983).

between biological transport (mostly downward transport by sinking biological detritus and subsequent metabolic release) and the large-scale features of oceanic circulation as shown in Figure 1.4.

The distribution of organic phosphorus in the ocean is less well known. In one oceanic study, Jackson and Williams (1985) reported on measurements of total dissolved phosphorus (acronyms are TDP or P_{td}) and dissolved inorganic phosphate (DIP or P_i) at several stations in the central Pacific Ocean (Figure 8.3). The differences between the two measurements were commonly about 0.1 to 0.2 μmol kg^{-1}, and are assumed to be a measure of the dissolved organic phosphorus (DOP). The occurrence of organic phosphorus at all depths suggests that much of this material is composed of fairly stable and long-lived compounds. Concentrations at depth are about the same in the Atlantic and Pacific (Paytan and McLaughlin 2007). A significant point is that while the DOP concentrations are quite small relative to the inorganic phosphate in deep water where the DIP is high, in surface waters where DIP is low the DOP may comprise more than half the total. The slightly greater absolute concentrations of DOP in surface waters suggests that some of this material is relatively labile, and should be considered in careful studies of the nutrient relationships of organisms living there. Indeed, phytoplankton living

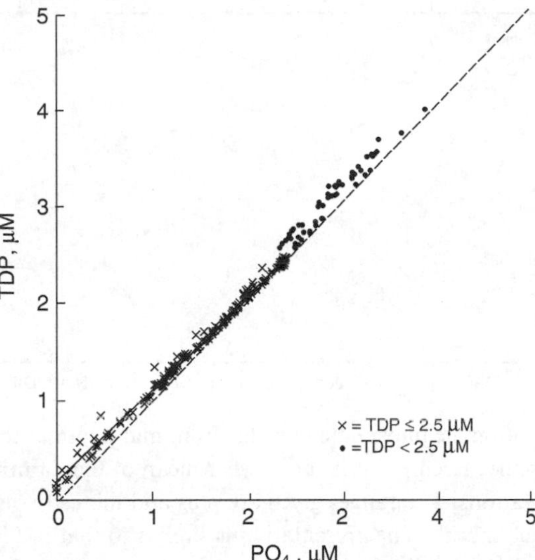

Figure 8.3 Relationship of total dissolved phosphorus (measured as phosphate after vigorous oxidation of the sample by high-intensity ultraviolet light) to inorganic phosphate in samples of water taken from the surface down to near the bottom at 15 stations in the north and south central Pacific Ocean. The dashed line is the 1:1 relationship (TDP = DIP). The distance by which any point falls above the line is a measure of the dissolved organic phosphorus in the water sample examined. (Jackson and Williams 1985.)

in low nutrient water may have, on their cell surfaces, increased concentrations of enzymes that hydrolyze phosphate esters.

8.1.4 Seasonal cycles

The classical studies on seasonal cycles of phosphorus in the ocean were carried out by Alfred Redfield, Homer Smith, and Bostwick Ketchum in the Gulf of Maine during the 1930s and by F. A. J. Armstrong and H. W. Harvey in the English Channel during the late 1940s. Armstrong and Harvey (1950) made repeated measurements at one station in the English Channel during the course of several years, and observed that in winter the water was relatively homogeneous from top to bottom (the depth was about 70 m), while in summer the water column was vertically stratified and surface phosphate concentrations were lower. Furthermore, during summer there was a marked increase in the fraction of the total dissolved phosphorus which was characterized as organic, while at the same time the concentration of inorganic phosphate decreased.

Figure 8.4, with data from Narragansett Bay, Rhode Island, shows that in shallow temperate coastal water there can be a dramatic seasonal change in the concentration of phosphate. No data for organic phosphorus accompany these data, so it is not known whether there may be a seasonal cycle to this component as well. There is still much to learn. The evidence (Pilson 1985) suggests that here the seasonally varying concentrations are controlled at least in part by changes in the rates of exchange between the sediments and the overlying water. The sediments release phosphate to the water above in the summer, and somehow it returns to the sediment in the winter.

Every region with seasonally varying temperature, light, and other conditions may be expected to show features of the seasonally varying concentrations of nutrients that are controlled by the changing balance of processes characteristic of the area.

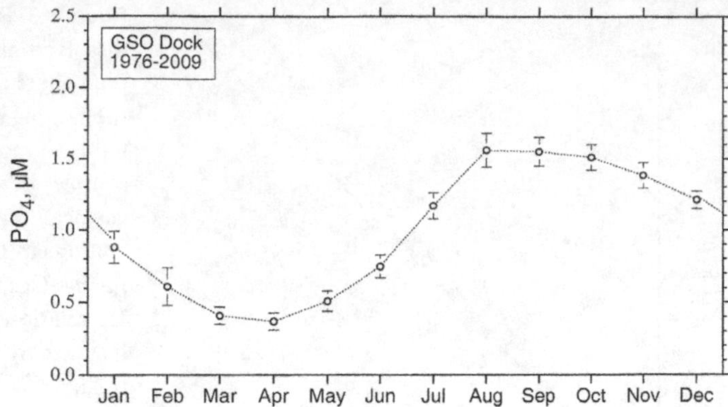

Figure 8.4 Concentration of inorganic phosphate in water from mid-depth at the GSO dock in Narragansett Bay, Rhode Island. Each point is the 33-year mean of the monthly means of approximately weekly observations. The ranges given are plus and minus two standard errors of the monthly means for each month. The presentation is similar to that in Pilson (1985) but includes a total of 33 years of data. Individual observations appear quite variable, within and between years, but the averaging of 33 years of data provides a compelling picture of the underlying seasonal cycle.

In no case do we yet have sufficient information to create a model that satisfactorily predicts such seasonally varying concentrations.

8.1.5 Phosphorus-poor regions

Many regions of the sea, notably the Mediterranean and the great central gyres of all the major oceans, are remarkably deficient in phosphorus (as well as in other nutrients) so very little can grow there. The water is exceedingly transparent, making it appear blue. It has been said that "blue is the desert color of the sea" (Sverdrup *et al.* 1942, p. 784).[1]

This is not to say that phosphate is always the most limiting nutrient, but it is certain that water with a low total phosphate concentration can grow only a low concentration of phytoplankton, because these plants cannot manufacture it and they cannot live without it.

In low-nutrient waters an odd problem arises. The occurrence of arsenic in the form of arsenate ions in seawater poses problems for both the analyst, and the plants living there. Arsenate is chemically very similar to phosphate, and it reacts with the molybdate reagent to form a blue-colored arsenomolybdenum compound with an absorption spectrum very similar to that of the phosphomolybdenum blue. Therefore, the usual analytical methods for phosphorus in seawater report arsenate as well. The concentration of arsenate is low, about 0.015 to 0.025 μmol kg^{-1} (Middleburg

[1] The sentence derives from one used in a number of German publications: *Blau ist die Wüstenfarbe des Meeres*. This in turn (Pilson 2010) derives from a metaphor originally expressed by Franz Schütt (1892) as "*... das reine Blau die Wüstenfarbe der Hochsee ist*".

et al. 1988), so this does not significantly affect the analyses of samples from productive or deep water. In oligotrophic surface waters, however, the phosphate concentration may be as low as 0.02 μmol kg^{-1}, or lower, and here the analyst, in reporting a value for phosphorus, may be unknowingly reporting an equal quantity of arsenate as phosphate. A procedure for eliminating the arsenate interference was described by Johnson (1971). When dealing with archived data where this or an equivalent procedure was not used, a simple first-order correction is to subtract 0.02 μmol kg^{-1} from the reported results.

The problem for plants also arises because of the chemical similarity of arsenate and phosphate. Arsenate competes with phosphate in many biochemical pathways; along with phosphate it is actively transported into the cells of at least some and probably all phytoplankton, and at almost the same rates, proportional to the concentration in the water. Arsenate is not a perfect substitute for phosphate, however, and when it is formed into biochemical analogs of phosphate compounds it disrupts some normal biochemical activities in cells, which is one cause of the poisonous nature of arsenic. When the phosphate concentration decreases to that of arsenate (such as may be the case in the central Sargasso Sea) the situation would seem likely to become serious. It is not fully understood how plants respond to this problem. The experiments of Johnson (1972) and Pilson (1974) led to the suggestion that bacteria and corals might take up arsenate, reduce it to arsenite (the trivalent form which is actually more toxic), and excrete it or allow it to escape from the cells. It is, however, a remarkable fact that many marine organisms incorporate arsenic into their biochemical composition to a far greater extent than do terrestrial organisms. Marine meat such as fish, shell fish, or crustacea, may have up to 10 times the concentration of arsenic as terrestrial meat (Cullen and Reimer 1989). The most abundant arsenic compound in crustacea and fish is arsenobetaine: $(CH_3)_3As^+CH_2COO^-$. This was the first arsenic compound certainly identified in marine organisms (Edmonds *et al.* 1977). Since then, dozens of additional compounds have been identified, including compounds of sugars, especially ribose, that incorporate arsenic into the structure (Francesconi and Edmonds 1998), fatty acids with dimethylated arsenic attached (Rumpler *et al.* 2008), and hydrocarbons with dimethylated arsenic attached (Taleshi *et al.* 2008). It is not known exactly how the most abundant compounds are synthesized or what their roles may be in the life of the organisms. It was at first thought that they were involved in mechanisms of detoxification, but it now seems more likely that at least some of them may play some essential biochemical role.

8.1.6 Phosphorite minerals

Phosphorite is a general term for sedimentary rocks that contain large amounts of phosphorous. Pure calcium phosphate can be made to precipitate from solutions of phosphate by adding calcium, to yield, for example: $Ca_3(PO_4)_2$ or $CaHPO_4$. These minerals are relatively soluble, however, and are rare in nature.

The apatites, the most abundant group of phosphorite minerals, are much less soluble. They have the approximate general formula: $Ca_5(PO_4)_3(X)$, where X may be:

Cl^-, F^-, OH^-. Many apatites also have partial substitutions such as CO_3^{2-} or SO_4^{2-} forPO_4^{3-}, and Mn^{2+} or Sr^{2+} for Ca^{2+}. The most abundant apatites are fluor-apatites and carbonate fluor-apatites. All intermediate gradations are possible. When carbonate is present as a major component the mineral has sometimes been given the special name: francolite. Most marine phosphorites can be called francolite.

The largest deposits of apatites have a marine origin but today are found on the continents. Mineable reserves of phosphate rock total about 65×10^9 t, of which most are in Morocco and western Sahara. In many cases it is not entirely clear how these deposits were formed, because present-day oceanic conditions are apparently not conducive to the formation of massive deposits. Apatite nodules and crusts are found in the ocean, however, covering fairly extensive areas in some continental-shelf regions. Those off California have the approximate general composition: $Ca_{10}(PO_4, CO_3)_6F_{2-3}$. Such phosphorites are commonly found in depths of 30 to 300 m. It is unclear whether all these deposits are actively forming today. If not, they were precipitated at some time in the fairly recent past. Phosphorites always seem to be found in regions of intensive coastal upwelling and low oxygen concentration, such as off Peru and Namibia, where they appear to be forming in the organic-rich and anoxic sediments. Goldhammer *et al.* (2010) presented clear evidence that bacteria in such sediments are in some way actively involved in the formation of apatite minerals.

A simple examination of the equilibria involved suggests that precipitation should be encouraged by low pressure, high phosphate concentration, and high pH. There is a problem with this, however, because high phosphorus concentrations are usually found in regions where a lot of biological destruction of organic matter (respiration) has occurred, so that the CO_2 concentration is elevated and the pH is therefore lowered. Further discussions are provided by Froelich *et al.* (1982), Schuffert *et al.* (1994), and Benitez-Nelson (2000).

8.2 Nitrogen

As with phosphorus, nitrogen is an absolutely essential ingredient of all living things, and is required in amounts commonly about 16 times greater than is phosphorus. The complex chemistry and biochemistry of nitrogen, however, introduce corresponding complexity into the oceanic behavior of this element. The large increase in global fluxes caused by humans, the multiple valence states, the atmospheric as well as terrestrial and oceanic sources and sinks of biologically active forms, and the currently inadequate knowledge of global fluxes, all contribute to both the importance and the difficulty of studying this element. Because of the extreme inertness of nitrogen when in the form of N_2, a major focus of interest has been the processes whereby N_2 is exchanged with other forms.

8.2.1 Chemical forms

About one third of all nitrogen on Earth is in the atmosphere (Table 8.3). There are two important stable isotopes, ^{14}N and ^{15}N, which in the atmosphere occur in the

Table 8.3 Estimates of the magnitudes of various reservoirs of nitrogen on Earth

	Gt of N
Atmosphere:	
N_2	3 877 000
N_2O	2.5
Mantle:	8 400 000
Crust:	
Igneous rocks	1 300 000
Sedimentary rocks	840 000
Ocean:	
N_2	22 200
NO_3^-	400
Organic N (DON)	60
Coal:	40
Biosphere:	30

Atmosphere and ocean data estimated from Appendix tables and other data in this text. Estimates for the mantle, and sedimentary and crustal rocks, have varied by more than a factor of 2; the values here are from Goldblatt *et al.* (2009), who regard the mantle value as a minimum estimate. Coal typically contains about 1.5% N by mass. According to Goldblatt *et al.* the inventory in the mantle is growing slowly, at the expense of the atmosphere.

ratio $^{14}N/^{15}N = 272 \pm 0.3$ (Faure 1986). Fractionation of $^{14}N/^{15}N$ in biological and chemical processes makes the isotopic ratio a useful tracer, at least in circumstances where the processes are not too complex.

In addition to gaseous N_2, nitrogen occurs in a wide variety of inorganic forms. The non-N_2 inorganic forms (and the organic forms as well) are included in the term *fixed nitrogen*. The inorganic forms of fixed nitrogen are perhaps better called *reactive nitrogen*, and in some literature they are. The oxidation states of inorganic nitrogen species range from a valence of –3 to +5 (Table 8.4). Representatives of all of these valence states do occur in nature, and most are important components of the natural geochemical cycles of nitrogen.

In seawater, most nitrogen is present as the dissolved molecular form, N_2. This gas is normally found at concentrations close to those appropriate for air saturation. At a temperature of 3 °C and S = 35‰, this is 580 μmol kg^{-1}. The total quantity dissolved in the ocean is therefore about 22×10^{12} tons (Table 8.3). Molecular N_2 is thermodynamically unstable in the presence of free oxygen, a situation that includes all of the atmosphere and most of the ocean. In the presence of oxygen, the stable form is the +5 oxidation state; in the presence of water this form becomes HNO_3, nitric acid. Nitric acid is a "strong" acid, so in all common circumstances the proton is completely dissociated, and in natural waters one finds NO_3^-, the nitrate ion. The reaction of N_2 with O_2 requires a very high input of activation energy, however, so throughout the atmosphere and ocean the two gases coexist without reaction, and the atmosphere does not explode. Only in lightning discharges, fires,

Table 8.4 Various forms of inorganic nitrogen

Valence	Formula	Name	Aqueous solution
+5	N_2O_5	Nitric anhydride	HNO_3 (nitric acid)
+4	N_2O_4	Nitrogen tetroxide	$HNO_3 + HNO_2$
+4	NO_2	Nitrogen dioxide	$HNO_3 + HNO_2$
+3	N_2O_3	Nitrous anhydride	HNO_2 (nitrous acid)
+2	NO	Nitric oxide	NO
+1	N_2O	Nitrous oxide	N_2O
0	N_2	Molecular nitrogen	N_2
−1	NH_2OH	Hydroxylamine	NH_2OH
−2	N_2H_4	Hydrazine	N_2H_4
−3	NH_3	Ammonia	NH_4OH

All the above forms occur as gases, or in aqueous solution. Some hydrate when in aqueous solution, as shown. All the forms listed above occur somewhere in nature, and most play important roles in the geochemical behavior of nitrogen.

N_2O_4, which is colorless, is a dimer of NO_2, which is brown, and as gases the two easily associate or dissociate according to conditions. NO_2 is responsible for the brown color of the fumes from nitric acid and the dimerization is responsible for the otherwise inexplicable visual structure of those fumes. In water they form the two acid species listed.

Nitric acid is a strong acid, completely dissociated in normal seawater. Nitrous acid is a weaker acid, but is also completely dissociated in seawater.

Nitrous oxide, N_2O, colloquially called "laughing gas," is widely used as an anesthetic and has certain attractive and interesting physiological effects. It is chemically very inert, so it has a very long residence time in the atmosphere (perhaps 120 years); its atmospheric concentration is currently increasing at about 0.25% yr^{-1}. The more proper name is "dinitrogen oxide."

Molecular nitrogen is extremely inert, requiring a high activation energy for reaction with either oxidizing or reducing agents.

Hydroxylamine is a very minor constituent in nature, occurring as an uncommon intermediate in biochemical processes. Hydrazine, a toxic substance used as a rocket fuel, is only known to be biologically produced by certain bacteria as an intermediate in the anammox reaction.

Ammonia in water is mostly in the form of ammonium ion:

$$NH_3 + H_2O \leftrightharpoons NH_4^+ + OH^-$$

At 18°C, $K_{NH3} = \dfrac{[NH_4^+][OH^-]}{NH_3} \approx 2.93 \times 10^{-3}$, At pH 8.2 , $\dfrac{[NH_4^+]}{[NH_3]} \approx 17.7$

and similar high-energy situations do N_2 and O_2 combine. Over a long enough time it seems likely that these situations would lead to the complete combination of all available oxygen and nitrogen; it is processes carried out by living organisms that return these gases back to the atmosphere and thus maintain Earth's atmosphere in a state far from thermodynamic equilibrium. Thus the biological processes responsible for transformations of nitrogen are crucial for the maintenance of the present environment of Earth and the possibility of life as we know it. Here we will consider some of the major transformations of global importance.

Because molecular nitrogen (N_2), also called *dinitrogen*, is so inert, the most important class of reactions includes those called *nitrogen fixation*, the various ways in which N_2 is reacted to make *fixed nitrogen*, by which is meant nitrogen in some chemically combined form. In lightning and in fires the products consist largely of the oxidized forms NO and NO_2, collectively known as NO_x. During biological fixation the first product is a reduced form equivalent to ammonia in oxidation state. Nitrogen fixation thus occurs in nature both inorganically and biologically; in addition, humans fix great quantities during the manufacture of NH_3 for fertilizers and other uses, and also inadvertently fix significant amounts during the combustion of fossil fuels.

8.2.2 Industrial nitrogen fixation

Humans fix nitrogen in the form of NH_3, to be used for fertilizer, explosives, plastics, and many other comparatively minor uses. The basis of the method now in use was invented by Fritz Haber in 1909 and Karl Bosch made it industrially practical by 1911. The great importance of this invention was recognized immediately (industrial production began in 1913) and, because of it, Fritz Haber was awarded the Nobel prize in chemistry in 1918. (Karl Bosch also received the Nobel prize in 1931 for his role in this and other advances.)

In the Haber process, as it is called (Table 8.5), H_2 reacts with N_2 at high temperature and pressure, in the presence of a catalyst, to form NH_3. In the modern industrial version of the process, the primary source of hydrogen is the methane in natural gas (Smil 2001). The simple equations in Table 8.5 show the ideal stoichiometry of the reaction, but do not convey the complexity of a large industrial activity carried forward in many parts of the world on a massive scale. In 2011, about 136×10^6 tons of nitrogen was fixed annually by modern versions of the Haber process. Production at this rate consumes the equivalent of 3 to 4% of the world's total natural-gas production.

Ammonia has many industrial uses, but the most important is for fertilizer. Some NH_3 is applied directly to the soil, and a greater amount is applied in the form of urea [$(NH_2)_2CO$], which is easy and safe to transport long distances, and has other advantages. The urea is synthesized by reacting CO_2 with ammonia. Some ammonia may also be combined with phosphate, sulfate, or nitrate, to be used as fertilizer. Some is oxidized to nitric acid, HNO_3, which has many uses, and some is used in the manufacture of plastics. About 87% of the ammonia fixed is used as fertilizer, so most of it is recycled in various ways within the ecosystem, and some must end up in the oceans.

According to Smil (1997, 2001) world agriculture could not support much more than one half the present world population without this nitrogen subsidy, so the increased nitrogen flux must continue for the foreseeable future.

8.2.3 Biological nitrogen fixation

Biological nitrogen fixation is confined to a very few types of organisms (Postgate 1998). No eucaryote (organisms with nucleated cells) can fix nitrogen. Some bacteria,

Table 8.5 Transformations of nitrogen

1. Modern Haber process:[a]
   a. $CH_4 + 2H_2O \rightarrow CO_2 + 4H_2$
   b. $N_2 + 3H_2 \rightarrow 2NH_3$
2. Further industrial transformations:
   a. $2NH_3 + CO_2 \rightarrow (NH_2)_2C{=}O$ (urea) $+ H_2O$
   b. $NH_3 + 2O_2 \rightarrow HNO_3 + H_2O$
3. Biological nitrogen fixation:[b]

$$N_2 + 8H^+ + 8e^- + 16ATP \rightarrow 2NH_3 + 16ADP + 16P_i + H_2 \quad (8.2)$$

4. Nitrification:[c] Some N_2O is also released

$$2NH_4^+ + 3O_2 \rightarrow 2NO_2^- + 2H_2O + 4H^+ + \text{energy} \quad (8.3)$$

$$2NO_2^- + O_2 \rightarrow 2NO_3^- + \text{energy} \quad (8.4)$$

5. Denitrification:[d] Some N_2O is also released

$$4HNO_3^- + 5CH_2O \rightarrow 5CO_2 + 7H_2O + 2N_2\uparrow + \text{energy} \quad (8.5)$$

6. Anammox:[e]

$$NH_4^+ + NO_2^- \rightarrow N_2\uparrow + 2H_2O + \text{energy} \quad (8.6)$$

[a] Both processes require high temperature, catalysts, and especially high pressure for 1b.
[b] Carried out by cyanobacteria such as *Trichodesmium*, bacteria, and some archaea; free-living and symbiotic; the reaction is very energy intensive; the nitrogenase enzyme requires Mo (some can use V) and Fe, and is very sensitive to oxygen.
[c] The first step is carried out by bacteria such as *Nitrosomonas* and by some archaea. The second step is carried out by other bacteria, such as *Nitrobacter* and by some archaea.
[d] Many bacteria, such as *Pseudomonas*, under low oxygen or anoxic conditions.
[e] This bacterial process is also denitrification, but is an autotrophic rather than heterotrophic process, in that no organic matter is oxidized.

some archaea, and some cyanobacteria can convert N_2 to NH_3. Among the most remarkable nitrogen fixers are the *Rhizobium* bacteria that live symbiotically in the root nodules of leguminous plants. Each partner in the symbiosis can live without the other, but N_2 is fixed only by the two forms living together (some strains of Rhizobia can be cultured in special circumstances to fix nitrogen on their own). A couple of hundred species or strains of procaryotic microorganisms (these have no separate nuclei in their cells) have been shown to fix nitrogen, in association with the roots of various kinds of plants or in other symbiotic associations, or free-living as some of the cyanobacteria, archaea, and bacteria. Evidence to date (Table 8.6) suggests that biological nitrogen fixation on land (about 144 Mt yr^{-1}), slightly exceeds present estimates of that in the oceans (135 Mt yr^{-1}). Cyanobacteria and/or bacteria are very active nitrogen fixers on coral reefs (Wiebe *et al.* 1975) and in sea-grass beds, but these occupy only a small area in the oceans. Of greater global importance are the

Table 8.6 Some estimated global rates of important nitrogen transformations and some individual fluxes

		Rate, as N (10^6 t yr^{-1})	
		Sub-totals	Totals
Nitrogen fixation			
Agricultural land		40	
Other land		100	
Fresh waters		4	
Total on land			144
Oceanic, pelagic		120	
Oceanic, benthic		15	
Total oceanic			135
Total biological			279
Lightning			5
Industrial (Haber process, 2011)		136	
Combustion		25	
Total direct by humans			161
Overall total			445
Denitrification			
In soils	Production of N_2O	11	
	Production of N_2	113	
In oceans	N_2O to the atmosphere	7	
	On continental shelves	250	
	Oceanic water column	150	
	In other marine sediments	50	
Total			581
Various transports:			
N_2O evasion from the ocean		5	
Rivers → sea	(all forms)	48	
Rain and dry-fall on land	NH_3	40	
	NO_x	25	
Rain and dry-fall on sea	NH_3	24	
	NO_x	23	
	DON	20	
N_2O → stratosphere		10	
Loss to marine sediments (burial)		25	

All estimates (except that for the Haber process) carry significant uncertainties, perhaps ± 20%, and in some cases rather more.

NO_x includes several oxidized forms of N; DON means dissolved organic nitrogen.

Data selected from Capone *et al.* (1997, 2009), Codispoti (2007), Duce *et al.* (2008), Galloway *et al.* (2004, 2008), Holland (1978), Seitzinger *et al.* (2006), USGS (2012).

active nitrogen-fixing cyanobacteria of the genus *Trichodesmium*, which are abundant in nutrient-poor warm surface waters (Carpenter and Romans 1991). The fixation by this alga alone has been estimated at possibly 80 Mt per year (Capone *et al.* 1997). There is a nitrogen-fixing bacterium, *Richellia intra-cellularis*, which lives symbiotically in some marine diatoms, and presumably provides a significant advantage to these diatoms. The discovery (Zehr *et al.* 2001) of abundant, very small, planktonic cyanobacteria that can fix nitrogen adds a major and previously unrecognized group of organisms to the list of oceanic nitrogen fixers. Evidence from Montoya *et al.* (2004) suggests that these minute cyanobacteria are major contributors to the pool of fixed nitrogen.

The fixation of N_2 requires a great deal of energy (Table 8.5), requiring for example, the ATP from the complete oxidation of one molecule of glucose for each two molecules of nitrogen fixed, as well as a source of electrons for the reduction. Some organisms appear to require less than the 16 ATP per molecule of dinitrogen, while others require more. The reaction is carried out by an enzyme complex called nitrogenase, and other enzymes are required to funnel energy to the complex (Postgate 1998). Nitrogenase is a complex of two proteins, each having sub-units. Each nitrogenase contains two atoms of molybdenum, and in addition there are 26 to 39 atoms of iron, depending on the species. (Some organisms can use vanadium instead of molybdenum.) The enzyme operates slowly, so many copies are needed, increasing the cost of supporting this process. Nitrogenase is sensitive to oxygen, so nitrogen-fixing organisms have elaborated several structures and mechanisms to prevent oxygen from reaching critical concentrations at the enzyme site. This is especially difficult for photosynthetic nitrogen fixers.

The nitrogen fixed by microorganisms ends up mostly as compounds of NH_3, primarily in amino acids and proteins. Some of this fixed nitrogen is transferred to the host if the fixers are living symbiotically. If the fixers are free-living, some of the fixed nitrogen leaks from the cells in the form of soluble amino acids, but probably most becomes part of the organic structures of these organisms and later passes into other biological pathways when the nitrogen-fixing organisms die or are eaten. On decay of the plants, or during their metabolism by herbivores, the organic nitrogen is released, mostly as NH_3 and urea. On land, ammonia is to some extent volatile and can enter the atmosphere, where it may be washed down to the ground or to the ocean again or be oxidized to N_2O_3, NO_2, or N_2O_5. This oxidation is also carried out by bacteria in soils, lakes, rivers, and the sea. The NH_3 and NO_3^- in aqueous solution may then be recycled through plants again on its way to the sea.

8.2.4 Other nitrogen fixation

In the atmosphere, nitrogen is fixed by electrical discharges. Every thunderbolt leaves behind a trail of NO_x (NO and NO_2, which are later oxidized to nitric acid), and perhaps other compounds as well, produced by the reaction of N_2 with O_2 (and possibly also with water vapor) under the conditions of very high temperature caused by the electrical discharge. The global production of fixed nitrogen by this process has been difficult to quantify, and estimates have varied by more than a factor of 20. Currently, estimates of about 5 Mt per year appear to be reasonable. Nitrogen is fixed in a similar fashion during the combustion of fossil fuels, as well as by biomass burning. Every automobile engine produces some nitrogen oxides, as do industrial power plants. The hotter they burn, the more they produce. Considerable effort is expended to reduce the discharge of these nitrogen oxides to the atmosphere. Through these processes, humans currently contribute about 25 Mt of fixed nitrogen to the atmosphere each year (Table 8.6).

8.2.5 Denitrification

Organisms that depend on oxygen for respiration cannot grow and may not survive in regions where O_2 is present in very low concentration or is absent. A few fungi are known to be able to use NO_3^- as an electron acceptor to oxidize organic matter, but none are known that can carry the reduction all the way to N_2, the final end point. Numerous species of bacteria and archaea can use NO_3^- as an electron acceptor and continue the reduction to N_2 as an end product (Zumft 1997). These bacteria and archaea are known to be ubiquitous in soils, in low-oxygen water, and in sediments throughout the world ocean, and were believed to be responsible for nearly all the loss of fixed nitrogen from the world's ecosystems. Only recently has it been found that many species of benthic foraminifera, and some related forms such as *Gromia*, can accumulate large concentrations of NO_3^- and use it as an electron acceptor for respiration (Risgaard-Petersen *et al.* 2006, Piña-Ochoa *et al.* 2010). These are the only eukaryotic organisms known that can carry the reduction of nitrate to N_2, and their abundance in sediments suggests that it may turn out that they are responsible for a significant fraction of the denitrification in the marine world.

The production of N_2 removes fixed or "reactive" nitrogen from the system and returns nitrogen to the atmosphere. Following is a schematic of the overall reaction:

$$4HNO_3 + 5CH_2O \rightarrow 5CO_2 + 7H_2O + 2N_2\uparrow. \quad (8.5)$$

In this equation, the CH_2O is a common and useful shorthand for organic matter, as this is the smallest unit representing approximately the chemical composition of carbohydrates, and comes close to representing the average composition of organic matter or biomass, with respect to the oxidation and reduction reactions it undergoes.

It is important to realize that the process described by Eq. (8.5) is a series of steps carried out by a very complex biochemical machinery. The process proceeds by several steps:

$$NO_3^- \rightarrow NO_2^- \rightarrow NO \rightarrow N_2O \rightarrow N_2.$$

Different enzymes or enzyme complexes are involved for each step (some contain molybdenum, others copper). If nitrous oxide reductase is not present, for example, then the process works up to the last step, but N_2O is released instead of N_2. Strains of bacteria like this have been found, and others deliberately created for experimental purposes. It also follows that if nitrate is not present, but one of the other oxidized forms is available, the latter can enter the pathway and be reduced. This works with some organisms, and some are known that can live with only organic matter and N_2O as an oxidant. Because there are separate processes involved in the cell, some intermediates in the series may diffuse out of the cell before being utilized. Nitrous oxide, being a small and uncharged molecule, can easily diffuse out of the cell. For this reason, whenever denitrification is occurring, there is some release of N_2O as well as N_2.

Denitrification processes occur in all regions where nitrate is present and oxygen is low or absent, such as in anoxic basins and sediments. They can also occur in certain areas of the open ocean where the O_2 concentration falls very low; at the present time the most extensive region of this type is the O_2-minimum zone in the Eastern Tropical Pacific. Goering and Cline (1970) incubated samples of water from this zone on shipboard and observed that nitrate disappeared after O_2 was exhausted; at first NO_2^- replaced the NO_3^- nearly quantitatively, but later NO_2^- also disappeared and no measurable oxidized form of fixed nitrogen remained.

In some sense the loss of N_2O from the system is not denitrification, because the N_2O is still a form of fixed nitrogen. However, N_2O is chemically quite unreactive, it cannot be used by plants, and much of it diffuses into the atmosphere, so it is effectively lost from the biologically active systems in soils or in the ocean. Thus, its loss is commonly included in reports of denitrification. In terrestrial soils N_2O is known to be produced as a sort of byproduct during both nitrification and denitrification (Schlesinger 1997). Its production during denitrification in marine sediments was first shown by Seitzinger *et al.* (1980, 1983), and evidence from the open ocean suggests that formation of N_2O in surface and subsurface waters is associated with the oxidation of ammonia to nitrate (Oudot *et al.* 1990). There is usually a flux of N_2O from the ocean to the atmosphere. This relatively unreactive gas has a residence time in the atmosphere of about 120 years. Nitrous oxide is an important greenhouse gas, and forms a rather uniform proportion of about 323 ppbv (parts per billion by volume) of the atmosphere; it is currently increasing at a rate of about 0.76 ppbv per year. Once in the atmosphere, the major loss appears to be in the stratosphere where ultraviolet light effects its destruction, and some of the reaction products participate in the destruction of ozone. Some reaction products are higher oxides of nitrogen that eventually return to the surface, but N_2 is also a product.

8.2.6 The anammox process

Denitrification as described above (Eq. (8.5)) has been known for a very long time (at least for bacteria), and was believed to be the only route by which nitrogen was returned to the unreactive dinitrogen state. A few investigators had imagined that nitrate or nitrite might oxidize ammonia, but evidence was lacking, and no organisms that could carry out the reaction were known (Strous and Jetten 2004). Then Mulder *et al.* (1995), investigating a fludized-bed sewage reactor, found evidence for the oxidation of ammonia under anaerobic conditions. Strous *et al.* (1999) identified the organism responsible as a planctomycete, a member of an odd, ancient, and not well-known group of bacteria. It carries out the following reaction:

$$NO_2^- + NH_4^+ \rightarrow N_2 + 2H_2O. \tag{8.6}$$

This reaction is the primary energy source for these bacteria, which use the energy to fix carbon and to grow. In the next few years it was discovered that these bacteria, *Candidatus sp*., are remarkably strange biochemically, and are geochemically significant (Kuenen 2008). They are the only organisms known that make and store hydrazine, a very toxic substance, as an intermediate in the reaction process. The hydrazine is stored in a vesicle inside the cell. The membrane around this vesicle is composed in part of a previously unknown class of lipids, the ladderanes (Damsté *et al.* 2002) that appear analogous to fatty acids and fatty alcohols but have a series of condensed cyclobutane rings at the non-polar end. Nothing like this had ever been seen anywhere in the world of biochemistry. The compounds are believed to contribute to an extremely impermeable membrane, imagined to be important in keeping the hydrazine under control. These bacteria are also unusually slow growing, doubling in about two weeks.

The geochemical importance of the anammox bacteria has been discovered to be quite important. The loss of nitrate that was believed for decades to be due to the classical denitrification reaction, with its associated consumption of organic matter, is now thought to be perhaps at least one-half due to the anammox reaction. The matter is under current active investigation (Lam *et al.* 2009, Zehr 2009).

The situation is so new that the writers in the current literature refer to the classical denitrification reactions (Section 8.2.5 above) as "denitrification," and the newly discovered pathways are distinguished as "anammox," even though the anammox reaction is certainly denitrification, in the sense that it is a route for the loss of fixed nitrogen from the ecosystem to the relatively unreactive dinitrogen pool. Probably the former process will be called *heterotrophic denitrification*, because organic matter is consumed, and the latter process will be called *autotrophic denitrification*, because organic matter is manufactured.

8.2.7 Biological nitrogen cycles

Our understanding of the broad features of the biological cycling of "fixed'" or "reactive" nitrogen in the ocean was initially formed largely through the classical studies of von Brand and Rakestraw, published in a series of papers in *Biological Bulletin* during the years 1937–1942. These investigators grew cultures of diatoms in large glass carboys of seawater, then placed the containers in the dark but in the presence of oxygen. Figure 8.5 shows the results of chemical analyses of the water over a period of several months, which revealed that a sequence of events took place as follows: ammonia was liberated during the decay of the plankton; the ammonia was oxidized to nitrite; the nitrite was oxidized to nitrate. If the experiment was interrupted by placing the carboys in the light, the ammonia was used up during growth of phytoplankton, but was released again if the carboys were again placed in the dark. The graphs given in Figure 8.5 are certainly idealized, and various experiments produced results differing in detail and timing from those shown here. Nevertheless, the general principles and interpretations appear to hold.

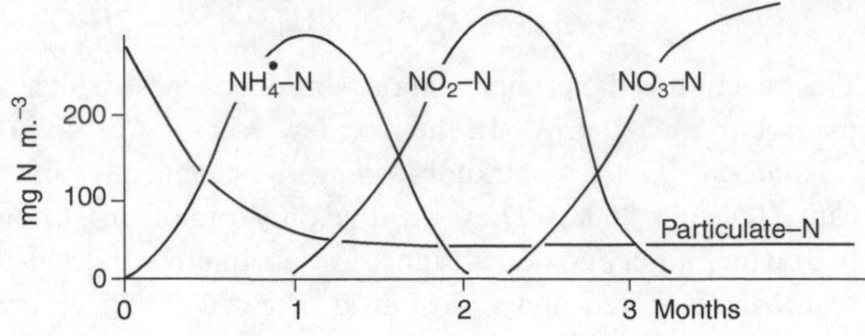

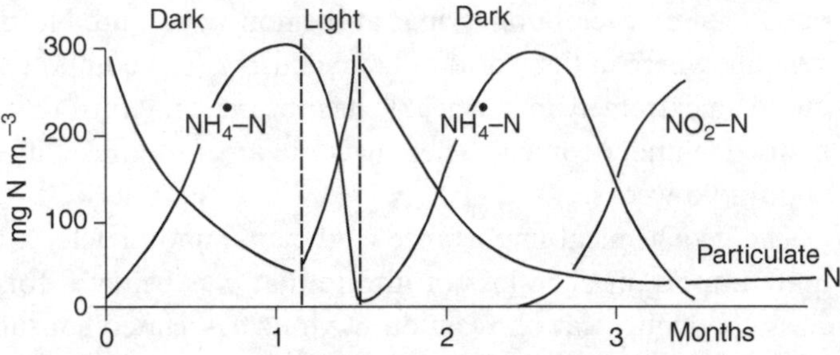

Figure 8.5 In a series of classical experiments during the 1930s, von Brand and Rakestraw stored suspensions of diatoms in seawater in the dark and observed the production of ammonia and its conversion to nitrite and finally to nitrate (upper graph). If the processes were interrupted by placing the experimental carboys in a lighted window, ammonia was taken up during the formation of organic matter as diatoms re-grew to their original density. When placed back in the dark, the sequence began as before (lower graph). From Harvey (1957).

- The first major nitrogenous product of digestion or decay of plant material is ammonia.
- Ammonia is oxidized to nitrite, NO_2^-, by one set of organisms (bacteria) which develop in numbers when the ammonia supply is adequate, and O_2 is present.
- Nitrite is oxidized to nitrate, NO_3^-, by a different set of bacteria which develop when the nitrite supply is adequate; again, O_2 must be present.

So far as is known, the above sequence applies in the ocean as well, except that in surface waters all processes would generally be expected to proceed simultaneously at rates varying locally depending on conditions. In surface waters, the phytoplankton are continually taking up nitrogen in its various forms, but most of these plants prefer ammonia, because it is in this form that it enters the biosynthetic pathways, and it is energetically costly to reduce nitrate to ammonia. Generally, phytoplankton will take up ammonia until it is reduced to a very low concentration and then turn to nitrite or nitrate. Nitrite is usually a very minor component of the pool of fixed nitrogen.

In water too far below the surface for phytoplankton to grow, the nitrogen released by respiration all ends up as nitrate (provided oxygen is present). Essentially all the inorganic fixed nitrogen in deep water is in the form of nitrate. This will not be used by phytoplankton until this deep water upwells again to the surface. Some nitrate must diffuse into sediments where it can be consumed. If a parcel of deep

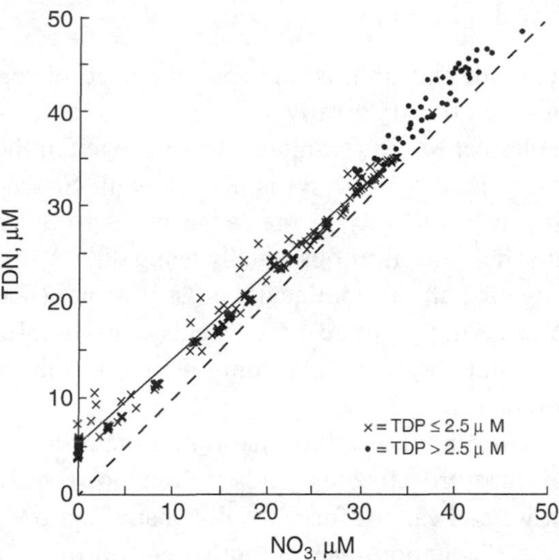

Figure 8.6 Dissolved organic nitrogen in seawater (Jackson and Williams 1985). Relationship between TDN (total dissolved fixed nitrogen, which includes nitrate and DON) and nitrate at various depths in the water column at stations located in the central basins of the Pacific Ocean from about 34° N to 76° S. The distance of each point upwards from the 45° slope is a measure of the concentration of DON. (The different symbols are used to identify points that fall in different parts of the range of total dissolved phosphorus, for additional evaluation. The fitted line refers only to the points identified with crosses.)

water becomes anoxic, or nearly so, the nitrate may be consumed by denitryifying bacteria in the water column.

8.2.8 Dissolved organic nitrogen

Important reservoirs of fixed nitrogen in the sea are found in the poorly characterized pool of compounds known as dissolved organic nitrogen, or DON, a situation analogous to that of phosphorus. DON is found everywhere in seawater; it appears to be synthesized mostly in the biologically rich upper euphotic layers of the sea, and to be degraded both there and much more slowly in the deeper layers. An extensive investigation into the oceanic distribution of DON was carried out in the Pacific by Jackson and Williams (1985), who obtained data throughout the water column off California and along a transect through the central basins from the North to the South Pacific (Figure 8.6). They showed that concentrations of DON generally ranged from 2 to 10 μmol kg^{-1}, with the higher values in the euphotic zone where often most of the fixed nitrogen was in the form of DON. Not much is known about the chemical composition of DON. In biologically productive surface waters, the excretion of urea by animals results in concentrations sometimes as high as 0.5 μmol kg^{-1} (with two atoms of nitrogen in each molecule of urea), but urea is taken up by phytoplankton and is easily degraded by bacteria, so it is not detected in deep water. Amino acids are certainly measurable in seawater, so they contribute as well, but concentrations range from 0.002 to 0.6 μmol kg^{-1}. Most of the DON in seawater remains uncharacterized, though evidence suggests that a substantial fraction is composed of material derived from the cell walls of bacteria (McCarthy *et al.* 1998). Nearly all of the DON in deep water and a considerable portion of that in surface water appears to be highly resistant to bacterial attack and very long-lived. Nonetheless, any complete evaluation of nutrient dynamics in the sea requires a serious consideration of the organic forms of both nitrogen and phosphorus.

8.2.9 Geochemical fluxes of fixed nitrogen

In contrast to the situation with phosphorus, there is little fixed nitrogen in continental rocks. Concentrations tend to vary greatly but average values (expressed as mol of N) are around 1 to 2 moles per ton for granites. Most nitrogen in these rocks is in the form of ammonia, but generally some N_2 is also present. Sediments typically contain rather more nitrogen, commonly in the range of 30 to 40 moles per ton (Goldblatt *et al.* 2009) with the concentrations usually being directly proportional to the clay content. An average for all the continental crust is around 7 moles per ton (Appendix Table J.2). Weathering of these rocks yields a negligible input to the oceans. Calculation shows that the rocks play only a minor role in the modern geochemical pathways of nitrogen.

In the past 30 years, a great deal of new information on the processes involved in the transformations of various nitrogen compounds and on global fluxes of nitrogen has become available, but we still cannot form a well-constrained budget. It is only within recent decades that the importance of both denitrification and nitrogen fixation in the marine environment has become evident, and only within the last decade has the remarkable anammox process been recognized as important. The rates of these processes are difficult to measure and extrapolation to obtain global fluxes requires some guesswork. Table 8.6 gives a tabulation of important fluxes, taken from several sources. In this table only the rates of industrial fixation and river transport are likely to be close to correct, and some of the other estimates may not be accurate to within a factor of two or more. Nevertheless, it appears that, in total, about 445 × 10^6 t of N_2 is fixed each year, nearly half caused by humans. Various marine environments have been shown to contain very active nitrogen fixers and the oceans may contribute about one-third of the global total. Present estimates of worldwide fixation are still likely to be underestimates.

The processes of denitrification (and burial of fixed nitrogen) must balance the total fixation of nitrogen, at least over a sufficiently long period of time. Estimates of the rates of these processes have been repeatedly revised, generally upwards, during the last several decades. Denitrification occurs in anoxic basins, but these are a very small part of the world ocean. Richards (1971) estimated the rate of denitrification in anoxic basins at a maximum of about 0.2 × 10^6t per year (0.2 Mt per year), a negligible pathway because their total volume and the total water flow through them is so small. In the eastern tropical north and south Pacific just west of Central America there is a large region of about 3.1 million square kilometers (bigger in area than the Mediterranean) under which there is a layer of water in which the oxygen is depleted almost or entirely to zero. The volume of water involved is about 1.1 × 10^{15} m^3, and forms a layer about 500 m thick. In this region the nitrate concentration is low, compared with concentrations expected on the basis of other constituents in the water. Until recently, this deficiency was ascribed to the process of heterotrophic denitrification, but now the anammox reaction has been shown to be at least as important. From concentration data alone it is not possible to calculate the rate at which this process occurs. Various approaches based on estimates of the flux of water through the region have yielded various rates, noted in Codispoti (2007); a plausible recent estimate is that denitrification in this region may amount to 75 Mt per year. Other globally important regions of this type are in the Arabian Sea and off the west coast of Africa, and Codispoti's estimate for the sum of these is 135 Mt per year.

Marine sediments are perhaps the most important locations where nitrogen is lost as N_2 from the pools of fixed nitrogen. Despite occupying a globally small area, only a few percent of the world ocean, sediments on continental shelves appear to be responsible for most of the ocean's denitrification. That this is the case must be due to both the generally high rate of organic-matter production over the shelves, and to the shallowness of the water, so much more of the organic matter fixed by phytoplankton falls to the sea floor and leads to anoxic conditions in the sediments, compared to the situation in deep water. The estimate that continental shelves (Table 8.6) may denitrify 250 Mt of nitrogen each year is surely uncertain, but it is so large compared to the flux of fixed nitrogen down rivers, that we can draw the conclusion that most of the 48 Mt of fixed nitrogen (natural and anthropogenic) carried towards the ocean by rivers is consumed in estuaries and on the shelves, and never makes it out to the major ocean basins. Indeed, the shelves must, on average, import nitrogen from the adjacent ocean. A considerable fraction (probably more than half) of the fixed nitrogen transported through the atmosphere to the sea is from anthropogenic sources, however, so the open ocean would appear to receive significant anthropogenic inputs.

Despite the various uncertainties, the general magnitudes of the important processes involved in global nitrogen cycles would seem reasonably well established. A consideration of the magnitudes of reservoirs and fluxes involved suggests that the global pool of fixed nitrogen might even be increasing. From Table 8.3, the total pool of fixed nitrogen amounts to about 660×10^9t. From Table 8.6, the annual global fixation rate is about 445×10^6t. This gives 1500 years for an apparent residence time of the fixed nitrogen pool, a remarkably short length of time, suggesting that the size of the total pool might respond rapidly to changes in rates of input or output. Since humans have recently increased the rate of fixation by about 50%, it seems plausible that the global pool of fixed nitrogen might be significantly increasing. When we start to think about this, it becomes evident that we do not know how sensitive the rate of denitrification might be to changes in the size of the pool of fixed nitrogen. We do not yet have quantitative evidence about the feedbacks that might operate to maintain the pool at a given size.

8.3 Silicon

Silicon is a major constituent of Earth and is always found combined with oxygen. Silicon dioxide, SiO_2, is commonly called *silica*. It occurs in this form mostly as the abundant crystalline mineral quartz, but most of the mass of Earth consists of silicate minerals, in which the SiO_2 is combined with various other elements. Indeed, expressed as SiO_2, silica makes up nearly 58% of the mass of Earth's crust. The layman does not often think of silica as a soluble nutrient. Nevertheless, numerous organisms both terrestrial and marine have an absolute requirement for this substance. It is an important oceanic tracer, it affects the physical properties of seawater to some minor extent, and by its presence or absence it exerts a considerable influence on the species composition of the plankton. If the concentration of dissolved silica in the ocean were reduced a 100-fold below its present concentration, the photosynthetic productivity of the ocean would probably be about the same as it is now, but there would be few diatoms and radiolaria, and in other ways the species composition of the plankton would be dramatically different; it is likely that the overall oceanic transport of carbon would differ in significant details.

8.3.1 Silica as a nutrient

Silica is classed as one of the important plant nutrients in seawater, even though not all plants appear to need it, and some animals do. The single-celled algae called diatoms (and a much less abundant group called silicoflagellates) make shells, called *tests* or *frustules*, out of precipitated amorphous silica (opal). For reasons not now entirely understood the diatoms often have a competitive advantage over other forms of phytoplankton (although, of course, only when silica is abundant), and in many productive regions they may be the dominant type of phytoplankton.

Among the animals, there is an important class of protozoa, the radiolaria, which make skeletons out of silica. They are unicellular forms related to the amoebae, and are in moderate abundance in open-ocean water. Some groups of sponges, primarily those that live in the deep sea, construct a framework or skeleton of silica, which is often remarkably intricate and beautiful.

In addition to the above uses as a major structural material for a variety of specialized but important forms of life, there is also the possibility that many other living things may need silica in minute traces for certain special biochemical reactions. For example, rats and chickens raised on silica-deficient diets exhibit severe abnormalities in bone formation, probably because of a failure of the biochemical reactions that make the organic matrix in which the bone develops (Simpson and Volcani 1981). While in a few such cases this requirement for silica has been demonstrated, available information is not adequate for generalization, and no such information is available for the larger marine organisms.

8.3.2 Occurrence and forms of silica

In the world as a whole, silica is found mainly as a constituent of the wide variety of silicate minerals. Most of this silica is combined with cations such as iron, magnesium and calcium, and aluminum is often included in the mineral structure. Free silica in rocks commonly occurs in the form of quartz, essentially pure SiO_2, the most stable crystalline form of the element. Amorphous silica, of several types, is produced through biological precipitation and infrequently by inorganic precipitation; this material lacks the crystallinity of quartz, and contains varying amounts of water in the structure.

Some of these silica materials are surprisingly soluble. Quartz itself is relatively insoluble, but some kinds of amorphous silica reach an equilibrium solubility in water of about 1800 μM at room temperature, or a concentration of 108 mg L^{-1} (expressed as SiO_2, the usual convention).

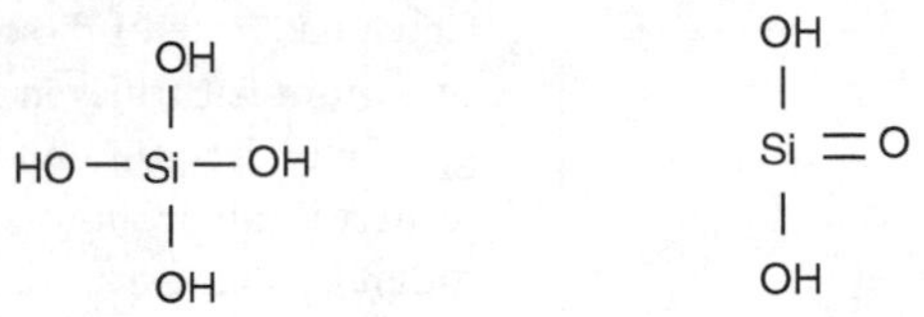

Silicic acid monomer

Silicic acid polymer

Figure 8.7 The form of silicic acid in solution is not certain. The silicic acid monomer with four hydroxyls ($Si(OH)_4$, or often written H_4SiO_4) is the most likely form, but the alternate form with two hydroxyls has been suggested, as is another form with six hydroxyls ($Si(OH)_6^{2-}$). In the polymeric form the value of n may be very large.

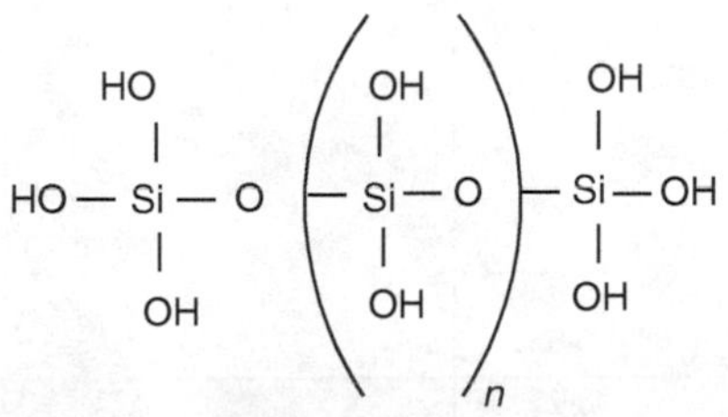

In solution at normal pH values SiO_2 is mostly present as the uncharged silicic acid. It is common, though perhaps not rigorously correct, to refer to this as dissolved silica, rather than as silicic acid. The structure in solution is not known for sure. Various formulas have been proposed (Figure 8.7), but the tetrahydroxy form, $Si(OH)_4$, is favored as the best representation.

Solutions of silica that are apparently supersaturated with respect to amorphous silica (i.e. concentrations above 1800 μM at 25 °C) can be prepared. In these solutions the silicic acid monomers combine to form long-chain polymers (Figure 8.7) reducing the monomer concentration to about 1800 μM. These long-chain polymers have not been well characterized, neither have the branched forms and rings that are also believed to occur (Alexander 1954).

When a solution containing polymers is diluted below a monomer concentration of 1800 μM, the polymers become unstable and dissociate into monomers. This process may be very slow, however, depending on the size of the polymers and conditions, and may take hours or weeks. In ordinary seawater dissolved silica should be exclusively monomeric, because the concentrations are seldom much above 200 μM.

Monomeric silicic acid is a very weak acid; the dissociation constant for the reaction:

$$H_4SiO_4 \leftrightharpoons H_3SiO_4^- + H^+ \quad (8.7)$$

has not been measured in seawater, but in solutions of a similar ionic strength it is about 3.9×10^{-10} (pK = 9.4), leading to the estimate that in seawater at

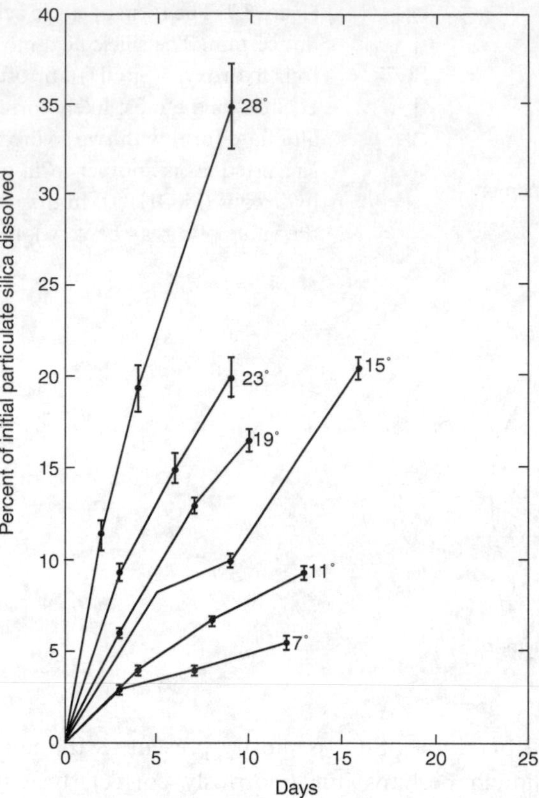

Figure 8.8 Rate of dissolution of diatom frustrules in seawater greatly undersaturated with respect to amorphous silica at different temperatures. (Lawson *et al.* 1978.)

pH 8.2 silicic acid may be only about 5% ionized, provided that not much of it forms ion pairs.

In the ocean the concentrations of silica span a very broad range, from less than 1 μmol kg^{-1} in surface waters of the Atlantic tropical regions to about 210 μmol kg^{-1} in deep water of the North Pacific. At the higher end of the range the mass concentration, about 13 mg kg^{-1}, is enough to have a barely significant effect on the density of seawater. Since silica in solution is mostly not ionized, it has a negligible effect on the conductivity of the water, and its presence, therefore, must cause the density to depart from that calculated according to the usual relationship from the salinity as measured by a conductive salinometer.

8.3.3 Solubility of silica

It has proved quite difficult to measure the solubility of silica at the low temperatures appropriate for reactions in soils, rivers, seawater, and sediments. The main problem is the extreme slowness of the reaction. Figure 8.8 shows that it can take many days for appreciable amounts of amorphous silica in the form of diatom frustules to dissolve. Rates of solution depend not only on temperature, but on the surface area

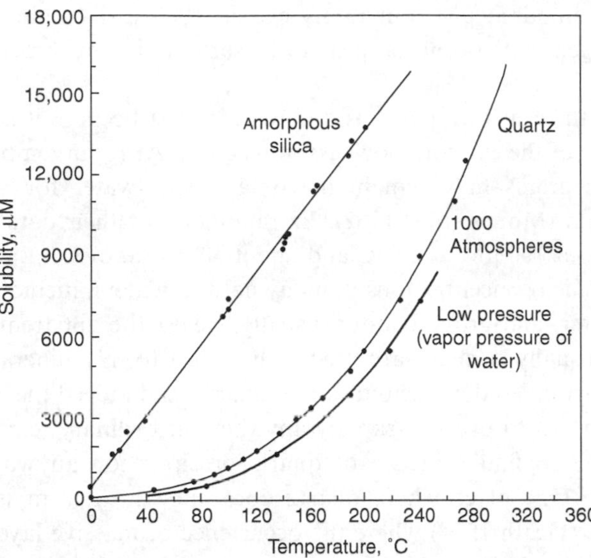

Figure 8.9 Solubility of amorphous silica (Alexander *et al.* 1954) and of quartz (Morey *et al.* 1962) in distilled water. The low-pressure data above 100 °C were obtained at an increasing pressure just greater than the vapor pressure of water at the temperature of measurement, so as to maintain the water in a liquid condition.

of the silica (lightly silicified diatom frustules dissolve faster than those more heavily silicified, and the specific surface area also varies with the intricacy of the structure), the presence of surface coatings, and the presence of adsorbed metal ions. Not only does it take a long time to reach equilibrium (even up to several months, depending on the conditions and the kind of amorphous silica used), but the equilibrium concentration seems not to be the same in all cases. Probably the nature of the solid surface, the degree of aging, and the presence of contaminants influence the apparent solubility. The best estimate is that the solubility of pure amorphous silica is about 1800 µM at 25 °C (Alexander *et al.* 1954), but often diatom frustules do not appear to reach this concentration in solution: Hurd (1983) reported values of about 1600 µM and about 900 µM in seawater at 25 °C and 3 °C, respectively. While the equilibrium solubility was not much affected by pH below values of about 8.5, the rate of solution was considerably pH-dependent, increasing about three-fold for an increase in pH from 6.7 to 8.3 (Hurd 1983). Solution rates were slower in seawater than in fresh water.

At higher temperatures amorphous silica is much more soluble, and the equilibrium is reached much faster. Figure 8.9 shows the solubility, in distilled water, of both amorphous silica and quartz as a function of temperature.

Increased pressure causes some increase in the solubility of both quartz (Morey *et al.* 1962) and amorphous silica (Jones and Pytkowicz 1973). The latter authors estimated the solubility of amorphous silica at 2 °C as 930 µM at 1 atm and 1170 µM at 1000 atm. The pressure effect may be of only minor consequence in the ocean. It increases the solubility of amorphous silica by about 25% at 1000 atm, the pressure in the deepest part of the ocean, but nowhere does the ocean seem to be saturated with respect to this form. While silica concentrations in the oceanic water column never greatly exceed 200 µM, the concentrations in sediment pore waters may reach several hundred μM, but rarely exceed 1000 μM (Schink *et al.* 1974). Even pore waters generally do not appear to be saturated with respect to amorphous silica.

The situation with regard to quartz is different, in part because it is so insoluble, and in part because of the extreme slowness of reaction. At room temperature in the laboratory, quartz grains may remain in contact with water for years without reaching equilibrium (Morey *et al.* 1962). Extrapolation of their data suggests that the solubility is about 100 μM at 25 °C and about 50 μM at 5 °C.

Since dissolved silica concentrations in many natural waters, including most deep seawater, river water, and ground water, usually exceed the apparent solubility of quartz, they must usually be supersaturated with respect to this mineral. There must be severe kinetic barriers to the precipitation of quartz, and indeed the crystallization of quartz is not known to proceed rapidly anywhere at ordinary temperatures and pressures. It is rare to find evidence of quartz precipitation anywhere, at Earth surface conditions. One place where quartz does appear to form is in very old deep-sea sediments (Heath 1974) where the occurrence of massive layers of chert, a microcrystalline form of quartz, in association with ancient deposits of amorphous silica suggests that this material slowly crystallizes over a long period of time (some tens of millions of years). The rate appears to be strongly sensitive to temperature, and chert deposits deep in the sediments are found where the temperature has risen above about 35 °C (Calvert 1983).

Almost all natural waters are undersaturated with amorphous silica, however, and consequently this material should dissolve. Amorphous silica is often abundant in marine sediments. Apparently it remains there because its dissolution is slow when pure, and the rates of dissolution are further slowed by surface coatings of organic matter or by adsorbed metal ions, such as those of aluminum, beryllium, iron, and gallium.

The possibility that other mineral phases besides quartz or opal might be effective in controlling the maximum observed concentration should be considered. For example, Wollast (1974) suggested that the magnesium silicate, sepiolite, could form by the reaction:

$$2Mg^{2+} + 3H_4SiO_4 + 4OH^- \leftrightharpoons Mg_2Si_3O_8 + 2H_2O. \qquad (8.8)$$

The solubility of sepiolite is very sensitive to pH (in fact to the 4th power of $\{H^+\}$), apparently being less soluble at high pH. The solubility of sepiolite is not well characterized, but it seems that in many parts of the ocean the water may be close to saturation with respect to this mineral. It has not, however, been observed as a component in marine sediment.

Other reactions which might control the concentration of silica in the oceans involve various clay minerals. Clays are complex alumino-silicates in which the general arrangement of aluminum and silicon atoms is in layers, along with oxygen atoms and hydroxyl molecules. Various other cations such as iron, magnesium, and potassium may be included. Figure 8.10 shows examples of the structures of several clay minerals. The properties of the various clay minerals are strongly dependent

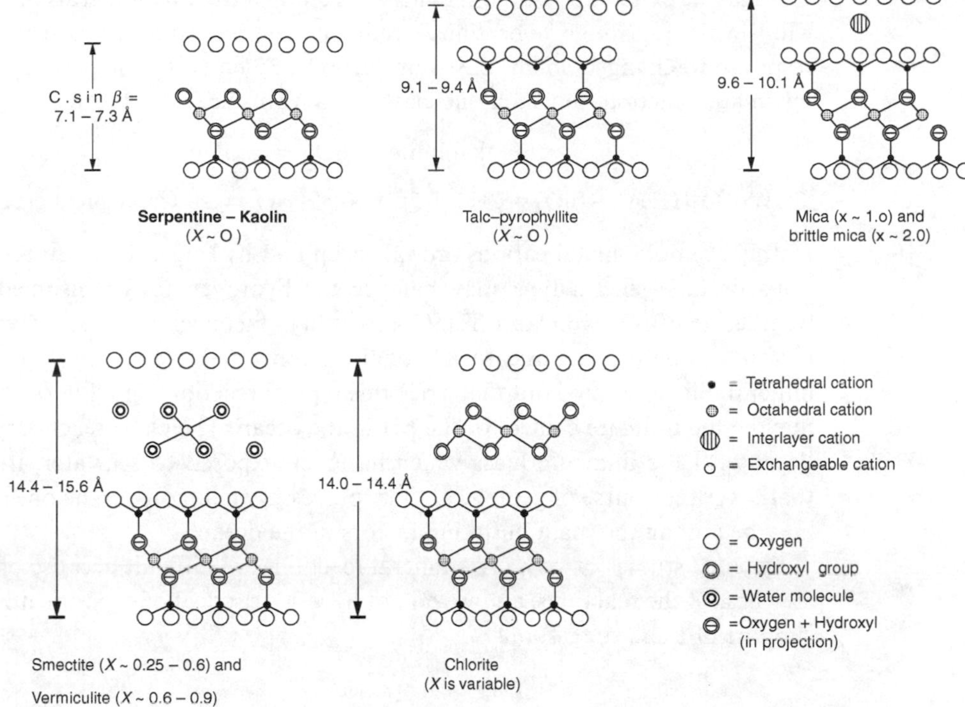

Figure 8.10 View of the structures of major groups of clay minerals. The clays are composed of sheets or layers in which the composition and the spacing between the layers varies between groups. Within the groups there are many variations in composition. The x values noted refer to the layer charge per formula unit, a parameter that need not concern us here. From Brindley and Brown (1980).

upon precise details of the arrangement and spacing of the atoms involved. Clay minerals always occur in very small particles, so even a small mass can have a very large surface area and a correspondingly great adsorptive capacity. The surfaces of clay minerals are composed of varied arrangements of aluminum, silicon, and other atoms, mostly combined with oxygen atoms.

Reactions with clays can be exceedingly slow, especially because this may involve changes in the mineral structure of the clay. Mackenzie *et al.* (1967) carried out experiments at 25 °C with several different kinds of clay minerals suspended in seawater with either no initial dissolved silica or elevated (430 μM) concentrations. Over a period of 9000 hours the clays in low-silica water released some silica, and in the high-silica water the concentration of silica decreased. While in some cases the concentrations were stable after 5000 hours, in other cases the concentrations were still changing after 9000 hours. In no case did the clays approach the same equilibrium solubility from both directions. It may be concluded that reactions are very slow, different clays behave differently, and that clays can release or take up silica depending on conditions.

It may be expected that the chemical structure of the clay minerals or their surfaces will slowly rearrange over time, gradually altering to form a new structure in response to changed conditions. For example, Sillén (1961) suggested the following schematic reaction, in which one clay reacts with silica and another clay is formed:

$$\text{kaolinite} + SiO_2 \leftrightarrows \text{zeolite} \tag{8.9}$$

$$3Al_2Si_2O_5(OH)_4 + SiO_2 + 2K^+ + 2Ca^{2+} + 9H_2O \rightleftharpoons 2KCaAl_3Si_5O_{16}(H_2O)_6 + 6H^+$$

In this example, metal cations are taken up and hydrogen ions are released. Reactions of this general type may balance the hydrogen ions consumed during the weathering of rocks on land (Eq. (7.46)), and in fact such reactions have been called *reverse weathering*. In this case, a kaolinite is converted to a different, more silica-rich mineral. Sillén pointed out that since this type of reaction is highly pH-dependent, it may be the ultimate control of the pH in the ocean. The clays react very slowly but, because of the immense mass which has been exposed to seawater, it seems likely that, over the course of geological time, processes analogous to the one shown above may be among the main buffering factors in the ocean.

Detailed studies of such clay-mineral reactions are difficult because of the extreme slowness of the reactions, and we do not have a great deal of experimental evidence in support of these suggestions.

8.3.4 Geochemical cycles of silica

There have been a number of attempts to construct a balanced budget for dissolved silica in the oceans. While the uncertainties are not so great as those for nitrogen, the budgets have still required considerable guesswork. Tables 8.7 and 8.8 summarize some of the estimated fluxes. More complete discussion may be found in Heath (1974) and Tréguer *et al.* (1995). The estimated accumulation rate of silica is the sum of estimates for pelagic sedimentation into the deep-ocean basins, which is probably fairly accurate, and for sedimentation on continental margins, which is poorly known. Most marine sedimentation occurs in estuaries, on shelves, on continental margins, and in the great delta deposits. It is difficult to measure the opal content of such neritic sediments, so possibly some of the unaccounted-for silica is in such sediments, masked by the large volume of detrital minerals. As mentioned before, clay minerals and other minerals can take up some of the silica released by dissolving opaline tests, so that might account for some of the missing silica.

In the above analysis, it has been assumed that the ocean is in steady state at the present time, but a speculative case can be made that it is not. Nutrient input by humans into most temperate rivers and lakes has increased the nitrogen and phosphorus concentrations there, but not the inputs of dissolved silica, the result being that in many places the additional growth of diatoms has depleted silica concentrations. Diatom frustules sink to the bottom of lakes and become buried, so the flux of dissolved silica to the oceans is reduced. For example, the concentration of dissolved silica in the Mississippi River has decreased by about 50% in historic times (Turner and Rabalais 1991). Billen *et al.* (1991) point out that this is true of many

Table 8.7 Estimated overall budget for silica (as SiO_2) in the world ocean

	10^{14} g yr^{-1}
River input	3.0
Air-borne dust	0.3
Sea-floor weathering	0.24
Hydrothermal leaching	0.12
Total input, and uncertainty	3.7 ± 1.2
Measured accumulation rate in sediments, and uncertainty	4.3 ± 1.1
Excess	~0.6, or 0

Inputs are calculated here for dissolved SiO_2, while the output is calculated as the accumulation rate of biogenic silica in sediments (Tréguer *et al.* 1995). The value for dust is most uncertain, as the fraction which dissolves has been only rarely investigated.

Table 8.8 Estimates of the silica budget in the water column in the world ocean

	10^{14} g yr^{-1}
Total ocean input (= permanent loss to sediments)	3.7
Diffusion from sediments (dissolving biogenic silica)	13.8
Total input to water column	17.5
Biological uptake to form diatom and radiolarian tests	144
Total particulate silica transported into the sediments	17.5
Dissolution within water column (= 144 – 17.5)	126

Derived from data in Nelson *et al.* (1995) and Tréguer *et al.* (1995). Steady state is assumed. Nelson *et al.* estimated that the biological uptake rate was uncertain by about $\pm 20 \times 10^{14}$.

other rivers and lakes as well, and note that the effect is greatly enhanced by the construction of dams on most of the world's rivers. The resulting impoundments provide opportunities for diatoms to grow and for the retention of their frustules in the sediments behind the dams. The imbalance between inputs to and outputs from the ocean noted in Table 8.7 is possibly not real, given the large uncertainties. Nevertheless, it is entirely possible that silica fluxes measured in the world's rivers today are somewhat less than they were one or two centuries ago.

The estimate of diffusion from the sediments is also uncertain. Typically, the dissolved silica concentration in the interstitial water of oceanic sediment increases with depth in the sediment, as shown in Figure 8.11. Whenever there is a gradient in the concentration of some substance in solution, there must be an associated diffusive flux. The source of the silica involved in this diffusive flux is probably the very slow dissolution of the frustules of diatoms and skeletons of radiolaria, buried in the sediment. The flux due to physical processes alone may be calculated from the diffusion constant (after corrections as noted in Figure 8.11) and the gradient:

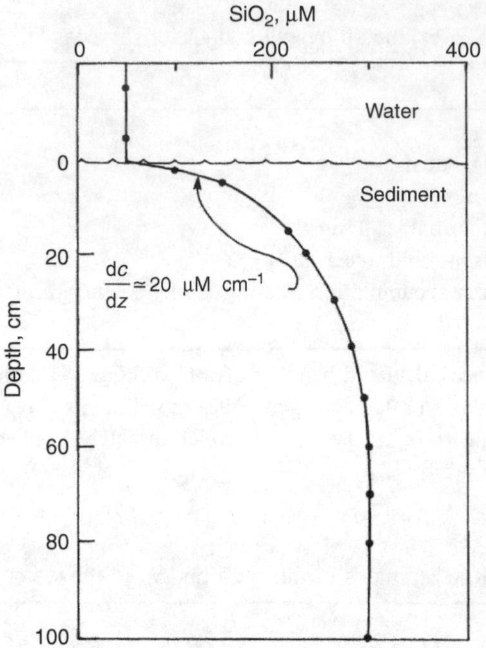

Figure 8.11 Characteristic shape of the vertical profile showing the concentration of dissolved silica in the interstitial waters of marine sediments. The flux upwards is calculated from the slope of concentration vs. depth near the interface (Schink *et al.* 1975, see also Wong and Grosch 1978), as follows: Flux(in units of mol $cm^{-2}s^{-1}$) $= \frac{\phi}{\theta^2} D \frac{dC}{dz}$ where:
ϕ is the porosity, this accounts for the fraction of sediment volume occupied by water;
θ is the tortuosity, this accounts for the increased diffusive path length around sediment particles; C is the concentration in nmol cm^{-3}, z is the depth in cm; and D is the molecular diffusion constant, this is about 4×10^{-6} $cm^2\ s^{-1}$, depending mostly on temperature.
After correction for porosity and tortuosity, the effective diffusion constant may be about 3×10^{-6} $cm^2\ s^{-1}$.

A crude estimate of the average gradient in marine sediments is about 20 μM cm^{-1}, from which the total flux from oceanic sediments can be calculated as follows:
$(3.0 \times 10^{-6}\ cm^2\ s^{-1})\ (20 \times 10^{-6}\ mol\ L^{-1})\ (10^{-3}\ L\ cm^{-3})\ (cm^{-1})\ (361 \times 10^6\ km^2)$
$(10^{10}\ cm^2\ km^{-2})\ (3.16 \times 10^7\ s\ yr^{-1})\ (60\ g\ mol^{-1}) = 4.1 \times 10^{14}\ g\ yr^{-1})$.

The main factor not accounted for above is the degree of biological mixing and irrigation of sediments, which, when present, greatly enhances all fluxes.

$$\text{Flux}(\text{mol cm}^{-2}\text{s}^{-1}) = \frac{\phi}{\theta^2} D \frac{dC}{dz} \tag{8.10}$$

As shown in Figure 8.11, this calculation gives a total diffusive flux from the sediments of the ocean of about 4×10^{14} g per year, based on estimated average gradient near the sediment–water interface of 20 μM cm^{-1}, a very poorly known value. Most estimates suggest that the total flux from the sediments is somewhat greater. Much of the diatom and radiolarian debris landing on the sediment surface must dissolve there before burial. In addition, biological mixing and irrigation of the sediments must greatly increase the total diffusive flux. The value of 13.8×10^{14} g per year as given in Table 8.8 is from Tréguer *et al.* (1995).

From Table 8.8, it appears that most of the silica that is biologically converted to opaline diatom frustules and radiolarian skeletons must be re-dissolved in the water column (or at the very top of the sediment) during or shortly after the sinking of the various particulate remains. We lack a quantitative understanding of why such a large fraction of the silica precipitated by organisms is apparently so readily dissolved, while some of the remaining silica is so slow to dissolve. A partial reason is that thin and fragile tests dissolve while more robust tests survive. This cannot be a complete explanation, because even some of the thinner tests are found in the sediments.

8.4. Other nutrients

The elements mentioned so far in this chapter, phosphorus, nitrogen, and silicon, are (in various literature sources) conventionally called the *nutrient elements*, or sometimes the *macro-nutrient elements*; it would, however, seem more sensible to call them *micro-nutrient* elements, because they are present in concentrations conveniently expressed in units of μmol kg^{-1}, or μmol L^{-1} = μM. There are, however, many other elements that are essential for life.

Consider the gross composition of living organisms. The major constituent is water, usually amounting to 70 to 95% of the total mass. Of the organic dry matter, about 45% is carbon, mostly in the form of the important classes of organic matter: carbohydrates, lipids, and proteins. Next comes oxygen, followed by nitrogen (mostly in proteins, some in nucleic acids and other minor compounds), sulfur, and phosphorus. Marine organisms generally contain considerable inorganic salt in the dry matter, consisting of sodium, potassium, chloride, magnesium, calcium, and sulfate.

In addition to these major and "micro" components, other elements are essential in trace amounts. For example, all living things contain iron, as this is an essential component of some proteins that are part of the respiratory pathway, so all organisms have an absolute requirement for iron. In addition, in photosynthetic plants iron is an essential component of the electron pathways in the photosynthetic apparatus. Iron is also required for many other enzyme systems. It is probable that copper, zinc, manganese, and cobalt, at least, are also essential for all organisms, because of their varied role in many enzymes. Some organisms require molybdenum, and some are believed to require boron. Perhaps these elements could be called *nano-nutrients*, because their requirements by marine organisms usually can be satisfied by nanomolar concentrations.

Many phytoplankton also have specialized requirements for various organic compounds, such as vitamins, which could therefore also be considered as nutrients. These are important to consider when dealing with the species composition of the plankton.

When considering the maximum amount of phytoplankton that can grow up in a parcel of water, however, the limiting nutrients are usually thought to be nitrogen and phosphorus, and silicon if diatoms are considered, and this appears to be the case over much of the ocean.

The concept of a limiting nutrient perhaps deserves more discussion. The idea was first popularized by Justus von Liebig in several editions of a seminal work on agricultural chemistry, from 1840 to 1862, as the Law of the Minimum.[2] The original idea was simple; for each species, nutrients are required in a fixed proportion, and when one of these nutrients is used up, the plant will cease to grow. If this nutrient is then added in a sufficient amount, the plant will continue to grow until another nutrient becomes limiting. In one sense this law is absolutely true. The basic biochemical structure of a cell contains, for example, phosphorus in the DNA and RNA, and in the energy-processing machinery. Without phosphorus, this cell cannot exist. In the absence of phosphorus nothing can grow. In nature, however, multiple factors are usually at work. Photosynthetic plants require light, they are affected by temperature and other factors, and multiple nutrient and other environmental limitations are probably common. With an abundance of all nutrients but one, and sufficient energy, plants can perhaps work harder to find and absorb the nutrient present in the lowest concentration. In a few cases some substitution of one element for another is possible, such as cadmium for zinc in the enzyme carbonic anhydrase. In addition, if there is not enough of some essential nutrient for growth and cell division, plants may still continue for a while to photosynthesize and produce carbohydrates and lipids, so the total biomass might still increase to some extent. Many ecological and physiological investigations deal with these intermediate situations and with the variable responses of different species.

Various trace metals have been studied with cultures of phytoplankton. It is well known that elevated levels of many metals are frequently toxic, but that adequate growth requires certain minimal concentrations. Only in recent decades could anything be said about the relation between trace-metal concentrations in the open ocean and the growth of plankton or larger organisms. Until three decades ago there was almost no evidence that the availability of trace metals ever limited the growth of plankton in the ocean, although it had often been speculated that the extreme insolubility of iron as ferric oxide, combined with its absolute requirement by all organisms, could lead to a shortage of iron. This speculation was confirmed when Martin and Gordon (1988) and Martin and Fitzwater (1988) presented strong evidence that, in the central gyre of the Northeast Pacific, iron is exhausted before nitrogen and phosphorus. Martin and Gordon showed that the amount of dissolved iron available from upwelling of deep water was only about 5% of the estimated

[2] The idea that various substances are required by plants in some necessary proportion, and that plants would cease to grow when one such substance is used up, antedates Liebig by some decades. However, in his book *Organic Chemistry in its Applications to Agriculture and Physiology*, published in 1840, and more forcibly in later works, Liebig gave the idea its clearest and most compelling statement. His powerful intellect, enthusiasm, and belligerence brought quantitative chemical analysis and rigor into the service of agriculture and physiology (Brown 1942).

requirement, based on the assumed ratio of one atom of iron to each 1250 atoms of phosphorus. Martin and Fitzwater found that adding iron in nannomolar concentrations to containers of seawater caused increased growth of phytoplankton. Nevertheless, plankton do grow there, and most of the available nitrogen and phosphorus are used. Remarkable as it may at first seem, it is now clear that in such regions the atmospheric transport of dust and its fallout onto the ocean surface can supply enough iron so that this element is not always limiting there. In one year, the fallout of dust provides about 60 μmol of iron onto each square meter of ocean surface (Duce and Tindale 1991). As Martin and Gordon showed, this is a sufficient supply for the plants that they can utilize essentially all the nitrate and phosphate supplied to the surface waters by upward mixing from below. One would normally think of iron in most of its likely forms as being quite insoluble. The iron in atmospheric dust is surprisingly soluble (up to 50% dissolves), apparently because chemical reactions in the atmosphere lead to reduction of the iron from the trivalent to the more soluble divalent state, and there may be other reactions that make even the trivalent form more soluble.

There are many parts of the ocean where the dust supply appears inadequate to provide enough iron. Concentrations of nitrogen and phosphorus are anomalously high in parts of the north Pacific, in much of the eastern Pacific equatorial regions, and over large areas of the off-shore waters of the Southern Ocean; these have been called "high-nutrient low-chlorophyll" (HNLC) regions. For many years there had been discussion about the causes of this situation. Martin and his co-workers provided compelling evidence in support of the speculation that the relative absence of iron carried in by rivers, combined with the very low atmospheric transport of dust to those regions, results in such a shortage of iron that the other nutrients are left partially unused. In such locations iron is therefore the limiting nutrient. The matter was of sufficient interest that the initial ship-board incubations in bottles was thought inadequate to assess the complex situation in the water column. Several major expeditions were mounted to fertilize multi-square-kilometer regions of the ocean surface with iron in several of the notable HNLC regions. Results from twelve such expeditions were summarized by Boyd *et al.* (2007). Varying amounts of iron were added, in one case as much as 2.8 tons of iron, as dissolved ferrous sulfate, discharged into large patches of ocean surface. The results all showed that adding iron increased the growth of phytoplankton; in some cases the patches were large enough that the blooms could be seen in satellite images. Detailed results varied according to the nature of the local ecosystems, and in any case the investigations could not be continued for more than a few weeks. The subsequent fate of iron in such experiments is still not very clear, but could be investigated during future expeditions.

Whether adding iron to the ocean in such regions would increase the total productivity of the ocean, and thus the draw down of atmospheric CO_2 is yet unclear, and certainly the results must vary from place to place. In some areas most of the "unused" nutrients in these HNLC regions would eventually be utilized anyway, as the water rich in phosphorus and nitrogen moves to regions where dust-fall supplies enough iron or the water mixes with near-shore water which is usually comparatively rich in iron. The higher concentrations of iron in near-coastal waters are due both to remobilization from sediments and from river input (Martin and Gordon 1988), and perhaps from greater dust-fall. In other places, especially in the Southern Ocean, this could not easily happen. The discussion goes beyond what we can usefully carry forward here.

As one last comment, it may be the case, however, that not all near-shore water contains sufficient supplies of iron. Duarte *et al.* (1995) provide evidence that, in some areas, sea grasses growing on carbonate sediments may be chronically short of iron. Iron fertilization alone considerably increased the growth rates of these plants.

8.5 Quantitative relationships

Phosphorus is the master variable which controls the availability of the others [nitrogen and oxygen] *Alfred C. Redfield 1958*

When the results of some of the early measurements of nutrient elements in the deeper water of major ocean basins were examined, it appeared that these elements were found in rather constant ratios to each other. Redfield (1934) showed that the linear relationship (Figure 8.12) had a slope of about 16 N to 1 P in mole units. He suggested that this should reflect the average composition of the organic debris that is metabolized and mineralized in the deep sea and, by extension, the average composition of the plankton that produces the organic debris. Subsequent work generally confirmed this suggestion and extended the ratios to include carbon, as well as the oxygen used for respiration or produced during photosynthesis.

While the overall chemical composition of organisms may vary considerably between species, or between individuals and life stages of single species, the average composition of marine organisms appears to be approximately uniform when summed over large areas of the ocean. Consequently, whenever photosynthesis exceeds respiration and the biomass of organisms is increasing in some large volume of water, the essential nutrients and carbon dioxide are taken up in approximately constant ratios to each other, and oxygen is produced. Conversely, when living matter is metabolized, the nutrients and carbon dioxide must (over a suitable time interval) be released in the same ratios, and oxygen is consumed. The ratios calculated by Redfield *et al.* (1963) were:

$$1\mathrm{P} : 16\mathrm{N} : 106\mathrm{C} : 138\mathrm{O}_2.$$

From these proportions, the composition of organisms, and the oxidation state of the carbon in them, could be expressed as a formula for average organic matter:

$$(\mathrm{CH_2O})_{106}(\mathrm{NH_3})_{16}(\mathrm{H_3PO_4}) \qquad (\text{formula wt.} = 3553.3\ \mathrm{g\ mol^{-1}}).$$

The 16:1 ratio is called the *Redfield ratio*, and by extension the other ratios included are also called Redfield ratios. For convenience, the formula given above may be

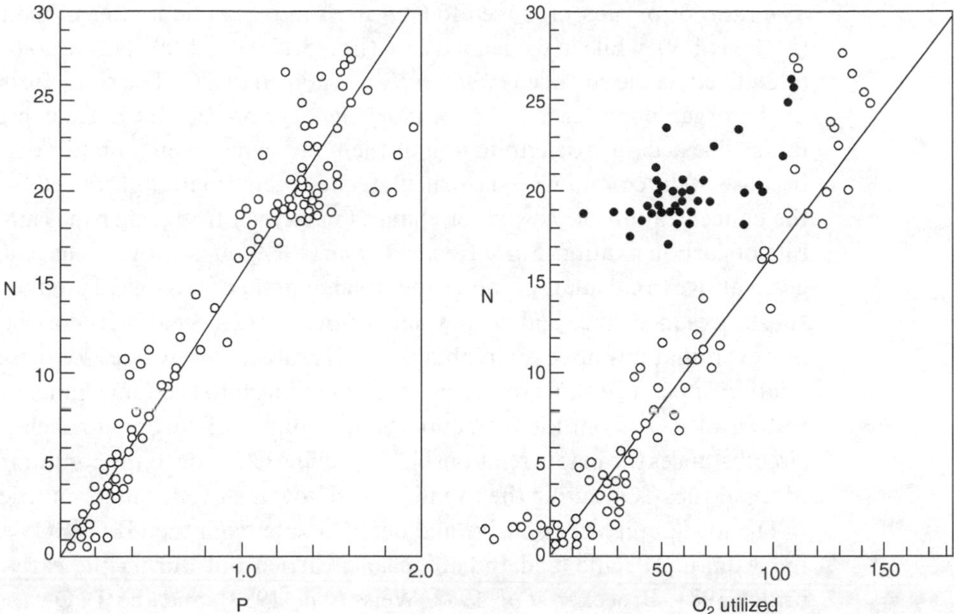

Figure 8.12 The first data showing the relationships between the concentrations of nitrate and phosphate and between the concentration of nitrate and the apparent oxygen utilization (AOU) in samples of water throughout the water column at open-ocean stations in the western Atlantic (Redfield *et al.* 1963). Concentrations are in mmol m^{-3}. These are slightly corrected presentations of the same data that Redfield (1934) had first used to show the relationship between the composition of plankton and the concentrations of nutrients in seawater. The filled circles are a subset of deep-water values; these will be discussed later.

called the *Redfield molecule*, though it does not, of course, represent an actual molecule. In this formula, the carbon in average organic matter is assigned a formal oxidation state equal to that in common carbohydrates such as starch or cellulose. The manufacture of this organic matter by photosynthesis and its oxidation during respiration can be represented as follows:

$$106\,CO_2 + 122H_2O + 16HNO_3 + H_3PO_4 \underset{\text{respiration}}{\overset{\text{photosynthesis}}{\rightleftarrows}} (CH_2O)_{106}(NH_3)_{16}(H_3PO_4) + 138O_2 \tag{8.11}$$

Note that the complete oxidation of this "molecule" requires 106 O_2 to oxidize the carbon; the nitrogen in organic matter is mostly in the oxidation state characterized here by listing it as ammonia, and thus requires an additional 32 O_2 to oxidize the ammonia to nitrate. It is important to recognize that if ammonia rather than nitrate is used during photosynthesis the forward equation would be:

$$\begin{aligned} &106CO_2 + 106H_2O + 16NH_3 + H_3PO_4 \rightarrow \\ &(CH_2O)_{106}(NH_3)_{16}(H_3PO_4) + 106O_2. \end{aligned} \tag{8.12}$$

The ratio of oxygen produced to CO_2 fixed into organic matter in the first case (Eq. (8.11)) is 1.30, while in the second case (Eq. (8.12)) it is 1.00. This ratio (O_2 produced)/(CO_2 fixed) is the so-called *photosynthetic quotient* or PQ. The ratio during respiration of the organic matter is the *respiratory quotient* or RQ. Both are expressed in mole units. The accurate determination of these ratios has been a matter of some interest, because of the common need to calculate the organic matter produced from changes in the concentration of oxygen, or changes in oxygen from other measurements of the rate of carbon fixation. Since Redfield *et al.* (1963), most workers have used the ratios given above to calculate or model the stoichiometry of biological uptake and release of substances in surface and deep waters of the world's oceans. It must be appreciated, however, that it is not easy to obtain entirely satisfactory values for these quantitative relationships. A careful evaluation requires taking into account changes in the concentration of CO_2 from the formation or dissolution of carbonate shells, and in some circumstances the bias in relationships involving CO_2 due to the continuing absorption of fossil-fuel CO_2 during the twentieth and now the twenty-first centuries.

The availability of extensive and detailed data from the GEOSECS expeditions to the Atlantic, Pacific, and Indian oceans, carried out during the early 1970s (Bainbridge 1981, Broecker *et al.* 1982, Weiss *et al.* 1983), and the TTO expedition in the North Atlantic (PACODF 1986), and several other large data sets, has made it possible to reevaluate these relationships. Several investigators, including Takahashi *et al.* (1985), Broecker *et al.* (1985), Anderson and Sarmiento (1994), and Körtzinger *et al.* (2001) plotted data along selected isopycnal[3] surfaces in the major ocean basins to evaluate the ratios in which nutrients are released and oxygen consumed in a more rigorous manner than had been possible before. Variations from region to region and depth to depth are small but probably real. The various investigations produced somewhat different values for some of the ratios, so it is clear that the values are not rigorously defined. For our purpose here the following modern revised equation is selected as a good basis for further calculation; in comparison with the classical Redfield equation, Eq. (8.11), it captures somewhat better the stoichiometric relationships characterizing the oxidation of organic matter in subsurface waters:

$$\begin{aligned}&(CH_2O)_{104}(CH_2)_{20}(NH_3)_{16}H_3PO_4 + 166O_2 \leftrightarrows \\ &124CO_2 + 140H_2O + 16HNO_3 + H_3PO_4.\end{aligned} \tag{8.13}$$

The major improvement from the classical ratios (Eq. (8.11)) is that there is a little more carbon per unit phosphorus, and a little more oxygen is required per unit carbon to oxidize the carbon part of the Redfield molecule. Since the oxidation state of the carbon in the original molecule was equivalent to that in carbohydrate, the apparent requirement for more oxygen requires that some of the carbon be associated with more hydrogen than is the case for carbohydrates. This is accommodated here by including the lipid component as 20 CH_2 units. The new formula (formula weight = 3774) is an

[3] The word *isopycnal* means constant density. For reasons having to do with the dynamical action of ocean currents and eddies, the depth at which a specified density is found varies across an ocean basin. A layer of constant density is called an isopycnal surface. Lateral mixing is extensive along such layers of isopycnal water.

improvement, and is in better accord with the observation that marine plankton organisms often contain considerable quantities of lipid materials such as triglycerides and waxes. The use of CH_2 units in the formula reflects this composition.

The range in published values for the oxygen-to-phosphate ratio for complete oxidation of the organic matter associated with one molecule of phosphate is 138 (the original Redfield ratio) to 175 (from Broecker *et al.* 1985). As we will see later, the choice of values within this range does have some effect on results from various calculations.

As pointed out by Laws (1991) the PQ associated with the production of carbohydrate alone is 1.00, for proteins it is about 1.04 if made from ammonia and 1.54 if made from nitrate, while for lipids it averages about 1.5. Because of the presence of proteins and the typically high contents of lipid materials, Laws' estimate of the PQ appropriate for the biosynthesis of most phytoplankton was 1.3 to 1.4 when growing on nitrate (and 1.1 to 1.2 when growing on ammonia). The RQ calculated from the ratios in Eq. (8.13) is 1.34. This ratio is lower than that observed by Hammond *et al.* (1996), who found an RQ of 1.45 for the remineralization of organic matter on and in the sea floor of the central equatorial Pacific. The higher carbon-to-nutrient ratios developing during remineralization have been found by many investigators, including Christian *et al.* (1997) who showed that the C/N and C/P ratios systematically increase with depth in sinking particles collected by sediment traps north of Hawaii, and by Hammond *et al.* (1996); both of these sets of observations suggested even higher carbon-to-nutrient ratios than are incorporated into Eq. (8.13).

The differences between these various recent estimates of the PQ and the RQ are possibly not significant, given the difficulties of obtaining realistic values; however, such as they are, they do suggest that the organic debris sinking to depth and metabolized there may have a slightly more lipid-rich character than the total organic matter synthesized at the surface.

While most phytoplankton prefer ammonia to nitrate as a source of fixed nitrogen (because it takes energy and reducing power to reduce nitrate to ammonia, which is the nitrogen form that enters biochemical synthetic processes), nevertheless most can probably use nitrate. With N, P, and CO_2 (plus Si and minor elements) available, most phytoplankton can carry out the complete synthesis of their biochemical structure. On the other hand, no organisms are known to be able to carry out the complete respiration as shown in Eq. (8.13). All forms of heterotrophic life appear to release nitrogen largely in the form of ammonia, or as various kinds of organic nitrogen compounds (such as urea) in which the nitrogen is in a reduced state. Only some species of bacteria carry out the oxidation of ammonia to nitrate. For this reason the initial stages in the respiration of organic matter (i.e. the metabolism of phytoplankton by zooplankton or the metabolism of zooplankton by fish) are perhaps best represented as follows:

$$\begin{aligned}&(CH_2O)_{104}(CH_2)_{20}(NH_3)_{16}H_3PO_4 + 144O_2 \rightarrow \\ &124CO_2 + 16NH_3 + 124H_2O + H_3PO_4.\end{aligned} \tag{8.14}$$

The released NH_3 might then be taken up again directly by plants carrying out photosynthesis, or oxidized first to NO_2^- and then to NO_3^- by the specialized bacteria that make a living at it. Note that according to Eq. (8.14) the RQ will be 1.16, while if the oxidation of ammonia to nitrate is included in the overall reaction the RQ becomes 1.34. In considering the various ratios to be expected in surface waters, therefore, it is important to note whether nitrate or ammonia is the dominant form of nitrogen participating in the reactions.

In addition, no known herbivorous or carnivorous organism is able to digest and metabolize every type of organic molecule in its normal food, so a considerable amount of organic matter is excreted as fecal pellets or as dissolved organic compounds. Varying amounts of dissolved organic compounds leak from all living organisms, and are often deliberately secreted as well. There is no reason to suppose that either the fecal pellets or the various forms of dissolved organic matter have the same composition as whole organisms, and in local areas these processes might significantly affect the ratios.

Using typical ratios embodied in the Redfield molecule it is possible to calculate the potential fertility of a sample of water. The potential fertility may be defined as the quantity of organic matter that could be produced by photosynthesis in a unit volume of water if it were brought from depth to the surface and illuminated there until the limiting nutrients are exhausted (Redfield *et al.* 1963). Commonly, carbon is about 45% of the ash-free dry weight of organisms. For example, a sample of seawater from about 500 m depth in the North Pacific could contain phosphate and nitrate at concentrations of 2 and 32 μmol kg^{-1} respectively (in which case neither nutrient is more limiting than the other). The calculation for 1 kg of this water is as follows:

$$(2\ \mu\text{mol P})\left(\frac{124\ \mu\text{mol C}}{1\ \mu\text{mol P}}\right)\left(\frac{12\ \mu\text{g C}}{1\ \mu\text{mol C}}\right)\left(\frac{2.2\ \text{g dry org.matter}}{1\ \text{g C}}\right) = 6547\ \mu\text{g}.$$

or about 6.5 mg of dry organic matter per kilogram of water. On a wet-weight basis this amounts to about 30 to 100 mg kg^{-1}. Over the oceans as a whole the concentration of plankton in surface water probably averages less than 1% of these values. The reason for the low average values is that for the most part the nutrients have been stripped out of surface waters, transported downwards, and released into deep water by the metabolic activity of organisms consuming the sinking plant and animal debris.

In the major basins of the open ocean, changes in the concentrations of nutrients and oxygen with depth exhibit relationships that are generally consistent with the ratios established above, especially from depths below a few hundred meters, but they often differ in details and complete explanations may not always be apparent. The nutrient and oxygen data from a GEOSECS station east of Barbados are shown in Figure 8.13. The nitrogen and phosphorus concentrations clearly parallel each other, while they both correlate inversely to the oxygen concentrations. The plot of N vs. P (Figure 8.14) shows that over the entire depth and concentration range the values all lie close to the 16:1 ratio line. When the adjacent points are serially connected, however, it is evident that there are small changes in the ratio with depth, such that different layers (water masses) show slight but consistent deviations from the 16:1 line. These small differences must in some way reflect the history of each water mass. Fanning (1992) plotted nutrient data from the GEOSECS survey and

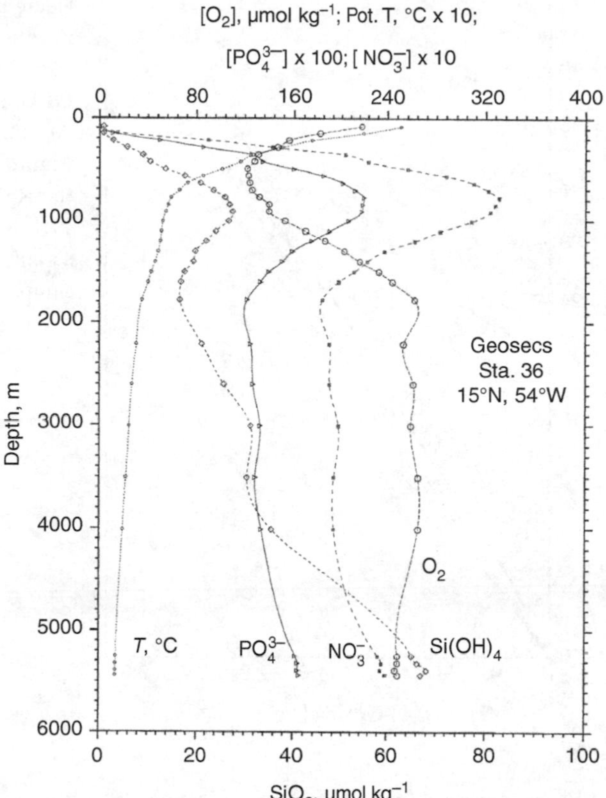

Figure 8.13 Nutrient, oxygen, and temperature profiles from a GEOSECS station in the North Atlantic Ocean east of Barbados, taken on October 11, 1972. Note that the depths of the phosphate and nitrate maxima coincide, while the upper silicate maximum is a little deeper, suggesting that silicate is dissolved from sinking particles a little more slowly than the other nutrients are remineralized. The increased silicate concentration near the bottom is due to the influence of Antarctic Bottom Water, which is especially rich in silicate. (Data from Bainbridge 1981.)

other sources to show in detail the variations throughout the world ocean. Anderson and Sarmiento (1994) used much of the GEOSECS data set to extract the ratios at which the nutrients are remineralized throughout the major ocean basins and over the depth range from 350 m to the bottom. Their evidence suggests small systematic differences between the ocean basins and some variation with depth.

The ratios of the major nutrients provide considerable insight into oceanic processes, as can be seen by comparing the values of these ratios in different regions. Figure 8.15 shows that there are distinct differences between the North Pacific and North Atlantic in the values of the ratio of nitrogen to phosphorus that characterize the near-surface waters of these central water masses and extend almost to the bottom. The extensive compilation of the N/P ratios for the three largest ocean basins by Fanning (1992) showed that the situation is more varied and complex than

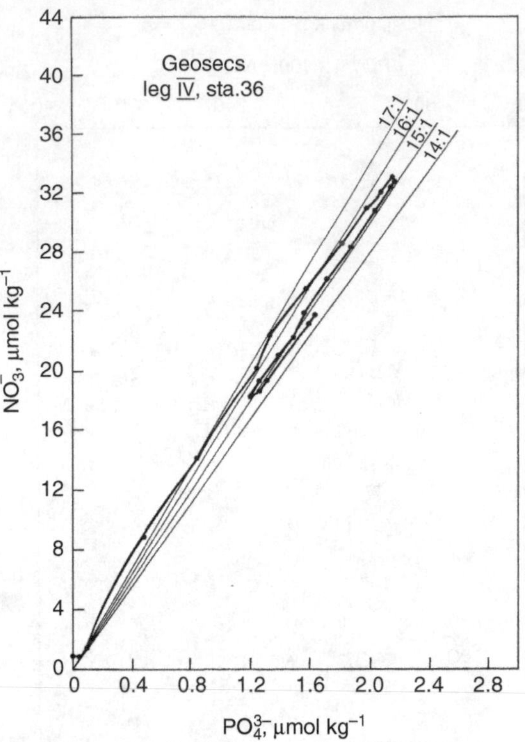

Figure 8.14 The concentration of nitrate plotted against the concentration of phosphate at a GEOSECS station in the North Atlantic east of Barbados (same as Figure 8.13). The light lines identify the Redfield ratio slopes from 14:1 to 17:1. The data points are connected serially according to the depth of each sample.

would be evident from Figure 8.15 alone, but the general observation still holds. Most of the Atlantic is characterized by higher ratios than are the Pacific and Indian oceans.

The reasons for this are still subject to speculation. It could be that the ratio is different in the organic matter delivered from the surface to depth in each major basin, but this seems unlikely. Another possibility is that the Pacific is a net source of phosphorus to the world ocean and the Atlantic a net sink, but this also seems unlikely. Much more likely is the hypothesis that the Pacific is a region where denitrification is more important than in the Atlantic. The mid-waters of the Pacific contain large regions where the oxygen concentration is reduced to zero and denitrification is known to take place. Thus, the Pacific could be a region of net denitrification in the world ocean. The deep water there is also "older," and has had more time to be exposed to sediments where denitrification can take place. It appears that the Pacific is a net sink for fixed nitrogen, and the Atlantic a net source. Because of this difference of the ratios in the major water masses the Atlantic must continually export nitrogen to the Pacific. The ratios are the same in the very deepest water of both oceans. This is presumably because the very deepest water in each case receives a considerable input directly from Antarctic Bottom Water.

If the Redfield ratio of 16N:1P indeed, on average, characterizes the synthesis of organic matter at the surface, then it would appear that organisms living in central water of the north Pacific are exposed to water mixing up from below that is richer in phosphorus and deficient in nitrogen, relative to the required ratio, while in the central

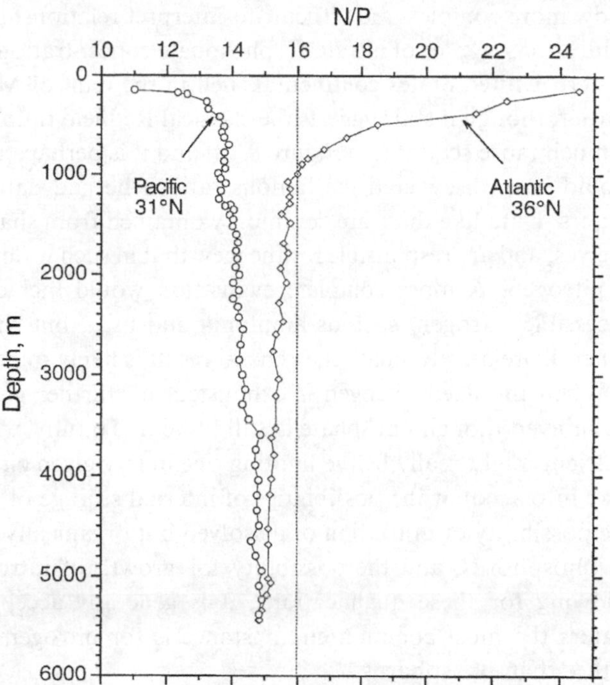

Figure 8.15 Ratios of nitrate to phosphate (molar units) in the water columns of the central gyres in the North Atlantic and North Pacific. Pacific data from GEOSECS Sta. 204 at 31° 22′ N, 150° 2′ W (Broecker *et al.* 1982); Atlantic data from TTO Sta. 241 at 36° 18′ N, 56° 27′ W (PACODF 1986). Data from the top 100 m were deleted because the concentrations there were too low to provide consistent estimates of the ratios, and the remaining data were corrected for the presence of arsenate by subtracting the arbitrary amount of 0.02 μmol kg^{-1}; this correction is the same as that used by Fanning (1992) in his evaluation of N/P ratios. The vertical dotted line identifies the classical Redfield ratio of ~16:1.

water of the north Atlantic the situation is reversed. It is still a matter of speculation whether this makes much difference to the species composition in the respective regions. If the North Atlantic exports fixed nitrogen to the Pacific, it should follow that there are more nitrogen-fixing organisms in the north Atlantic; on the other hand, the relative shortage of fixed nitrogen in the north Pacific would suggest that nitrogen-fixers should have some advantage there. Indeed, Karl *et al.* (1997), using data from a 7-year time series of measurements at a station 100 km north of Hawaii, have demonstrated active nitrogen fixation there. The rates they measured were large enough to supply the fixed nitrogen necessary for up to half the new production (see Chapter 11 for a definition of new production). It is possible that the atmospheric input of anthropogenically derived nitrogen, which is much more important in the North Atlantic than in the North Pacific, could have had some influence on the surface water of the Atlantic, but it seems unlikely that this could have had much impact in the deep water. A full exploration of these questions is yet to come.

In contrast to the central regions of the oceans, where the Redfield ratios are clearly evident, the near-shore coastal and estuarine regions that border the ocean basins show more complex and difficult-to-interpret relationships between the nutrient concentrations. A plot of nitrate vs. phosphate concentrations measured in water over the eastern United States continental shelf in the Gulf of Maine (Figure 8.16) has some scatter, though it still suggests the classical Redfield ratio of 16N:1P. Frequently there is much more scatter (see Figure 8.18) and it is perhaps unlikely that Alfred Redfield would have discovered the famous ratio if he had data from only shallow coastal waters. Data like these are commonly obtained from shallow waters over continental shelves, and are responsible for the view that in such waters the major limiting nutrient is nitrogen. A more complete evaluation would include other forms of fixed but accessible nitrogen, such as ammonia and urea, but since their concentrations are generally relatively small, the general result is likely to be nearly the same.

When the fixed nitrogen is exhausted, no further growth of phytoplankton will occur even though phosphate is still present. To fully evaluate the question of which nutrient might really be the limiting one in any given circumstance, it is necessary to take into account the possibilities of internal storage of nutrients by phytoplankton, the possibility of utilization of dissolved but organically combined forms of nitrogen or phosphorus, and the possibility for growth of nitrogen-fixing organisms. Even allowing for these qualifications, it is generally accepted that in shallow coastal waters the most common circumstance is for nitrogen to be the limiting nutrient, rather than phosphorus.

Further information may be obtained by examining the seasonal changes of nitrogen and phosphorus in coastal waters. The phosphorus concentrations in Narragansett Bay (Figure 8.4) and the inorganic fixed nitrogen concentrations (Figure 8.17) show a surprising scatter when plotted against each other (Figure 8.18). It appears that the processes that control the concentrations of these nutrients in this location must operate rather independently on each nutrient. The N/P ratio is always below the Redfield ratio (Figure 8.19), and exhibits a minimum in the summer. This occurs even though the fresh water entering the upper end of the bay carries a ratio of $N/P \approx 12$ and the seawater entering the other end has a ratio $N/P \approx 5$. It seems that the low values in the bay must be caused by an active denitrification process in the sediments of Narragansett Bay, first shown to be important there by Seitzinger *et al.* (1980, 1983). From the evidence in Figures 8.17 and 8.19 it may be concluded that the process of denitrification is most active in the summer during times of warm temperatures. This active denitrification is believed to be important in most and perhaps all near-shore sediments (Seitzinger and Giblin 1996, Codispoti 2007). It is likely that nutrient concentrations in near-shore shallow marine waters everywhere are strongly affected by active sediment–water exchange processes.

8.6 Initial nutrients

The relationships between oxygen and the nutrient elements nitrogen and phosphorus in the deep sea provide information on the nature of the water that descends into the deep. In evaluating these relationships it is necessary to know both the concentrations of

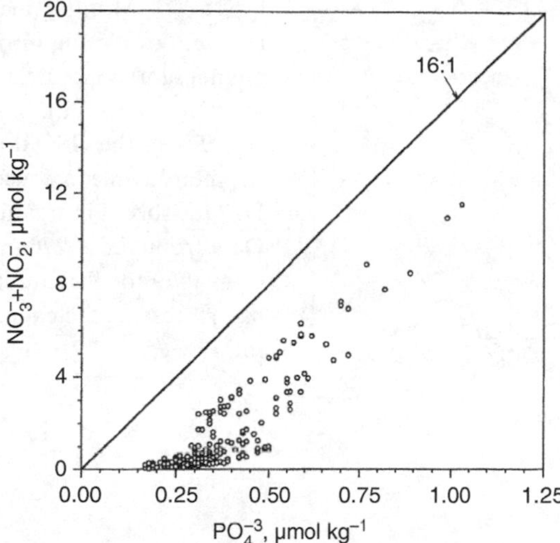

Figure 8.16 Relationship of nitrate (plus nitrite) to dissolved inorganic phosphorus several stations in the Gulf of Maine near 70° W,44° N in early June, 2003. From http://grampus.umeoce.maine.edu/nutrients/#Data.

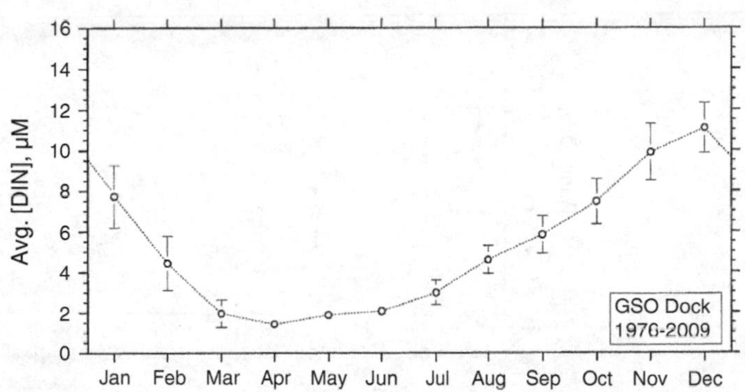

Figure 8.17 Concentration of dissolved inorganic fixed nitrogen (DIN, which is the sum of $NO_2^- + NO_3^- + NH_4^+$) in water from the GSO dock, Narragansett Bay, Rhode Island. Each point is the 33-year mean of the monthly means of approximately weekly observations. The ranges given are plus and minus two standard errors of the means of the monthly means. The presentation is similar to that in Pilson (1985) but more data are included. Compare Figure 8.4.

oxygen present in the water and also the quantity of oxygen used or produced in association with the consumption or production of the nutrients. In analyzing the situation in a sample of deep water, one measures the concentration of nitrate, phosphate, and oxygen. The N/P ratio will probably be about 15. The amount of oxygen that has been consumed by organisms metabolizing the organic matter that was in the water originally and the generally much greater amount that has rained down from the surface, and the nutrients released during this metabolism, are calculated as follows:

1. Assume that when this sample of deep water was last at the surface (probably in some high latitude region) the oxygen concentration was in equilibrium with the atmosphere.

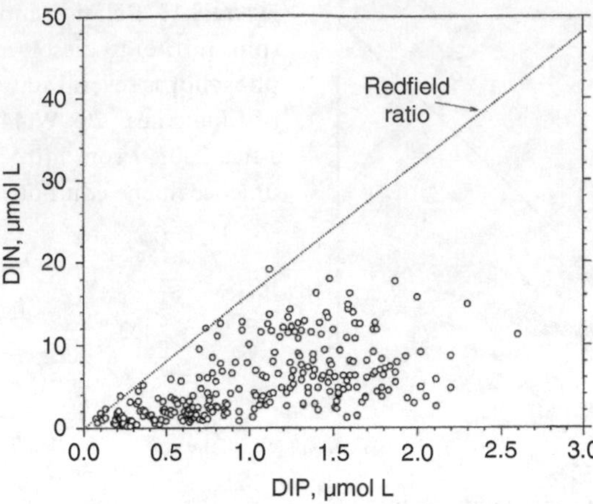

Figure 8.18 Monthly means of DIN (the sum of nitrate, nitrite, and ammonium) concentrations in water samples from southern Narragansett Bay, RI, at the GSO dock, plotted vs. the monthly mean concentrations of DIP (dissolved inorganic phosphate). Data from about 270 months of data from 1979 to 1999 are included. The line represents the canonical 16:1 ratio.

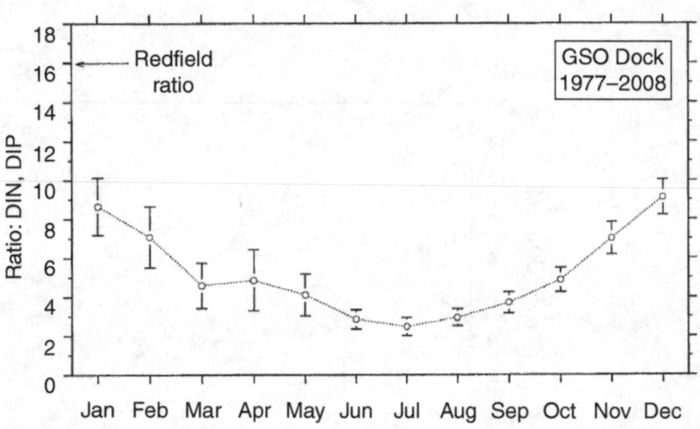

Figure 8.19 Ratios of the monthly mean concentrations of DIN to dissolved inorganic phosphate (DIP) in the same water samples as those used for Figures 8.4, 8.17, and 8.18. The 32-year means are plotted, and the error bars are twice the standard error of the overall mean for each month.

2. From the potential temperature and salinity determine the air-saturation concentration of oxygen in the water. (This value should probably be adjusted for bubble injection, but this has seldom been done; the uncertainties involved have commonly led investigators to ignore the correction.)
3. Subtract the observed concentration from the air-saturation concentration. This gives the apparent oxygen utilization (AOU).
4. Calculate, based on the Redfield ratio (Eq. (8.13)), the concentrations of nitrate and phosphate that must have been released.

If the seawater left the surface stripped of its nutrients, with negligible concentrations of N and P, then, based on Eq. (8.13), the ratio of AOU to phosphate should be about 166 O_2:1 P. In deep water, however, there is always more phosphate and nitrate than can be accounted for by the AOU, using the above ratio. Because of this,

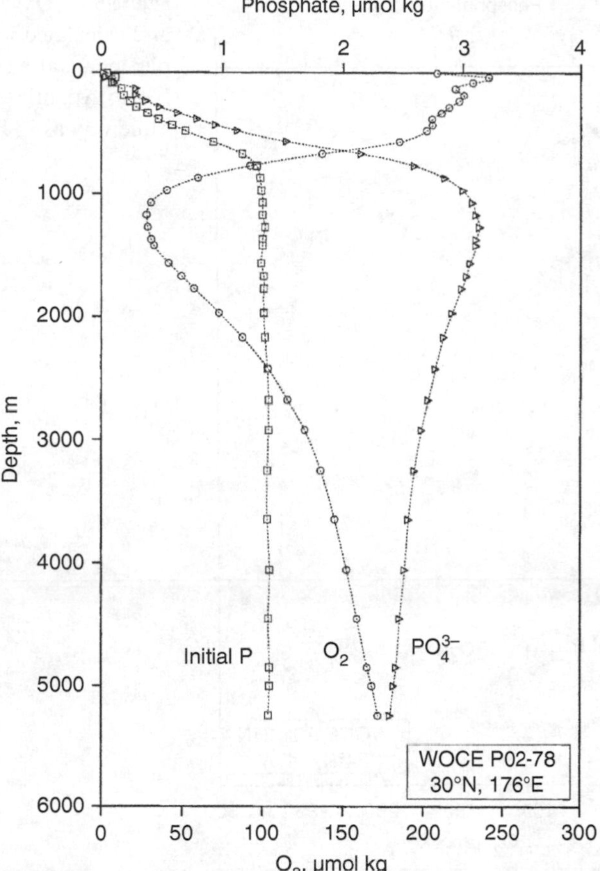

Figure 8.20 Oxygen, phosphate, and calculated initial phosphate at a station in the North Pacific. Initial phosphate is that phosphate not produced by the mineralization of organic matter and the associated consumption of oxygen, calculated in this case using 166 as the ratio of oxygen to phosphate. Note the remarkable uniformity in the concentration of initial phosphate in water below about 900 m.

it seems that much of the water sinking in the regions that generate the deep water of the world ocean is, in fact, not completely depleted in nutrients prior to equilibration with the atmosphere and sinking. The first observation of the existence of "initial" nutrients (in this case NO_3^-) was shown in Figure 8.12. In that figure all the deep-water samples (filled circles) have more NO_3^- than the Redfield ratio would predict. Redfield *et al.* (1963) gave this additional or extra nutrient the term *preformed nutrients*, but the term used here is *initial nutrients*, because these must be present in the water when it initially sinks. Such deep water is formed at high latitudes in the wintertime, when biological productivity is minimal, some unused nutrient is present, and there must often be some mixing between surface and deep water during this process.

The initial phosphate is estimated as follows:

$$\text{Initial P} = \text{Measured P} - (\text{AOU}/166). \tag{8.15}$$

Similarly, the initial nitrate is estimated as follows:

$$\text{Initial } NO_3^- = \text{Measured } NO_3^- - 16 \times (\text{AOU}/166). \tag{8.16}$$

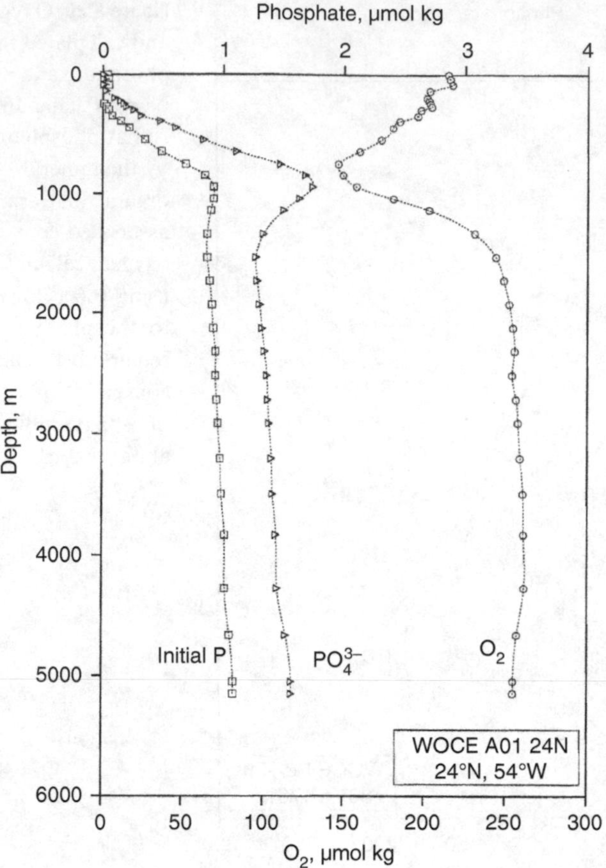

Figure 8.21 Oxygen, phosphate, and calculated initial phosphate at a station in the North Atlantic, calculated the same way as in Figure 8.20. ■

It is important to note that the magnitudes of the estimated initial nutrient concentrations depend directly on the magnitudes of the numerical coefficients in Eq. (8.13).

The initial nutrients display a remarkable uniformity of concentration in deep water. Phosphate is especially uniform (Figures 8.20, 8.21), with only a barely detectable increases in concentration from 1000 m to the bottom in either the Atlantic or Pacific. There is somewhat more initial phosphate in the Pacific than the Atlantic.

An important point about the initial nutrients is that they constitute the ultimate reservoir of nutrients dissolved in the ocean that have not been utilized to make organic matter. This must be the pool of nutrients that could possibly be in some way accessed to increase the organic production in the ocean to reduce the atmospheric concentration in glacial times, or in modern times to absorb and transport some of the excess anthropogenic carbon dioxide.

At this time I must say that I do not fully understand the distribution of the initial nutrients. I do not understand how it has come about that the initial phosphate in the Pacific below 1000 m should be so uniform. According to the simple scenario described above, this entire mass of deep water would have been formed by water sinking with nearly saturated oxygen concentrations and a phosphate concentration of about 1.3 μmol kg^{-1}, or by a mixture of waters that yield the observed relationships. It is not clear, at least to me, where or how this happens, or what combination of processes can yield the results shown in Figures 8.20 and 8.21. This needs further investigation.

Summary

Plants require many elements in order to grow. Those that are abundantly present in seawater are not generally considered in the category of nutrients. The three elements commonly called nutrients are phosphorus, nitrogen, and silicon. In addition, a number of micro-nutrient elements, such as iron, copper, cobalt, and zinc, are present, and needed, in very small amounts.

Nearly all the phosphorus in seawater is present as phosphate in its various ionized forms. Phosphorus is nearly always present in concentrations less than about 3 μmol kg^{-1}, and its salts are quite insoluble, so it is generally believed to be the substance that, over long timescales, sets an upper limit to the amount of plant life that can grow in the sea. The ionized phosphate species are mostly ion-paired with various cations. Small amounts of phosphorus are found in combination with organic compounds, and this becomes significant in surface water where the inorganic forms are present in very low concentration. The low concentrations of phosphorus in surface water are due to the downward transport of organic debris and release at depth. The structure of large-scale ocean circulation leads to dramatically different concentrations in deep water of the different major ocean basins.

Gaseous nitrogen is unavailable as a nutrient to all organisms except those that can fix nitrogen. Most fixed nitrogen in the sea is in the form of nitrate, other common forms are ammonia, and nitrite; these can be used by various plants and bacteria. The exchanges between the several forms of nitrogen are of considerable consequence. The continuing loss of fixed nitrogen through denitrification by several processes that yield N_2, as well as lesser amounts of N_2O, requires that nitrogen fixers continually make up the difference, though these activities take place in different places. Denitrifying organisms are also important on a long geochemical timescale, as without them the world's oxygen might all be consumed by reaction with nitrogen,

Silicon, in the form of silicate, is required by diatoms and some other plants and animals in order to form their shells or other support structures, and it is likely that it is important as a trace nutrient for many other organisms. Careful budgets have shown that while vast amounts of amorphous silica, mostly as diatom tests, accumulate in the sediments, most of the silica formed into biological structures dissolves in the water column and is eventually recycled.

Iron is the most studied micro-nutrient element because it is essential for all organisms but has a very low solubility at the normal pH of seawater. Many areas of the surface ocean have some amount of the nutrient elements present but reduced growth of plants because there is not enough iron. Experiments have shown that addition of iron to the water results in blooms of phytoplankton, using up the unused nutrients.

Phosphorus and nitrogen are present in deep water in ratios normally within the range of 1P:14N to1P:16N. Average values in plankton are generally about 1P:16N, and this ratio is the original Redfield ratio. It is living organisms that are responsible for maintaining this approximate ratio in the deep ocean, as well as causing deviations from it in many special circumstances. While there is some variability, the ratio holds so widely that it is used as the basis for many biogeochemical calculations, and observed departures from the ratio may be used as a starting point for examining the processes involved. The ratios of nutrient changes to changes in the concentration of oxygen and carbon during respiration are somewhat more variable, but provide valuable information about many biogeochemical processes. A suitable set of ratios to use in the circumstance that the nitrogen used during photosynthesis is in the form of nitrate, and the end product of respiration is also nitrate, is the following 1P:16N:124C:166O_2. and by extension this is also called the Redfield ratio, as modified, and specifies approximately the average composition of organic matter that is oxidized in water below the mixed layer.

Departures from the modified Redfield ratio help to elucidate the nature of various processes. For example, the repeated finding of N:P ratios far below 16:1 in near-shore water highlights the importance of denitrification in shallow-water sediments. Similarly, the finding that in deep water the concentrations of phosphate and nitrate are considerably greater than would be expected from the respective Redfield ratios to AOU, highlights the situation that somewhere there is a significant sinking into the deep sea of water that carries a considerable concentration of both oxygen and nutrients.

SUGGESTIONS FOR FURTHER READING

Benitez-Nelson, C. R. 2000. The biogeochemical cycling of phosphorus in marine systems. *Earth-Science Rev*. **51**: 109–135.

This is a thorough and well-expressed review of the subject. Highly recommended.

Paytan, A. and K. McLaughlin. 2007. The oceanic phosphorus cycle. *Chem. Rev*. **107**: 563–576.

This most recent review of the field brings the subject almost up to date, and provides an excellent entrée to the literature.

Brandes, J. A., A. H. Devol, and C. Deutsch. 2007. New developments in the marine nitrogen cycle. *Chem. Rev*. **107**: 577–589.

The marine nitrogen cycle is a most active field of investigation, and the subject is well covered in this review, up to the time of publication.

Postgate, J. 1998. *Nitrogen Fixation*, 3rd edn, Cambridge, Cambridge University Press.

Thorough review of the processes involved in nitrogen fixation.

Westheimer, F. H. 1987. Why nature chose phosphates. *Science* **235**: 1173–1178.

A nice discussion of the importance of phosphates in the chemistry of life.

Redfield, A. C. 1958. The biological control of chemical factors in the environment. *Am. Sci.* **46**: 205–221.

In this classic paper, Redfield elaborates on his view that organisms control the N/P ratio in the ocean.

Open University. 2005. *Marine Biogeochemical Cycles*. Elsevier Butterworth Heinemann, Oxford and Boston, MA.

This book has a wealth of colorful and detailed illustrations. It lacks references for the sources of the material.

9 Trace metals and other minor elements

Knowledge of the chemical speciation of trace elements in seawater is, from a practical standpoint, at an undeveloped stage.

BURTON 1975

In the earlier discussion of the major elements in seawater, the arbitrary division between major and minor elements was set at a mass concentration of one milligram per kilogram, equal to one part per million. This is a convenient separation, because only salts present in concentrations somewhat greater than 1 mg kg^{-1} can have a detectable effect on the density and related physical properties of seawater. Most of the elements present in smaller concentrations are not conservative in seawater; by examining the distributions of these elements we may learn much about the biological and geochemical processes in the ocean. Furthermore, the analytical difficulties of measuring or even detecting many of the substances present in small concentrations in seawater often pose special problems that must be addressed during any consideration of these elements.

9.1 Analytical considerations

In approaching the analysis of any substance present in seawater at a concentration of, for example, around 1 μg kg^{-1}, a value common for some important elements, it should first be realized that the substance is present in a matrix consisting of 35 000 000 μg of a complex mixture of salts and other substances. This fact alone can impose severe difficulties. Furthermore, when dealing with concentrations at this level or below, it is often not appreciated that equivalent or even much greater quantities of the substance to be analyzed may be introduced during collection at sea, as contaminants in the reagent-grade chemicals required for the analysis, in dust from the air, and on the walls or in the very composition of ordinary laboratory glassware and other equipment.

Among the metallic elements that attracted the early interest of seawater analysts were iron, cobalt, copper, cadmium, and gold. Numerous papers were published on the concentrations of these and other similar elements in seawater between 1900 and

Table 9.1 Analysis of trace metals in seawater: the state of the art in 1970

Metal	Concentrations, $\mu g\ kg^{-1}$			Current estimates of the "true" values
Cadmium				
Lab A[(a)]	0.03			
Lab B	2.1	0.5	0.7	0.0001–0.12
Cesium				
Lab A	0.29	0.29	0.28	
Lab B	0.332	0.305	0.310	
Lab C	0.18	0.2	0.17	0.29
Cobalt				
Lab A	0.014	0.014	0.013	
Lab B	0.037	0.036	0.037	
Lab C	0.510	0.448	0.427	
Lab D	0.037	0.040	0.038	0.0006–0.006
Copper				
Lab A	1.35	1.55	1.42	
Lab B	7.5	6.3	6.3	
Lab C	27.4	2.21	2.81	
Lab D	0.9	0.8	1.0	
Lab E	15.0	15.0	14.4	0.06–0.4

These data are extracted from the results of a trace-element intercalibration study, reported by Brewer and Spencer (1970). Results are from three identical samples.

(a) The laboratory coding is different for each element. As an interesting observation on human nature, many analysts would not participate unless the identifications of results with analysts were kept confidential.

1975. That their results differed greatly from place to place and appeared to show various and random patterns did not deter several generations of analysts from continuing to measure and publish.

During preparations in 1969 for the upcoming GEOSECS Expeditions, it was thought desirable to measure a suite of trace metals in the samples to be collected. As the published results were disparate, an intercalibration exercise was planned. At stations in the Pacific Ocean and Caribbean Sea, large water samples were carefully collected in plastic containers, acidified to prevent loss to the walls of the containers, divided into portions, and then sent to several investigators for analysis. Surely, every investigator who agreed to participate must have thought that he or she was competent to do the job and would produce first-class results; furthermore, knowing it was an intercalibration exercise, each analyst would no doubt have been especially careful. An extract of the results (Table 9.1) shows, however, that in those days it was not possible to achieve consensus on the true values for the concentrations of any of the several elements listed. In most cases each investigator reported good consistency within his or her own data from analyses of the different bottles, but the variation in the reported values from investigator to investigator was as much as 20- to 30-fold: a dismal record.

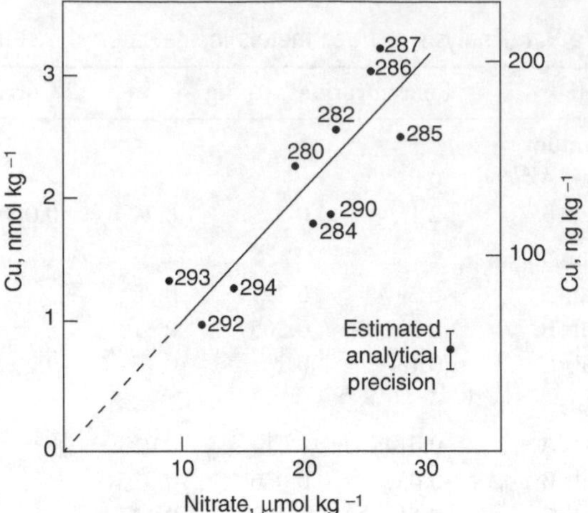

Figure 9.1 The first really believable data on the concentration of copper in seawater. Boyle and Edmond (1975) took great precautions to avoid contamination of water samples collected from surface waters during a transect across an oceanographic front south of New Zealand, and they used a very sensitive and specific technique for analysis. The finding that the concentrations of Cu appeared linearly related to the concentrations of nitrate showed for the first time that it would be possible to make oceanographic sense out of the copper data, and provided some confidence that the values were at least approximately correct.

The results of that intercalibration exercise were an embarrassment to the community and, while some chemists no doubt continued to believe that *their* own numbers were right, most observers concluded that consistent and true data were not going to be easy to get. There are dozens of papers in the literature from the first seven and a half decades of the twentieth century where some value is reported for the concentration of a transition metal dissolved in seawater. The majority of these reported values bear little or no relationship to the true concentrations.

Most investigators concluded that data on the concentrations of trace substances in seawater should all be taken with a grain of salt (so to speak) unless there is very good reason to believe them; they developed the view that perhaps *no* trace-element data could be considered entirely trustworthy until the methods and data had been duplicated in more than one laboratory. As a very powerful test, the data must make oceanographic sense. If the data are widely scattered from station to station, especially in the same water mass, then often they might as well be thrown out. On the other hand, confidence in their validity would be greatly strengthened if the data plot in some smooth way against other significant oceanic variables.

The first real breakthrough in attempts to learn the true concentrations of these dissolved metals in seawater came in 1975 with the publication of a paper by Ed Boyle and John Edmond showing that their data from measurements of copper in surface waters south of New Zealand made sense when plotted against another oceanographic variable, in this case nitrate (Figure 9.1). The evidence of a relationship between the concentrations of copper and a nutrient suggests that this element is biologically transported, and is also consistent with its status as an essential constituent element in living organisms.

Heartened by these results, and encouraged by Patterson (1974) in the use of ultra-clean techniques for both collection and processing, many investigators, in a remarkably short time, succeeded in collecting large amounts of excellent data on the concentrations of numerous trace elements in different parts of the world ocean.

Lest the reader think that the existence of really excellent data, such as are reported below for most of the trace elements, means that the analytical problems have been solved, an additional comment should be made. A great deal of time and material effort is required to achieve reliable results for some of the important elements (such as cadmium, cobalt, copper, iron, lead, and nickel) that are present at very low concentrations in the central water masses of the oceans. The real concentrations are so vanishingly small that the quantities present are easily swamped by contamination. An undertaking to make such measurements involves much planning and attention to detail, and is not to be entered into lightly.

9.2 Various patterns of distribution

Since the mid 1970s, several investigators have shown that not only are most trace elements present in smaller concentrations than previously believed, but many have complex non-conservative distribution patterns within the sea. These patterns have often led to new insight into oceanographic processes, and a great deal of interesting biogeochemical information has been extracted.

Table 9.2 provides a summary of much of the available concentration data for the minor elements. In this table there is a brief notation as to the type of distribution that appears to characterize each element. Only a few of these minor elements exhibit a conservative distribution (that is, they are found in a constant or nearly constant ratio to the salinity); these elements are not taken up by the biota or are taken up in such small amounts as to have a negligible effect on their distributions. For example, lithium, rubidium, and cesium, which along with sodium and potassium are members of group 1 in the periodic table, behave like the latter two more common elements, and appear to be conservative. Surprisingly, molybdenum seems conservative. It occurs in seawater as the molybdate ion, which is similar in shape to the sulfate ion, so this may provide a partial explanation for the evidence that it is not strongly transported by organisms or other sinking particles.

Uranium is radioactive and accordingly was detected early and can be measured accurately. Because it forms a stable complex with carbonate ions in seawater, uranium is relatively soluble and in the oceans it is the most abundant of the heaviest 35 elements. Somewhat surprisingly, uranium appears conservative, being found in a constant ratio to the salinity.

Those elements labeled as having a nutrient-like distribution are likely to be incorporated into body materials, transported downwards when organisms or their debris sink, and be released into the water when the organic matter is metabolically destroyed. Particle-reactive elements are those that appear to stick to the surfaces of

Table 9.2 Concentrations of the minor constituents in seawater

Element			Concentration (units per kg)			
Number	Symbol	Name	Approx.	mean	Range	Type of distribution
3	Li	Lithium	25	μmol	—	Conservative
4	Be	Beryllium	20	pmol	4–30	Nutrient; scavenged
7	N	Nitrogen (NO_3)	30	μmol	<0.1–45	Nutrient
13	Al	Aluminum	10	nmol	0.1–40	Mid-depth minima
14	Si	Silicon	100	μmol	<1–200	Nutrient
21	Sc	Scandium	15	pmol	8–20	Surface depletion
22	Ti	Titanium	200	pmol	4–300	Surface depletion
23	V	Vanadium	30	nmol	20–35	Slight surface depletion
24	Cr	Chromium	4	nmol	2–5	Nutrient
25	Mn	Manganese	0.5	nmol	0.2–3	Depletion at depth
26	Fe	Iron	1	nmol	0.1–2.5	Depletion at surface. and depth
27	Co	Cobalt	40	pmol	10–100	Depletion at surface, and depth
28	Ni	Nickel	8	nmol	2–12	Nutrient
29	Cu	Copper	4	nmol	0.5–6	Nutrient; scavenged
30	Zn	Zinc	6	nmol	0.05–9	Nutrient
31	Ga	Gallium	20	pmol	2–50	Complex, scavenged
32	Ge	Germanium	70	pmol	<7–115	Nutrient
33	As	Arsenic	23	nmol	15–25	Nutrient
34	Se	Selenium	1.7	nmol	0.5–2.3	Nutrient
37	Rb	Rubidium	1.4	μmol	—	Conservative
39	Y	Yttrium	250	pmol	80–300	?
40	Zr	Zirconium	200	pmol	12–300	Nutrient; scavenged
41	Nb	Niobium	3.9	pmol	2.8–3.9	~Conservative, surface depletion
42	Mo	Molybdenum	100	nmol	92–105	~Conservative
44	Ru	Ruthenium	(20	fmol)	?	?
45	Rh	Rhodium	0.8	pmol	0.3–1.0	Scavenged, nutrient ?
46	Pd	Palladium	0.6	pmol	0.2–0.6	Surface depletion
47	Ag	Silver	25	pmol	0.5–45	Nutrient, complex
48	Cd	Cadmium	0.7	nmol	0.001–1.1	Nutrient
49	In	Indium	(0.1	pmol)	0.05–0.15	Scavenged ?
50	Sn	Tin	(4	pmol)	(1–12)	Depletion at depth ?
51	Sb	Antimony	(1.2	nmol)	?	?
52	Te	Tellurium	0.6	pmol	0.4–1.7	Scavenged
53	I	Iodine	0.4	μmol	0.2–0.5	Nutrient
55	Cs	Cesium	2.2	nmol	—	Conservative
56	Ba	Barium	100	nmol	32–150	Nutrient
57	La	Lanthanum	30	pmol	8–57	Surface depletion
58	Ce	Cerium	20	pmol	16–26	Surface depletion
59	Pr	Praeseodymium	5	pmol	1–8	Surface depletion
60	Nd	Neodymium	25	pmol	5–40	Surface depletion
62	Sm	Samarium	4	pmol	1–6	Surface depletion

Table 9.2 (*cont.*)

Element			Concentration (units per kg)			
Number	Symbol	Name	Approx.	mean	Range	Type of distribution
63	Eu	Europium	1	pmol	0.3–1.7	Surface depletion
64	Gd	Gadolinium	6	pmol	2–9	Surface depletion
65	Tb	Terbium	1	pmol	0.2–1.5	Surface depletion
66	Dy	Dysprosium	8	pmol	2–12	Surface depletion
67	Ho	Holmium	2.5	pmol	0.5–3	Surface depletion
68	Er	Erbium	8	pmol	2–10	Surface depletion
69	Tm	Thulium	1	pmol	0.3–1.5	Surface depletion
70	Yb	Ytterbium	7	pmol	1.5–11	Surface depletion
71	Lu	Lutetium	1	pmol	0.2–1.8	Surface depletion
72	Hf	Hafnium	0.3	pmol	0.12–0.8	Surface depletion
73	Ta	Tantalum	0.12	pmol	.09–0.29	Surface depletion
74	W	Tungsten	45	pmol	42–67	Conservative
75	Re	Rhenium	40	pmol	—	Conservative
76	Os	Osmium	46	fmol	± 3	~Conservative ?
77	Ir	Iridium	0.7	fmol	0.5–0.9	???
78	Pt	Platinum	1	pmol	0.54–1.64	Surface depletion
79	Au	Gold	50	fmol	20–200	variable
80	Hg	Mercury	2	pmol	0.5–12	Complex, scavenged
81	Tl	Thallium	60	pmol	—	Conservative
82	Pb	Lead	10	pmol	5–175	High in surface water
83	Bi	Bismuth	150	fmol	24–500	Depletion at depth
92	U	Uranium	13.6	nmol	± ~ 1–2%	Conservative

Name	Abbrev.		Fraction	Number of chemical units	
mole	mol	=	1 mol	6.022×10^{23}	atoms or molecules
millimole	mmol	=	10^{-3} mol	6.022×10^{20}	atoms or molecules
micromole	μmol	=	10^{-6} mol	6.022×10^{17}	atoms or molecules
nannomole	nmol	=	10^{-9} mol	6.022×10^{14}	atoms or molecules
picomole	pmol	=	10^{-12} mol	6.022×10^{11}	atoms or molecules
femtomole	fmol	=	10^{-15} mol	6.022×10^{8}	atoms or molecules

This list omits the gases and some radioactive elements. Modified from Bruland (1983) with data for Al (Orians and Bruland 1985, 1986); Ag (Pacific only, Martin *et al.* 1983); Au (Falkner and Edmond 1990); Bi (Lee *et al.* 1986); Co (Saito and Moffett 2002); Ga (Orians and Bruland 1988a, b); Hg (Gill and Fitzgerald 1988, Mason and Fitzgerald 1993); In (Pacific only, Amakawa *et al.* 1996); Ir (Anbar *et al.* 1996); Mo and W (Sohrin *et al.* 1987); Nb, Ta, Hf, W (Pacific only, Sohrin *et al.* 1998); Os (Chen *et al.* 2009); Pd, Pt, Ir and Ru (Goldberg 1987); Re (Anbar *et al.* 1992); Rh (Pacific only, Bertine *et al.* 1993); Te (Lee and Edmond 1985); Ti (Orians *et al.* 1990); Y, La, Pr, Nd, Sm, Eu, Gd, Tb, Dy, Ho, Er, Tm, Yb, Lu (Pacific only, Zhang *et al.* 1994); and Zr (Pacific only, McKelvey and Orians 1993).

Where appropriate, concentrations are normalized to salinity 35‰. Values in brackets are not well established. Additional information and graphical presentations of many distributions can be found in Nozaki 1997.

sinking particles and thus to be swept out of the water column at usually all depths and carried to the bottom. The label 'surface depletion' is self-explanatory, but the reason for this distribution may not always be apparent due to inadequate information. There are many variations on these themes. Every element has its own charm and may be of some geochemical or biological interest; here we discuss a few notable or characteristic examples.

9.2.1 Cadmium

The distribution of cadmium in the sea follows a pattern startlingly close to that of phosphate (Figure 9.2). While it might have been expected that elements like cadmium could be, to some extent, scavenged from surface waters by sinking particles, the closeness of this relationship was utterly unexpected. Furthermore, the difference between the major ocean basins (Pacific and Atlantic) previously observed for nutrients is apparent for cadmium as well. This extraordinary correspondence between these two elements caused considerable discussion. In the case of the nutrient distributions our understanding of the relation between nitrogen and phosphorus comes about because (after Redfield) we can readily imagine that average marine organisms are generally similar in composition, with respect to the essential components of their biochemical structure, and the two nutrients are transported and released in similar ratios. It was generally believed, however, that cadmium had no biochemical function and, indeed, it is rather toxic. The evidence in Figure 9.2 and similar data led to a search for a biological function. In 1995, Lee *et al.* showed that cadmium seems to be a useful nutrient for at least some phytoplankton. It appeared to act as part of the active center in some forms of the enzyme carbonic anhydrase, just as zinc does, and the cells actively control the internal concentration of cadmium. Subsequently Lane *et al.* (2005) discovered and purified a cadmium-containing carbonic anhydrase enzyme from a marine diatom, and Xu *et al.* (2007) showed that cadmium substitutes, at least partially, for zinc in zinc-limited cultures of a coccolithophore. Thus, cadmium is a micro-nutrient, although, so far, it has not been proven to be essential. The observed distribution of cadmium compels us to believe that it must be taken into the tissues of marine organisms in near proportionality to the total mass of tissue and regenerated at about the same rate as phosphate (and nitrate) during decomposition. Why it should appear to follow phosphate so closely is, however, still somewhat of a mystery.

A practical benefit from the observation shown in Figure 9.2 is that in some circumstances cadmium may be used to estimate the concentration of phosphate. A little cadmium is incorporated into the calcareous tests of foraminifera, in proportion to the concentration in the water. The concentration of cadmium in such tests from various parts of the sedimentary record thus provides an estimate of the concentration of phosphate in the water in which they grew. In turn, this provides some knowledge of the patterns of oceanic circulation at times in the past (Boyle 1988).

Further evaluation of larger and geographically more widely distributed data sets (de Baar *et al.* 1994) has, however, demonstrated that the relationship shown in Figure 9.2 is not simple. As can be seen already in this figure, the slope of the line

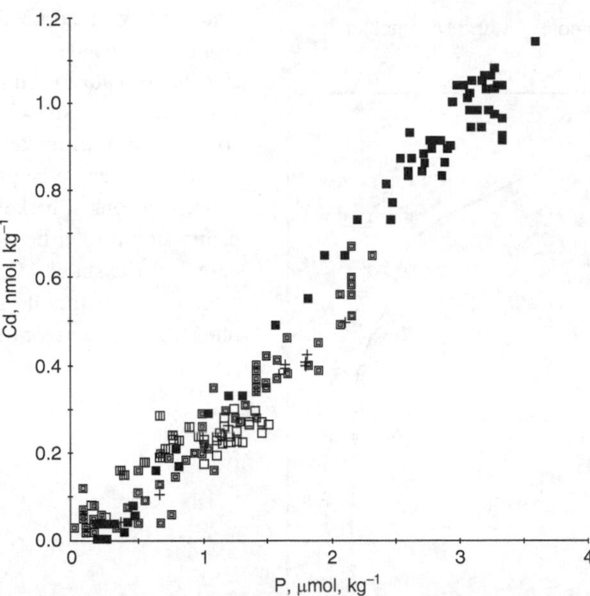

Figure 9.2 Relationship of the concentration of cadmium to that of phosphate in open ocean samples taken below the surface mixed layer. The data come from a variety of regions: solid squares are from the northeast Pacific, partially filled squares from the North Atlantic and the Arctic, and crosses are from the Gulf of Mexico. From Boyle (1988).

(the Cd/P ratio) changes with the phosphate concentration and thus with the depth and age of the water. The ratio Cd/P may be less than 0.5×10^{-4} in some surface waters and as much as 3.5×10^{-4} in deep North Pacific water. The present ratio in different deep-water masses may vary, so the extraction of precise paleoceanographic information from the sediment record is not straightforward.

9.2.2 Nickel

Nickel is another element that shows generally nutrient-type profiles (Figure 9.3). In this case the correspondence to phosphate is very poor and to silicate a little closer, but evidently biological uptake and the release of nickel from sinking particles follows somewhat different dynamics than either nutrient. Nickel is an essential trace element for many organisms, as it is a component of the active site of hydrogenases and some other enzymes.

9.2.3 Selenium

An interesting and roughly nutrient-type distribution (Figure 9.4) has been reported for selenium, but in this case two different oxidation states are present, Se(IV) and Se (VI), as well as a small amount of some form of selenium that can only be measured after exposure to a strong oxidizing agent (so it is presumably combined in some way with organic matter). Organic selenium is most abundant near the surface and was not detected in the deepest samples. Both inorganic forms are always present, even though only Se(VI) is thermodynamically stable in the presence of oxygen. The situation is not well understood, but Cutter and Bruland (1984) suggested that when organic matter is metabolized the form released is Se(IV), which then is kinetically

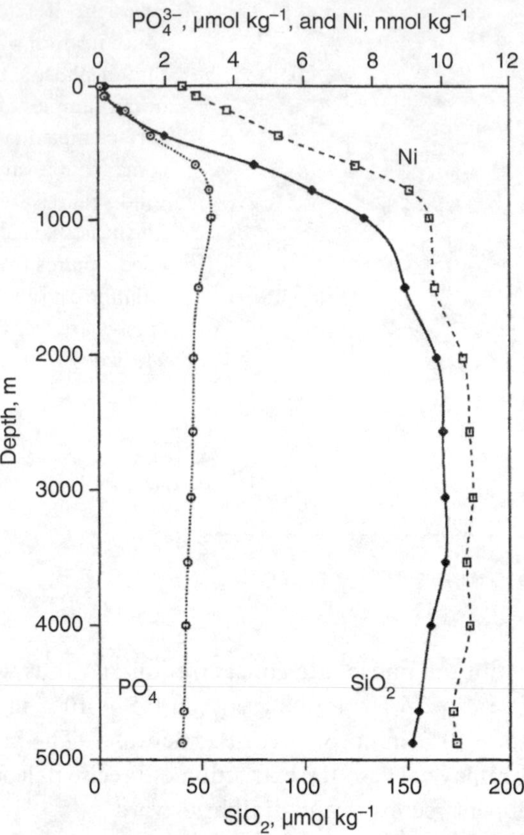

Figure 9.3 Vertical profile of nickel, along with profiles of phosphate and silicate, at a station in the North Pacific (32° 41′ N, 145° 0′ W). The concentration of nickel shows a general nutrient-type profile. The concentrations of nickel are not exactly proportional to either nutrient plotted here, but the shape of the profile resembles that for silicate more than phosphate. Data from Bruland (1980).

hindered from further oxidizing, having perhaps a mean life of 1100 years. Additional facts to be taken into account when trying to comprehend the chemistry of selenium in seawater are that it is an essential nutrient for the growth of all phytoplankton that have been examined, that only the Se(IV) form usually seems to be taken up, and that in one experiment where this has been examined the phytoplankton excreted dissolved organic selenium compounds (Hu *et al.* 1996). Another aspect of the distribution of selenium that is somewhat mysterious is that, although presumably it is associated with biochemical compounds and reactions in the soft tissues of organisms, its distribution in the sea appears to correlate much better with dissolved silica than it does with phosphate.

9.2.4 Manganese

Sometimes the distributional patterns of an element may reflect an important local process. The vertical distribution of manganese in the central North Pacific (Figure 9.5a) apparently involves more processes than are required to explain those trace elements previously considered, but all concentrations are relatively low, less than 0.1 to 1.0 nmol kg^{-1}. However, near the Galapagos hydrothermal vent area concentrations in the water column are over 15 nmol kg^{-1} (Figure 9.5b).

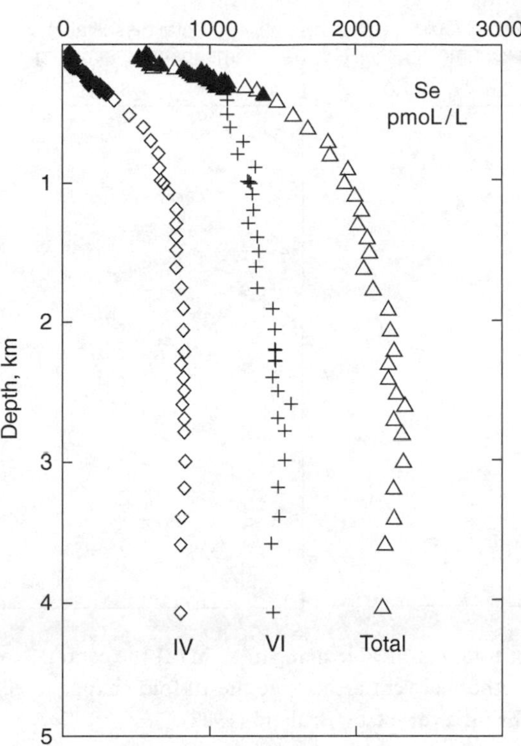

Figure 9.4 Vertical profiles of selenium in two oxidation states Se(IV), Se(VI), and total selenium at a station in the eastern North Pacific Ocean (28° N, 122° W). From Measures *et al.* (1983).

At depths around 2500 m, comparable to those of the major vents, the water for some distance around is strongly labeled by manganese, which is dissolved from the hot basalt rocks and carried up with the venting fluids. Note that in this figure the substance analyzed is the *total dissolvable manganese*, which means that it could, and probably does, include both dissolved and particulate forms brought into solution by the addition of acid used in the analysis. In its oxidized form (the +4 oxidation state), manganese is quite insoluble and forms particles or attaches to other particles. The high concentrations seen in Figure 9.5b can only be maintained by continuous injection into the water. The particles must be continually sinking out. A startlingly different distribution will be seen when we consider the Black Sea, in Chapter 12.

9.2.5 Lead

Profiles of lead in the North Atlantic and North Pacific (Figure 9.6) show a distribution unlike that of any other element. The concentrations are highest in surface water, and decrease dramatically with depth. Furthermore, the concentrations are highest in the Atlantic. These distributions are evidence that most of the lead must have entered the oceans from the atmosphere. This and other evidence shows that the primary source of the atmospherically transported lead is from a variety of releases by humans, mostly the exhausts from cars using leaded gasoline. The higher concentrations in the Atlantic were due to the high rate of burning of leaded gasoline in

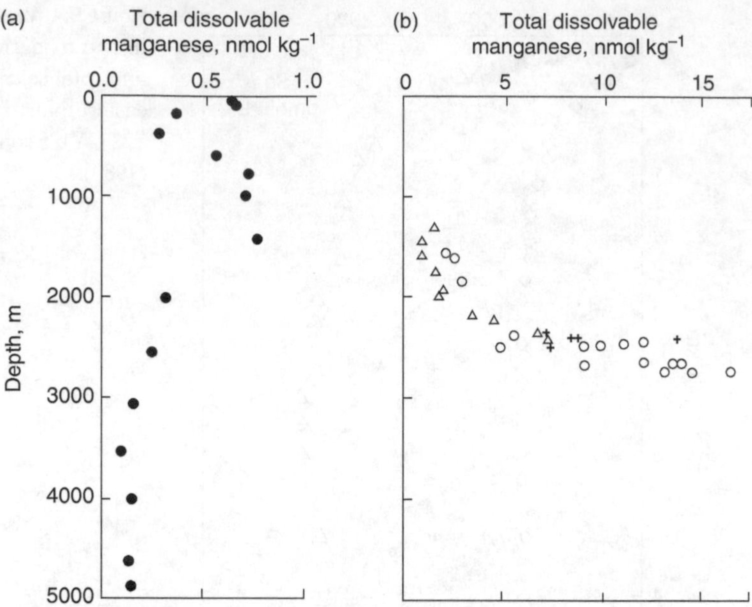

Figure 9.5 Vertical profiles of total dissolvable manganese in (a) the central North Pacific and (b) near the Galapagos hydrothermal vent area. Note the 10-fold change of concentration scales between the figures. (From a review by Bruland (1983).)

countries bordering the North Atlantic, compared to the Pacific, as well as to the greater volume of the Pacific and the average direction of the winds. The decrease from year to year near the surface in both profiles is due to the phasing out of leaded gasoline, and the surface concentrations have continued to decrease (Figure 9.7).

9.2.6 Aluminum

Vertical profiles of aluminum (Figure 9.8) also show consistent surface enrichment in the Atlantic and Pacific (but not the Mediterranean), followed by a mid-depth minimum and an increase with depth. In this case, it is believed that the distributions are natural, not influenced by humans, but the processes involved are not understood. The surface enrichment appears to be due to the dissolution of aluminum from clays in atmospherically transported dust, with scavenging at mid-depths and release at greater depths, but the causes of the latter processes are obscure. The decrease in concentration from the Atlantic to the Pacific is ascribed to continued scavenging in the older waters of the Pacific, combined with a lower input of dust (Orians and Bruland 1985). The Mediterranean is mysterious.

9.2.7 Germanium

The oceanic chemistry of germanium is both remarkable and mysterious. Being next below silicon in group 14 of the periodic table, its chemistry is in some ways very

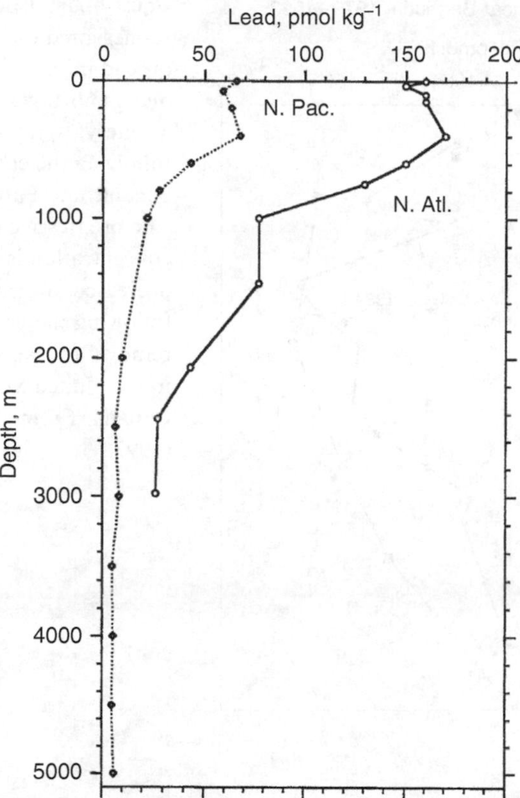

Figure 9.6 Vertical profiles of dissolved lead in the North Pacific (1800 km northeast of Hawaii) and in the North Atlantic (250 km northwest of Bermuda) measured in 1977 and 1979 respectively (Data from Schaule and Patterson 1981, 1983). The Pacific data represent perhaps the first accurate measurements of lead in the open ocean; they were made using the ultra-clean techniques pioneered by Clair Patterson. The size of the anthropogenic signal delivered from the atmosphere to surface waters and the downward transport of lead from the surface due to sinking particles are both evident. ■

similar to that of silicon, but it also shows striking differences. Although at least some organisms do discriminate against it (Froelich *et al.* 1989), germanium is generally taken up at the ratio Ge/Si $\approx 0.7 \times 10^{-6}$ into the silica tests of most organisms that make them. Correspondingly, it is released when these tests dissolve and it appears in the water column to show a profile almost exactly like that of silica (Figure 9.9), again in the ratio of 0.7×10^{-6}.

Unlike silicon, however, germanium is also found in organic combination in the sea. The compounds so far identified are monomethylgermanium and dimethylgermanium. Each of these is present in the ocean in greater concentrations than that of the inorganic form, and each appears to be conservative in the open ocean and in estuaries. Among all the elements, this situation is unique to germamium. While deep-water concentrations of the inorganic germanium are about 100 to 110 pM, the concentration of dimethylgermanium is about 120 pM and that of monomethylgermanium about 330 pM. Everywhere, the open ocean has about 450 pM of these organic germanium compounds. River water carries only very low and barely detectable concentrations of these compounds, and in estuaries the concentrations show conservative mixing just like the salt from seawater (Figure 9.10). Nevertheless, the sources of these compounds appear to be on the continents. Lewis *et al.* (1989) observed no methylation of inorganic germanium in experimental cultures of algae and fungi, but they discovered that some inorganic germanium is methylated in a

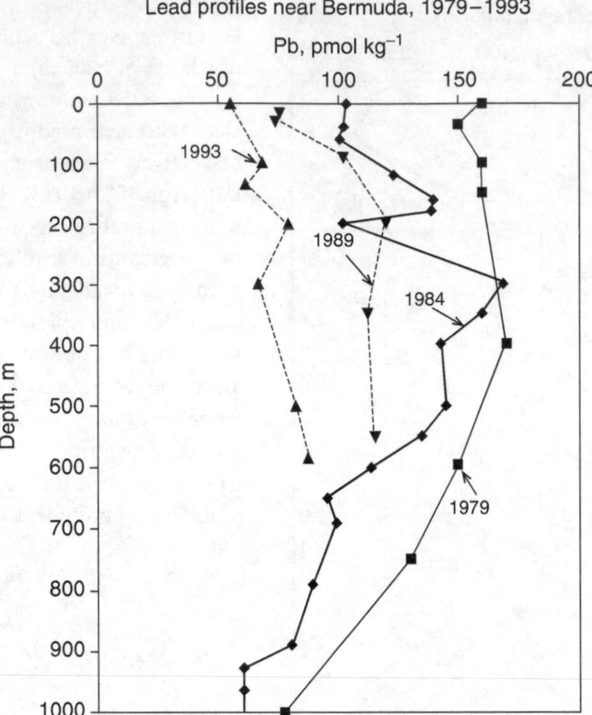

Figure 9.7. Ed Boyle at MIT re-measured the concentrations of lead in the upper 1000 meters northwest of Bermuda at intervals in the years following the early work of Schaule and Patterson in 1979. The progressive decrease in concentration is ascribed to the decreased anthropogenic input following the near total removal of lead from gasoline in the United States and Canada. From Wu and Boyle (1997).

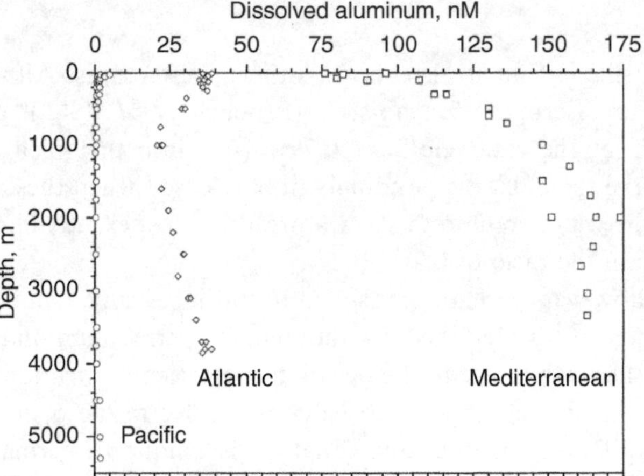

Figure 9.8 Vertical profiles of dissolved aluminum in the North Pacific (28° N, 155° W), North Atlantic (41° N, 64° W), and Mediterranean (34° N, 26° W). Modified from Donat and Bruland (1995).

sewage treatment plant, and they found methyl germanium compounds in water draining from anaerobic swamps where there is active production of methane. This suggested that bacterial formation of methylated germanium compounds may occur in terrestrial anaerobic environments and may account for the trace amounts observed in rivers. If the source of these compounds is indeed terrestrial, their residence times in the sea must be millions of years, because the river input is so

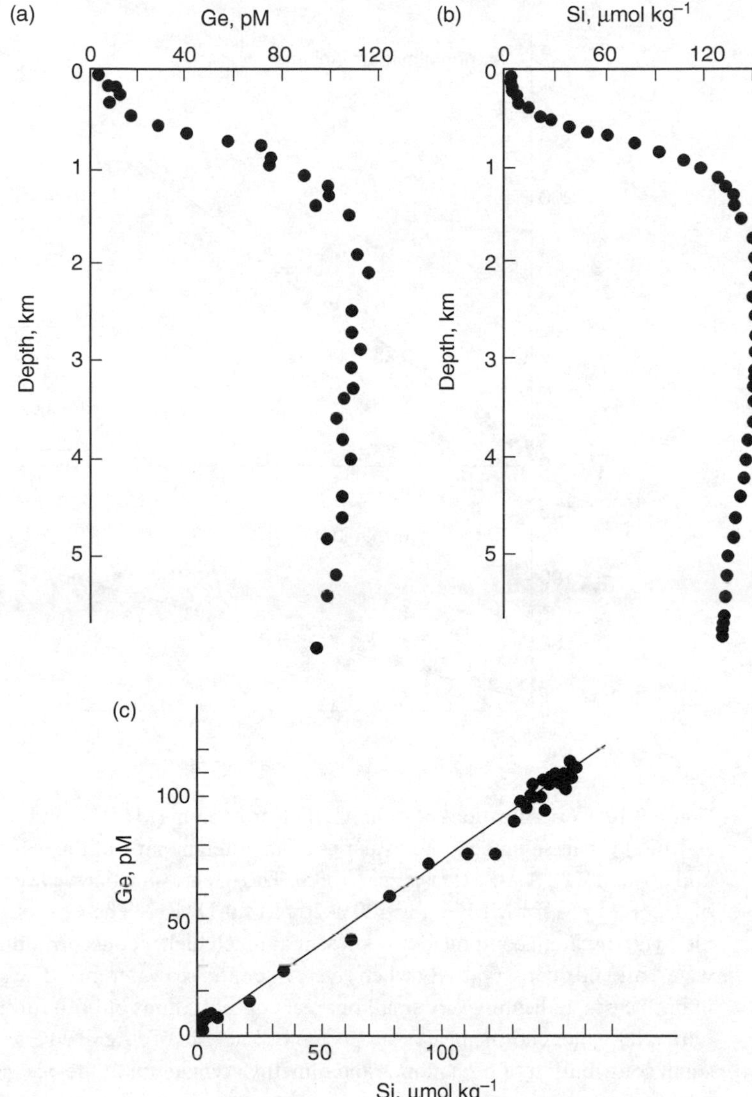

Figure 9.9 Vertical profiles of (a) dissolved germanium and (b) dissolved silicon (Froelich and Andreae 1981) in the western North Pacific (GEOSECS Sta. 227, about 1700 km northwest of Hawaii), and (c) a plot of germanium versus silicon. The slope of this line, $Ge/Si = 0.7 \times 10^{-6}$.

small. In any case, their concentrations are so uniform in the sea that their residence times must be very long. Lewis *et al.* (1989) also investigated the sinks for these compounds. So far, the major clue appears to be their discovery that decreased concentrations of methylated germanium compounds are present in marine anoxic basins, accompanied by increased concentrations of inorganic germanium. The evidence therefore suggests that these compounds may be bacterially synthesized in

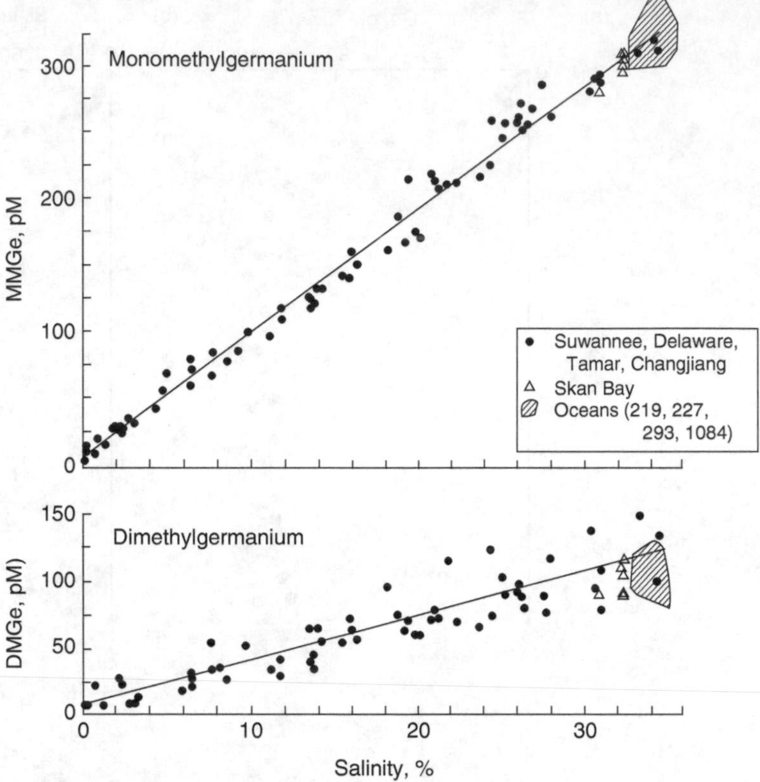

Figure 9.10 Concentrations of monomethylgermanium (MMGe) and dimethylgermanium (DMGe) versus salinity in four estuaries from different parts of the world, in several ocean areas, and in Skan Bay, Alaska (Lewis *et al.* 1985). These regression lines give concentrations at $S = 35‰$ of 330 ± 15 pM for MMGe and 120 ± 20 pM for DMGe. These graphs show that the two methylgermanium compounds are present at much higher concentrations in the sea than in fresh water, and mix conservatively when river water and seawater mix. The slight positive intercept in both cases, indicating very small but real concentrations of both forms in the zero-salinity pure river water end-member is supported by analyses of several rivers, showing that there is a small contribution of both mono- and dimethylgermanium to the ocean from rivers.

terrestrial anoxic environments, and bacterially destroyed in marine anoxic basins and possibly marine anoxic sediments. On a global scale these rates appear to be exceedingly slow. We currently have no evidence on how constant the concentrations of these methylated forms may be over geological time.

The concentration of inorganic germanium, or at least the ratio of germanium to silica, does vary somewhat. While the oceanic Ge/Si ratio is currently about 0.7×10^{-6}, evidence from opal (mostly diatoms, which seem to faithfully record the concentration in the water) preserved in deep-sea sediments shows that during each of the last five glacial periods the ratio of Ge/Si was about 0.6×10^{-6} and in one case as low as 0.5×10^{-6} (Bareille *et al.* 1998) There is yet no certain explanation for these changes, but it is clear that various geochemical fluxes carry different Ge/Si ratios; rivers are variable, with ratios ranging from 0.3 to 1.2×10^{-6}, while hydrothermal fluids from the mid-ocean ridges generally carry high ratios, around 8–13 × 10^{-6}, and diffusional fluxes from continental-shelf sediments also differ from the oceanic average (McManus *et al.* 2003). As the relative proportions of these various fluxes may vary over time, so may the resultant ratio observed in the ocean.

9.3 Mercury, an interesting special case

This element is selected for somewhat extended treatment because it provides examples of complexity in its chemistry and biogeochemical cycling, it has numerous anthropogenic and natural sources, and it has potential for causing harm (Fitzgerald 1989). It is, however, difficult to analyze at environmental concentrations and much is yet not known about important aspects of its behavior.

Mercury is generally rare in Earth's crust; the usually quoted average concentration is 0.085 ppm (parts per million) but, being concentrated in its ores, it is readily obtainable and easily separated from the ore. Mercury was one of the few elements known in ancient times and it has been used in several ways for at least 3500 years. Its primary ore, mercuric sulfide, HgS, was known to the ancient Greeks as *cinnabar*, a word of earlier oriental origin and still in use. Cinnabar is red in color, soft, and easily ground into a paste, so perhaps its first use was as the pigment also known as vermilion. Simple roasting of the ore yields the element as a vapor, which can be readily condensed as the liquid metal.

$$HgS + O_2 \rightarrow Hg\uparrow + SO_2\uparrow. \tag{9.1}$$

This and related processes were evidently parts of a technology developed long ago. Mercury vapor is very toxic, however, causing neurological damage, kidney damage, and death; and probably the work was normally assigned to slaves.

During the long history that mercury has been known and used, its properties were important in the development of ideas about natural substances. It was known to the Romans as *argentum vivium*; the equivalent in English is *quicksilver*. It influenced the development of alchemical and chemical thought, and the early technology of chemical practice. The sense that mercury is almost alive perhaps comes from the observation that if a small amount is spilled on the ground or on a countertop it breaks up into numerous small droplets rolling in every direction and it is difficult to scoop up again. This behavior is caused by its very high surface tension (six times that of water). The high surface tension prevents it from wetting glass; this property, its great density (about 13.5 times that of water), low vapor pressure, and apparent insolubility led to its use in thermometers, barometers, and many other kinds of chemical apparatus. These properties and the uses for mercury materially advanced the early practice and development of physics and chemistry.

Mercury occurs in three oxidation states: Hg^0, Hg^+, and Hg^{2+}. The elemental form is occasionally found associated with its ores as liquid droplets but, as noted later, is otherwise widespread in nature in very low concentrations. Singly charged Hg^+

occurs uncommonly in nature as Hg_2Cl_2; this compound is known as *calomel* and among its many uses are some in medicine and as a component of electrodes for measuring electromotive force (e.g., pH electrodes). Doubly charged Hg^{2+} is the stable valence state in the presence of oxygen (or as the very insoluble HgS). The cation Hg^{2+} forms very strong bonds with many substances; in saline solutions such as seawater the predominant form in solution is probably $HgCl_4^{2-}$, a tetrachloride complex, and some Hg^{2+} is also associated with other ions and with organic matter.

In ocean and lake water, in soils, and in marine sediments, Hg^{2+} is strongly adsorbed onto particles and organic matter, but some is also taken up by bacteria and other biota. Under mildly reducing conditions, such as are found in marine sediments, bacteria methylate mercury to form monomethylmercury (CH_3Hg^+), and dimethylmercury (CH_3HgCH_3). The latter compound appears to be less stable, and is much less often found in measurable amounts. Divalent mercury is also reduced to elemental mercury (Hg^0) by bacteria, and in surface waters can also be photochemically reduced (Fitzgerald *et al.* 2007).

Monomethylmercury is relatively stable in natural waters, occurring in seawater as CH_3HgCl. It is accumulated biologically, and concentrated up the food chain (Mason *et al.* 1996). This substance is especially toxic to mammals. Fish in many lakes have accumulated levels of monomethylmercury sufficient to present a toxic hazard if eaten in large amounts. This accumulation is natural and indeed universal, but where the concentrations of mercury have been enhanced by some process, the situation can become serious, as it has in several parts of the world. The worst example of toxicity in humans occurred in the 1950s around Minamata Bay, Japan. Large amounts of mercury were discharged by an acetaldehyde plant into the bay. The mercury and its compounds were converted by sediment bacteria into monomethylmercury, which became concentrated in the fish and shellfish that formed an important local food supply, but of course this was not known at the time. Cats died, about 200 000 people became ill, more than 100 people died, and more than 2200 babies were born with severe and permanent mental retardation and physical deformities.

The Minamata disaster, as well as the known toxicity of other forms of mercury, led to an increasing level of geochemical and regulatory interest in this element, resulting in a considerable research effort, with hundreds of papers and several symposia and reviews devoted to the subject, as well as a recent global assessment (UNEP Chemicals Branch 2008). The research effort was at first hampered by the generally low environmental background concentrations of mercury and the several forms in which it occurs, leading to considerable difficulty in measuring these. Many details of the natural cycling of mercury are yet unclear.

Elemental metallic mercury has an appreciable vapor pressure at ordinary temperatures (Appendix Figure D.1) and should be present in the atmosphere, from several sources. Very small amounts of elemental mercury are sometimes found in association with its ores. Small amounts are found in volcanic vapors, and droplets have even been observed at hydrothermal vents (Stoffers *et al.* 1999). As mentioned earlier, biochemical processes, widespread in nature and probably

Table 9.3 Anthropogenic emissions of mercury (t yr^{-1})

Burning coal	878
Metal production (non-gold)	200
Large-scale gold mining	111
Artisanal gold mining	350
Cement	189
Chlor-alkali industry	47
Waste incineration	125
Dental amalgam (cremation)	26
Total	1930

Data from the United Nations Environment Program, estimated for the year 2005. (UNEP Chemicals Branch 2008.)

due mostly to bacteria, convert the soluble salts of mercury into the elemental form Hg^0. For this reason, soils and even natural vegetation (Hanson *et al.* 1995), emit small amounts of mercury vapor. Currently, mercury is mined at an officially reported rate of about 1960 tons per year, and sold as the liquid metal. Some of this mercury must eventually escape to the atmosphere, although the amount is not well known. Mercury and its salts are used for many purposes; the largest use is as a catalyst for industrial processes (AMAP/UNEP 2008) followed by artisanal gold mining, manufacture of batteries, dental amalgam, and lamps. Incineration of municipal and other wastes thus releases some mercury into the atmosphere. Coal contains variable amounts of mercury, generally around 0.5 ppm. The burning of coal transfers some of this to the atmosphere, the fraction emitted depending in part on the extent of flue gas cleanup; currently, burning coal contributes most of the anthropogenic mercury to the atmosphere. A recent estimate (Pacyna *et al.* 2006) put the release in the year 2000 from coal combustion at 1400 tons per year and the total anthropogenic emissions at 2200 tons per year. Other estimates (Table 9.3) suggest slightly lower values and a total emission of about 1930 tons per year.

Numerous measurements of mercury in the atmosphere show that it is detectably present everywhere, at an average concentration of about 1.5 ng m^{-3} (UNEP Chemicals Branch 2008). This mercury is believed to be over 90% in the form of elemental Hg^0. This global average concentration corresponds to a total atmospheric burden of about 5900 t or 29 Mmol.

With these approximate atmospheric data available, we can ask: What is the equilibrium concentration of mercury vapor dissolved in seawater relative to that in the atmosphere? The only data on the solubility of the vapor in seawater appear to be those of Sanemasa (1975). A derivation of the Henry's law constant from his data is given in Appendix D, and shown in Figure 9.11. Calculation of the expected equilibrium concentration in surface seawater (Box 9.1) shows that usually this should be in the range of 17 to 46 fM, depending on the temperature of the water. (The uncertainty in the reported values of the Henry's law constant, the lack of more

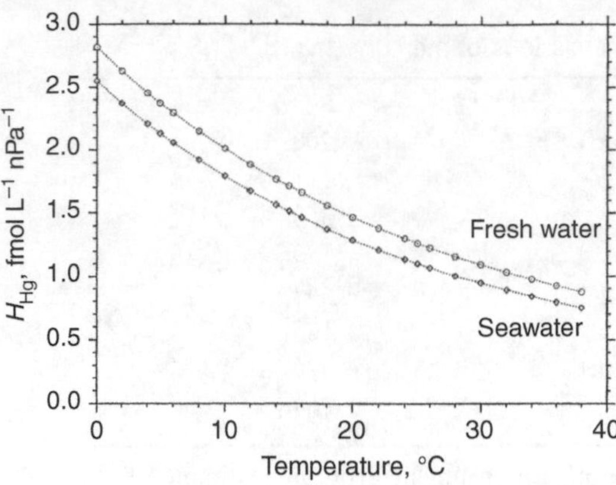

Figure 9.11 Solubility of mercury vapor in distilled water and in seawater. Here the Henry's law constant, H_{Hg}, is given in units of fmol L^{-1} nPa^{-1} (= fM nPa^{-1}), convenient units when calculating the solubility in water relative to typical atmospheric concentrations. These relationships were calculated from data in Sanemasa (1975); see Appendix D. Sanemasa did not specify the salinity of the seawater used; a reasonable assumption is that it was 34‰, typical for surface seawater near Japan.

refined measurements relative to the salinity, and the difficulty in making precise analytical measurements at these low concentrations, suggest that at this time it is not worthwhile to consider the relatively small effect of salinity variations on the solubility. It is sufficient to obtain these approximate values.)

Next, we can ask: What are the observed concentrations of Hg^0 in seawater? There have been only a few measurements but, such as they are, they tell a compelling story. Many of the data have been obtained by Bill Fitzgerald and several co-workers at the University of Connecticut. Kim and Fitzgerald (1986) measured [Hg^0] in surface water during a 1984 cruise along the equator (93° W to 155° W) in the Pacific Ocean and found values ranging from about 70 fM to about 220 fM over a temperature range of 26 to 19 °C. At this time, the concentrations they measured in the atmosphere were about 1 ng m^{-3}, so the equilibrium concentrations in the water there should have been about 12 to 16 fM (cf. Box 9.1). Evidently, the seawater in that region was considerably supersaturated relative to the atmospheric concentration, and Hg^0 must have been diffusing into the atmosphere. During cruises in the North Atlantic, Mason *et al.* (1998) found concentrations of Hg^0 in surface water ranging from about 200 to 1550 fM, with an average value of 650 fM in the region between Newfoundland and Iceland. The temperature of the water was commonly about 5 °C, so the equilibrium concentration in the water should have been about 40 fM. Again, in this region the partial pressure of elemental mercury in the surface mixed layer was much greater than in the air above, and mercury vapor must have been diffusing into the atmosphere. Additional information comes from Andersson *et al.* (2008) who found average concentrations of about 220 fM in a transect across the Arctic Ocean. The supersaturation so far appears to be general, and the ocean appears to be a source of mercury vapor to the atmosphere.

There are not enough measurements to make a well-founded global estimate of the evasion flux of mercury from the ocean to the atmosphere. Nevertheless, Mason *et al.* (1994) and Hudson *et al.* (1995) found a way to incorporate useful estimates of this evasion rate into global models of the cycling of mercury. Mason *et al.* (1994) observed that the concentration of Hg^0 was often related to the

Box 9.1 Calculation of the equilibrium concentration of mercury vapor in seawater

[1] Partial pressure (*pp*) of mercury vapor in air (assume dry air):
Average concentration in air (UNEP Chemicals Branch 2008):

$$\frac{\text{mass concentration in air}}{\text{mol.wt.of Hg}} = \frac{1.5 \times 10^{-9}\text{g m}^{-3}}{200.54\text{g mol}^{-1}} = 7.48 \times 10^{-12}\ \text{mol m}^{-3}$$

Number of molecular volumes in 1 m^3 at standard atmospheric pressure and 18 °C:

$$\frac{1000\ \text{Lm}^{-3}}{23.89\ \text{L mol}^{-1}} = 41.86\ \text{mol m}^{-3}$$

Number of moles of mercury per mole of air (this ratio is called the *mixing ratio* or the *mole fraction* of mercury in air):

$$\frac{7.48 \times 10^{-12}\ \text{mol m}^{-3}}{41.86\ \text{mol m}^{-3}} = 1.79 \times 10^{-13}$$

At 18 °C, dry air at sea level exerts a pressure of 101 325 Pa – VP H_20 = 99 293 Pa, *pp* of Hg = $(1.79 \times 10^{-13}) \times 99\,293$ Pa = 1.78×10^{-8} Pa, or 17.8 nPa

[2] Concentration of mercury in seawater in equilibrium with the atmosphere:
(From Figure 9.11 or from Appendix D, Table D.9)
H_{Hg} = 2.55 fM nPa^{-1} at 0 °C which yields [Hg^0] = 45.9 fM
= 1.37 fM nPa^{-1} at 18 °C which yields [Hg^0] = 24.4 fM
= 0.95 fM nPa^{-1} at 30 °C which yields [Hg^0] = 16.5 fM

Calculated for equilibrium with the modern (ca. 2010) atmosphere.

concentration of chlorophyll *a* and thus to the total concentration of living material in the water. While there is evidence that bacteria can reduce mercury in the water from the Hg^{2+} valence state to the elemental Hg^0 form, mercury can also be both reduced and oxidized photochemically (Fitzgerald *et al.* 2007). The photochemical effects are mediated in some way by dissolved organic matter, and the concentration of dissolved organic matter is greater in regions of high productivity. The concentration of Hg^0 at any time must be the net result of several competing processes. Mason *et al.* (1994) reasoned that the evasion rates in the tropical Pacific, calculated from gas exchange equations, could be scaled to the whole ocean according to the biological productivity in the water where the Pacific measurements had been made and the global ocean productivity. Taking the global productivity as 23 Gt of carbon per year, they estimated that from the ocean alone there is a total evasion rate of 10 Mmol of Hg^0 per year. Since the standing stock of mercury vapor in the atmosphere is about 29 Mmol (as calculated above), this is a very significant input to the atmosphere. Their picture of the global (ocean and

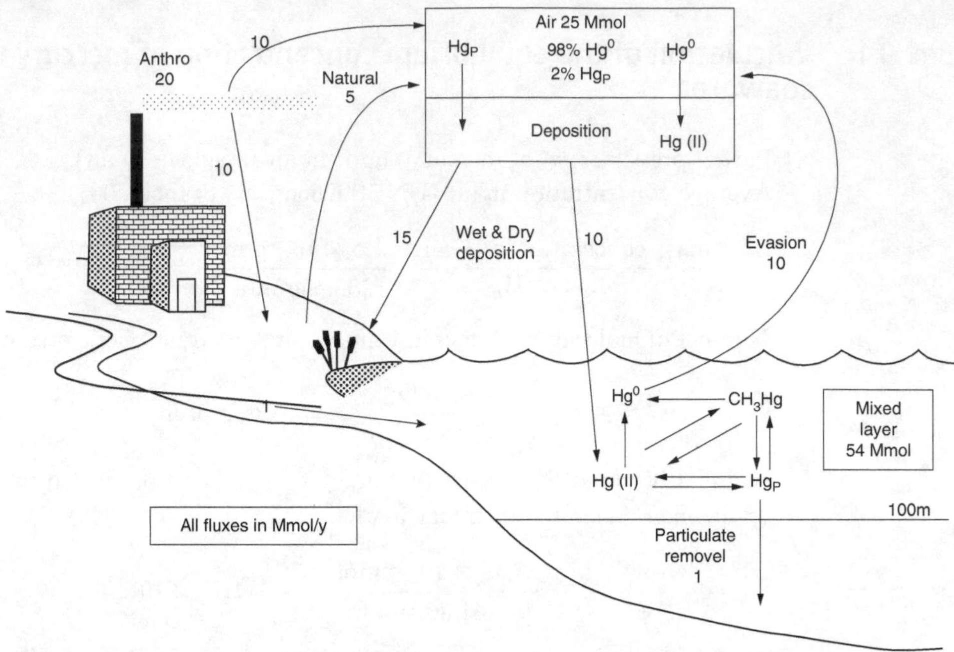

Figure 9.12 A sketch of the global cycle of mercury, with emphasis on the role of the atmosphere (Mason *et al.* 1994). None of the fluxes or concentrations is really well known but, as shown here, they give an impression of the present magnitudes, and the relationships between them. A considerable portion of the estimated anthropogenic emission of 20 Mmol from incinerators, coal burning, and the smelting of ores is in particulate form (Hg_p) or as Hg^{2+} compounds easily rained out, so it is believed to fall to ground close to the sources. Here the anthropogenic emissions are apportioned 50% to local fallout and 50% to more distant mixing into the atmospheric reservoir. The input from rivers to the ocean is estimated to be 1 Mmol per year, while atmospheric deposition to the ocean surface by wet and dry fallout is estimated to be 10 times greater. Mercury vapor in the atmosphere is slowly oxidized to the ionized form which makes it particle-reactive and also more soluble in rain water, so deposition to the ocean surface as well as to land depends on this conversion. Re-emission of Hg^0 vapor from the ocean surface approximately balances the atmospheric input to the ocean.

terrestrial) atmospheric cycling of mercury is shown in Figure 9.12. In this modeled estimate the annual fluxes to and from the atmosphere (each equal to 25 Mmol per year) are equal to the total mass of mercury in the atmosphere, suggesting that the residence time of mercury vapor in the atmosphere is approximately one year. Several adjustments made by Hudson *et al.* (1995) do not substantially change the picture. Some 40% of the flux to the atmosphere is anthropogenic and another 40% is from the ocean. Furthermore, since mercury appears to be recycled rather readily from the oxidized to the elemental state, a portion of the evasion rate from the ocean must be influenced by the anthropogenic mercury deposited onto the surface of the ocean. This may be true of the land as well. The ultimate sink for mercury is the transport and burial of mercury-containing particles in marine sediments.

Presumably, this rate has also been accelerated, although evidence is still lacking. Lake and bog sediments do show a several-fold increase in concentration during the last century, however, so these provide a record and at least a temporary sink for mercury deposited on land.

The total mercury in the surface layers of the ocean as well as deeper in the water column tends to be found at concentrations of about 1 to 5 pM (1000 to 5000 fM), with higher values near the coast and in polluted areas. The several forms present (Hg^0, CH_3Hg^+, CH_3HgCH_3, and *reactive* species [probably ionic species such as $HgCl_4^{2-}$]) occur in varying ratios to one another in different places, suggesting that there is an active interchange between them mediated by the biota (e.g. Mason and Fitzgerald 1990; Mason *et al.* 1993, 1998).

There is much yet to be learned about the complex natural and human-enhanced pathways and reactions of this interesting metal.

9.4 Speciation

An important consideration in the study of trace elements in seawater is the form in which they exist in the water. We have already seen examples of elements that can be found in different valence states (selenium and mercury); several others (e.g. iron, manganese, copper) also undergo oxidation and reduction (mediated biologically or inorganically or both) that influence their behavior and biological effects. Many elements dissolved in seawater are present as various inorganic complexes. The mercuric ion mentioned in Section 9.3 is present mostly as $HgCl_4^{2-}$ and a number of other elements also occur mostly as chloride complexes (e.g. $AgCl_3^{2-}$, $AuCl_2^-$, $PtCl_4^{2-}$), while others are present as carbonate complexes or as complexes with hydroxyl ion (Byrne *et al.* 1988).

We have seen examples of elements that may be found with methyl groups attached (such as the extremely stable mono- and dimethylgermanium species $CH_3Ge(OH)_3$ and $(CH_3)_2Ge(OH)_2$, and the less-stable and biologically recycled methylmercury species noted above), but a number of other elements (e.g. arsenic, antimony, selenium, and tin) may also be found in various methylated forms, perhaps most commonly produced in anoxic environments (Donat and Bruland 1995).

Certain metals (e.g. iron, cobalt, copper, zinc) are known to be quite strongly complexed with organic matter in seawater. Among the most important and much studied of these is copper, and here we will examine the situation with regard to this element as an example that provides insight into a variety of considerations that will also apply to several other elements. We will examine iron later.

Concentrations of total dissolved copper in seawater vary considerably, but most observations in the open ocean are encompassed by the range of 0.5 to 5.5×10^{-9} mol kg^{-1} (Figure 9.13). Copper is an absolutely essential trace element for all living things because of its role in certain enzyme systems. It is also inherently very toxic, and many organisms have elaborate mechanisms to control their internal concentrations and excrete unnecessary copper (as do genetically normal humans; a single

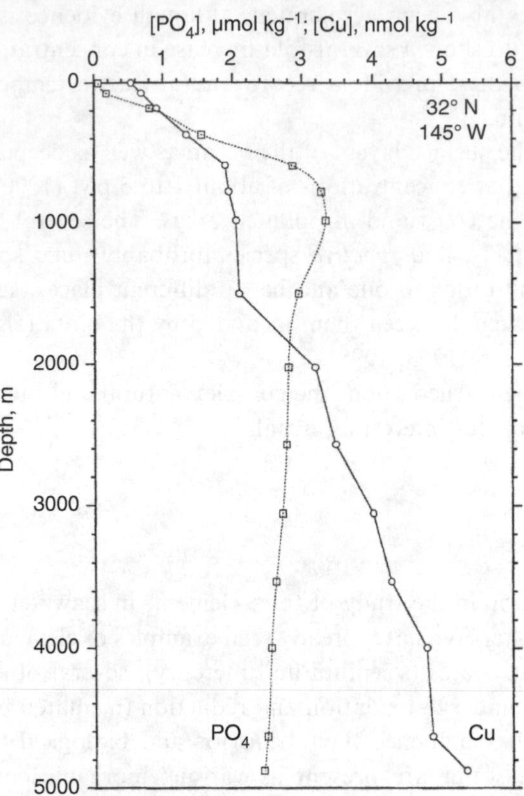

Figure 9.13 Typical distribution of copper in the water column in the central North Pacific. The vertical distribution of phosphate is also shown, in order to provide a picture of the vertical distribution of nutrients at that station, and to show that the concentration of copper does not follow that of phosphate or other nutrients. Copper is, however, stripped out of surface water and carried by transport on particles down to depth where it is released. This distribution is, in detail, quantitatively unexplained, but suggests multiple pathways of downward transport, and perhaps a source by diffusion out of bottom sediments. (Data from Bruland 1980.)

gene defect can lead to copper toxicity from the amounts consumed in an ordinary diet). Many experiments have been done to examine the toxicity of copper to phytoplankton. These have been interpreted to show that the toxicity is due to the concentration or the activity of the free copper ion in solution. An example of data on copper toxicity is shown in Figure 9.14, where the calculated copper ion activity is plotted (rather than the concentrations). Other data from Sunda *et al.* (1987) have shown that some zooplankton are also damaged by copper toxicity at similar concentrations. To compare data such as those in Figure 9.14 with concentrations found in the surface ocean, we need to calculate the activity of copper ion corresponding to the observed concentrations. Surface-water concentrations in the central North Pacific can be about 0.5×10^{-9} mol kg^{-1}. Other work (e.g. Bruland and Franks 1983) shows that in the central North Atlantic the concentration of copper can be about 1×10^{-9} mol kg^{-1}. Nearer the coast, in water over the continental shelf, the concentration may be four or five times greater. Using the data in Byrne *et al.* (1988) it can be calculated that approximately 90% of the copper ion in seawater at 15 °C and normal pH is complexed with inorganic anions, mostly with carbonate and some with hydroxyl ion. This leaves about 10% free (and more at lower temperatures). For simplicity, we can take the activity coefficient of free copper ion to be about 0.21 by analogy with calcium ion. Therefore the activity of the copper ion in surface seawater is about 0.2 × 10%, or about 2% of the actual concentration. This means that the activity of

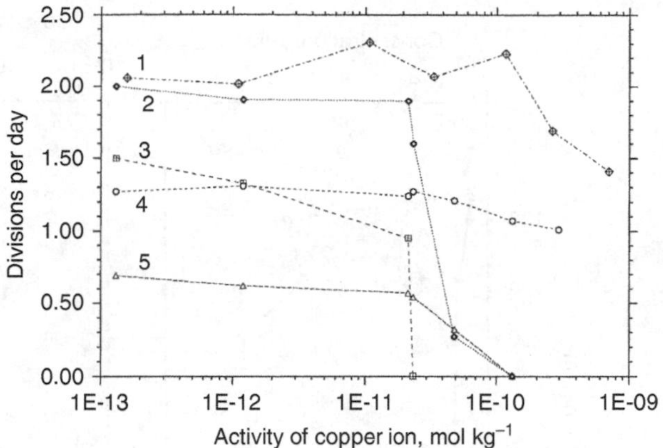

Figure 9.14 Growth rates of five species of marine phytoplankton (belonging to three different major groups) exposed to several different concentrations of free copper ion. The calculated activity of the copper ions is plotted, rather than the total concentration of copper or the concentration of the free ions. [1] *Skeletonema costatum*, a coastal diatom; [2] *Asterionella glacialis*, a coastal diatom; [3] *Synechococcus* sp., an oceanic photosynthetic cyanobacterium; [4] *Emiliania huxleyi*, a cosmopolitan coccolithophorid; and [5] *Cyclococcolithina leptopora*, an oceanic coccolithophorid. The most sensitive species could not grow at copper ion activities as low as 2.4×10^{-11} mol kg^{-1}. (Data selected from Brand *et al.* 1986.)

copper ion in water from the central North Atlantic should be about 0.02×10^{-9} mol kg^{-1} $= 2 \times 10^{-11}$ mol kg^{-1}. Thus, even in the central gyre, the concentration of copper in surface seawater seems to be surprisingly close to the toxic concentration that completely prevents some phytoplankton species from growing at all, as shown in Figure 9.14. Near the coast, the water could be lethally toxic to at least some forms, while in the central North Pacific the copper might be expected to exert at least a depressing effect on some species. Possibly the situation is worse than expressed so far, because the phytoplankton species in culture and used for the experiments described are likely be tougher than others that have not been successfully cultured. In upwelling areas another factor should be considered; the upwelled water may have elevated levels of copper, as well as of many other substances, so the situation is likely to be worse again. From what we have seen so far, it appears likely that most of the surface water of the ocean could be marginally or lethally toxic to many of the species of phytoplankton that grow there. Clearly, there must be some factor that we not have considered.

Several investigators have found that the electrochemically measured activity of copper ion in surface seawater is undetectable or is much less than would be expected from the concentration present, even allowing for the reduction in activity caused by the inorganic complexes noted earlier. It appears that the copper ion is complexed very strongly by at least two kinds, or two classes, of organic substances, called *organic ligands*. These compounds have not been isolated, and their existence is deduced only from their effect on the apparent activity of the copper ion. The chemical nature of these organic ligands is therefore yet unknown (so it is not known if there are two

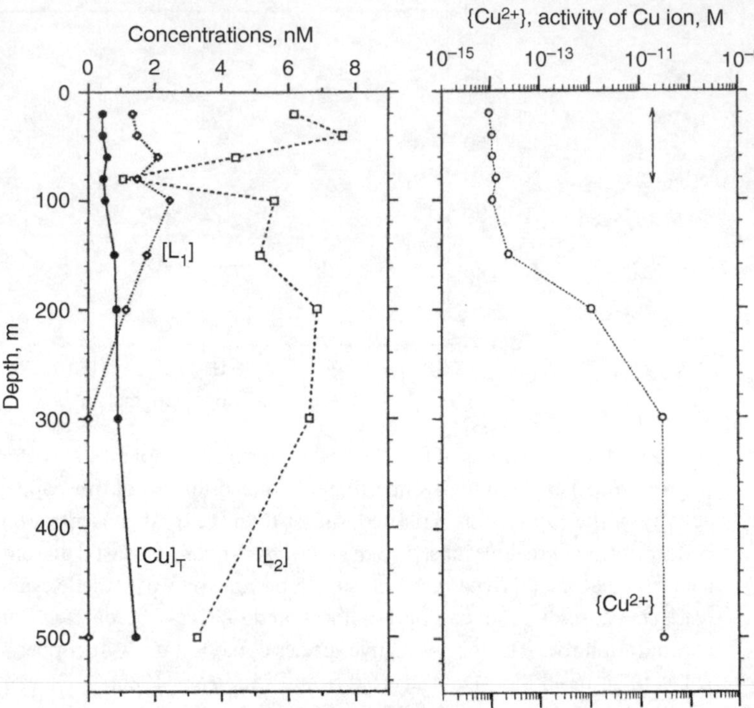

Figure 9.15 Measured concentrations of total dissolved copper $[Cu]_T$ and calculated values for other parameters at a station in the North Pacific (33° N, 139° W) occupied during the summer of 1987. From a series of titrations with small additions of dissolved copper, and electrochemical measurements of free and easily dissociated copper ion, Coale and Bruland (1990) calculated the concentrations and the association constants of two classes of organic ligands that bind copper ions. One ligand or class of ligands, L_1, binds copper extremely strongly; this ligand was present at concentrations somewhat greater than the total copper itself in surface water, but was undetectable at 300 m and at greater depths. The other ligand or class of ligands, L_2, was generally present at concentrations greater than those of L_1 but still in the nM range; this ligand was found at roughly similar concentrations down to as deep as the measurements were made, about 1400 m. Using the data shown here, the association constants of several complexing agents, and the estimated activity coefficient of free copper ion, Coale and Bruland calculated that the activity of free copper ions $\{Cu^{2+}\}$ was as low as 10^{-14} M in surface water. Note the log scale for the right-hand graph; due to the presence of L_1, which binds nearly all the copper in the surface water but is not present in deeper water, the activity of copper ion varied by three orders of magnitude over the 500-m depth range observed here, while the total concentration of copper varied by only about a factor of two. The small line with arrows is placed at the value for the activity of copper ion shown to be lethally toxic to some species of phytoplankton (see Figure 9.14). (Data from Coale and Bruland 1990.)

compounds or two classes of compounds), but their existence has a dramatic effect on the proportion of copper in seawater that is free to react with other substances or to affect phytoplankton. The binding between copper ions and the organic ligands is exceedingly strong, so that in surface seawater the activity of copper ions is greatly reduced, and more than 99% of the copper is effectively unavailable. The ligand known as L_1 binds copper ion most strongly and is observed to be present only in surface water (Figure 9.15). The ligand known as L_2 binds copper, but not so strongly; this ligand is observed to be present in water down to at least 1400 m, with little change in concentration over this depth range (Figure 9.15). From these observations it is concluded that L_2 is quite stable, while L_1 is produced in surface water but is not stable enough to survive for the length of time necessary to mix it down into deep water.

What are we to make of these observations? First of all, it is assumed that the organic ligands must be made by phytoplankton (or possibly by bacteria or other organisms) in the surface water. One might have imagined that phytoplankton would protect themselves from the toxicity of the copper in seawater by developing internal mechanisms to detoxify and excrete copper that is absorbed in excess of the small essential amount needed of the element. Some phytoplankton, fairly resistant to copper, may have done so. Why others would spend energy making and excreting enough of a highly specific compound with an extremely high affinity for copper, to react with and sequester all the copper in the surrounding water, is somewhat of a puzzle. An organism doing this would seem to be protecting not only itself but also its competitors. Another possibility is that these ligands are synthesized to help excrete copper, so that they occur in seawater because they are, in some sense, waste products. We will know more when the sources, chemical nature, and something of the dynamics of these compounds have been worked out.

On the basis of experiments with added chelators (organic chemicals that complex in a specific way with metal ions) biologists have believed for a long time that organic complexes of trace metals are somehow important in the growth of phytoplankton. It seems likely that the results of early experiments showing this (e.g. Johnston 1964) were in fact due simply to the effect of an added chelator reversing the toxic effects of metal contamination caused by the lack of metal-clean techniques at the time.

Perhaps more relevant to conditions in the sea is the early observation of Barber and Ryther (1969) that recently upwelled and nutrient-rich water in the eastern tropical Pacific would not support the growth of phytoplankton very well until it had been conditioned in some fashion by the addition of chelators or zooplankton extract, or by an increase in the natural concentration of organic matter. This observation seems consistent with the recent data on a natural copper-specific organic ligand found only in surface water and presumably made there, and with the evidence (Figure 9.15) that the activity of copper ion in water from deeper than several hundred meters is about the same as that shown to be lethally toxic to at least some sensitive phytoplankton.

Some other metals are also complexed with organic matter in seawater, but, except for iron, studies have perhaps not gone as far as they have with copper.

9.5 Iron, another special case

The trace element that has commanded the most attention in recent years is iron. This metal is abundant on Earth and essential for all life, but in its normally oxidized form ($Fe(OH)_3$) it is quite insoluble. This insolubility, and thus its presumed

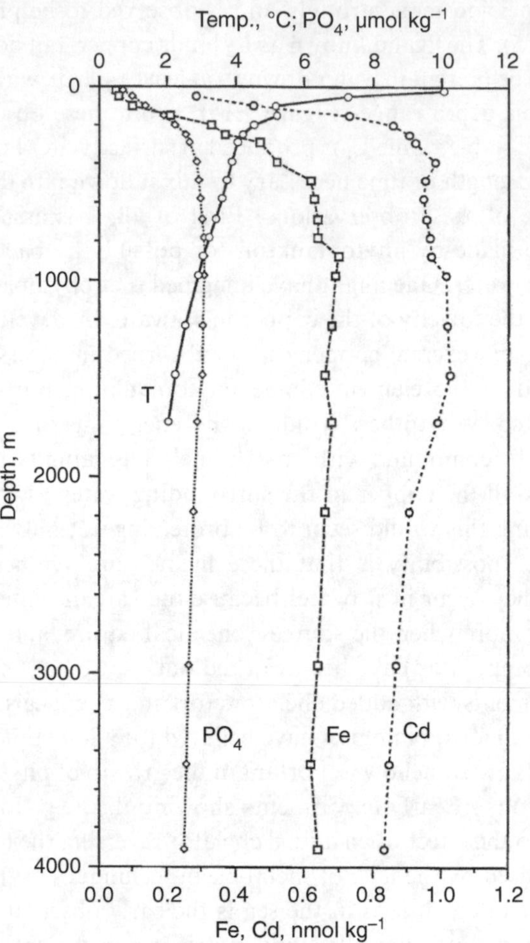

Figure 9.16 Data from Station "Papa" in the eastern North Pacific (50° N, 145° W) collected August 5, 1987 (Martin *et al.* 1989). Relative to the phosphate, the concentration of iron drops more sharply in the uppermost layers of water. In this region, Martin *et al.* showed that growth of phytoplankton could be enhanced by the addition of iron to samples of surface water.

unavailability to phytoplankton, has worried biologists for many decades (e.g. Gran 1931, Harvey 1937), since long before it was possible to make accurate assessments of its concentration in seawater. How could phytoplankton ever get enough iron to supply essential needs?

Iron exists in seawater in both ferrous (Fe^{II}) and ferric (Fe^{III}) oxidation states. In the presence of oxygen the ferric state is the stable condition. Ferric iron occurs in vanishingly small concentrations as Fe^{3+}. It has a strong affinity for OH^- ions; the $Fe(OH)_1^{2+}$, $Fe(OH)_2^+$ and $Fe(OH)_3^0$ and $Fe(OH)_4^-$ species all may exist, and some of the iron may also form ion pairs with chloride and other anions in seawater (Byrne *et al.* 2002). It is a challenging task to measure the effective solubility of iron in organic-free seawater. Relative to solid $Fe(OH)_3$, the amount of iron in solution at pH values near 8.0 and at 25 °C is thought to be about 0.1 nmol kg^{-1}, or perhaps a little less. However, concentrations considerably greater than this are found in seawater (Figure 9.16). Direct measurement of the solubility in surface seawater results in estimates of about 0.3 nmol kg^{-1} (Liu and Millero 2002). The solubility appears a little greater under the conditions found at depth: colder temperatures and lower pH, where dissolved iron may be found at concentrations of about 0.6 nmol kg^{-1}. It appears that most of the iron in seawater is complexed with organic ligands (Bruland and Rue 2001).

All organisms absolutely must have some minimum amount of iron in their cellular machinery; the amount required varies with the biochemical processes involved. For example, organisms that fix nitrogen do this with a complex of enzymes that contain iron, so they must have much more iron in their cellular machinery than organisms without this capacity. A typical concentration of iron in marine bacteria is about 1 atom of iron for each 125 000 atoms of carbon (Tortell *et al.* 1996). Several diatom species growing with NH_3 as their sole source of nitrogen had an average of about 1 atom of iron for each 310 000 atoms of carbon, while the same species growing with NO_3 as their sole source required twice that cellular concentration, presumably because of the additional iron-containing enzymatic machinery (e.g. nitrate reductase which contains iron) required for the reduction of nitrate to the biochemically usable reduced form (Maldonado and Price 1996). In the case of both the bacteria and phytoplankton studies the organisms were grown under conditions of strict iron limitation, so the ratios quoted reflect minimum requirements observed in these cultures. Calculations by John Raven (1988) based on the known components of the necessary biochemical machinery suggested that minimum requirements for rapid growth of phytoplankton with all enzymes working at maximum efficiency might be 1 atom for each 42 000 atoms of carbon if growing with NH_4^+ as the source of nitrogen, and 1 atom of iron for each 26 000 atoms of carbon if utilizing NO_3^-. Fixing N_2 as the source of this element requires much more iron; the need would be about 1 atom of iron for each 410 atoms of carbon. In fact, nitrogen-fixing cells grow quite slowly and the nitrogen-fixing machinery is slow. In one study the observed carbon-to-iron ratio in *Trichodesmium* (perhaps the most important nitrogen fixer) collected at sea averaged 3100 atoms of carbon for each atom of iron (Berman-Frank *et al.* 2001), so these cells are extremely iron-rich. As a further complication, in the presence of excess available iron some phytoplankton species may exhibit luxury uptake and increase their iron stores by sometimes at least 10-fold, allowing for future growth if iron becomes limiting.

The very low maximum solubility of the inorganic species of iron, and the biological stripping of iron from surface waters and downward transport along with carbon and other nutrients, lead to the situation where considerable areas of the ocean do not have enough iron to support the production of enough biomass to use up all the nitrogen and phosphorus in surface water. Such parts of the ocean have been termed "high-nutrient low-chlorophyll" (HNLC) regions. The definitive evidence that this phenomenon is caused by lack of iron was presented by John Martin and colleagues (Martin and Fitzwater 1988, Martin *et al.* 1989) who provided some of the first accurate measurements of dissolved iron in seawater (e.g. Figure 9.16). They incubated surface water from several stations in the eastern North Pacific with and without added iron. The added iron greatly increased the growth of phytoplankton.

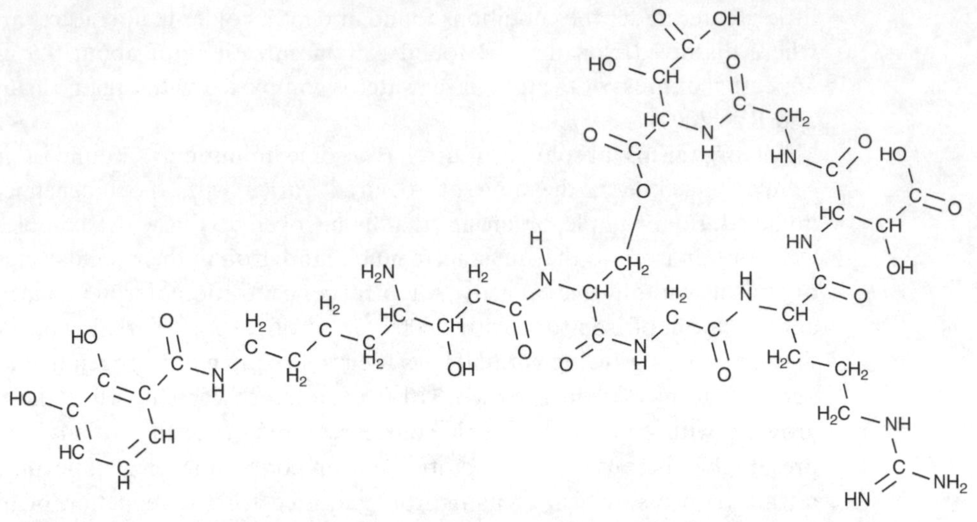

Figure 9.17 This compound, named Alterobactin A, is an example of a siderophore, an organic ligand with one of the highest affinities for iron of any yet reported (Reid *et al.* 1993). It is produced by a marine bacterium.

In sediments where there is no oxygen some ferric iron is reduced to the more soluble ferrous oxidation state. This soluble form can diffuse into the overlying water, and in addition some comes down rivers in colloidal form or combined with organic matter, so that near shore and in shallow water there is generally an adequate supply (Johnson *et al.* 1999). Regions of the ocean that receive inputs of wind-blown dust receive enough iron dissolving from the dust to supply the needs of phytoplankton, so most of the other nutrients can be used up (Jickells *et al.* 2005). Far from land, in regions with little dust fall, the input of iron is often insufficient.

The acute shortage of iron over considerable areas of the ocean means that competition for this essential nutrient is intense. Some organisms, of which bacteria are the best known in this regard, secrete substances called *siderophores*; the name comes from the Greek word for iron. Siderophores are complex organic molecules that have an extremely great affinity for iron, hold atoms of iron very tightly bound, and are generally thought to be specific for iron. There are several types of these compounds (Butler 2005), and different bacteria secrete different forms. Bacteria may then acquire iron by recognizing and absorbing their own siderophore–iron complex; some bacteria have evolved receptors to recognize the siderophore–iron complexes of their competitors. Probably most of the iron in surface water is complexed by these specialized organic ligands. Figure 9.17 shows an example of a siderophore. It must be quite energetically costly to make and then release into the water compounds with such an elaborate structure. There are now dozens of known marine siderophores (Vraspir and Butler 2009), in addition to many known from terrestrial sources.

Evidence so far suggests that eukaryotic phytoplankton do not secrete siderophores; instead, they rely on iron receptors and associated transport machinery on their cell surfaces (Sunda 2001).

The large ratio of carbon to iron in phytoplankton, combined with the demonstration that it is iron that limits the growth of phytoplankton in HNLC regions, led to the realization that very little iron added to the water there would allow the phytoplankton to fix a comparatively enormous amount of carbon. In turn, such an addition would increase the flux of particulate carbon from the surface down to some depth. The effect would be to remove CO_2 from surface water and so from the atmosphere. At a seminar at Woods Hole in 1988 John Martin is famously reported to have said: "Give me half a tanker of iron and I'll give you an ice age." While a considerable exaggeration, this captured the idea; the quote was endlessly repeated, and many people were led to plan experiments to test the idea. A dozen or more experiments have been carried out in the ocean, on what would heretofore have been considered a massive scale (de Baar *et al.* 2005). Initial experiments were in the eastern Equatorial Pacific Ocean, and later in the Southern Ocean and in the subarctic North Pacific Ocean, all regions demonstrably short of iron. In one experiment in the Southern Ocean (Boyd *et al.* 2000), 8.6 tons of ferrous sulfate ($FeSO_4 \cdot 7H_20$) was dissolved on shipboard in large tanks and then discharged into a 50-square-kilometer patch of ocean. The resulting plankton bloom was easily visible from space, via satellites measuring ocean color. Experiments could not be followed long enough to discover what happened to the carbon fixed, due to limitations of ship time, and also because the patches disperse over time. In general, however, the results did not suggest that the procedure would result in long-term sequestration of enough carbon to be worthwhile, but much biologically important information was acquired. In a later chapter we will learn that generally most carbon fixed anywhere in the surface mixed layer is remineralized before it sinks very deep into the ocean.

The availability of iron to phytoplankton is strongly affected by the pH of the medium, because of complex changes in the relative amounts of the various inorganic and organic species of iron. Lower pH leads to a decreased capacity to take up iron (Shi *et al.* 2010). This leads to the suggestion that as the ocean becomes more acidic in the future, the regional extent of apparent iron limitation may increase.

9.6 Trace elements in sediments

It is beyond our scope here to discuss the occurrence of trace elements in sediments in any detail, but a few general statements may be useful.

For most elements, the mass-weighted average concentrations in all marine sediments must be fairly close to the crustal average. The variations from place to place are considerable, however, as some sediments may be composed largely of riverborne detrital minerals (e.g. in the great deltaic fans), airborne dust (as in parts of the central Pacific), biological carbonate deposits (the foram and coccolith oozes, and coral reefs), biological silica deposits (such as the diatom or radiolarian oozes), highly organic-rich deposits (such as are found in the anoxic basins), or chemical precipitates (such as the manganese crusts and nodules, and the deposits around hot vents and cold seeps). Each of these deposits displays a characteristic but variable composition of major and minor elements, and with our new and increasing knowledge of trace-element distributions within the water column there will be new analyses and interpretations of their distributions in sediments.

Some elements that are present in extremely low concentrations in seawater, such as iron and aluminum, are major components of the sediments. Other elements are rare everywhere, but most are present in much higher concentrations in sediment minerals than in seawater. The occurrence of many elements in relatively high concentrations in sediments does not necessarily eliminate the difficulties in obtaining accurate analyses, however. It is, in many cases, crucially important that investigators in different laboratories carry out periodic intercalibrations or use certified reference materials where these are available (Wangersky 2000, Loring and Rantala 1988).

One type of deposit on the sea floor is of special interest for the curious marine chemist. The so-called manganese crusts and nodules occur widely in different parts of the ocean, but it is the manganese nodules that have commanded the most attention ever since they were first seriously studied after collection by the *Challenger* expedition. Many parts of the ocean floor, generally in places with very low sedimentation rates, are thickly carpeted with these remarkable objects. They are typically 1 to 10 cm in diameter, sometimes irregular but often roughly spherical; when cut in half they commonly appear to be composed of concentric layers, easily visible, with the initial deposit often beginning on some bit of inorganic or organic debris such as a shark tooth (Cronan 2001). While there is a wide range of composition, the dominant components are manganese and iron in approximately 1:1 ratios, on average. Thus, they should probably be called manganese–iron nodules. Depending on location, they generally contain high concentrations of copper, cobalt, nickel, and a number of other elements of interest. The concentrations are high enough that for the last half-century there has been considerable commercial interest in trying to figure out how to harvest them, both for the manganese and for the copper, cobalt, and nickel, and several other metals that would be obtained as byproducts.

Some manganese–iron crusts can grow fairly rapidly, especially if they are near a source of manganese, such as adjacent to the hot fluids coming out of mid-ocean vents, but the manganese nodules on the sea floor appear to grow extraordinarily slowly. Rates calculated by radioisotope dating are in the range of millimeters per million years. There are regions of the sea floor many thousands of square kilometers in area densely carpeted with nodules about the size of a baseball, which are growing very very slowly by accreting manganese and iron (as well as copper, etc.) out of overlying seawater that has at most only nanomolar concentrations of these elements. The common appearance of concentric layering in the nodules suggests that the conditions for growth have changed periodically over time, but there is no evidence about what these changes might consist of. The nodules are fascinating to contemplate, but not easy to study.

Summary

The rich tapestry of patterns in the distributions of many of these minor elements in the sea and the insight they provide into oceanic processes have become apparent only since modern, highly sensitive, analytical techniques and ultra-clean procedures were effectively combined, beginning in the mid 1970s. It is only since then that the data on the distribution of these elements achieved oceanic consistency. There are many elements for which there are now quite satisfactory data, but our understanding of what controls their concentrations is still feeble, and much of what we think we know is guesswork derived simply from examining the nature of the distribution. Some regions are poorly sampled, and there are several elements for which very few data yet exist. Some elements are found in more than one oxidation state or in various chemical combinations. Some have distributions that could not have been predicted; the processes involved are not well documented or are largely unknown. Some are so strongly complexed with various (still uncharacterized) organic ligands that their availability to organisms is only a tiny fraction of the analytically determined total concentration. Human activities have dramatically changed the concentration of lead in seawater, and have probably affected the concentrations of some other elements. One element (mercury) is volatile and actively cycles between atmosphere, land, and ocean. Iron is essential for all life, but it is so insoluble that over considerable areas of the ocean its concentration is low enough that for many organisms it may be the primary limiting nutrient. Many marine bacteria synthesize and secrete siderophores; these are substances that have a very great affinity to complex iron, and enable the bacteria to extract iron from the exceedingly low concentrations in seawater.

Manganese–iron nodules are found abundantly in certain regions of the deep sea where sedimentation rates are especially low. They must grow by absorbing a suite of trace metals from the overlying water, but the controlling processes are not well understood.

There is certainly much more to learn, and one wonders if there are more real surprises yet waiting.

SUGGESTIONS FOR FURTHER READING

Bruland, K. W. and M. C. Lohan. 2006. Controls of trace metals in seawater. In *The Oceans and Marine Geochemistry*, ed. H. Elderfield (*Treatise on Geochemistry*, vol. 6, ed. H. D. Holland and K. K. Turekian). Elsevier, New York, pp. 23–47.

This is the best general modern summary of the inputs, outputs and controls on the concentrations of trace metals in the ocean. Ken Bruland has been one of the main contributors to the field.

Turner, D. R. and K. A. Hunter, eds.. 2001. *The Biogeochemistry of Iron in Seawater*. Wiley, New York.

There has been an immense amount of work on iron is the ocean, and a great deal of the information on inputs, concentrations, analytical methods and thermodynamics of iron is included in this book.

de Baar, H. J. W., P. W. Boyd, K. H. Coale, *et al.* 2005. Synthesis of iron fertilization experiments: from the iron age to the age of enlightenment. *J. Geophys. Res.* **110**: C09S16.

Most of the information from the big experiments on fertilizing patches of ocean is summarized here.

Cronan, D. S., ed. 2000. *Handbook of Marine Mineral Deposits*. CRC Press, Boca Raton, FL.

This book covers more than just manganese nodules, but there is a lot on this subject.

Nozaki, Y. 1997. A fresh look at element distribution in the North Pacific. *EOS* **78**(21): 221.

Yoshiyuki Nozaki prepared a periodic table in which the space for each element is replaced with a small graph showing the vertical distribution of that element in the water column.

Vraspir, J. M. and A. Butler. 2009. Chemistry of marine ligands and siderophores. *Annu. Rev. Marine Sci.***1**: 43–63.

This review includes structures of the many marine siderophores discovered to date, and an entrée into the literature on the siderophores.

10 Radioactive clocks

. . . every radioactive mineral can be regarded as a chronometer registering its own age with exquisite accuracy. ARTHUR HOLMES 1913

The presence of natural radioactive isotopes of numerous elements in the ocean and throughout Earth provides us with a wealth of accurate clocks by which it is possible to determine the rates of many important processes. Their evaluation has provided us with an assured sense of the timescales of many geological and oceanic phenomena. The supplemental introduction of anthropogenically produced radioactive isotopes has in many cases provided opportunities to obtain additional information.

The existence of radioactivity, and what we have learned from the radioactive elements, has profoundly affected our comprehension of the world around us, but all this has come about only recently. A little over 100 years ago, the age of Earth was estimated by physicists (e.g. Lord Kelvin in 1897) as being only 20 to 40 $\times$ 10^6 years, because the laws of physics did not allow that it could take much longer than that to cool from a molten ball to its present state. Geologists by then knew that Earth must be much older, but saw no escape from the laws of physics. Then radioactivity was discovered, and the consequences were rapidly explored by Becquerel, Marie and Pierre Curie, Rutherford, Soddy, Thompson, Strutt, Holmes, and others. It was a wonderfully productive time in science (see Faure 1986 for a brief review and entrée to the history). The radioactive elements present in the composition of Earth provided a source of heat sufficient to keep the interior hot for the necessary length of time and also provided several accurate clocks by which the geological timescale could be calibrated. By 1913, Holmes had established that some ancient rocks were as much as 1.6 $\times$ 10^9 years old, and our general appreciation of the true age of Earth was pretty well settled. Subsequently, the age of Earth and of at least some other bodies in the Solar System has been established at around 4.6 $\times$ 10^9 years, with relatively small uncertainties, all by observing several of the natural radioactive clocks. With regard to the oceans, our knowledge of the rates of renewal of deep water and the rates of transport and sedimentation of many substances has come in large part from investigating these clocks.

10.1 Radioactivity

This section presents only a brief review of the nature of radioactivity and of the necessary equations for important calculations. Further background and data may be obtained from Friedlander *et al.* (1981), Faure (1986), and Firestone and Shirley (1996), which have been drawn on for the treatment here.

The atomic number of an element is the number Z of protons in the nucleus of the atom. The chemical nature of an element, hence its location in the periodic table and its name, depend on the number of electrons in the electron shells around the nucleus, and this in turn depends on the number of protons in the nucleus. Except for the lightest form of hydrogen, all nuclei also contain N neutrons, and the mass number A is the sum of the number of protons and the number of neutrons ($A = Z + N$). For every element, the number of neutrons in the nucleus is a variable. When an element occurs in several forms, each with a different number of neutrons and therefore a different mass number, each form is called an *isotope* of the element. If an isotope of an element has too many or too few neutrons for the number of protons, the nucleus is not stable and spontaneously transforms (directly or through several intermediates) into a stable nucleus. In this transformation process there is always a detectable release of energy in one or more of several possible forms. Because normally the number of protons has changed, the product will be a different element with different chemical properties.

Some of the ways in which a nucleus can spontaneously transform (Figure 10.1) are as follows.

a. *Emission of a negative electron (β^- particle).* In this case a neutron in the nucleus becomes a proton, the atomic number increases by 1 unit but the atomic mass number does not change. A neutrino is emitted at the same time. There is more than one kind of neutrino, but it is not useful to distinguish them here. Often the resulting nucleus is not in its rest state, still having some excess energy, so as it settles down it may emit energy in the form of one or more γ-rays as well.
b. *Emission of a positive electron (β^+ particle).* In this case a proton becomes a neutron, the atomic number decreases by 1 unit, and the atomic mass number does not change. As with β^- emission, a neutrino is emitted, and the product nucleus may give off energy as γ-rays before reaching its rest state.
c. *Electron capture.* In this case an electron is captured from those in orbit around the nucleus. This process occurs only with nuclei where there is a deficiency of neutrons. Usually an inner K-shell electron is involved. The electron enters the nucleus and combines with a proton to form a neutron, so the number of protons and the atomic number decrease by 1, the number of neutrons increases by 1, and the atomic mass number does not change. Here also a neutrino is emitted. Depending on the element, the nucleus may have extra energy after the capture, and a γ-ray may be emitted. The missing space for an electron in the inner shell will be filled by an electron from an outer shell dropping down. The energy involved in the latter electron shift is released as one or more X-rays.

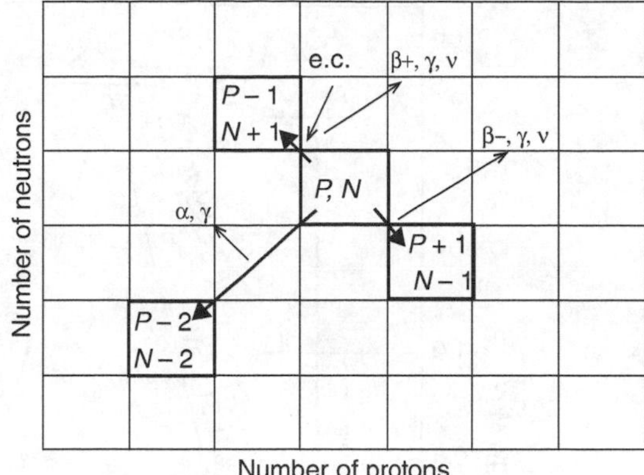

Figure 10.1 Radioactive decay schemes. Imagine a radioactive element, with *P* protons and *N* neutrons in the nucleus. There is only a limited number of ways that it can undergo a nuclear transformation. In alpha decay, an alpha particle (= a helium nucleus) escapes, resulting in a loss of two neutrons and two protons. The most common transformation process is "beta decay"; an electron escapes from the nucleus, as a neutron is transformed into a proton. In the least common process a proton is transformed into a neutron with either the escape of a positron (a positively charged electron) or the entrance of an electron (e.c., electron capture) into the nucleus. In most of these reactions one or more γ-rays are emitted, and in the reactions involving electrons or positrons a neutrino (ν) is also emitted. ■

d. *Alpha decay*. Many nuclides (mostly heavy ones with an atomic number greater than 58) can decay by emission of an alpha (α) particle. This particle, with 2 neutrons and 2 protons, is the nucleus of the helium-4 atom. The emission of an α particle reduces the atomic number by 2 and the atomic mass number by 4 units. Sometimes the resulting daughter nuclide still has excess energy, so there may also be an emission of γ-rays. Alpha decay results in the production of atoms of two elements from each event, the major daughter product and a helium atom.

Characteristics of the different forms of energy and particles released during nuclear transformations vary greatly. High-frequency electromagnetic radiation, such as γ-rays, can, depending on the frequency, travel some distance in solids before being absorbed. If not shielded, this radiation can be detected at a distance, as can X-rays. In contrast, β^- particles can travel only a short distance, depending on their energy. The weaker ones cannot penetrate the walls of a glass container, or even human skin. Alpha particles can travel only very short distances in solids, and not far even in gases. However, being relatively heavy, they cause a great deal of ionization and other chemical effects within that short distance.

Some radionuclides may undergo spontaneous transformation by more than one route, yielding two different products. The ratio of the two products formed is known as the *branching ratio*. An excellent example is provided by potassium-40 (Figure 10.2), which also illustrates the variety of processes that may be involved in an apparently simple nuclear transformation. This nuclide undergoes transformations by the first three of the four mechanisms listed above. It can emit a β^- particle

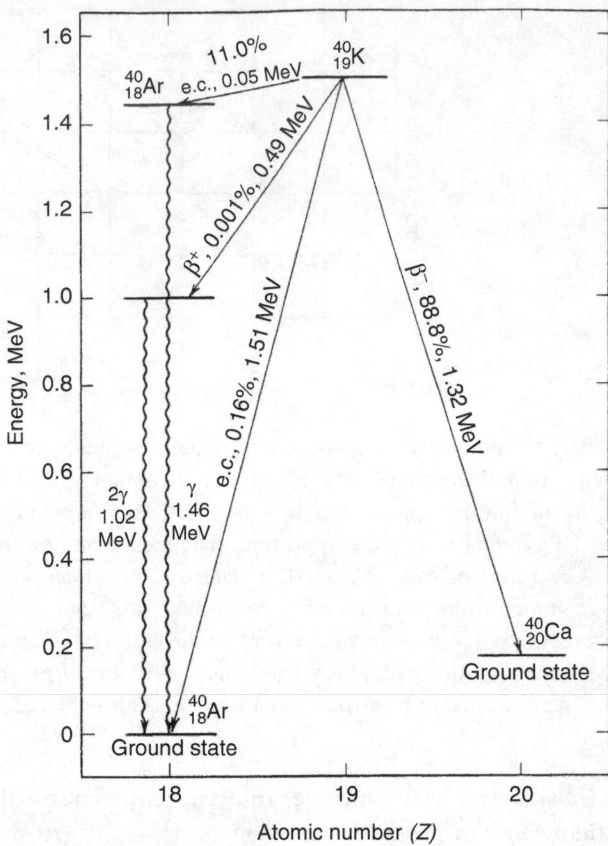

Figure 10.2 Decay scheme for the transformation of ^{40}K to ^{40}Ca and to ^{40}Ar. (From Faure 1986.) The decay to ^{40}Ca involves the emission of a β^- particle and a neutrino, for a total energy release of 1.32 MeV (= 2.11×10^{-13} J per atom of ^{40}Ca). MeV means million electron volts. The second branch of the decay scheme, the transformation to ^{40}Ar, can proceed by three pathways, of which the two electron-capture pathways are the most important. In most of the e.c. decays a 1.46 MeV γ-ray is also emitted. The energy release is 1.51 MeV (= 2.42×10^{-13} J per atom of ^{40}Ar). The decay constant for the first branch, $\lambda_1 = 4.962 \times 10^{-10}$ per year, while the decay constant for the second branch, $\lambda_2 = 0.581 \times 10^{-10}$/yr. The overall decay constant $\lambda = \lambda_1 + \lambda_2 = 5.543 \times 10^{-10}$/yr, and therefore the half-life of ^{40}K is 1.25×10^9 yr. The total energy released, the weighted average of the energy in the two branches, is 1.34 MeV or 2.15×10^{-13} J per atom of ^{40}K.

(negative beta particle or electron) and end up as ^{40}Ca. In a tiny fraction of its transformations, ^{40}K can emit a β^+ particle (positron) to become ^{40}Ar in an excited state, which then emits a γ-ray to become stable ^{40}Ar. In 11.2% of its transformations, ^{40}K undergoes electron capture to become ^{40}Ar in either the ground state or in an excited state that emits a γ-ray to become ^{40}Ar in the ground state. The overall result is that 88.8% of ^{40}K is transformed to ^{40}Ca and 11.2% to ^{40}Ar. This latter branch forms the basis of one of the most important techniques for dating ancient rocks. If the parent potassium atom is part of the mineral structure in some rock, the argon formed by radioactive decay, although it is a gas, is usually trapped in the mineral, and will stay there until the mineral is melted or strongly heated. The amount of ^{40}Ar relative to the amount of ^{40}K remaining is a direct measure of the age since the mineral was formed.

From the information above it is evident that to measure the total amount of radioactivity in a sample it may be necessary to measure the production of α particles, β particles, γ-rays, or X-rays. Selecting the measurement technique appropriate for a given isotope requires more information than can be presented here. Many techniques are now remarkably sensitive, so that in especially favorable circumstances one can identify and actually count the individual atomic transformations one by one, although for statistical certainty a large number is usually needed. Most detectors are less than 100% efficient, and there may be some background counts, so hundreds to thousands of counts are usually needed as a minimum.

Many different units have been used to express the amount of radioactivity, and for historical or special reasons several of these retain common usage. The basic unit is the nuclear transformation of one atom in one second. This unit is known as the *becquerel* (symbol Bq) after an early pioneer in the field. The next common unit is the *curie* (symbol Ci) and was originally defined as the number of disintegrations per second in 1 gram of radium; now it is defined as 3.7×10^{10} Bq. The units μCi and pCi are commonly used, for practical reasons. For empirical work with counters whose efficiency is not known, or not accounted for, it is common to report counts per minute or per second (cpm or cps), values of which are generally less than the number of nuclear transformations.

The spontaneous transformation of each radioactive atom is in each case a matter of statistical probability and is not influenced by any external conditions normally found at the surface of Earth. (The process of electron capture is increased slightly by very high pressures, such as are found in the center of Earth, and other reactions may be influenced by extreme conditions, such as may be found in the Sun.) Accordingly, the number of disintegrations observed in a given mass of some radioactive element is solely a function of the number of atoms present and of the probability that any one will undergo transformation in a given time. This probability is called the *decay constant*, and is a characteristic of each nuclide. The basic radioactive decay law is expressed in this simple relationship

$$\frac{dN}{dt} = -\lambda N. \tag{10.1}$$

where N is the number of atoms present, λ is the decay constant (for example, 1×10^{-4} per year or 3.17×10^{-12} per second), and t is time in the same units as λ. This equation is the definition of a first-order reaction, that is, the rate depends only on N, and not on any other substance present.

Equation (10.1) is integrated to yield various useful forms of the equation for radioactive decay:

$$\ln\frac{N}{N_0} = -\lambda t, \tag{10.2}$$

where N_0 is the number of atoms present at $t = 0$, and N is the number remaining at time $= t$. Thus,

$$\frac{N}{N_0} = e^{-\lambda t}$$

and

$$N = N_0 e^{-\lambda t}. \tag{10.3}$$

The decay constant λ is the most useful way of characterizing the rates of radioactive decay for each element, but for convenience in comprehending the timescales involved it is a common practice to report the half-life, $T_{1/2}$, obtained by setting $N = 0.5N_0$ in Eq. (10.2).

$$T_{1/2} = \frac{\ln 0.5}{-\lambda} = \frac{0.693}{\lambda}. \tag{10.4}$$

The half-life is the time taken for one half of the atoms initially present to undergo transformation.

It is also sometimes useful to know the mean lifetime of radioactive atoms, which is not the same as the half-life but is always longer. The relationship is

$$\tau = \frac{1}{\lambda}, \tag{10.5}$$

where τ is the mean lifetime of all the atoms initially present.

In many situations, the focus of interest is on the amount of daughter product atoms produced. In the simplest case, the daughter atoms are stable and do not undergo further transformations. If the number of daughter atoms, produced by the decay of a parent are designated D^*, then

$$D^* = N_0 - N.$$

From Eq. (10.3),

$$D^* = N_0 - N_0 e^{-\lambda t},$$

so

$$D^* = N_0(1 - e^{-\lambda t}). \tag{10.6}$$

By further substitution,

$$D^* = Ne^{\lambda t} - N = N(e^{\lambda t} - 1).$$

If the total number of atoms of the daughter product, those formed by radioactive transformation plus those initially present at time $t = 0$, is designated as D, then

$$D = D_0 + D^* = D_0 + N(e^{\lambda t - 1}). \tag{10.7}$$

We now have two equations that can be used in dating. Suppose, for example, as with ^{14}C in marine shells, we can reasonably assign a concentration value to N_0, the concentration when the shell was formed. Then we can measure N, the present concentration, by its present radioactivity and solve Eq. (10.2) for t, the age of the shell.

Conversely, it may be easier to measure the concentration of the daughter atoms and also to assign a concentration value to D_0, the initial value of daughter product. An example would be the initial value of ^{40}Ar in a mineral that contains potassium. The potassium concentration can be measured and the ^{40}K concentration can be calculated from that. The D_0 of ^{40}Ar initially present is generally small and can be evaluated on the basis of where the mineral formed and the other argon isotopes present. Then an estimate of D^*, the corrected concentration of ^{40}Ar, can be entered into Eq. (10.7) and the equation solved for t, the age of the mineral. In practice, a variety of considerations must often be dealt with to achieve a well-established date, but the principle is compelling, and in favorable circumstances so are the calculated dates.

When the radioactive parent is a long-lived element such as ^{238}U, and the initial daughter products are radioactive and relatively short-lived, and there has been no chemical separation of the parent and daughter for several times the half-life of the daughter, the activity (the number of disintegrations per unit time) due to each daughter is nearly equal to that of the parent. Then the isotopes are in *secular equilibrium*, and

$$N_1\lambda_1 = N_2\lambda_2 = N_i\lambda_i, \qquad (10.8)$$

where $N_1\lambda_1$ refers to the parent nuclide and $N_2\lambda_2$, $N_i\lambda_i$, refer to a series of one or more daughter nuclides. In several cases we can learn something of interest to seawater chemistry by examining the extent to which the concentrations of the daughters depart from secular equilibrium.

10.2 Radionuclides in seawater

In total, there are known about 280 stable nuclides and about 1700 unstable and therefore radioactive nuclides. Of the latter, about 67 occur naturally on Earth. Here we are concerned with those radioactive nuclides present in the highest concentration and those that have properties that make them useful as tracers of oceanographic processes.

Most of the radioactivity in seawater (Table 10.1) is contributed by potassium-40; in 1 kilogram of seawater ($S = 35‰$) about 743 atoms of ^{40}K disintegrate every minute. Since the half-life of ^{40}K is 1.28×10^9 years, only about 3.5 half-lives of this isotope have passed since Earth was formed. The present whole-Earth inventory of ^{40}K is therefore about 8% of that present when Earth condensed. The next elements of importance are rubidium and uranium. Unlike the decay of ^{40}K, that of ^{87}Rb is simple and produces only stable ^{87}Sr as a final product. The three isotopes of uranium all decay by alpha emission and all produce radioactive daughter nuclides that decay in turn. Some of these will be discussed later.

Potassium is a relatively minor component of sea salt, but it is very important to the physiology of most cells. Many organisms maintain potassium as the main cation inside their cells, so that even in terrestrial organisms potassium is often more concentrated there than in seawater. For example, the overall concentration in the average human male is reported to be 48 mmol of K per kg of body weight, and in the

Table 10.1 Concentrations of quantitatively important, naturally occurring, and nearly conservative radioactive nuclides in seawater

Nuclide	Concentration, kg^{-1}	Decay mode	Half-life, yr (λ [yr^{-1}])	Activity, kg^{-1}
^{40}K	0.01171% of total K = 1.196 μmol	β^- (89.3%), e.c., β^+ (10.7%)	1.277×10^9 (5.428×10^{-10})	12.39 Bq 743 dpm
^{87}Rb	27.84% of total Rb = 0.39 μmol	β^-	4.81×10^{10} (1.36×10^{-11})	0.109 Bq 6.52 dpm
^{238}U[(a)]	99.3% of total U = 13.5 nmol	α	4.468×10^9 (1.55×10^{-10})	0.040 Bq 2.40 dpm
^{235}U[(a)]	0.720% of total U = 97.9 pmol	α	7.038×10^8 (9.85×10^{-10})	1.84×10^{-3} Bq 0.11 dpm
^{234}U[(a)]	0.0062% of total U = 0.85 pmol	α	2.455×10^5 (2.82×10^{-6})	0.046 Bq 2.76 dpm

Data from Tables 4.1, 9.2; Firestone and Shirley 1996; U from Chen *et al.* 1986; ^{87}Rb from Nebel *et al.* 2010; $^{234}U/^{238}U$ from Anderson *et al.* 2010.

All values are normalized to $S = 35‰$. The natural abundance of each isotope, as percent of the total for each element, is on the mole % basis, the same as the % of total number of atoms. For each isotope, both the half-life, in years, and the decay constant (λ) are given, and the activity is expressed both as Bq kg^{-1} and as disintegrations per minute per kilogram (dpm kg^{-1}). To convert activities in Bq kg^{-1} to pCi kg^{-1}, multiply by 27.03

(a) The activity given for each isotope of uranium refers to the first decay only. Uranium-235 decays through a series of 11 steps to ^{207}Pb, so if all daughter products remained in the same water for a long enough period of time to reach secular equilibrium the activity of the whole series would be $11 \times 1.84 \times 10^{-3} = 2.02 \times 10^{-2}$ Bq kg^{-1}.

Uranium-238 decays to ^{234}U in 3 steps and ^{234}U to ^{206}Pb in 11 steps (some steps also include some minor branches). If all the uranium isotopes were in secular equilibrium the total activity due to uranium would therefore be 0.58 Bq kg^{-1}. However, some of the intermediate nuclides, especially thorium, are insoluble and are rapidly lost from the water column, so subsequent decay steps occur mostly in or near the sediments. For the water column as a whole, the total radioactivity due to uranium and its daughters is therefore reduced but variable, depending on local conditions.

female 39.4 mmol of K per kg (Miller and Marinelli 1956). These values were determined by measuring the radioactivity of 12 male and 3 female people, with counters set to measure the γ-rays from the decay of ^{40}K. In the typical 70-kg human male, the radioactive disintegrations of potassium amount to 58.3 Bq kg^{-1}. This amounts in total to 4080 Bq, meaning 4080 radioactive disintegrations per second inside the typical adult human male. For typical 55-kg human females, the rate is 2630 per second.

Similarly, the radioactivity that is due to the potassium in seawater can be measured directly. However, the measurement of the other radioactive nuclides in seawater, present as they are in very small amounts, usually requires extensive chemical separation to concentrate the elements involved, with due attention paid to possible contamination and to the special techniques required for very low-level counting.

Table 10.2 Partial list of non-conservative radioactive nuclides in seawater

Nuclide	Decay mode	Half-life	Activity, dpm kg^{-1}	Total mass on Earth, kg
		Produced by cosmic rays		
^{3}H	β$^-$	12.32 yr	0.04	7
^{7}Be	e.c.	53.2 d	0.5	3×10^{-3}
^{10}Be	β$^-$	1.51×10^{6} yr	1×10^{-3}	4×10^{-5}
^{14}C	β$^-$	5730 yr	0.32	48 800
^{14}C (atmosphere only)			847	
^{32}Si	β$^-$	153 yr	3×10^{-5}	0.5
^{32}P[(a)]	β$^-$	14.26 d	1×10^{-3}	50×10^{-6}
^{33}P[(a)]	β$^-$	25.34 d	3×10^{-3}	100×10^{-6}
		Produced by decay of uranium		
^{234}Th	β$^-$	24.10 d	–	–
^{230}Th	α	75 380 yr	–	–
^{226}Ra	α	1600 yr	0.068	–
^{222}Rn	α	3.824 d	–	–
^{210}Pb	β$^-$	22.2 yr	–	–
		Produced by humans (bomb testing, other activities)		
^{90}Sr	β$^-$	28.78 yr	1	–
^{137}Cs	β$^-$	30.07 yr	0.5	–
^{239}Pu	α	24 110 yr	1×10^{-3}	–
^{3}H	β$^-$	12.32 yr	–	~6[(b)]
^{14}C	β$^-$	5730 yr	–	1560[(c)]

Data from Burton 1975, Lal and Lee 1988, Broecker *et al.* 1995, Firestone and Shirley 1996. Total mass of ^{14}C calculated from ^{14}C of reservoirs, agrees with Usoskin and Kromer (2005).
Concentrations of some of these nuclides vary greatly, and the activities listed (as dpm kg^{-1} of seawater) are only approximate guides.

[(a)] It might at first seem silly even to list ^{32}P and ^{33}P, with the implication that they could be oceanographically useful, when the total inventory on Earth is so small, about 51 mg for ^{32}P. Simple calculation shows, however, that this small amount provides 190 atoms on each square centimeter on Earth or about 25 000 atoms per cubic meter in the upper mixed layer of the ocean. This was sufficient for Lal and Lee (1988) to show that it is indeed possible to collect the phosphate from a few cubic meters of seawater and measure the ^{32}P in it, as well as that in zooplankton collected from several hundred cubic meters, and show the age difference between the two, which is useful to know.

[(b)] About 100 kg of tritium was released by bomb explosions up to 1963, but by 2010 more than 94% of this had decayed.

[(c)] Of this total, roughly 61% is now in the ocean, 20% in the atmosphere, and the remainder in the terrestrial biosphere.

The radioactive nuclides noted in Table 10.1 are, with one exception, all *primordial*, meaning here that they were part of Earth during its initial formation. The exception is ^{234}U, which is produced by decay from primordial ^{238}U. The radioactive nuclides listed in Table 10.2 are all produced either by cosmic rays in the atmosphere, by the decay of primordial uranium, or by nuclear explosions in the atmosphere. (Small, and on this scale negligible, amounts of ^{3}H and ^{14}C are also released from industrial, medical, and research production and use of these isotopes.) Because of their different sources, different decay rates, and different chemistries, each of these isotopes provides a tracer for different processes in the ocean, and an opportunity to establish the rates of some of these processes. Here we will discuss some examples from the series of uranium daughter products and from carbon-14.

10.3 The uranium series

Uranium-238 is the most abundant isotope of this element, and for this reason the decay series beginning with this nuclide is the most important. Nineteen daughter products are formed from ^{238}U (Figure 10.3), with the last one being stable ^{206}Pb. Some of the intermediate daughter products are so short-lived that no geochemical process can act on them effectively before they are transformed into the next element in the series. Useful information can be obtained, however, by examining the concentration and behavior of at least ^{234}Th, ^{230}Th, ^{226}Ra, ^{222}Rn, and ^{210}Pb. Three of these are discussed here.

10.3.1 Thorium-234

Uranium is apparently conservative in seawater, being relatively soluble and present in a constant or nearly constant ratio to the salinity. The element thorium, however, is quite insoluble, and *particle-reactive*. Any thorium introduced into seawater is rapidly attached to the surfaces of particles, especially mineral particles but also organic particles, that may be present in the water (Figure 10.4). As these particles sink, the thorium is removed from the water column in a process called *scavenging*. This removal can be measured experimentally in large mesocosm tanks (Figure 10.5). More importantly, it can be measured in the ocean by noting the deficiency of ^{234}Th in the water column, relative to the amounts that should be present from the radioactive decay of ^{238}U, as described below.

In the case of a very long-lived parent and a short-lived daughter, where both nuclides remain together in the same place, the concentration of the daughter eventually reaches a steady state so that the rate of radioactive decay of the daughter equals the rate of production by radioactive decay of the parent.

From Eq. (10.8) and the data in Tables 10.1 and 10.2, and Figure 10.3, the expected concentration of ^{234}Th in surface seawater at $S = 35$ ‰ is calculated to be 2.00×10^{-19} mol kg^{-1}, which should be decaying at a rate of 2.4 atoms per minute in each kilogram of seawater. In the central gyres of the open ocean, where biological productivity is quite low, the concentrations of ^{234}Th (determined by concentrating the thorium from several kilograms of seawater and measuring the radioactive emissions) are found to be close to this expected value. However, in biologically productive regions, and in coastal waters where the sediment load can be significant and which are often also biologically productive, the concentrations are much reduced. In these latter situations the ^{234}Th is lost from the water by two processes,

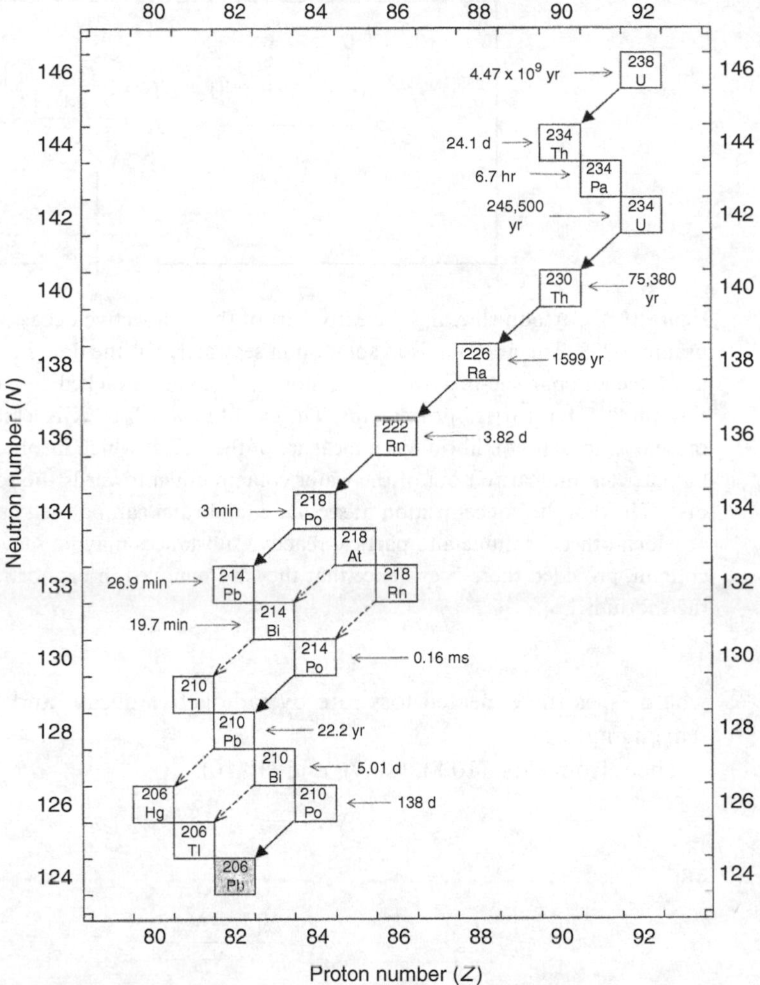

Figure 10.3 The decay of radioactive ^{238}U to stable ^{206}Pb. Each decay involves either the emission of an α particle, resulting in a nucleus with two less neutrons and two less protons, or emission of a β particle, resulting in a nucleus with one less neutron and one more proton. Several of the branches are not quantitatively important and are indicated by dashed lines. Many of the 20 nuclides noted are either very short-lived or are part of quantitatively insignificant pathways. Half-lives range from milliseconds (ms) to minutes (min), hours (hr), days (d), or years (yr).

radioactive decay and scavenging by particles that sink out of the surface water. At steady state, the total loss flux is given by

$$\text{loss rate} = N_{2a}\lambda_{2a}, \tag{10.9}$$

where the subscript *a* refers to the actual concentrations present and the overall total loss rate constant. The overall rate constant is the sum of two rate constants:

$$\lambda_{2a} = \lambda_{2r} + \lambda_{2s}, \tag{10.10}$$

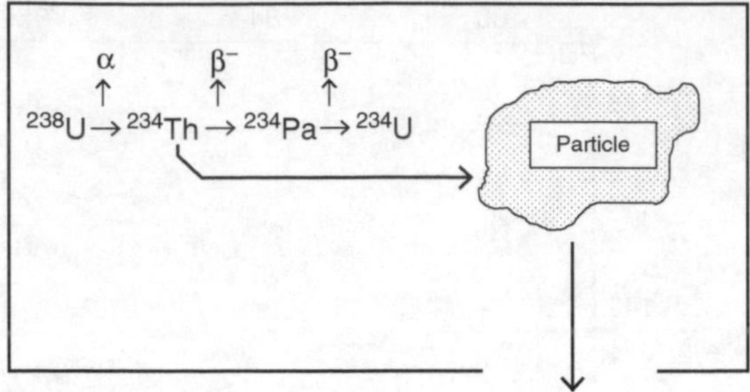

Figure 10.4 Cartoon showing the early part of the radioactive decay series beginning with uranium-238. The uranium is in solution in seawater, but the daughter isotope thorium-234 is insoluble and particle-reactive. Some thorium becomes attached to particles; as the particles sink this ^{234}Th is carried downwards. The extent to which there is less thorium than should be present at secular equilibrium is a measure of the rate at which insoluble thorium is scavenged by particles and carried out of the water column down towards the sediments. The departure of ^{234}Th from the concentration at secular equilibrium can be used as a measure of the rate at which other insoluble and particle-reactive substances may be scavenged from the water column, provided there is evidence that they are removed in proportion to the removal of the thorium.

where λ_{2r} is the expected loss rate by radioactive decay, and λ_{2s} is the loss rate by scavenging.

Then, from Eqs. (10.8), (10.9), and (10.10):

$$N_1\lambda_1 = N_{2a}(\lambda_{2r} + \lambda_{2s})$$

and

$$\lambda_{2s} = \frac{N_1\lambda_1 - N_{2a}\lambda_{2r}}{N_{2a}}. \tag{10.11}$$

If the ^{234}Th concentration is measured in a sample of water, and the salinity is known so that $N_1\lambda_1$ is also known, the only remaining unknown is λ_{2s}, the scavenging rate constant for thorium. Values of this scavenging rate constant have been measured in many places (e.g. Figure 10.6). In the example given, the half-lives of thorium relative to the scavenging process range from perhaps 2 to 20 days in Narragansett Bay, a relatively shallow (8 m) arm of the sea, while in the New York Bight (depth around 50 m) the half-life is typically longer. There is an especially strong seasonal dependence observable in Narragansett Bay, with the scavenging rates being ten times faster in the summer than in the winter. The interpretation of these results is that more sediment particles are resuspended from the sediments in the shallower region, and that both sediment resuspension and chemical uptake processes are fastest in the warm summer season. Additional scavenging due to biologically produced particles is also likely to be involved.

The importance of such observations is that from them we can learn the rates at which particles (biological or inorganic) and particle-reactive substances are carried downwards from surface seawater, perhaps into the sediments, and can begin to

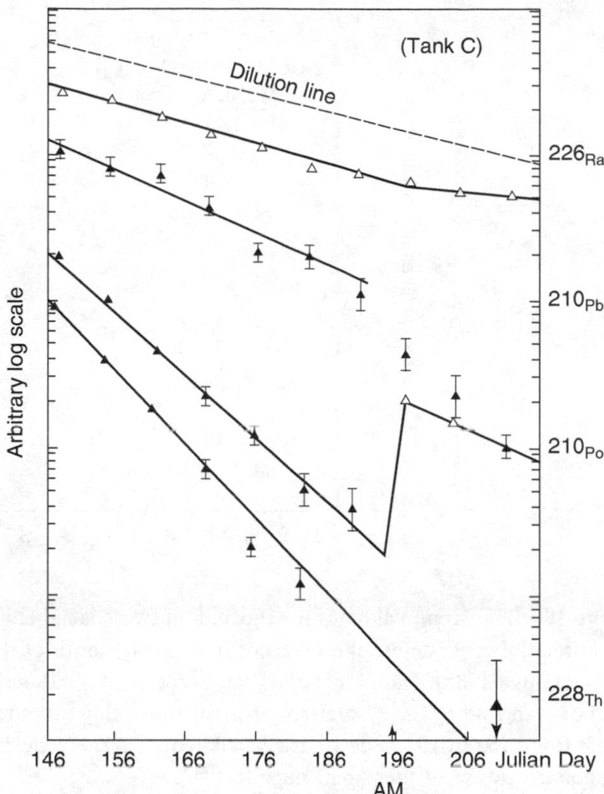

Figure 10.5 Loss of radioisotopes from the water column of an experimental ecosystem tank at the Marine Ecosystems Research Laboratory (MERL), University of Rhode Island. Small amounts of these isotopes were added to the tank on day 146 and the concentrations remaining in the water column were measured during the subsequent 70 days (Santschi *et al.* 1980b). Here we are not concerned with the change in conditions introduced on day 196, only with the loss rate of thorium under normal operating conditions. Several such experiments yielded the scavenging rates shown in Figure 10.6.

understand the factors that affect and control these rates. From this it is also possible to guess the rates at which some anthropogenically introduced metals and organic chemicals, if particle-reactive, will be removed from seawater (Broecker *et al.* 1973; Hinga 1988) and carried to the sediments. Such chemicals should behave somewhat like thorium (Santschi *et al.* 1980a).

10.3.2 Radon-222

Radon is a noble gas, with no chemical reactivity under the conditions in seawater. While the thorium daughters of ^{238}U are very insoluble and are largely scavenged to the sediments, radium is more soluble and, when ^{226}Ra is formed in the sediments from ^{230}Th (Figure 10.3), some diffuses into the overlying water and mixes back into the ocean again. In addition, some ^{226}Ra is carried into the ocean by rivers.

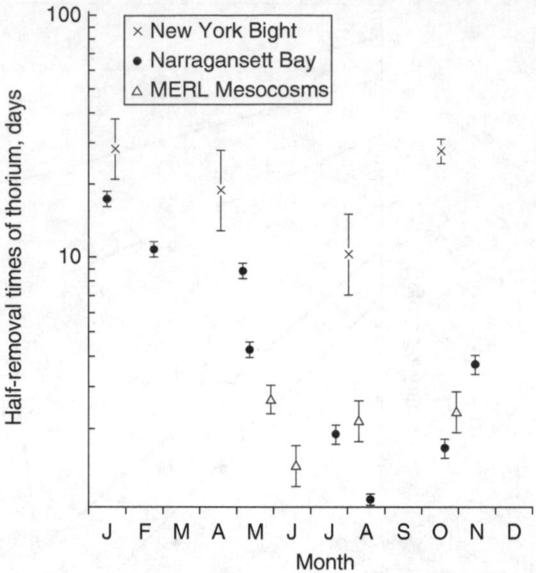

Figure 10.6 Half-removal times for thorium in two coastal environments and from experimental ecosystems, plotted according to the month of the year in which the data were taken (Hinga 1988). Data for the MERL experimental mesocosms are from experiments such as shown in Figure 10.5. Data from Narragansett Bay (about 8 m deep) and from the New York Bight (50 to 100 m deep) are calculated from the deficiency in natural ^{234}Th relative to the concentration of the parent nuclide ^{238}U.

The ^{226}Ra in seawater is therefore a relatively constant source of ^{222}Rn. The ^{222}Rn in turn decays and, with a half-life of only 3.824 days, it is an important tracer for processes acting on a timescale of a few days.

The oceanic process most studied by the use of radon measurements is air–sea gas exchange. Radon diffuses into the air from soils and rocks, but its lifetime is sufficiently short that air over the ocean far from continental influences has only very low concentrations, and radon in the surface layers of seawater tends to diffuse into the air (Figure 10.7). In a manner analogous to the calculation of the scavenging rate for thorium, it is possible to calculate a loss rate for radon to the atmosphere and then to solve for the gas exchange coefficient, e_G (Chapter 5). To show how this calculation is done, we can use the example in Figure 10.7. As with the U/Th pair discussed earlier, the parent ^{226}Ra is very long-lived relative to the daughter ^{222}Rn, which allows us to use the simplified Eqs. (10.8) and (10.11). Again let N_1 be the parent and N_2 the daughter, and assume that the water is in steady state relative to gas exchange processes. From Figure 10.7, the activity of the parent and daughter in secular equilibrium, Eq. (10.8), would be (converting to units per cubic meter, for convenience),

$$N_1\lambda_1 = N_2\lambda_2 = 68 \text{ dpm m}^{-3}.$$

The observed rate at which radon disappears is, from Eq. (10.10),

$$\lambda_{2a} = \lambda_{2r} + \lambda_{2s},$$

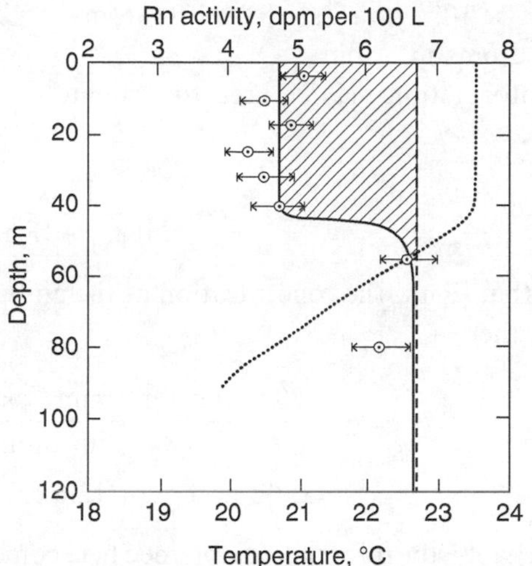

Figure 10.7 Concentrations of radon-222 measured (Broecker and Peng 1982) in surface water at GEOSECS station 57 (24° S, 35° W). The dotted line is temperature, indicating that the depth of the upper mixed layer is about 40 m. The dashed line is the expected steady-state concentration of ^{222}Rn if no loss is occurring (except by radioactive decay). The measured radon data are shown by circles, and the deficiency by the shaded area. Such data can be used to estimate the natural rate of gas exchange in this region, under the prevailing weather conditions averaged over the previous several days. Note the remarkably low concentrations of radioactive substance that can be measured. The observed value of 4.8 dpm per 100 L corresponds to a concentration of only about 382 atoms of ^{222}Rn in each liter of seawater.

where (in this case) λ_{2s} is the rate of escape through the surface of the water. Then, from Eq. (10.11) and the observed decay rate (48 decays min^{-1} m^{-3}, as shown in Figure 10.7) in the upper part of the water column,

$$\lambda_{2s} = \frac{N_1\lambda_1 - N_{2a}\lambda_{2r}}{N_{2a}} = \frac{(68 - 48)\text{ atoms min}^{-1}\text{m}^{-3}}{N_{2a}},$$

while from Eqn. (10.1) and the decay constant calculated from data in Figure 10.3, we can estimate the concentration of radon (N_{2a}) in each cubic meter of the upper mixed layer,

$$N_{2a} = \frac{48\text{ atoms min}^{-1}\text{m}^{-3} \times 1440\text{ min d}^{-1}}{0.1813\text{ day}^{-1}} = 3.81 \times 10^5\text{ atoms m}^{-3},$$

$$\lambda_{2s} = \frac{20\text{ atoms min}^{-1}\text{m}^{-3}}{3.81 \times 10^5\text{atoms m}^{-3}} = 52.5 \times 10^{-6}\text{min}^{-1}.$$

The flux of radon from the surface may be estimated by noting that the upper mixed layer, the volume from which the radon has escaped, is here about 40 m deep, so that under 1 m^2 of surface there is a column of water containing 40 m^3 from which the radon has escaped. The total escape flux, therefore, is:

$$\text{flux} = \lambda_{2s}\, N_{2a} \times \text{total volume affected}$$

$= (52.5 \times 10^{-6}\ \text{min}^{-1})\ (3.81 \times 10^5\ \text{atoms m}^{-3} \times 40\ \text{m})$
$= 800\ \text{atoms m}^{-2}\ \text{min}^{-1}$.

This information may be used to evaluate the exchange coefficient, according to Eq. (5.8),

$$e_G = \frac{\text{Flux}}{[\text{Rn}]_I - [\text{Rn}]_w}.$$

Assume that $[\text{Rn}]_I$, the concentration at the air–water boundary, is negligible at this location, then

$$e_G = \frac{800\ \text{atoms m}^{-2}\text{min}^{-1} \times 1440\ \text{min d}^{-1}}{3.81 \times 10^5\ \text{atoms m}^{-3}}$$
$$= 3.02\ \text{m d}^{-1}, \text{or } 12.6\ \text{cm hr}^{-1}.$$

Remarkably, the information obtained here comes from measuring the concentration of a gas in the upper mixed layer amounting to only a few hundred atoms per liter.

Measurements such as this have been made in many parts of the ocean. Combined with data on the wind speed and temperature of the water they provide large-scale estimates of the gas exchange rates over those parts of the ocean where such measurements are possible. By adjusting according to the Schmidt number, it is possible to scale the gas exchange rates to other gases under a variety of conditions (Chapter 5) and thus, for example, to provide estimates of the exchange rates of carbon dioxide or oxygen between the atmosphere and the surface ocean.

10.3.3 Lead-210

The ^{222}Rn formed from ^{226}Ra in water, soils, and sediments decays with a half-life of 3.824 days to ^{218}Po, which is very short-lived, having a half-life of only 3 minutes. The next daughter nuclides also have very short half-lives, until the nuclide ^{210}Pb, which has a half-life of 22.2 years. This half-life, coupled with the chemistry of lead, makes ^{210}Pb a most useful nuclide for examining rates of sedimentation in lakes and in marine areas with relatively fast sedimentation rates.

In the sediments, there is always some ^{226}Ra, the concentration of which can be measured by observing its radioactivity. The radium decays to radon, and the latter decays through several quick steps to ^{210}Pb. After some time, these isotopes should be in secular equilibrium (allowing for the possibility that some radon may diffuse out of the sediment, depending on the depth interval measured), and the activity of ^{210}Pb supported by the ^{226}Ra should be equal to that of the ^{226}Ra in the sediments. There are other sources of ^{210}Pb, however. One is from the ^{226}Ra in the water above. Any ^{222}Rn formed there (or that came from the sediment) that had no chance to escape to the atmosphere will decay to ^{210}Pb, which is quite insoluble and particle-reactive (like thorium) and is carried down to the sediments. In addition, in lake and near-shore marine areas, there is a source from the air. Radon gas, formed in soils, can escape to the atmosphere, especially if the soil is porous. With a half-life of 3.8 days the radon can be carried in the atmosphere over all lakes, estuaries, and

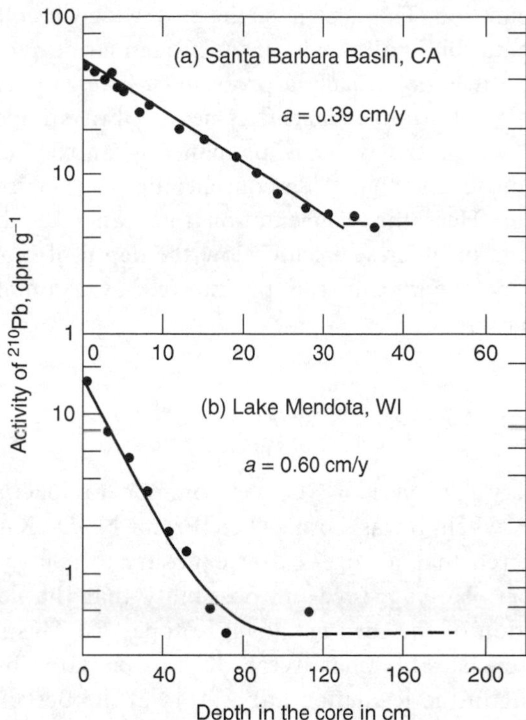

Figure 10.8 Concentration of unsupported lead-210 with depth in sediment cores from the Santa Barbara Basin, California, and from Lake Mendota, Wisconsin (Faure 1986). These regions were selected to show relatively simple regimes of sedimentation, nearly constant through time and not confused by bioturbation. At depth in the sediment, all the ^{210}Pb is derived through a series of radioactive decays from ^{226}Ra, which can be measured and the amount of supported ^{210}Pb calculated. The excess ^{210}Pb or unsupported ^{210}Pb is the additional ^{210}Pb derived from ^{222}Rn originating from some other region. The usual assumptions are that the radon enters the atmosphere from soils in the region and decays to ^{210}Pb, which falls out of the atmosphere and ends up in the sediments; and that the annually averaged rate of this process is constant, or nearly so.

At some depth in the sediment the unsupported ^{210}Pb will have decayed to the point that this excess can no longer be distinguished from the constant background provided by sources within the sediment. The depth at which this occurs will be deeper in regions with rapid sediment accumulation than in regions with slow accumulation. The slope of the activity of unsupported ^{210}Pb with depth is a measure of (a), the rate of sedimentation.

continental-shelf areas, at least. The ^{222}Rn in the air decays to ^{210}Pb, and this is then scavenged from the atmosphere by rain or dry fallout and, being insoluble and particle-reactive, it ends up in the sediments. This provides *excess* ^{210}Pb or *unsupported* ^{210}Pb, excess in concentration to the amount supported by decay of the ^{226}Ra in the sediments. Once it is buried by new layers of sediment, no further addition from outside occurs and the ^{210}Pb decays with a half-life of 22.2 years (Figure 10.8). By measuring how the concentration of ^{210}Pb changes with depth in the sediments, it is possible, in favorable circumstances, to calculate the rate of sedimentation. This is perhaps the most widely used technique to estimate the rate of sedimentation in all rapidly sedimenting marine and fresh-water areas.

While it is not usually a problem in lakes, the effect of animals in stirring and mixing the bottom sediments is nearly always important in marine environments, except where the water is permanently anoxic. In most marine sediments it is common to find that ^{210}Pb concentrations in the upper layers are variable or even uniform. Here the sedimentation rate can often be estimated by examining the gradients in older sediment below the depth of biological mixing. Conversely, the extent of the mixing can be inferred by examining the details of the isotope distributions.

10.4 Carbon-14

The very existence of ^{14}C, the common radioactive isotope of carbon, was not known when I was born. Then Franz N. D. Kurie, working at Yale in 1934, discovered that nitrogen, after exposure to neutrons, produced tracks in a cloud chamber. He suggested the possibility that the neutrons might have caused the formation of the then unknown isotope ^{14}C. This suggestion was later confirmed, and a measurable quantity of ^{14}C was prepared by Martin Kamen in 1940, then working in the Radiation Laboratory at the Berkeley campus of the University of California. Among the many dramatic advances that followed (all within my lifetime) was the enormous development of biochemistry and molecular biology made possible by using ^{14}C for detailed tracking of carbon flows in biochemical processes.

By the late 1930s, it had been suggested that ^{14}C might be produced by neutrons derived from cosmic rays hitting the atmosphere. The term *cosmic ray* is a general expression for a variety of energetic particles that bombard Earth from space. Some are galactic (a few may be extra-galactic) in origin, while most are from the Sun (however, those from the Sun are not energetic enough to make ^{14}C). Cosmic rays consist mostly of protons with a wide range of energies. Some α particles and a few heavier nuclei also contribute; the latter can be extremely energetic. When energetic cosmic rays hit nuclei in atoms of O, N, Ar, Kr, or other atmospheric components, the result may be a variety of broken pieces of the nuclei, called *spallation* products, including neutrons that may in turn also interact with various atoms to yield additional products. Prominent among the products of these reactions are the radioactive isotopes ^{3}H, ^{10}Be, ^{14}C, ^{26}Al, ^{32}Si, ^{36}Cl, ^{39}Ar, and ^{81}Kr.

Most of the carbon-14 is produced in the atmosphere by neutrons that react with nitrogen:

$$^{14}N + n \rightarrow {}^{14}C + p.$$

In this reaction a neutron hits the nucleus of ^{14}N, is stopped and incorporated into the nucleus, and a proton is ejected. The mass number does not change, but the atomic number is decreased by one unit. The ^{14}C formed as a product of the reaction is radioactive, with a half-life of 5730 ± 40 years, and decays in the following manner:

$$^{14}C \rightarrow ^{14}N + \beta^- + \text{a neutrino}.$$

The free radioactive carbon atoms formed in the atmosphere must eventually be oxidized to $^{14}CO_2$, although details of the reactions involved are not well known. The atmospheric $^{14}CO_2$, in turn, mixes with the other reservoirs; and evidence from the distribution of $^{14}CO_2$ suggests that CO_2, and its tracer $^{14}CO_2$, exchanges from the atmosphere into the surface water of the oceans in less than 5 years. It is taken up into the biosphere as plants incorporate CO_2 into organic matter.

By 1946, Willard Libby, at the University of Chicago, had calculated that the production rate in the atmosphere, now estimated to average about 16 000 atoms m^{-2} s^{-1}, would be great enough that the isotope could be detected in natural samples, and by the next year, 1947, it had been done. By the early 1950s, it was determined that the concentration of $^{14}CO_2$ absorbed from the atmosphere into plant tissue was such that the disintegration rate in each gram of carbon in the modern plant tissue was 13.56 dpm (= 2.71 Bq mol^{-1}). This rate could, with care, be measured accurately. Libby suggested that, after an organism dies, the carbon incorporated from CO_2 into its structure should undergo no further exchange with the atmosphere, as long as the biological structure remains intact. The concentration of ^{14}C within, for example, dried wood, should change only by radioactive decay. He suggested that the age of biological materials could be ascertained by measuring the ^{14}C content. His leadership in establishing the technique of using ^{14}C for dating samples of carbon-containing materials was recognized by the award of the 1960 Nobel prize in chemistry. Dating by ^{14}C has been a standard technique ever since.

The concentration of ^{14}C in atmospheric carbon dioxide is controlled by production-rate processes, by its own radioactive decay, by processes through which the CO_2 in the atmosphere exchanges with carbon in other reservoirs that may have different concentrations of ^{14}C, and, during the last hundred years, by human-caused effects. The flux of galactic cosmic rays is not known to vary on an accessible timescale. The changing output of solar cosmic rays and associated variations in the magnetic field certainly cause changes in the numbers of galactic cosmic rays that reach the upper atmospheres of the Earth. The magnetic field of the Earth varies over time, and this variation changes the trajectory of at least some of the galactic cosmic rays and the numbers that hit the atmosphere. There is some variation in the production rate due to these effects, over short and long time periods, though the magnitudes involved are not easy to evaluate (Usoskin and Kromer 2005).

Before the effects caused by humans, concentrations of ^{14}C in the atmosphere were also affected by the rates of the various exchange processes and the magnitudes of other reservoirs where ^{14}C can reside long enough to be reduced in concentration by radioactive decay. For example, the rate of erosion of limestone exposed on land might vary, and there could be changes in the size of the standing stock of organic matter in trees and in soils, and especially in the large amount of CO_2 dissolved in the ocean. All these effects can change the concentration of ^{14}C in atmospheric carbon dioxide.

Humans have caused and are still causing significant changes, which fortunately do not affect the dating of archeological relics or geological processes occurring more than about a hundred years ago. The burning of fossil fuels in significant amounts has diluted the atmospheric concentration of ^{14}C, because the fossil fuel has lain underground for so long that all ^{14}C has long since disappeared. This dilution of atmospheric CO_2 with ^{14}C-free carbon is called the *Suess effect*, after Hans Suess who first suggested that the phenomenon could be significant. The Suess effect is apparent in wood from trees growing subsequent to about 1880, up until the 1950s. For this reason, wood from trees harvested in the middle of the nineteenth century was used to establish the natural atmospheric concentrations of ^{14}C as a baseline to be used in dating earlier objects (Box 10.1). The explosion of nuclear devices in the atmosphere during the two decades following the first atomic bomb (in 1945) introduced enough ^{14}C to completely override the Suess effect and to approximately double the atmospheric concentration (Table 10.2); most of this excess is now in the ocean and it provides a tracer of several oceanic processes.

The availability of wood dated accurately to within one year by counting tree rings makes it possible to examine the ^{14}C content of the atmosphere when the tree ring was formed. From such evidence (Figure 10.9) it is clear that the specific activity of the CO_2 in the atmosphere has varied significantly. This is enough to cause errors of several hundred years in the assigned dates of older archaeological objects if the effect is not corrected for. These changes may be great enough to influence the conclusions that are drawn about terrestrial or oceanic exchange processes calculated from modern distributions of CO_2. (Conversely, the changes can be examined to evaluate the magnitudes and nature of the effects that caused them.)

When samples of ocean water are collected, the CO_2 extracted, and the concentration of ^{14}C in the carbon measured, it is found that everywhere the CO_2 in the ocean is "older" than the CO_2 in the atmosphere (Figure 10.10). That is, there is less ^{14}C per unit of total carbon than is found in the atmosphere. Usually, the ^{14}C content is expressed in Δ-notation (Box 10.1), because this gives the measured concentration of ^{14}C in the carbon sample analyzed (corrected for isotopic fractionation) but does not specify the age. An apparent age could be calculated, based on the assumption that, when it was isolated from the atmosphere, the water sample started with a ^{14}C content in equilibrium with the atmosphere, and applying the radioactive decay equation (such as Eq. (10.2)). As we will see, however, the surface ocean is not generally in equilibrium with the ^{14}C in the atmosphere, because the carbon in surface ocean water has a lower specific activity. The actual "age" depends on definition and model assumptions.

Among the major ocean basins, the oldest CO_2 is found in the deep water of the North Pacific. The distribution within the ocean basin is complex (Figure 10.11), and carries much information about the details of deep circulation. The information is difficult to sort out, however, for several reasons: the complexity of the distribution, the natural variation in atmospheric concentration so that it is not easy to assign precisely the atmospheric composition when the water was in contact with the atmosphere, and the fact that in addition to the ^{14}C present at the time of sinking, the deep water also has a contribution of ^{14}C released from particles sinking directly from the surface. Finally, the best modern data all come from a time subsequent to

Box 10.1 Notes on calculation and reporting ^{14}C activity or concentration

Wood grown in nineteenth century is the original basis for the assignment of an absolute international standard of activity. That wood grew before there was a measurable influence of the Suess effect caused by the burning of fossil fuels. Samples of this wood had, when measured shortly after the middle of the twentieth century, an activity that, corrected for radioactive decay since it grew, amounted to an average of 13.56 disintegrations per minute in each gram of carbon (dpm g^{-1} C). In SI units this becomes 0.226 Bq for each gram of carbon. This is the activity that wood would have had if it had grown in 1950 in an atmosphere similar to that of the middle nineteenth century. The reference generally cited, if often incorrectly, for the source of the measurements is Karlén *et al.* (1964).

The National Bureau of Standards (now NIST) prepared a standard from oxalic acid that was carefully intercalibrated with the wood samples for both the activity of ^{14}C and for $\delta^{13}C$. Because most natural samples have undergone some isotopic fractionation during the process of formation from the CO_2 in air, it is necessary to account for this by correcting to some standard value by using the value for ^{13}C. The fractionation of ^{14}C relative to ^{12}C is theoretically not exactly twice that with ^{13}C, but it is close enough that the factor of 2 is commonly used. The activity of the wood standard relative to that of the oxalic acid standard is given by:

$$A_{ON} = 0.95A_{Ox}\left(1 - \frac{2(19 + \delta^{13}C)}{1000}\right),$$

where A_{ON} is the normalized activity (generally in dpm g^{-1} or Bq g^{-1}) to be assigned to the standard, A_{Ox} is the measured activity, and $\delta^{13}C$ the value for the standard, the last two both measured in the laboratory during the calibration process. If, as expected, the $\delta^{13}C$ measurement is the same as when the standard was established, namely –19‰, then this correction disappears. Subsequently prepared standards are also calibrated to this standard.

The measured activity of a sample (A_s) is similarly corrected according to the degree of isotopic fractionation by normalizing relative to the ^{13}C content of the nineteenth-century wood, which had a $\delta^{13}C$ value of –25‰. The parameter A_{SN} is the normalized sample activity:

$$A_{SN} = A_S\left(1 + \frac{2(25 + \delta^{13}C)}{1000}\right).$$

By arbitrary agreement, time zero for assigning dates is 1950; this honors the year of publication of the first successful dating of an object.

Sometimes there may be more than one model that can be used for assigning an age to a sample, and with some kinds of samples (e.g. those of ocean water) it may not be appropriate to assign a specific age; in these cases it is common to list the ^{14}C in $\delta^{14}C$ units, which can also be adjusted for isotopic fractionation, after measuring the $\delta^{13}C$, to yield $\Delta^{14}C$ units:

$$\delta^{14}C = \left(\frac{^{14}A_S}{^{14}A_{ON}} - 1\right)1000;$$

$$\Delta^{14}C = \delta^{14}C - 2(\delta^{13}C + 25)\left(1 + \frac{\delta^{14}C}{1000}\right);$$

This fractionation-corrected $\Delta^{14}C$ unit (Stuiver and Polach 1977) is used to express the ^{14}C concentration in ocean water samples and other geophysical measurements.

One of the confusing features of the practice of dating with ^{14}C is that the results are commonly reported in terms of "conventional radiocarbon age" or "radiocarbon years". By convention, this is the age calculated on the basis of a half-life of 5568 years. This half-life was an early estimate, whereas the true half-life is currently (and has been since 1962) believed to be 5730 years. The reported age in "radiocarbon years" must be corrected by a calibration graph or algorithm to get the true age in "calendar years." In some sense this is not as odd as it seems, because the relationship of age to ^{14}C content has in any case changed over time, due to changes in the ^{14}C content of the atmosphere, and some correction generally has to be made, in any case.

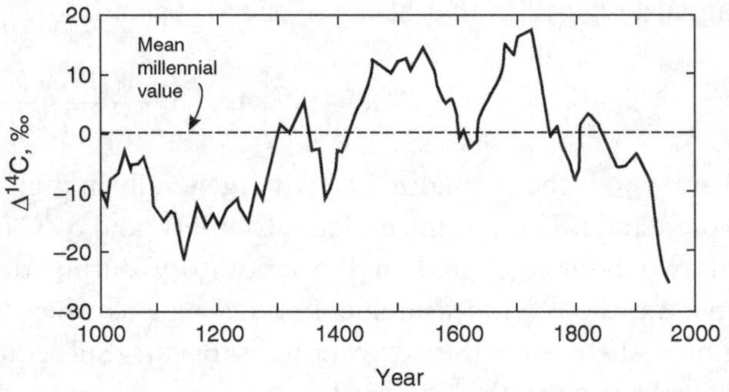

Figure 10.9 Ratio of $^{14}C/C$ expressed in $\Delta^{14}C$ notation (Box 10.1) for atmospheric CO_2 during the last 1000 years (figure from Broecker and Peng 1982). Samples of wood, dated accurately by counting tree rings, were analyzed for ^{14}C, and the values corrected for radioactive decay to the standard year 1950. Evidently, concentrations of ^{14}C in atmospheric CO_2 have varied through time. The record here terminates just as significant testing of nuclear explosives in the atmosphere began. The drop since 1900 is due to the Suess effect. The reasons for the other variations are not well known. Some evidence suggests that they are due to variations in the output of cosmic rays and other forms of energy by the Sun, as well as changes in Earth's magnetic field, but other causes include changes in the relative sizes of some reservoirs of carbon dioxide. ■

atmospheric testing of nuclear weapons, so that surface water also has a large amount of recently added anthropogenic ^{14}C.

The deep-water ^{14}C apparent ages of up to 2300 years (Figure 10.10) do not reflect the time since that water was at the surface, however, because surface water always

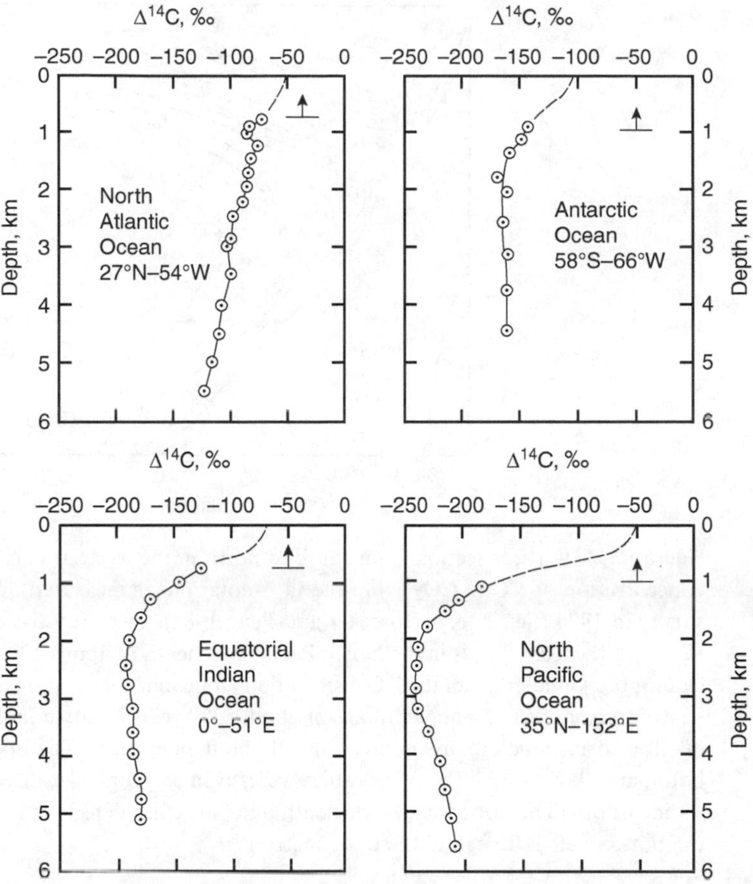

Figure 10.10 Vertical distributions of ^{14}C in the CO_2 of seawater at stations in each of four major oceanic regions, expressed in the delta notation (Broecker and Peng 1982). The upper portion (approximately the upper 1 km) of the water column at each station is represented by a dashed line. This is the region affected by artificially increased ^{14}C concentrations originating from nuclear explosions in the atmosphere, and the dashed lines are meant to suggest the concentrations probably present before the atmospheric testing of nuclear devices began. The zero point on the scale corresponds to the ^{14}C content of the pre-industrial atmosphere. A $\Delta^{14}C$ value of –250 corresponds to an apparent age of about 2400 years. The difference between deep water in the North Atlantic and the North Pacific suggests that it must take about 1000 years for water from the deep North Atlantic to make its way to the North Pacific. ■

shows a surprisingly large apparent age. Some evidence for this comes from the few samples collected and analyzed before the onset of nuclear contamination. The best evidence, however, comes from analyzing accurately dated samples of coral (Figure 10.12). Evidently, apparent ages of several hundred years are common, and vary from place to place. Surface water is replaced at a sufficient rate, by upwelling and general mixing with water below, that it usually does not come into isotopic equilibrium with the atmosphere. The extent of equilibration varies from place to place.

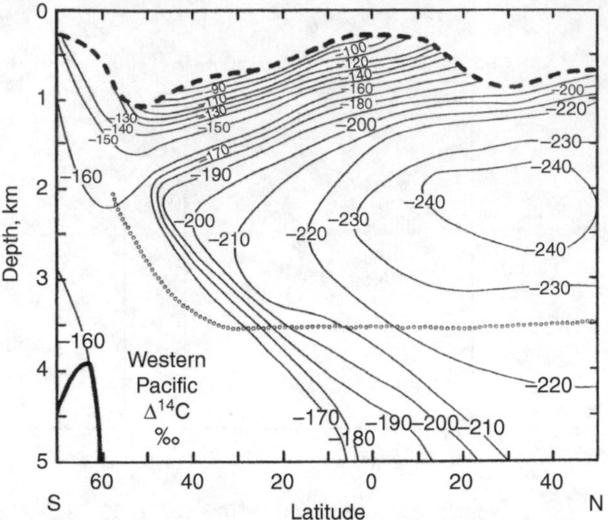

Figure 10.11 Vertical section from south to north in the western Pacific Ocean, showing the concentration of ^{14}C in CO_2 expressed in Δ-notation, as measured during the GEOSECS survey in 1974 (figure from Broecker and Peng 1982). Evidently the oldest water shown here, that with the least ^{14}C, is in the North Pacific at a depth of about 2200 m. The age of the water cannot be evaluated from the ^{14}C distribution alone, but for purposes of comparison a $\Delta^{14}C$ of –240 corresponds to an apparent age of about 2300 years, relative to the ^{14}C standard. A value of –160 corresponds to an apparent age of about 1400 years. The upper dashed line is a boundary above which the $\Delta^{14}C$ values were, even 35 years ago, affected by bomb-produced radiocarbon. (The dotted line is the depth of a particular density surface, and the heavy line at the bottom left is a peak in the bathymetry.)

The water sinking at the regions of major downwelling, or in regions where mode water is formed, may begin its trip already apparently several hundred years old, relative to the atmosphere. Broecker and Peng (1982) estimate that water sinking in the Norwegian Sea probably had a pre-bomb $\Delta^{14}C$ of –67, corresponding to an apparent age of about 550 years, and water sinking around Antarctica had a pre-bomb $\Delta^{14}C$ of about –175, corresponding to an apparent age of 1600 years. Taking these estimates into account, and along with the estimates of the input of fresher carbon to the deep sea by particulate transport (Figure 1.4) and the volume-weighted average $\Delta^{14}C$ for the world ocean below a depth of 1500 m, they arrived at an estimate of the water-replacement time for this deep water of 670 years. For the whole ocean below 1500 m, this may be currently about the best average value. Some parts of the deep water, especially in the Atlantic, are certainly much younger than this and some, such as in the North Pacific, are considerably older.

Given the very large reservoir of CO_2 in the deep sea, all of it carrying less ^{14}C than the CO_2 in the atmosphere, it follows that changes in ocean circulation leading to changes in the rate at which the deep reservoir is replaced would in turn lead to changes in the ^{14}C content of atmospheric CO_2. Some portion of the atmospheric changes that took place in earlier centuries (Figure 10.9) could have been due to

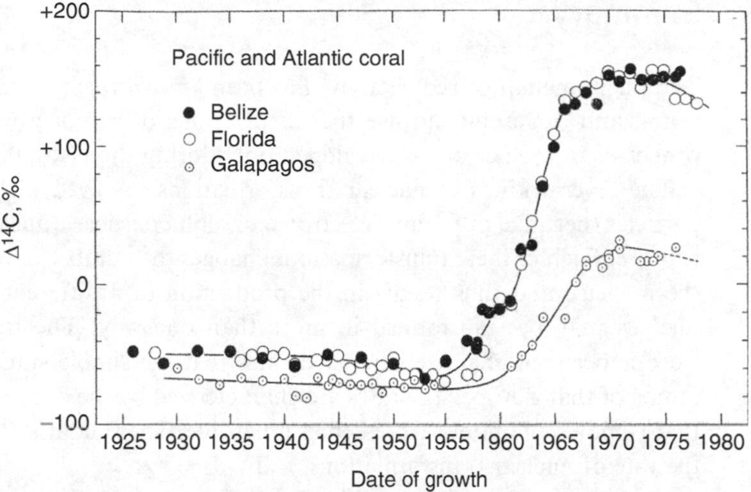

Figure 10.12 Concentration of ^{14}C (expressed in delta notation) in corals growing at two locations in the Caribbean Sea and at the Galapagos Islands in the eastern tropical Pacific. Because the corals investigated lay down annual bands, it is possible to establish an accurate date for the age of the coral samples. Such corals provide integrated average values for ^{14}C in the carbonate of surface waters during the period of their growth. Significant bomb-produced ^{14}C contamination began shortly after 1955. The GEOSECS data shown in earlier figures were collected in the early 1970s.

These data, collected for her dissertation at SIO by Ellen Druffel (1980), provide compellingly clear evidence on several important points:

a. Apparent ^{14}C ages of surface waters vary naturally from region to region.

b. A region strongly affected by upwelling, such as the Galapagos area, has an apparently older age ($\Delta^{14}C$ of –50 corresponds to an "age" of about 400 years, and –70 corresponds to an "age" of about 600 years).

c. Data subsequent to about 1956 show the effect of bomb-produced ^{14}C.

d. There are strong regional differences in the signal from bomb-produced radiocarbon in surface waters.

climate-induced changes in ocean circulation and the rate of deep overturn. Indeed, even discounting the recent Suess effect and the recent input of bomb-produced ^{14}C, it is an open question and a matter of speculation whether the ocean is now, or has ever been, completely in steady state with respect to the apparent age of its deep water.

Since the surface water of the ocean varies from place to place in the extent to which it departs from equilibrium with the isotopic composition of the atmosphere, dating of objects formed in surface seawater, such as mollusk shells found in kitchen middens, must take into account the apparent age of the water they came from. In the literature this apparent age is often called the "reservoir effect." Even in any one location the surface water may have varied in the degree of departure from equilibrium with the atmosphere due to local variations in currents and upwelling and mixing with deeper layers of water. Dating of such samples therefore requires some oceanographically informed judgment.

Summary

The phenomenon of radioactivity has been known for little more than one hundred years, and our ability to use the distributions of radioactive elements to examine processes in the ocean has developed only during the last half of that time.

The several kinds of nuclear transformations involved include the emission of β particles (negative or positive electrons) or alpha particles (nuclei of ^{4}He), and electron capture. Each of these transformations changes the number of protons and neutrons in the nucleus, and thus results in the production of a different element. Sometimes a nucleus may be transformed in more than one way. The transformations of each isotope occur at characteristic fixed rates that are simple statistical properties of the atoms of that isotope. The rates are characterized by the decay constant, λ, a property that is more commonly expressed as a half-life. The fundamental equation, describing the rate of nuclear transformations, is $dN/dt = -\lambda t$.

Important radioactive nuclides in seawater include ^{40}K, ^{238}U, ^{235}U, ^{234}U, ^{234}Th, and ^{14}C. Many others have been exploited in various ways to study ocean processes.

Important processes that have been quantified by exploiting several of the nuclides that are part of the decay series originating from ^{238}U include air sea exchange rates (using ^{222}Rn), sedimentation rates (using ^{210}Pb), and particle transport of carbon and other substances downward through the water column (using ^{234}Th).

The distribution of ^{14}C in the sea provides some of our best information about the timing of the formation of major water masses in the ocean and their relative ages, and on the rates of air–sea exchange.

The slowness with which the radiocarbon in the atmosphere equilibrates with the surface ocean, and the consequent apparent age of surface waters everywhere, means that the apparent age of surface seawater in each location must be evaluated when using radiocarbon to date objects that were formed in surface water.

SUGGESTIONS FOR FURTHER READING

Those who wish to learn something of the extensive development of tracer techniques in general, and of isotopic methods in particular, that have greatly increased our knowledge of the rates and details of ocean processes may wish to examine the following books:

Broecker, W. S. and T. H. Peng. 1982. *Tracers in the Sea.* ELDIGIO Press, Lamont-Doherty Geological Observatory, New York.

This classic provides an original and clear exposition of the use of various tracers and simple calculations to gain insight into ocean processes. It was a creative tour de force, and provided new insights into the way the ocean operates for generations of students. The book is awkwardly organized, especially the index.

Faure, G. 1986. *Principles of Isotope Geology*, 2nd edn. John Wiley & Sons, New York.

This is a standard reference describing the theory and application of isotopic tracers to a variety of earth science problems.

Sarmiento, J. L. and N. Gruber. 2006. *Ocean Biogeochemical Dynamics.* Princeton University Press, Princeton, NJ.

A detailed description of the use of tracers and modern models to describe the functioning of the present ocean. The best and most thorough exposition currently available.

11 Organic matter in the sea

. . .organic chemistry appears to me like a primeval forest of the tropics, full of the most remarkable things. FRIEDRICH WÖHLER 1835 IN LETTER TO J. J. BERZELIUS

The presence of significant concentrations of organic substances dissolved or suspended in seawater has been known or guessed at for a very long time. As knowledge of organic chemical structures and techniques to isolate and identify such compounds developed, so did attempts to learn about the organic substances in seawater. Even in the nineteenth century, the existence of these substances in the sea caught the imagination of biologists and chemists. The importance of organic matter in affecting the chemical and biological attributes of seawater has become increasingly evident. However, the overwhelming amount of salt, the recalcitrant nature of much of the organic matter, and the low concentrations of recognizable individual organic compounds have all combined to leave us with yet a poor understanding of this evidently complex mixture of substances. Some organic matter, such as simple sugars and amino acids, must be recycled rapidly because these substances are easily absorbed and metabolized by bacteria. However, estimates of the age of deep-sea organic matter, from its content of ^{14}C, show that some of it must be more than 6000 years old, long antedating the record of human observations of seawater. The nature of this very old organic matter is still a matter of speculation.

Even though we do not know the structures of much of the organic matter in seawater, it is still possible to trace some individual compounds with exquisite sensitivity. Much information can be obtained by examining the chemistry of these compounds and their isotopic composition; additionally, by examining organic substances in marine sediments, it is even possible to infer a great deal about the climate and patterns of productivity in years past.

11.1 Historical note

As we saw in Chapter 1, the recorded history of observations on the organic matter in seawater began with Aristotle (about 330 BCE), who attributed a fatty quality to seawater and said that, especially in hot weather, a fatty oily substance formed on the surface, due to its lightness, having been excreted by the sea, which has fat in it. This sea-surface slick material has been collected and studied and, indeed, a part of the substance is fatty in nature, more concentrated than in the water column below (Garrett 1967; Hunter and Liss 1981).

In 1892–1894, the Austrian chemist Konrad Natterer observed that prolonged heating of dried seawater caused the residue to give off a substance which smelled like acrolein (acrylic aldehyde). This material is known to be formed from the glycerol portion of ordinary fats.

Acrolein has a biting, acrid, and unmistakable odor, familiar to many from accidents in the kitchen. Thus, the smell of acrolein was fairly strong evidence for the presence of a fatty substance or at least of glycerol in some form in seawater, and one wonders whether even Aristotle could have made the connection.

Natterer also extracted dried sea salts with ethyl alcohol and obtained a brown liquid. This material smelled of old fat and gave other chemical evidence for the presence of lipid material. By this extraction technique he found about 2 mg of organic matter in each liter of seawater collected from the open Mediterranean and up to 20 mg L^{-1} from water collected near the coast of Greece. These samples are quite likely to have been contaminated; nevertheless, as will be seen shortly, these are roughly the concentrations that we might measure today.

In 1909, the German chemist August Pütter suggested that some organisms living in seawater could only obtain sufficient nourishment to sustain themselves by absorbing organic food directly from solution. This idea, known as *Pütter's hypothesis*, focused a great deal of attention on dissolved organic matter and stimulated much controversy and continuing research. While certainly many bacteria absorb and metabolize some dissolved organic matter, it appears that some invertebrate animals also can absorb small amounts of amino acids and other dissolved organic substances (Stephens 1982).

An investigation by the German chemist Kurt Kalle (1933) of the optical absorbance of seawater as a function of wavelength led him to identify the absorbance of light in the blue and near-ultraviolet spectral regions with the presence of organic matter. An absorbance in this region is characteristic of yellow-colored compounds, so Kalle called the organic material *gelbstoff*. For years thereafter, whenever a marine chemist was asked, "What is the organic matter in seawater?" he or she could knowingly reply, "Gelbstoff," and thereby satisfy at least some listeners. About the same time, the physiologist August Krogh discovered small amounts of organic nitrogen in seawater. Redfield *et al.* (1937) reported on the vertical and seasonal variations in the concentration of phosphorus bound in some organic form, in the Gulf of Maine, providing a further indication of the complexity of marine organic matter and showing that the concentrations of at least this fraction of the material varied seasonally due to biological activity.

The development of sensitive bioassay techniques in the 1950s led to the quantitative estimation of several organic compounds at remarkably low concentrations and to some early appreciation of the great variety of compounds that may be present. Subsequent and continuing development of sensitive chromatographic techniques, and especially gas chromatographic and mass spectrometric techniques, have made it possible to acquire a wealth of information on some individual compounds and their isotopic composition, even when they are present in very low concentrations.

11.2 Primary production

Phytoplankton floating in the sunlit surface layers make most of the organic matter in the sea. The photosynthetic formation of this organic matter is called *primary production*, because it is the first step in the production of nearly all the food for other marine organisms. Relatively much smaller amounts of organic matter enter the sea from rivers, from the atmosphere, from photosynthesis by larger fixed algae along the shores, and by bacterial chemosynthesis on parts of the sea floor.

Most of the organic matter in the sea is included within the operationally defined fraction called *dissolved organic matter* (DOM), usually measured as *dissolved organic carbon* (DOC); all of this is ultimately derived from living organisms.

Marine phytoplankton are mostly single-celled algae and bacteria, remarkably diverse in morphology, evolutionary history, and biochemical behavior. All the photosynthetic forms convert sunlight into controlled chemical energy, which is used to synthesize the organic constituents of life from carbon dioxide, water, and the other nutrients (Fogg 1975, Falkowski and Raven 2007). Some of this organic matter is leaked or excreted from the cells in soluble form, some is lost when the phytoplankton cells are broken while being eaten by zooplankton, and some is lost during feeding, death, and decay of subsequent participants in the foodweb. Bacteria are ubiquitously present everywhere in the sea; they consume dissolved organic matter and may also produce or leak some as well.

The rate of primary production (the rate at which organic matter is synthesized) is technically difficult to measure accurately, and indeed requires careful definition. All techniques yield empirically calibrated approximations. Rates vary laterally, vertically, and temporally, sometimes in each case by orders of magnitude, so it is difficult to obtain an entirely satisfactory total for the whole ocean.

Several common terms are defined as follows:

Gross production: Total amount of carbon fixed by phytoplankton (CO_2 converted to organic matter) per unit time; expressed per unit volume or per unit area.

Net production: Total carbon fixed, as above, corrected for carbon respired by the phytoplankton themselves.

New production: That fraction of the net production for which the nutrients are supplied from outside the region (or some defined "box") where the production takes place. (The contrasting term is not, as might be expected, *old production*, but rather *recycled production*, that production supported by nutrients recycled within that defined box; this is a poor term because it is not the production that is recycled.)

Export production: That fraction of the net production that is not consumed and regenerated as inorganic carbon and nutrients in the surface layer, but instead is transported out of this region, generally downward by sinking particles of organic debris and often whole phytoplankton cells. Export production may also include the harvest of fish in areas where this is important. At steady state, export production should equal new production. As we will see later, the concentration of dissolved organic matter is higher in surface water than in deeper water, so if surface water is subducted into deeper regions, the dissolved organic matter carried along is also a component of export production.

The measurement of gross production is important for a quantitative understanding of the biochemistry of photosynthesis and plant metabolism. However, it is the net production that results in an increase of biomass and a yield of food usable by other organisms, and it is the net production that is usually intended when measurements are made and reported.

Net primary production is usually measured by the carbon-14 method introduced by Steemann Nielsen (1952). In this technique, ^{14}C-labeled sodium bicarbonate is added to a water sample and the sample is incubated under specified conditions of temperature, light, and time. The sample is then filtered, and the ^{14}C-activity is measured on the filter pad. After correction for dark uptake, the resulting value is converted to an estimate of the amount of carbon fixed into particulate organic matter, presumably phytoplankton. Depending on the length of the incubation, the result is more or less self-correcting for respiration, but may also include some zooplankton production, and certainly misses any organic carbon that is manufactured by the cells but released in dissolved form and passes through the filter. With additional or different processing the dissolved organic product can be measured as well. Procedures to take account of the changing temperature and light regime with depth, as well as changes of efficiency throughout the daily cycle, are usually necessary. Despite all these corrections, the result still may be a somewhat uncertain approximation of reality. At the present time, however, this procedure is the basis for most of our knowledge of the net primary production in the ocean.

Another approach is to measure the change of oxygen concentration in light-incubated bottles, along with dark-incubated bottles for respiration. While there are uncertainties with this method as well, it is argued that the total net organic production is measured more accurately than by the ^{14}C technique, provided one has an accurate estimate of the Redfield ratio of carbon to oxygen. In most open-ocean regions the oxygen technique is, however, not sensitive enough for routine use. Furthermore, the proper ratio of oxygen released to carbon fixed is uncertain and indeed is certainly variable.

The simplest equation for photosynthetic production of organic matter from carbon dioxide and water using the energy from sunlight is

$$CO_2 + H_2O + \text{energy} \rightarrow CH_2O + O_2. \quad (11.1)$$

The formula CH_2O is a convenient one-carbon shorthand for the sugar glucose ($C_6H_{12}O_6$), a primary product of photosynthesis that enters the biochemical pathways to the rest of biochemistry; it also captures the oxidation state of the carbon at this stage. In Eq. (11.1) the ratio $O_2/C = 1$. This equation would hold approximately if all the material formed were carbohydrate, but, of course, the biochemical transformations in plants also yield lipids, proteins, and all the other substances needed to form their living structure. From the discussion in Chapter 8 it is evident (Eqs. (8.13), (8.14)) that a value for the O_2/C ratio more in accord with oceanic distributions would be about 1.16, yielding the following stoichiometry,

$$124CO_2 + 124H_2O \rightarrow (CH_2O)_{104}(CH_2)_{20} + 144O_2. \quad (11.2)$$

This ratio of $144/124 = 1.16$ is based on carbon alone, and would still hold if NH_3 were the nitrogen source, while with NO_3^- as the source of nitrogen the ratio would be $166/124 = 1.34$. Indeed, no generally applicable ratio can account for local variations in the dominant species and their biochemical condition at any particular time. Nevertheless, there are often circumstances where the simplicity of measurement and straightforward interpretation of the results make the measurement of changes in $[O_2]$ the method of choice.

The term new production originated with Dugdale and Goering (1967), and was most explicitly related to large-scale processes by Eppley and Peterson (1979). It had long been understood that phytoplankton production must depend on some combination of nutrients released by grazers (regenerated nutrients) and nutrients carried upward into the photic zone by mixing or upwelling. Dugdale and Goering, however, devised a technique that they supposed would measure the two independently. Reasoning that the nitrogen input from below is in the form of nitrate while the nitrogen excreted by grazers is mostly in the form of ammonia (nitrifying bacteria that convert ammonia to nitrate must be present everywhere in the ocean, but evidence shows that nitrification proceeds very slowly in most surface seawater), they proposed that the rate of nitrate uptake by phytoplankton would be a measure of production supported by input of new nutrients to the system. A sensitive nitrogen-15 tracer technique was used to make the necessary measurements of rates of uptake of NH_3 and NO_3. They suggested that the uptake rate of nitrate was a measure of new production, in contrast to recycled production supported by ammonia. Whether or not these measurement techniques can really apportion the total net production in this way is debatable. Indeed, to the extent that some phytoplankton can migrate vertically to acquire nutrients, especially nitrate, from deeper in the water column, the procedure becomes problematic. Still, the concept of new production took hold and led to more quantitative thinking about primary production and the total transport of substances in the sea.

The phytoplankton-based ecosystem in the sunlit upper layers of the sea (the *photic zone*, also called the *euphotic zone*) is characterized by rapid grazing and metabolism of the organic matter produced. In any one region, the system is never truly in steady state, responding as it does to vagaries of wind, turbulence, weather, season, and biotic factors, but there must be a continuing, though certainly not constant, rain or drizzle (Berger *et al.* 1989) of particles sinking from the surface. These particles carry nutrients down from the surface, and the stocks in the surface layer would be depleted if nutrients were not also continually resupplied by vertical and lateral advection and mixing.

An additional complication arises if some of the organic matter is released in dissolved form along with organically combined nitrogen and phosphorus. Adequate measurements of these forms are tedious and, as will be described later, few satisfactory data exist. Some of this dissolved organic matter may be taken up and metabolized only slowly, and the rest may be lost from the upper layers only by transport with the water to a region where water sinks to depth. There is some reason to think that this pathway may be quantitatively important, but it is an aspect of export production for which we need more evidence.

The concept of new production refers to the proportion of net production supported by imported nutrients, while the concept of export production refers to the rate at which organic matter is exported from the system. Over a suitably long time of averaging, the two rates must be the same (when based on the same system with the same boundaries). Both terms are used, however, because the ecosystem may not be in steady state during an interval of measurement and for some time one rate may well exceed the other.

The concept of export is not restricted only to the photic zone, and it is possible to speak of export through, for example, the 1000-m level.

Most measurements of primary production in the ocean have been made by the carbon-14 technique. Some 9000 of these data were combined by Berger *et al.* (1988, 1989) into a general map showing the variation throughout the world ocean. Even with this number of measurements, however, gaps in the coverage remain. Berger *et al.* (1988, 1989) filled in the gaps using much more abundant data on the concentration of phosphate at 100 m and an empirical relationship between production and phosphate concentration. The result (Figure 11.1) is a distribution throughout the ocean much like that predicted from earlier data, but more detailed and quantitative. There are areas of very low production in the central gyres of all oceans, and much higher production along the coasts and in the equatorial current systems (Table 11.1). Maps such as these provide useful estimates of the relative magnitudes and spatial distribution of rates of production. As treated by these authors, the available data provided an ocean-wide total net primary production of about 27 Gt of carbon per year.

The geographical distribution is well captured by the map as shown but the true value of total global net primary production is almost double the estimate given above. Many of the earlier measurements by the carbon-14 method were biased low because the reagents used were probably contaminated by trace metals at high enough concentrations to depress the phytoplankton activity before it was possible to make accurate analyses of these troublesome substances (Fitzwater *et al.* 1982). After examining this problem, Martin *et al.* (1987) suggested that the true value must be around 50 Gt per year.

Several authors have applied models that incorporate satellite-derived chlorophyll concentrations along with chlorophyll measurements and corresponding carbon-14

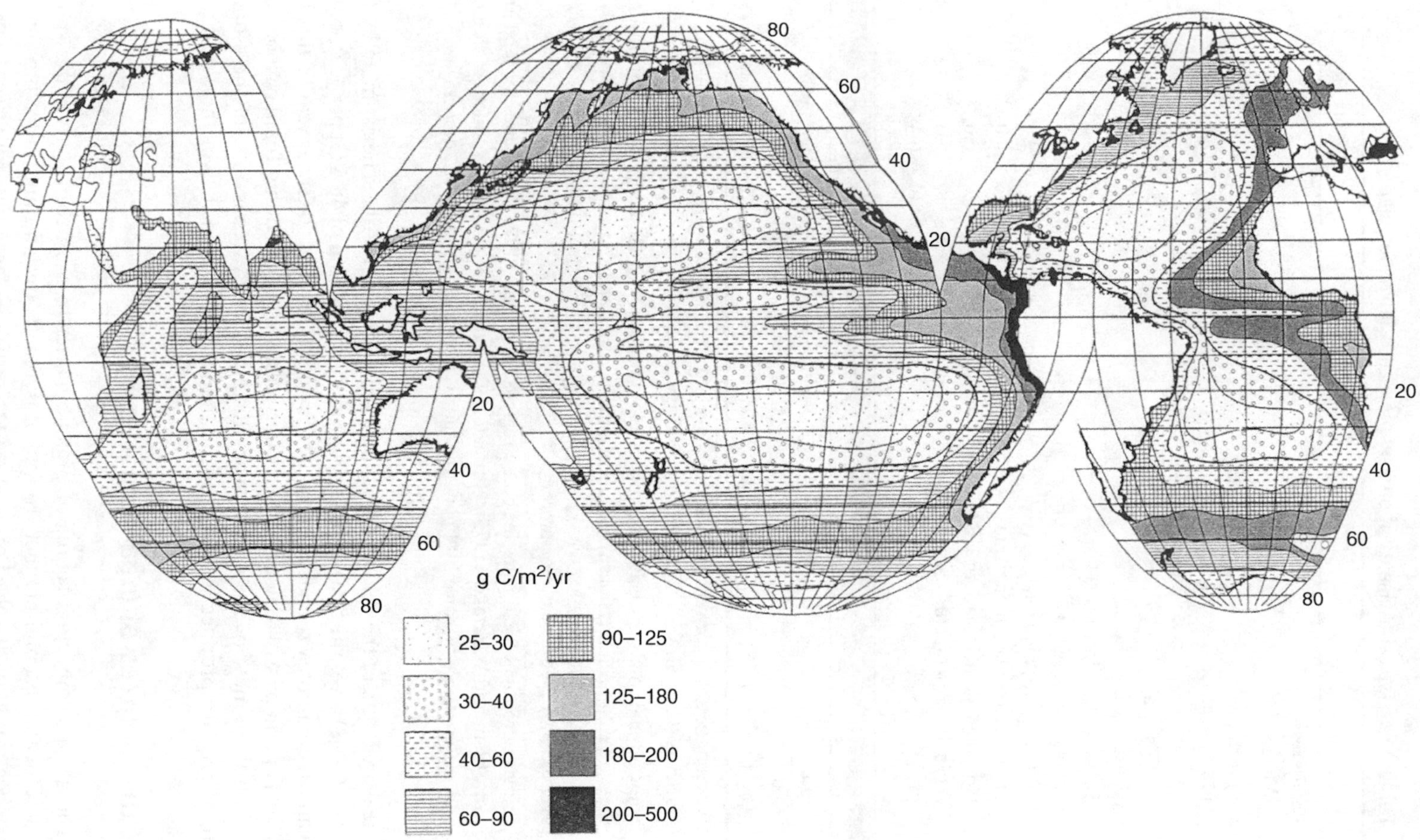

Figure 11.1 Map of the distribution of net primary production in the world ocean, combining about 9000 measurements, mostly by the ^{14}C method, with estimates based on the phosphate concentration in regions without productivity data. From these data Berger *et al.* (1988, 1989) estimated the world total to be 27 Gt per year, expressed as carbon. (Berger *et al.* 1989, with the coding key from Berger *et al.* 1988.)

Table 11.1 Typical estimates of primary production in various places

	C, mol m^{-2} yr^{-1}	C, g m^{-2} day^{-1}
Land areas		
Wheat	11	0.37
Maize	73	2.4
Mature spruce forest	37	1.2
Tropical rain forest	82	2.7
Land average	31	1.0
Ocean areas		
Sargasso Sea, low value	1.5	0.05
Sargasso Sea, mean	6	0.2
Peru Current, high value	275	9.0
Peru Current, common value	150	5.0
Continental shelf, eastern USA	18	0.6
Artificial enrichments	500	16.5
Ocean average	11.5	0.38
Totals	**C, Tmol yr^{-1}**	**C, Gt yr^{-1}**
Land	5000	60
Ocean	4160	50
World	9160	110

Where necessary, conversion from dry matter to C was done using the relationship: dry matter ÷ 2.2 = C.

Some annual totals, such as the high value for the Peru Current, are possibly unrealistic, as they are simple conversions from unusual daily values and shown for comparison only. Values for individual land area types are averages from a large range that can vary from the average by a factor of two or more.

Primary production estimates for ocean regions from Fogg (1975). Estimates for land areas from Lieth and Whittaker (1975). Estimates of the total net primary production on land and sea are those adopted by Eglington and Repeta (2006) and other authors.

measurements in the field to estimate global total production. The estimate by Field *et al.* (1998) gives the total oceanic net primary production of 48.5 Gt per year, but this value is commonly rounded to give the convenient value of 50 Gt per year (Table 11.1).

It is perhaps worth noting that at our present rate of adding fossil carbon to the atmosphere, now over 9 Gt per year, this input now exceeds 15% of the estimated total primary production on land or in the sea, and exceeds 8% of the world total.

11.3 Other sources of organic matter

The second important source of organic matter to the ocean is river transport of some of the residual product from photosynthesis on land and in fresh water. The concentration of dissolved organic matter in river water varies from river to river by

Table 11.2 Sources of organic carbon to the ocean (global totals)

	C, Tmol yr^{-1}	C, Gt yr^{-1}
Primary production in the sea		
Table 11.1	4160	50
River input[(a)] (Meybeck 1982)		
DOC	18	0.215
POC	12	0.143
Total	30	0.36
Atmospheric input[(b)] (Duce 1989)	4	0.05

(a) River inputs were calculated by taking estimates of the volume-weighted average concentrations of DOC and POC in river water (5.75 and 3.83 mg L^{-1}, respectively) and multiplying by the world total river flow, 37 400 km^3 yr^{-1} (Baumgartner and Reichel (1975). These values are not different from those in a recent review (Eglington and Repeta 2006: DOC, 0.2; POC, 0.15 Gt yr^{-1}).

(b) Some of the atmospheric input is recycled marine organic matter, and some is certainly pollen and other kinds of terrestrial detritus, but the proportions are not well known. Duce (1989) assigned an uncertainty of ± 50% for the atmospheric input.

nearly two orders of magnitude, and analytical coverage of the world's rivers is poor. There are, however, fairly good relationships between type of drainage basin and the concentrations of dissolved, particulate, and total organic carbon (DOC, POC, and TOC, respectively) in rivers. Meybeck (1982) used these relationships to estimate a weighted average DOC concentration for all rivers of 5.75 mg L^{-1}, expressed as carbon. His estimate of the average concentration of POC was 3.83 mg L^{-1}, though this value is less certain. Taking the average flow of the world's rivers to the ocean as 37 400 km^3 per year, this gives a total discharge of DOC and POC into the ocean of 0.215 and 0.143 Gt per year, respectively (Table 11.2). Over a wide range of drainage basins, the discharge of TOC was often just a little less than 1% of the total net primary production in the basin. The age of this material varies greatly from river to river and also temporally within a given river (Eglington and Repeta 2006). In some cases carbon-14 dating shows that the organic matter is apparently recently synthesized, while in other cases it appears to be many hundreds of years old. In the latter case it seems that the organic carbon is derived from soil organic matter that has aged a considerable time in the soil. It is also likely that some particulate carbon in rivers is derived from the breakdown and erosion of rocks carrying resistant organic matter from ancient sediments.

Much of the river-borne particulate matter and the associated POC is likely to be trapped by sedimentation in estuaries or on the deltaic fans near the mouths of rivers. Some of the POC adsorbed on the surfaces of particles may be desorbed as the river water mixes with seawater, however. Some very high-molecular-weight DOC appears to flocculate and settle out on first mixing with seawater, but evidence that over 90% of the DOC exhibits conservative mixing in estuaries (e.g. Mantoura and Woodward

1983) suggests that most DOC does get carried into the ocean. It is still uncertain how much of the DOC from rivers escapes the coastal zone to mix into the rest of the ocean.

The only other significant source of organic matter for the ocean is the atmosphere. Rain water contains a surprising amount of organic matter. Concentrations are quite variable, but Duce (1989) estimated that average concentrations of carbon in oceanic precipitation might be 5 to 15 μmol kg^{-1}. Precipitation on the oceans is about 385 700 km^3 per year (Baumgartner and Reichel 1975), so the input of organic carbon from rainfall may be 2 to 6 Tmol per year (0.03 to 0.07 Gt per year). Unlike the situation with river input, however, where it is clear that essentially all of the organic carbon originates in terrestrial ecosystems, some of the organic matter in oceanic rain must be recycled from the ocean itself. It is not known what fraction is recycled oceanic material, but presumably most of it is. The composition of this material is poorly known.

In sum, the available evidence suggests that, although the estimates are uncertain, the land-derived annual input of organic carbon into the ocean is probably close to 18 Tmol from rivers plus possibly 2 to 4 Tmol from rain and dry fallout. The total from terrestrial sources may therefore be about 20 Tmol per year, or about 0.5% of the primary production in the ocean.

The riverine and atmospheric inputs might possibly be more important contributors to marine organic matter than their rather small fluxes relative to primary production would at first suggest. Much of river-borne DOC is the residue left after extensive biological use and alteration, and may be relatively resistant to bacterial attack. It may be only coincidence, but it should be noted that by analyzing certain fractions of marine organic matter for traces of chemical structures of a type derived only from terrestrial sources, Meyers-Schulte and Hedges (1986) were able to show that perhaps 0.5% of this marine organic matter was derived from terrestrial sources.

Atmospheric carbon entering the sea is largely particulate (at least the fraction so far measured is) and has survived transport through the air with its potential for photochemical attack and other oxidative processes. Some of the atmospherically transported carbon consists of pollen and other resistant material. In some samples, a large fraction of the carbon in atmospheric particulate material is *black carbon*, a term used to describe a substance that might also be called soot, the product of incomplete combustion in numerous processes from forest fires to diesel engines. Black carbon is thought to be mostly elemental carbon in a poorly characterized physical state. Such material could be very resistant to oxidation (Duce 1989); while little is known of the resistance of black carbon in the sea, it seems likely that much of it could end up permanently sequestered in the sediments.

11.4 Fate of the primary product

After satisfying the respiratory needs of the phytoplankton itself, most of the remaining (net) production goes into growth of the phytoplankton, and a variable fraction (possibly 10% on average) leaks out or is excreted as dissolved organic carbon. The phytoplankton particles and the DOC then support most of the rest of the food web in the sea (Lee and Wakeham 1989). A significant fraction, up to 27% in one series of experiments (Copping and Lorenzen 1980), of the phytoplankton carbon ingested by zooplankton can also be lost or excreted as DOC. Much, but apparently not quite all, of the DOC is rapidly consumed by bacteria, which in turn are eaten by protozoa and other zooplankton. Most (perhaps 80 to 90%, but varying with location) of this growth and respiration of plants, bacteria, zooplankton, fish, and other animals takes place in the upper (photic) layers of the water column.

Some fraction of the phytoplankton themselves, and of the organisms that eat them and their fecal products, and other detritus, sinks below the sunlit surface layers to be consumed below. The fraction of production that sinks below the regions where photosynthesis is possible varies with circumstances, tending to be greater the higher the production, and the value measured varies with the depth level chosen. A common representative value, however, is 10% of net primary production sinking below 200 m (Figure 11.2). Sinking particles carry down nutrients, many trace metals, and other substances. Information on the biological processing and even repackaging of the particles as they sink can be gained by examining the changing ratios of various substances to each other and to carbon, and the changing concentrations of individual organic components in particles collected in sediment traps at different depths.

It is clear that the rain of sinking particles varies widely both spatially and temporally (Berger *et al.* 1988, 1989; Honjo *et al.* 2008). In general, the exported production is a considerably higher fraction of the total net primary production in productive waters than in non-productive regions, with values varying from above 50% to below 10%. In particular, sudden blooms may yield a lot of sinking particles as the bloom comes to an end. As particles sink, they carry bacteria and protozoa that also respire, and in turn they too are captured, eaten, and partially or totally metabolized. The rate of capture is high near the surface where many organisms are supported by this food, and decreases with depth as the availability of particles becomes less and less. The net effect is that the flux of sinking particles decreases exponentially with depth (Figure 11.3).

At least eight equations relating the flux of sinking particles to the mean annual production and the roughly exponential fall-off with depth have been developed (Bishop 1989). An equation developed by Betzer *et al.* (1984, as modified by Berger *et al.* 1988) illustrates the relationships described:

$$J_{(z)} = \frac{0.409 \times \mathrm{PP}^{1.41}}{z^{0.628}}, \qquad (11.3)$$

where $J_{(z)}$ is the flux of particles exported downwards through some depth, z, and PP is the net primary production.

Both $J_{(z)}$ and PP must be in the same units; common units are g m^{-2} yr^{-1}. This equation fits some of the data sets well but, as shown by Bishop (1989), the fit over many data sets is not especially good. The important points are that the rate depends to some variable extent on the primary production (and to a power greater than 1) and that the decrease with depth is roughly exponential. Both vary spatially and

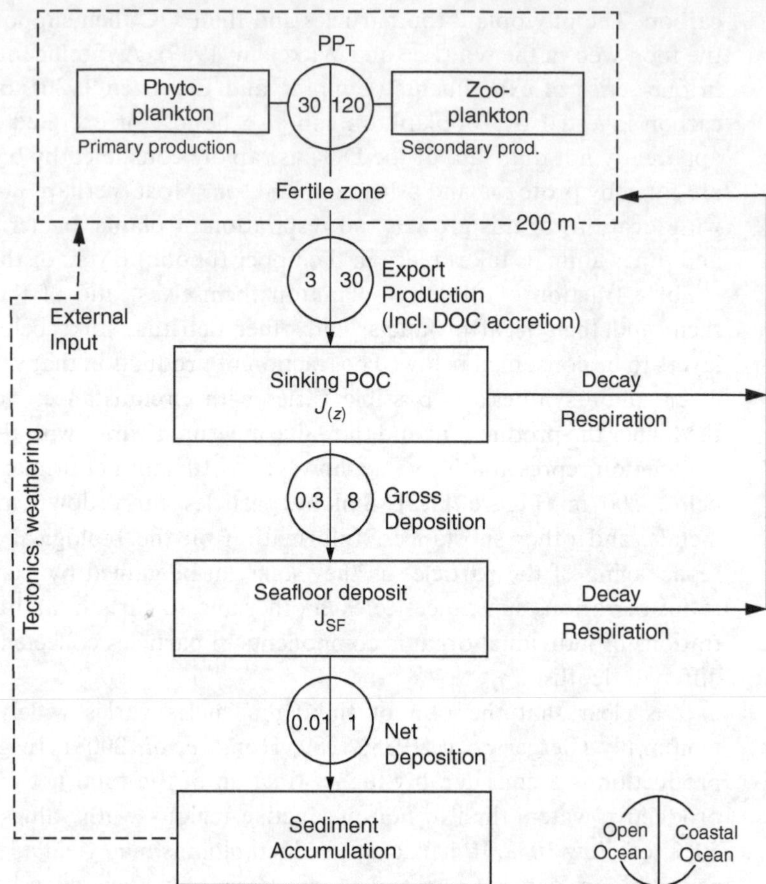

Figure 11.2 Schematic representation of the cycle of marine carbon. (Figure redrawn from Berger *et al.* 1989.) Here for convenience the upper layer where rapid recycling takes place and most of the marine food web is concentrated is taken as the upper 200 m, and the net primary production (PP_T, here in units of C, as g m^{-2} yr^{-1}) is divided into two segments. On the left are typical open-ocean values; on the right are typical values for the coastal ocean. True rates may be almost double those given here, as this figure was prepared before the new estimates deriving from Martin *et al.* (1987) and subsequent authors, but the relative proportions still hold. Everywhere, the values must be locally variable. In the open ocean, an average of perhaps 10% of net primary production sinks below the upper region, and is termed export production. Of this sinking particulate matter, about 90% is consumed on the way down through the water column and 0.1% or even less of the original net primary production finally accumulates in the sediment. The rest is consumed on or in the sediments. Return fluxes of carbon are mostly due to the complex processes of mixing and upwelling of deep water. It should be borne in mind, however, that the dissolved forms of organic carbon may play a role that is not negligible with respect to the fluxes given here. On the right, in the coastal ocean a larger fraction is consumed by benthic organisms, depending on the depth of the water column, and a larger fraction is buried. Data from the coastal zone are more variable and the averages less well defined.

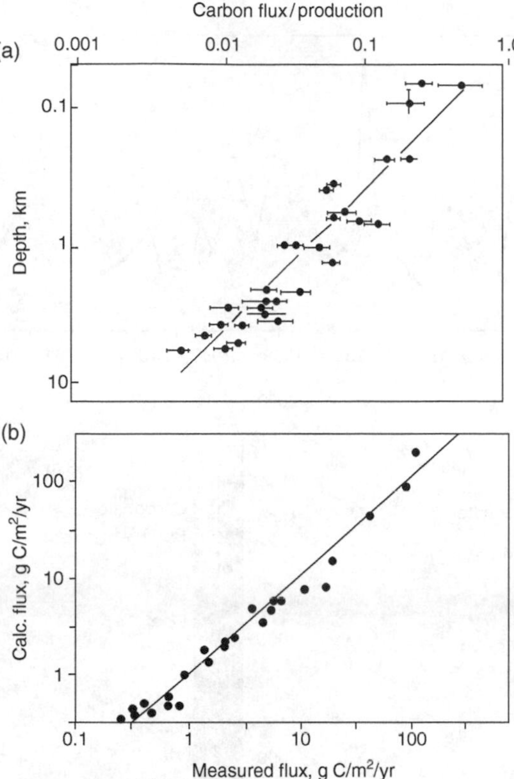

Figure 11.3 Sediment traps have been deployed at many depths in various parts of the ocean. The material accumulating in these traps has provided a great deal of information on the downward rain of particles from the surface to the sediments below, and the chemical changes as these particles sink through the water column (a). The dominant effect is that the mass of carbon in sinking particles decreases exponentially with depth (Suess 1980). Betzer *et al.* (1984) re-evaluated Suess' data; from the results they developed Eq. (11.3). Panel (b) shows their fit of the Suess' data to this equation. (Figures from Berger *et al.* 1989.)

temporally. None of the other equations fit all available data sets particularly well. There are many confounding problems. The value of PP used may not reflect the production that yielded the particles, as there will commonly be a temporal and even spatial offset between the various measurements. As reflected in Eq. (11.3), many data sets suggest an exponential increase in export production with an increase in PP, so it is not generally appropriate to use an arithmetic average of seasonal data on PP to get an annual average of export production. The extreme seasonality or episodic nature of PP in some areas makes the problem worse. In addition, different types of biological communities may export differing fractions of the net PP. Even quite deep in the water column the flux of sinking particles may show a strong seasonality and inter-annual variability (Figure 11.4), and under regions with strong seasonality the range with season may be a factor of 50 or more, as illustrated by the extreme cases in that figure. Estimates of the world total export production are therefore quite uncertain, but Martin *et al.* (1987) estimated the world total export production (roughly assignable to a depth of 100 m) to be 7.4 Gt per year. These values serve usefully as the basis of many other calculations. Bishop (1989) showed that an equation from Martin *et al.* (1987) fits much of the available sediment-trap data fairly well if the starting point is the measured flux at 100 m, so this equation can then be used to estimate the flux at various depths in the water column below:

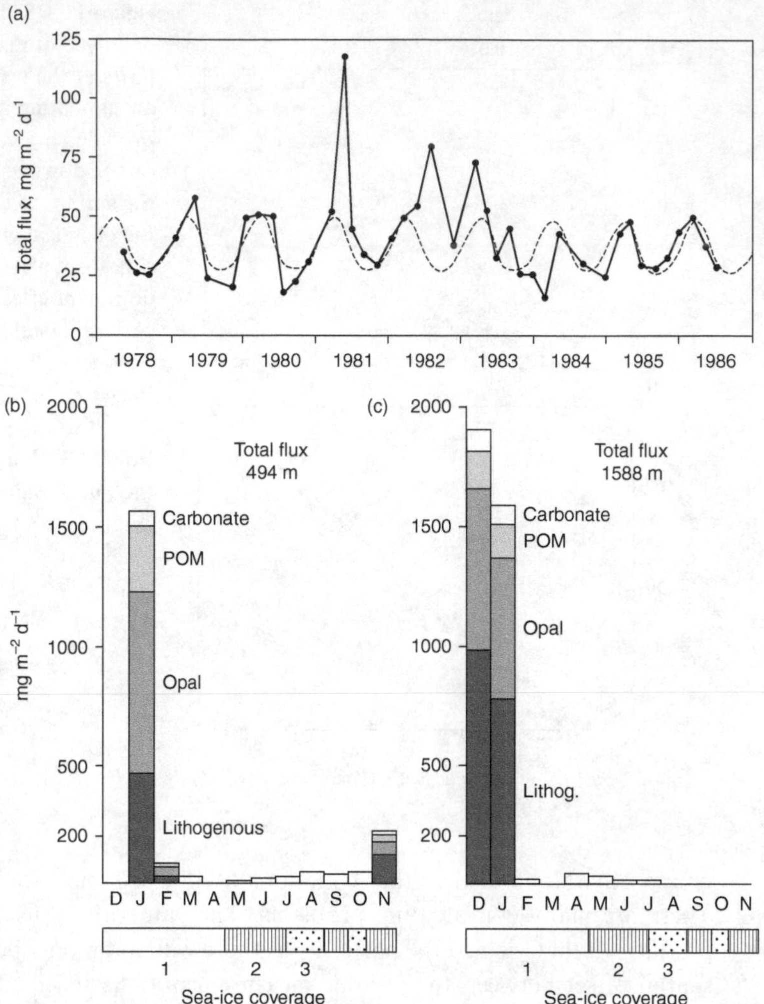

Figure 11.4 (a) Eight-year record of total particle flux into sediment traps deployed at 3200 m in the Sargasso Sea southeast of Bermuda. Each point is the average over a 2-month interval. The dotted line is a smoothed curve representing a idealized timing of the flux (redrawn from Deuser 1987). (b) and (c) Flux of total particles and major components at two depths in the Bransfield Strait just north of the Palmer Peninsula, Antarctica (redrawn from Wefer *et al.* 1988): the coding for sea ice coverage is 1: ice-free; 2: ice edge moving in; 3: ice-covered. Lithogenous means particulate material derived from rocks. Important features of sediment-trap records illustrated here are the locally variable and sometimes dramatic seasonality, and the large inter-annual variability, even in regions such as the Sargasso Sea.

$$J_{(z)} = J_{(100)}\left(\frac{z}{100}\right)^{-0.858}, \tag{11.4}$$

where $J_{(z)}$ is the flux at some depth, $J_{(100)}$ is the measured flux at 100 m, and z is the depth in meters. For example, a carbon flux of 1.70 mol m^{-2} yr^{-1} at 100 m, the world average from Martin *et al.* (1987), would yield a flux of 0.236 mol m^{-2} yr^{-1} at a depth of 1000 m.

Table 11.3 Calculation of the global fluxes of carbon and oxygen into deep water (below 1000 m) of the world ocean (excluding marginal seas)

Area of ocean below 1000 m[(a)]	314.9×10^{12} m^2
Volume below 1000 m[(a)]	98.16×10^{16} m^3
Input of sinking carbon (see text)	0.236 mol m^{-2} yr^{-1}
Carbon ultimately buried in sediments[(b)]	0.005 mol m^{-2} yr^{-1}
Carbon respired below 1000 m	0.231 mol m^{-2} yr^{-1}
O_2 used by carbon (C × 1.43) × area	**1.039×10^{14} mol yr^{-1}**
O_2 input, calculated for air-saturated water	
[O_2] at saturation, mean T&S below 1000 m	0.342 mol m^{-3}
Average [O_2] in water below 1000 m[(a)]	0.174 mol m^{-3}
Average apparent AOU in deep water	0.168 mol m^{-3}
Advection of new deep water	30×10^6 m^3 s^{-1} (30 Sv)[(c)]
Total input of oxygen (at saturation)	3.24×10^{14} mol yr^{-1}
Corresponding O_2 utilization rate	1.59×10^{14} mol yr^{-1}
O_2 input, plausibly more realistic calculation	
[O_2] of input water[(c)]	0.285 mol m^{-3}
Total input of oxygen	2.70×10^{14} mol yr^{-1}
Corresponding O_2 utilization rate	**1.05×10^{14} mol yr^{-1}**

(a) Estimates of the areas, volumes, and volume-weighted average oxygen concentrations at various depths or for the whole of the world ocean (except marginal seas) from data in Levitus (1982).

(b) Open-ocean values from Berner (1982), allocated to the area used here.

(c) The rate of 30 Sv for the advection of surface water to the deep ocean is a common conventional value, as is the assumption that this water is saturated with oxygen. The latter assumption is not valid for the purpose of this calculation. The oxygen concentration in the Greenland–Iceland overflow water entering the North Atlantic is about 311 mmol m^{-3}, or about 91% of saturation, and this is possibly characteristic of the northern source of deep water. In the Antarctic, it is difficult to find Antarctic Bottom Water (ABW) with concentrations much above 260 mmol m^{-3}. Most of the coldest water in the ocean is formed under the ice in the Weddell Sea in the wintertime. This water is characterized by [O_2] ≈ 327 mmol m^{-3}, or about 88% of saturation with oxygen at its temperature of about –2 °C (Weiss *et al.* 1979), and it mixes with intermediate waters having an even lower concentration during the formation of ABW with a concentration of about 260 mmol m^{-3}. Other sources of ABW are less well characterized. For the purpose of rough calculation here I take the average concentration of oxygen characterizing the ~ 30 Sv of water entering both north and south deep-ocean basins as 285 mmol m^{-3}.

A test of whether this global flux estimate derived from sediment-trap data is realistic may be made by comparing it to an independent estimate of the oxygen flux into deep water (Table 11.3). Given the great uncertainties in both these calculations, it is reassuring that the results are at least within the same order of magnitude; and this provides some assurance that both calculations reflect rates that are reasonably close to reality.

The export of new production from the surface might still be underestimated here. However, the world average export of carbon was used, while the average production over only the regions with water depths greater than 1000 m must be less than the world average because the very productive coastal zones would then be excluded. In so far as the sediment-trap data can be trusted, the calculated global flux of carbon is therefore likely to be a little too high. Walsh (1989) has, however, calculated that a significant part of the production over the shelves is transported off the shelves and down the slopes into deep water, the flux of carbon amounting to $\sim$ 0.07 mol m^{-2} yr^{-1}, averaged over the area of the deep sea. Thus, some contribution from the shelves is appropriately included in the simple calculation above.

On the other hand, as mentioned before, the highly episodic nature of the loss of particles from the photic zone may lead to underestimates of the downward flux. This may be especially the case in high latitudes, where few data have been collected. Of course, really big particles (such as dead whales, etc.) are not caught in sediment traps, and these must have a proportionately greater impact the deeper the bottom on which they land.

Lastly, there is direct evidence that sediment traps may underestimate the flux of even small particles. Buesseler (1991) compiled data from several studies (a total of 51 observations) where the ^{234}Th deficit in upper waters was measured at the same stations where ^{234}Th was measured in the material collected by sediment traps deployed at depths from 10 to 300 m. Results varied widely, from much more collected in traps to much less. The median value for the amount of material collected by sediment traps was 0.66 of the deficit estimated by measuring the concentration of ^{234}Th in the water column above. This suggests that, on average, sediment traps missed about one third of the settling particles; whether this is indeed so, and whether the discrepancy also applies to sediment traps deployed deeper, is not really known. It seems at least possible that the correspondence between the fluxes of settling carbon and trapped carbon might be even worse.

While many adjustments can be thought of, and more complex and detailed models developed, the calculations in Table 11.3 nevertheless suggest that our knowledge of either the global export production or of the replenishment of deep water is unlikely to be in error by as much as a factor of 2.

Up until very recently it was assumed that essentially all the export of organic matter from the regions of primary production is due to sinking particles. New information on the distribution of dissolved organic matter within the oceans, however, has provided evidence that some recently synthesized organic matter may escape rapid utilization, and thus may be carried down in a dissolved form in sinking water masses, where it is only slowly consumed by bacteria. Whether this route for transport of organic matter into the deep sea is significant in the overall balance is uncertain. Dissolved organic matter is discussed in a later section.

Some of the sinking particulate matter goes all the way to the bottom. Most of the organic carbon that gets there is metabolized by the benthos, the organisms that live on and in the sediments at the bottom of the sea, waiting for food to fall from above. Some organic matter is permanently buried, however. The amount buried depends on both the rate of input to the sediments and on the overall rate of sedimentation of

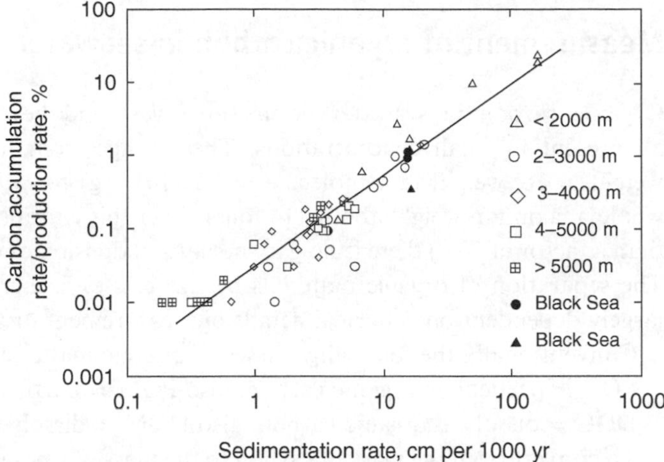

Figure 11.5 Rate of organic carbon accumulation in sediments expressed as a percentage of the net primary production in surface waters above the sediment, plotted against the total sedimentation rate (Calvert *et al.* 1991). The equation for the regression line is $A = 0.028S^{1.25}$, where A is the normalized accumulation rate of organic carbon (as percent of production) and S is the total rate of sediment accumulation in cm per 1000 yr. From these data it is evident that the rate of carbon burial in sediments is controlled by both the input of organic carbon and by the total rate of sediment accumulation.

mineral substances. If the organic particles can sit for a very long time on the surface of the sediment, they will eventually be consumed by benthic organisms, but if they are rapidly buried, there is a much better chance for long-term preservation deep in the sediments. Calvert *et al.* (1991) summarize data to show that in shallow, rapidly sedimenting regions as much as 10% of primary production may be buried in the sediments, whereas in deep, slowly sedimenting regions an amount of carbon equivalent to perhaps only 0.01% of primary production may end up permanently buried (Figure 11.5).

In evaluating the total burial rate of organic carbon in marine sediments Berner (1982) showed that most is buried on continental shelves and in the vast, rapidly sedimenting deltaic fans, while only a very small fraction of the total is in deep-sea sediments. Berner estimated that at the present time the total organic carbon in surficial sediments was accumulating at a rate of 0.157 Gt per year. He estimated that about 20% of this is metabolized by bacteria in the sediments before final burial deeper in the sediment column. Correcting for this, the final burial rates in shallow shelf and deltaic sediments amounted to about 0.110 Gt per year, while the accumulation rates in deep-sea sediments amounted to about 0.016 Gt per year, for a total in the ocean of 0.126 Gt per year. This rate is only 0.25% of the total primary production. In fact, the fraction is likely to be somewhat less than this, because the accumulation of organic carbon in shelf and deltaic sediments must contain a significant component of the particulate organic matter and even some of the dissolved organic matter discharged from rivers, and the riverine flux alone exceeds the total burial rate of organic carbon in marine sediments.

11.5 Measurement of organic carbon in seawater

Organic carbon in seawater occurs in a vast and bewildering range of forms, all present in small concentrations. These range from methane (CH_4, molecular weight = 16; weight of 1 molecule = 2.7×10^{-23} g) to *Balaenoptera musculus* (blue whale, maximum weight about 114 tons). Over this enormous range in mass (greater than a factor of 10^{30}) there is no gap, no natural division or cut-off in size of particles. The separation of organic materials by size classes is therefore quite arbitrary and largely dependent on empirical details of measurement or separation technique.

Conventionally the following classes of organic matter are distinguished:

POC = particulate organic carbon, also POM = particulate organic matter; and

DOC = dissolved organic carbon, also DOM = dissolved organic matter.

To convert from the mass of carbon to the mass of dry salt-free organic matter it is usual to use a factor of 2.2, the same as that commonly used for terrestrial biomass. This factor is equivalent to assuming a molecular weight of 26.4 g per mole of C, not far from the composition of Redfield organic matter given earlier (~27.6 g per mole of C).

The largest classes of POC are usually collected with harpoons. Smaller-size classes are collected by hook and line or by dragging a net through the water or by filtering through membranes of various sizes. (Net and membrane sizes range from a mesh size of about 20 cm for large fish and porpoises down to about 10 μm for phytoplankton and 0.5 μm for bacteria.) Intermediate-size classes may be collected by both techniques, as well as by other approaches.

In measuring the POC in seawater, the usual practice is to consider that this is the material retained on a glass fiber or silver filter with a nominal pore size of about 0.50 μm. Such filters can also trap some particles smaller than the nominal pore size (and can also adsorb some DOM), so the cut-off is not abrupt. Once filtered, however, the measurement of the mass of organic material on the filters is straightforward, making due allowance for problems of blanks and contamination, and for the removal of inorganic carbonates. The problem of having enough material to analyze is usually solved by filtering a sufficiently large volume of water. The method now most commonly used to measure the carbon on the filter is dry combustion in a furnace at 600 to 800 °C, followed by the determination of the CO_2 formed, using a gas chromatograph or an infrared-absorption gas analyzer. Other methods of quantifying the CO_2 may also be satisfactory. Observed concentrations of POC range from 5 to 10 μg kg^{-1} (0.4 to 0.8 μmol kg^{-1}) in deep water to as much as several hundred μg kg^{-1} in productive surface water.

Dissolved organic carbon is defined operationally as material that passes through a filter, usually one with a nominal pore size of about 0.5 μm. The material that passes through such a filter includes substances from the smallest molecules through colloidal sizes up to small particles, including viruses and even some bacteria. Most of the material does appear to be in true solution, however. Since the POC is usually such a tiny fraction (generally about 1 to 2%) of the total organic carbon in most samples of seawater, it is a frequent practice not to bother filtering prior to measurement of DOC, except in phytoplankton-rich surface waters. Some of the DOC is volatile, but the proportion is not exactly known and no method that is currently applied measures all of it. There is no evidence to suggest that the volatile material is anything more than a negligible fraction (<5%) of the total DOC, although individual components may be of significance for themselves.

In contrast to the relative ease with which POC may be measured, it is much more difficult to measure the DOC in seawater. Attempts to extract and weigh it have proven unsatisfactory, because no extraction technique gets it all and salt is also extracted. All modern approaches to measuring total DOC in seawater depend on the oxidation of the organic matter and the measurement of the resulting CO_2. In every case the first step must be the removal of inorganic CO_2, which is generally accomplished by acidification and then stripping out the released CO_2 by bubbling with compressed N_2 or O_2 gas. It is at this stage where any volatile DOC is lost.

In one of the earliest methods (Krogh 1934; Plunkett and Rakestraw 1955), a sample of seawater was treated to remove both CO_2 and chloride (which interfered with the analysis), then dried and the dry salts heated with sodium dichromate in sulfuric acid. The resulting CO_2 was captured and titrated. This technique was tedious and difficult, and produced scattered results for the concentration of organic carbon, generally in the range of 100 to 200 μM.

The DOC may also be oxidized by autoclaving with persulfate (Menzel and Vaccaro 1964), by high-intensity ultraviolet irradiation (Armstrong *et al.* 1966), or by dry combustion of the solids at high temperature (Starikova 1970; Sharp 1973; Sugimura and Suzuki 1988). The CO_2 produced by these oxidation techniques has been detected and measured by sweeping the gases out of the solution into either an infrared gas-phase photometer or a gas chromatograph with a thermal-conductivity detector.

The procedure developed by Menzel and Vaccaro (1964) was the first convenient method that appeared to give reproducible results, and it is still in use, as are the ultraviolet-oxidation technique and the modern version of the dry-combustion technique. All now give roughly similar results, though this was not always so. Intercalibration exercises at first yielded highly discordant results, with values differing by more than a factor of two between techniques and investigators, and the results of one such exercise took up the whole of the January 1993 issue of the journal *Marine Chemistry*. Eventually, Sharp (1993) and Peltzer *et al.* (1996) reported procedural details of the techniques that have now led to generally acceptable values. It should be noted that there is no easy way to prove the correctness of any value.

11.6 Concentration and age of marine organic matter

While the methodological problems described earlier have left us with some small uncertainty as to the concentration of dissolved organic carbon in any given sample of seawater, the general ranges are not too much in dispute; concentrations (expressed as C) range from about 40 to perhaps as high as about

200 μmol kg^{-1} in the open ocean, with the highest values of course being found in productive surface waters and the lowest values in the oldest deep water. The data from Sharp (1993) suggest a value of about 40 μmol kg^{-1} as representative of deep water in the equatorial Pacific Ocean. For the purpose of making representative calculations, in this book I use a world ocean average concentration of 50 μmol kg^{-1}.

Given the commonly used value of 2.2 to relate the mass of organic matter to the mass of carbon, the carbon concentration of ~50 μmol kg^{-1} suggests that the average total concentration of dissolved organic substance in the ocean amounts to about 1.3 mg kg^{-1}. Concentrations several times this high are commonly present in productive surface waters.

The total mass of organic carbon dissolved in seawater (assuming an average concentration of 50 μmol kg^{-1}) is about 830 Gt, a value larger than the total standing stock of trees and other terrestrial biota. The total mass of DOC is also about 17 times the total annual primary production in the ocean (cf. Table 11.2). Since much of the organic production is already accounted for in supporting the rest of the food chain, the fraction of organic production that goes into replacing the DOC must be rather small and, as a minimum, one could imagine that the DOC would on average have to be at least several hundred years old. Although various investigators have produced somewhat different results, most data sets agree that in deep water below 1000 m the concentrations of DOC appear fairly uniform with depth and location. Given that nearly all the organic matter in the ocean must be manufactured at the surface or delivered to the surface, and an average turnover time of the ocean of around 1000 years, it would seem reasonable that the average age of the organic carbon in deep water must be of comparable magnitude and probably greater than one turnover time of the ocean.

The first measurements of the ^{14}C content of deep organic carbon suggested an age of about 3500 years (Williams *et al.* 1969). Later, with better and more sensitive techniques, the ^{14}C age of organic carbon in the deep Pacific was measured again and found to be on average about 6000 years (Williams and Druffel 1987). The carbon on which the measurements were made was extracted from seawater as CO_2 after oxidation of the organic matter with high-intensity ultraviolet light (Figure 11.6), and about 40 μmol of carbon was released from each liter of sample. There remains some possibility that this technique misses a fraction of the carbon. If this is so, there is no evidence on the age of the missing carbon.

The old age and rather uniform distribution with depth of the carbon in deep water (Figure 11.6) allows us to draw some conclusions and also raises a number of perplexing questions. Since the age as measured must be an average including some carbon that is fairly young, some of the carbon must be significantly older, and thus it must be extraordinarily resistant to the processes that might remove or degrade it. Any carbon-containing molecules that old must have circulated from the deep water to the surface and back again a number of times. At the surface these carbon molecules have been exposed to a high level of biological activity, but they have still resisted bacterial attack. We have no information on what characteristics of the structures would lead to such resistance. It has been suggested that if these molecules

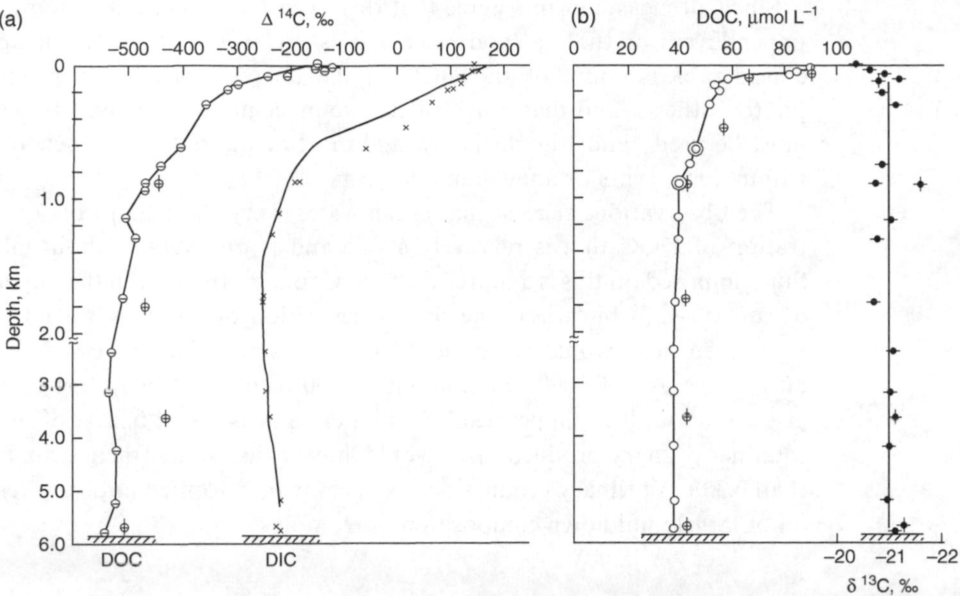

Figure 11.6 Measurements of the ^{14}C content of organic carbon in seawater to estimate its age (Williams and Druffel 1987). The concentration of ^{14}C is given as $\Delta^{14}C$; see Box 10.1 for the definition of $\Delta^{14}C$. After removal of the DIC, the DOC was oxidized with high-intensity ultraviolet light, the resulting CO_2 collected, and the ^{14}C determined.

(a): Concentration of ^{14}C in DOC and DIC. The values for DOC at the surface of about −150‰ correspond to a ^{14}C age of about 1310 yr BP, and the value of −540‰ at a depth of 5710 m corresponds to a ^{14}C age of about 6240 yr BP. From these data it is commonly stated that deep-water DOC is about 6000 years old.

(b): Concentration of DOC, measured by ultraviolet oxidation, and the concentration of ^{13}C in the DOC, given in the usual δ notation.

The samples represented by crossed circles appear to be due to unresolved analytical artifacts; for our purpose here they may be disregarded. They are, however, a warning that when dealing with DOC in seawater, even our best procedures in the hands of the most accomplished investigators may yield data that are yet not fully understood.

are so resistant to bacterial action then possibly their ultimate fate is to be degraded by photochemical attack at the surface, and experimental evidence on the degradation of organic molecules by ultraviolet light is consistent with that suggestion (Mopper *et al.* 1991). Calculations by these authors led them to suggest that, if photodegradation is the primary fate or *sink* for the very resistant organic material, this may be the rate-limiting process controlling its loss from the ocean.

The most resistant molecules have also failed to adsorb to sinking particles and be carried down to burial in the sediments, although it is known that all surfaces introduced into the sea quickly become coated with marine organic matter. This is, however, another route by which old organic carbon could be removed permanently from the ocean and must be the fate for some of it.

Since all measurements agree that there is a higher concentration of DOC in the upper layers of the sea, and since this is on average younger, it appears that a considerable amount of the DOC in the upper ocean is the product of recent photosynthesis, and that much of this younger material is used by bacteria. Some must be used almost immediately, and most of the rest of the new organic carbon within a few years or a few hundred years.

The observations suggest that ocean water everywhere has a background concentration of DOC that is relatively stable and is on average about 6000 years old. Superimposed on this is a more labile DOC found primarily in the upper layers of the ocean. If we (arbitrarily) take the concentration of the resistant DOC as 40 μmol kg^{-1}, then there would be about 665 Gt of this material, expressed as carbon. With an average age of 6000 years, about 1/6000 of the material must be replaced each year, or about 0.11 Gt per year. This latter value is about 0.22% of the world ocean total net primary production. It would appear that some fraction that is a little less than 0.3% of primary production is somehow transformed into very resistant material of largely unknown composition.

11.7 Nature of marine organic matter

The chemical nature of much of the organic matter in seawater is still largely unknown. Small amounts of each of many possible organic compounds have been measured in seawater, but in sum they do not add up to more than 30 to 40% of the total. Of the remaining material, some general chemical characterization has been carried out, but the sources, the structural nature, and the fates of this material are not well understood.

Most of the organic matter in the ocean passes through a 0.5 μm filter and is conventionally considered to be in the dissolved form (DOC or DOM) although, in fact, some of the material passing through these filters is in colloidal-sized particles of organic material (Wells and Goldberg 1992). The particulate matter caught on filters generally amounts to between 1 and 10% of the total organic matter, the higher percentages being found only in regions with phytoplankton blooms. Except when they are concentrated in dense swarms, living organisms, the best characterized fraction, commonly make up less than 1% of the total organic carbon (Table 11.4). The marine biota, in all their rich diversity, are the source of nearly all the organic matter in the sea, and some of that diversity must be reflected in the complexity of the non-living organic matter.

The DOM in rivers is already quite weathered and composed mostly of resistant material because much of it has survived passage through soils and lakes for usually some considerable time. This has caused speculation that river-borne DOM could contribute to the resistant organic matter in the ocean. There are several distinct signatures possessed by terrestrially derived organic matter that can be used to evaluate this speculation. One of these is the isotopic composition of the carbon. Terrestrial organic matter is generally characterized by a low ratio of the isotopes ^{13}C to ^{12}C; the $\delta^{13}C$ value is generally close to –28‰. In contrast, the $\delta\,^{13}C$ value for

Table 11.4 Concentrations of several classes of organic carbon in seawater

	Range, μM	% of TOC
Dissolved organic carbon (DOC)	40–250	90–99
Particulate organic carbon (POC)	1–30	<1–10
Colloidal organic carbon	1–3	<1–3
Volatile organic carbon (VOC)	1–5	<1–5
Living organisms		<0.1–10

Approximate calculation for dissolved organic matter (DOM): DOM/DOC = 2.2 by weight.

Therefore, DOM range = ~1 to ~6 mg kg^{-1}. The estimate of colloidal organic carbon (particles that can pass a 0.5 μm filter) is derived from Wells and Goldberg (1992), who investigated surface water off southern California.

marine organic matter is generally close to –22‰, about the same as that of marine organisms. This evidence shows that most of the marine organic matter must derive ultimately from marine organisms, but if only a small fraction were originally terrestrial, this would not shift the isotopic signal enough to be detected. Another characteristic of terrestrial organic matter is the presence of chemical structures derived from lignin, a substance not formed by marine plankton. Opsahl and Benner (1997) extracted organic matter from both Atlantic and Pacific seawater and found concentrations of lignin-derived phenolic compounds amounting to 27 and 11 ng L^{-1}, respectively. By setting these values in a ratio to the concentrations of such compounds in river water, they estimated that the perhaps 2.4 and 0.7% of the DOM in the two oceans might be derived from terrestrial sources. The estimate must be quantitatively uncertain because not all river-borne DOM is derived from lignin, and only a portion of the oceanic DOM can be extracted by the available procedures. It is clear, however, that some small fraction of marine DOM is ultimately terrestrial in origin and the rest must be marine.

In the following sections, we take a brief look at the POC and then briefly examine the nature of the dissolved substances in the ocean.

11.7.1 Marine snow

Many kinds of inorganic particles are found suspended in the ocean or, usually, sinking at varying rates. These include clays and other minerals from dust-fall and river input, shells of marine mollusks and protozoa, and coccoliths. Since all of these particles are coated with marine organic matter of unknown composition, they constitute some portion of the particulate organic matter in the sea. The largely organic particles that are microscopically recognizable on filters include sinking phytoplankton cells, exoskeleton molts from marine crustacea, and fecal pellets from a wide variety of organisms. These fecal pellets may incorporate all of the above classes of particles. Microscopic examination of filters also reveals a large amount of amorphous debris, often with recognizable particles in it.

In 1952, Dennis Fox and his colleagues published their idea that there is, suspended in the sea, a very finely dispersed mud or sludge of organic and inorganic composition; for this they coined the term *leptopel* from Greek words meaning "thin mud." The term did not catch on.

When looking through the windows of a bathysphere deep under the surface in the sea near Japan, Suzuki and Kato (1953) observed much flocculent material, evidently non-living, reflecting light, and thus easily seen against the blackness beyond. Recalling the evocative words of Rachael Carson (1951), who thought of the sediments on the ocean floor as slowly accumulating under an endless snowfall, Suzuki and Kato proposed the name *marine snow* for this suspended but presumably slowly sinking flocculent material. They suggested that it was identical with the marine leptopel of Fox *et al.* The term marine snow stuck and is still widely used (Alldredge and Silver 1988) though less poetic scientists often speak of *organic aggregates* or *particulate material.*

Everywhere in the ocean one can find some concentration of organic aggregates or marine snow. These particles have a varied composition and origin. Many marine organisms secrete slimes of mucous material that may coagulate and stick together, along with cast-off exoskeletons of marine crustacea, fecal pellets, abandoned gelatinous "houses" of appendicularians, and similar debris. Any dust, clay particles, pollen, or other atmospheric particles falling through the ocean may be trapped or stuck to these sheets of mucus and other aggregate materials. The aggregates provide surfaces for the attachment of bacteria and other microbes. As they sink slowly through the water column, the aggregates carry these organisms down with them, and they become small islands of high biological activity, contributing to the total metabolism deep in the water column. This slowly sinking dilute snowfall of organic debris provides food for deep-living plankton organisms and, to the extent that it gets to the bottom, food for the benthos as well.

Marine snow is difficult to sample in its native state. When seen by divers or through the windows of a submersible, it appears as particles, sheets, and filaments ranging in size from the smallest visible point of reflected light up to occasionally even a meter across. Much is broken up during bottle sampling, and then appears only as particulate matter when the water is filtered. Probably some of the organic slime can pass through the usual filters, so it is then classified as dissolved. Visual observations and careful sampling by divers have been crucial in its investigation (Alldredge and Silver 1988). It is not known how much of the material collected in sediment traps or on filters from bottle sampling or by deep pumping was originally in this loosely aggregated form, but presumably marine snow makes up a significant fraction of the reported particulate matter in seawater.

The common presence of blobs and irregular sheets of mucus-like material will have two effects on smaller sinking particles. First, it will tend to trap them and provide an opportunity for attack by bacteria and small grazers, thus tending to encourage the metabolism of these particles high in the water column. Second, as more particles are caught, they will tend to be clumped into larger masses, which will eventually sink much faster so that some clumps will be taken further towards the bottom than might otherwise be the case. Little is known about the relative importance of these two effects and how they may vary with depth.

The concentrations of filterable organic aggregates were found by Riley *et al.* (1965) to be rather uniform with depth in the Sargasso Sea below a few hundred meters. They interpreted this to suggest that some fraction of these aggregates is formed *in situ* (otherwise they should decrease with depth due to filtering by organisms), and suggested that this might be due to adsorption of DOM, resulting in the growth of the particles. Another route to the formation of particles is due to bubbles near the sea surface. As with other surfaces in the ocean, the gas–water interface surrounding each bubble can become enriched with organic matter. When a bubble comes to the surface and breaks, some of this surface skin can be compressed, forming a small organic particle. Sutcliff *et al.* (1963) found that filtered seawater forms particles when bubbles are passed through. In another process, small particles can stick together to make larger particles. Wells and Goldberg (1992) showed electron micrographs of the apparent agglomerations of colloidal-sized particles, which could eventually result in macroscopic particles. In sum, there is evidence that some of these particles can grow in seawater by incorporating smaller particles and/or DOM.

11.7.2 Vitamins

Among the remarkable early results from studying marine microbial physiology was the discovery that many marine algae and bacteria have absolute requirements for certain vitamins, while others can synthesize and possibly release these essential components of the biochemical machinery. Small amounts of several of these substances are nearly always detectable in seawater by the use of bioassay techniques (Swift 1980), and organisms acquire these essential nutrients from extremely low concentrations in the surrounding solution.

Three prominent vitamins are biotin and thiamine (Figure 11.7) and vitamin B_{12} (Figure 11.8). These compounds presumably get into the water by leakage from the cells of the organisms that make them, or by loss of cell material when predators eat the organisms. Another route is from the excreta of fish and other organisms, because it has been shown that, for example, vitamin B_{12} is present in the gut contents of fish, presumably synthesized by bacteria there (Cowey 1956). The widespread occurrence of both the capacity of some organisms to synthesize these vitamins, and the absolute requirement for the growth of other organisms, suggests the possibility of many kinds of symbiotic arrangements. One imagines that there may be many other compounds in a similar situation. The varying concentrations of growth factors, such as these vitamins, found in surface waters and their no doubt varying rates of synthesis and consumption must considerably influence the extraordinary variety of species and the patchiness in distribution of the phytoplankton (Cowey 1956; Swift 1980).

Some of these compounds appear to be quite stable in seawater. Menzel and Spaeth (1962) measured biologically useful concentrations of vitamin B_{12} as high as 0.1 ng L^{-1} at a depth of 2600 m in the Sargasso Sea and 0.15 ng L^{-1} at 500 m, even more than the 0.04 ng L^{-1} they measured in the surface mixed layer. Concentrations as low as 0.01 ng L^{-1} were detectable, which is about 7 fM

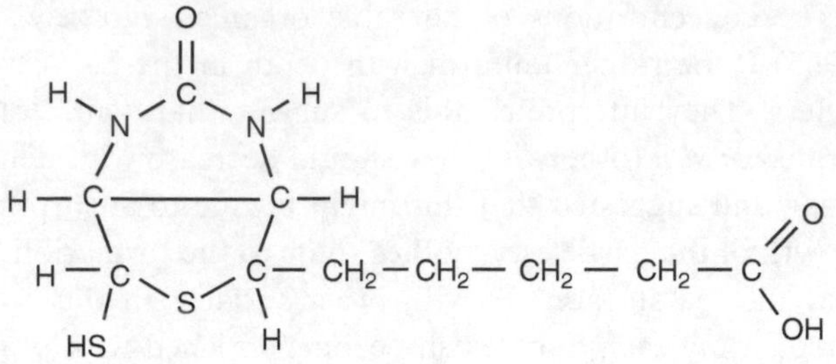

Figure 11.7 Among the organic compounds detected in seawater with bioassay techniques are these two vitamins. Both are essential components of the biochemical machinery involved in the metabolism of carbohydrates, but are not synthesized by mammals or many other higher organisms, nor by some bacteria and algae. Thus, they are essential nutrients for these organisms, and must be supplied from outside. They are synthesized by many other bacteria and algae, and these organisms must be the sources of these vitamins in seawater. Reported concentrations of biotin in seawater range from undetectable up to about 10 ng L^{-1}. Concentrations of thiamine, also known as vitamin B_1, in seawater range from undetectable up to about 490 ng L^{-1}.

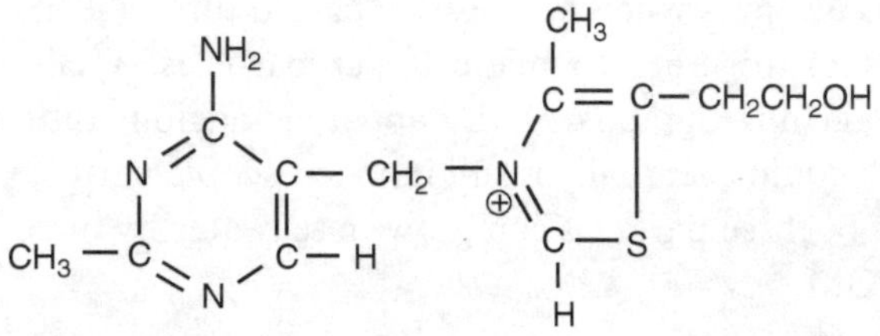

(fM $= 10^{-15}$ mol L^{-1}). It is quite remarkable that such a complex molecule as this (Figure 11.8) remains dissolved and undegraded in seawater for extended periods of time, perhaps hundreds of years.

11.7.3 Volatile hydrocarbons

Among the volatile organic compounds in seawater, the smaller members of the hydrocarbon series are interesting because of the concentrations present, the fact that in general not much is known about where they come from or to what extent they are utilized in the ocean, and also because they are usually supersaturated in surface waters, so that the ocean is a continuing source of these compounds to the atmosphere.

The smallest organic molecule, methane (CH_4), is the most abundant of the series, with concentrations commonly around 2000 pM (Table 11.5). Concentrations of ethane, the next in the series of saturated hydrocarbons, are much less (around 15 pM) and concentrations continue to drop for the higher members of the series.

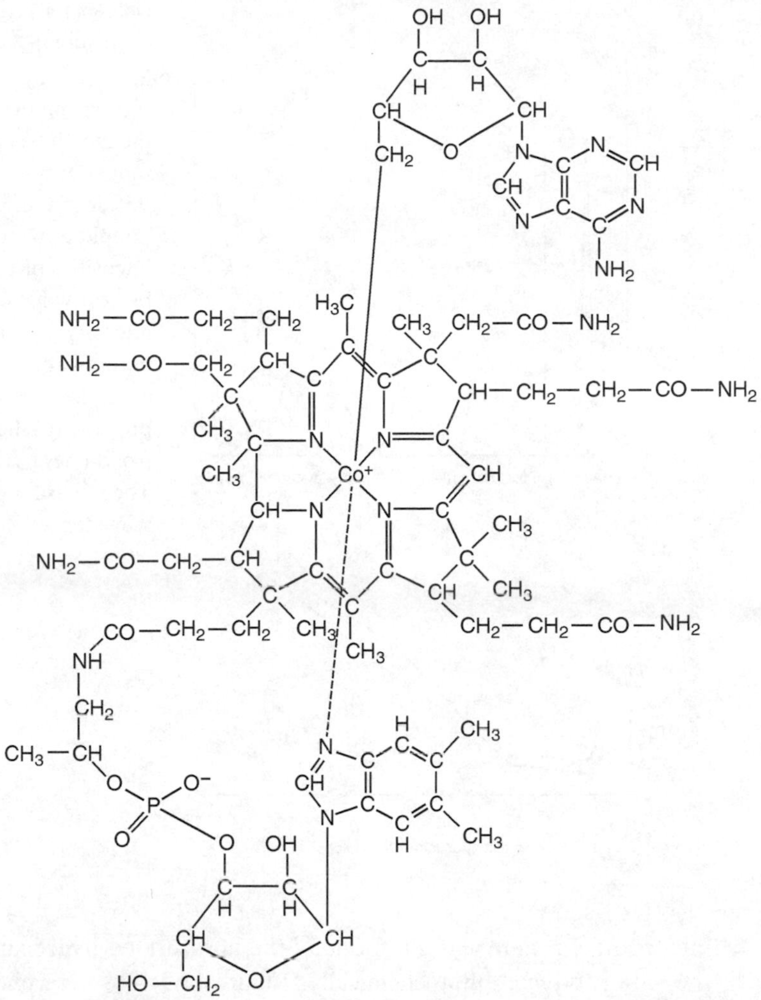

Figure 11.8 Structure of coenzyme B_{12} (5′-deoxyadenosylcobalamin). The form shown here is the metabolically active structure. The 5′-deoxyadenosyl unit (the group at the top of the diagram) is attached through the 5′ carbon atom to the cobalt atom. This is the only direct carbon-to-metal bond reported in biological systems. Only bacteria make vitamin B_{12}. The form of vitamin B_{12} found dissolved in seawater probably does not have the 5′-deoxyadenosyl unit attached; instead there is probably a hydroxyl unit. In the latter form, with a molecular weight of 1346, vitamin B_{12} is the largest and most complex molecule identified as dissolved in seawater (other variants may also exist). Biologically active concentrations in the ocean have been measured as low as 0.01 ng kg^{-1} (~7 fM); this concentration, however, is still several billion molecules per liter.

At comparable carbon numbers, the small unsaturated hydrocarbons are present in surface water at concentrations several times greater than those for the saturated series, though why this should be is not known. Many of these compounds are more concentrated in water with abundant phytoplankton, so biological activities are

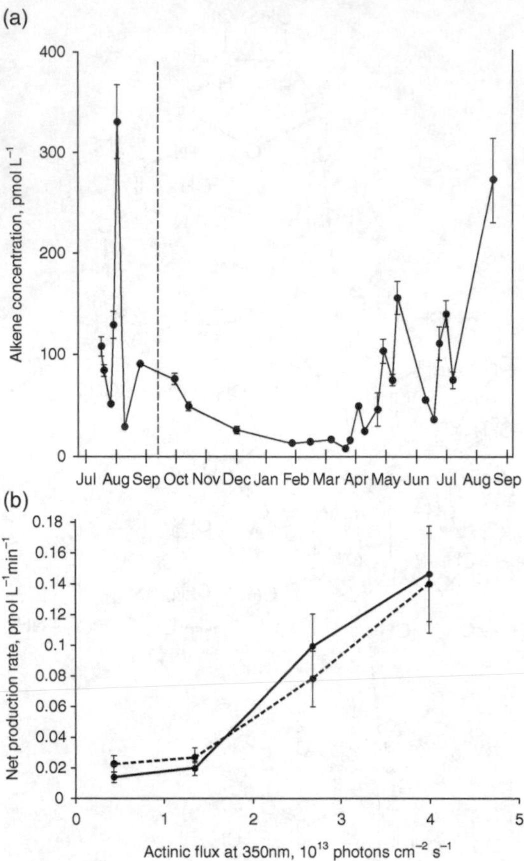

Figure 11.9 Gist and Lewis (2006) measured the concentration of propene in seawater collected monthly at two stations on the coast of England facing the North Sea during one annual cycle (upper graph, a). In an experiment (lower graph, b) they exposed a large sample of seawater to a range of intensities of ultraviolet light (the highest value being of about the same intensity as seawater is exposed to naturally during the wintertime at mid-day), and measured the rate at which propene (dashed line) and ethylene (solid line) were produced in the water. There was no production when seawater was kept in the dark.

important. In many cases (especially the unsaturated hydrocarbons) it appears that they are produced photochemically (Figure 11.9). In these photochemical reactions organic matter in seawater is in some way converted by the action of ultraviolet light into these light hydrocarbons, and probably into other substances as well. This might be one route by which the old organic matter in the ocean is degraded to smaller compounds and eventually eliminated.

11.7.4 Oxides and sulfides

Carbon monoxide (CO) is a prominent constituent of the volatile organic gases in seawater (Table 11.5). It is produced biologically; some macroalgae make CO to pressurize their gas bladders, used as floats. It is formed during photochemical destruction of dissolved organic matter. It is also produced by a variety of invertebrates with gas-filled floats, such as the siphonophore Portuguese man o'war that floats on the surface, and some other remarkable deep-living colonial forms; in each case it appears to pressurize the bladder, although other gases are also always present.

Dimethyl sulfide (DMS) is an important product from marine phytoplankton. It arises as a breakdown product from dimethyl sulfoniopropionate, a compound

Table 11.5 Partial list of naturally occurring volatile organic substances detected in seawater, along with representative concentrations, in pM (= 10^{-12} M)

Name	Formula	Surface	Deep
Methane	CH_4	2000	1000
Ethane[(a)]	CH_3CH_3	15	
Propane[(a)]	$CH_3CH_2CH_3$	8	4
Butane[(a)]	$CH_3CH_2CH_2CH_3$	3	
Ethene (Ethylene)[(a)]	$CH_2{=}CH_2$	100	
Propene[(a)]	$CH_2{=}CHCH_3$	25	14
1-Butene[(a)]	$CH_2{=}CHCH_2CH_3$	20	
Isoprene[(b)]	$CH_2{=}C(CH_3)CH{=}CH_2$	5	
Acetylene[(a)]	$CH{\equiv}CH$	10	
Carbon monoxide[(c)]	CO	12 800	
Carbonyl sulfide[(d)]	COS	30	
Dimethyl sulfide[(e)]	$CH_3{-}S{-}CH_3$	2600	100
Bromoform[(f)]	$CHBr_3$	8	4
Dibromochloromethane[(f)]	$CHBr_2Cl$	0.5	1
Bromodichloromethane[(f)]	$CHBrCl_2$	0.6	1
Dibromomethane[(f)]	CH_2Br_2	3	1
Chloroiodomethane[(f)]	CH_2ICl	2	0.5
Methyl iodide[f)]	CH_3I	3	0.5
Diiodomethane[(f)]	CH_2I_2	2	0.5

The concentrations are given for the molecular substance, not just the carbon.

(a) Production processes unknown, probably some is photochemical. Based on estimated flux for Atlantic only from 35° S to 35° N, the global fluxes to the atmosphere are extrapolated for comparison as follows (all values in Gmol yr^{-1}): ethane 20, propane 6, n-butane 2.7, i-butane 0.6, ethene 44, propene 18, 1-butene 11, acetylene 1.8; total all non-methane hydrocarbons ≈ 100.

(b) Production probably biological, by many kinds of phytoplankton.

(c) Production is photochemical and biological. Ocean-to-atmosphere flux ≈ 6 Tmol yr^{-1}.

(d) Production is photochemical. Ocean-to-atmosphere flux ≈ 13 Gmol yr^{-1}.

(e) Production is by phytoplankton. Ocean-to-atmosphere flux ≈ 1500 Gmol yr^{-1}.

(f) The halomethanes appear to be all biologically produced, with possibly some photochemical or other inorganic processes contributing. Macroalgae are certainly the source for some; apparently phytoplankton are for others.

Some of these substances show complex vertical distributions, even in the upper mixed layer, indicating active production and/or consumption. Concentrations and fluxes in this list should be considered as only approximations to more complex pictures. Some vary by at least a factor of ten from the value noted here.

Surface water values for C_2 to C_4 hydrocarbons from the central Atlantic from Plass-Dülmer *et al.* 1993. Values for methane and deep-water values for C_2 to C_4 from Swinnerton and Linnenbom 1967, who were the first to report concentrations of C_1 to C_4 gases in seawater. Values for COS from Andreae and Ferek 1992, for CO from Erickson 1989, for DMS from Andreae and Barnard 1984, for isoprene from Broadgate *et al.* 1997 and for the halomethanes from Moore and Tokarczyk 1993.

apparently used by many phytoplankton to help regulate intracellular osmotic pressure. It is generally supersaturated in seawater relative to the atmosphere, so, as it diffuses upwards, it is a significant source of sulfur to the atmosphere. In the troposphere it is fairly easily oxidized; on breakdown it releases sulfur that is eventually oxidized to sulfate, thus contributing both to acidification of rain and to the numbers of cloud condensation nuclei. Dimethyl sulfide can be found at a concentration of around 100 pM even in very deep water of the Atlantic Ocean. It is not known if it is formed there, but, if not, it must be quite stable to biological or chemical attack deep in the water column.

Carbonyl sulfide (COS) appears to be formed largely or entirely by photochemical processes in surface waters. It is also present in concentrations greater than air saturation, so there is a net flux of COS into the atmosphere, and is the most abundant sulfur–carbon compound in the atmosphere. While the flux from the ocean is much less than that of DMS, COS is stable to attack in the troposphere, so it eventually reaches the stratosphere where the more vigorous oxidation conditions eventually break it down. Accordingly, COS is a source of sulfate aerosol in the stratosphere and contributes to the steady-state background of stratospheric haze.

11.7.5 Volatile halocarbons

Marine algae make an impressive variety of chlorinated, brominated, and iodinated organic compounds (Faulkner and Anderson 1974).

Among these are a number of halogenated methanes (Table 11.5), which appear to be fairly stable compounds in the ocean environment and, based on evidence from the North Atlantic, are probably present in detectable concentrations everywhere in the ocean (Moore and Tokarczyk 1993). Figure 11.10 is a typical example of a section between Greenland and Labrador, showing the concentrations of dibromomethane. The higher concentrations in shallow water near the coasts supports direct evidence that this compound is synthesized by attached macroalgae. Concentrations of 0.1 to 0.2 ng L^{-1} were observed in bottom water at 34° N, as far south as measurements were made. It will be interesting to learn if this and similar compounds continue to be detectable in older water further south and into the Pacific.

Some of these natural halogenated volatile compounds contribute to the chlorine and bromine in the atmosphere, and those that are sufficiently stable to last for considerable lengths of time in the troposphere contribute to the chlorine and bromine in the stratosphere.

The functions of these and related compounds are assumed to be involved with deterring herbivores or with other aspects of the chemical warfare in the sea.

11.7.6 Fatty acids and other lipids, amino acids, and sugars

The first organic compounds in the sea that were identified and directly measured by chemical techniques were the fatty acids. After solvent extraction from seawater and some chemical processing Williams (1961, 1965) employed one of the earliest

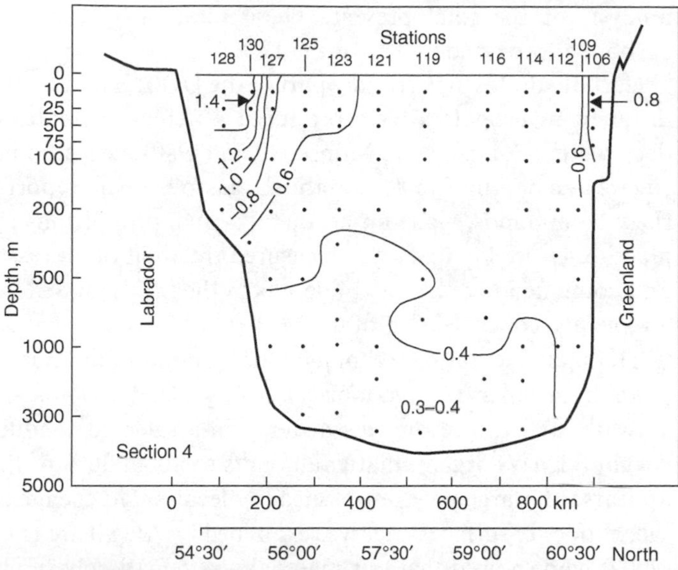

Figure 11.10 Concentrations of dibromomethane (CH_2Br_2) along a section from Labrador to Greenland (Moore and Tikarczyk 1993). The concentration units are ng L^{-1}. The high values in shallow water near the coasts suggest that the sources are the benthic biota living in shallow water. Other evidence points to attached macroalgae. The concentrations present in deep water could be due to production there, but it is far more likely that they indicate that this compound is relatively stable in deep water and is able to last there for at least several hundred years. ■

commercial models of the gas chromatograph to separate and quantitate several fatty acids. He measured the concentrations of dissolved fatty acids in deep water (> 1000 m) at several stations in the Pacific; he found several individual fatty acids present, but the total concentration of all of them totaled only about 3 μg L^{-1}. Expressed as carbon, this amounts to 0.1 to 0.2 μM, much less than 1% of the DOC.

Several investigators have examined the concentrations of amino acids in seawater. They are present in such low concentrations that contamination of the samples with, for example, a fingerprint or even by reagents such as hydrochloric acid is a serious problem. Reported concentrations have dropped as investigators apply ever more stringent techniques to eliminate interferences and contaminations. It appears, however, that these compounds are present and detectable, but the concentrations of dissolved free amino acids are only 1 to 15 nM in deep water (Lee and Bada 1975), or less than 1 nM (Druffel *et al.* 1992). Treatment of dried sea salt with hot 6 N HCl or autoclaving whole seawater samples with the same concentration of HCl yields additional amino acids released from the dissolved organic matter, presumably by hydrolysis of polymers such as proteins or peptides, and possibly from humic substances, viruses, and other colloidal particles. The hydrolyzable amino acids amount to about 10 to 100 times the concentration of dissolved free amino acids. Lee and Bada (1975) identified 11 individual amino acids, for an overall total of about 160 nM, while Druffel *et al.* (1992) found only about 100 nM by an analysis of the total present. These totals are equivalent to 0.4 to 0.7 μM of carbon, approximately 1% of the DOC.

Carbohydrates are certainly part of the DOC, but, as with many other compounds, different investigators have produced different estimates of the concentrations in deep water. For example, Mopper *et al.* (1980) measured individual free monomeric sugars at a depth of 1500 m in the Sargasso Sea and reported a value of 0.74 μM for these compounds, made up in roughly equal proportions of fructose, galactose, and glucose. Druffel *et al.* (1992) measured the total of all dissolved carbohydrates, free and combined together, at 14 depths in the Sargasso Sea below 400 m and obtained an average concentration of 0.35 ± 0.07 μM.

The introduction of techniques for ultrafiltration (Buesseler *et al.* 1996) made it possible to separate large-molecular-weight and colloidal-sized organic matter from seawater and to prepare this material in a salt-free condition. This high-molecular-weight (HMW) organic matter amounts to about 30% of the total DOM. Much of it appears to be large polymers of sugar molecules. The chemical nature of these polymeric saccharides in surface water was examined by Aluwihare *et al.* (1997) and Repeta *et al.* (2002) who showed that this material was remarkably similar in water samples from both the Atlantic and Pacific and, even more remarkably, very similar substances were present in river and lake water. This polymeric saccharide material was composed of seven different monosaccharides attached to each other with both linear and branching linkages, and also containing nitrogen and acetate groups. Apparently similar material was present in old cultures of the marine diatom, *Thalassiosira.* Their estimate is that this polysaccharide substance makes up 70% of the HMW DOM in surface seawater, and perhaps 20% of the total DOC. Given the similarity of this material from the two oceans, we may suppose that it is a relatively common and stable component of surface-water DOC, but most of it must eventually disappear, because of the evidence that polymeric saccharides are not abundant in deep water. This is an area of active research, and no doubt new surprises and insights will keep appearing.

In general, all these common types of organic compounds that are recognizable products of ordinary biochemistry (and a few others not mentioned here) probably amount in total to less than 30% of the total DOC measured in deep seawater, and a somewhat higher percentage in surface water.

All of the classes of compounds mentioned in this section are present in higher concentrations in surface water than in deep water. This is entirely expected, because they must be synthesized mostly in productive surface water, and many are excreted or leak from phytoplankton and other organisms. In these regions of active supply they are consumed rapidly by bacteria. Only minuscule amounts survive into deep water, although it is also possible that there is some very slight leakage into deep water from the comparatively few organisms living there. Another source for the concentrations found there is the bodies of living organisms themselves. Many marine bacteria are extremely tiny and can pass through the filters normally used to separate particulate and dissolved phases. There are in addition viruses that can pass the filters, and the bodies of very soft-bodied larger organisms may be crushed on filters and leak some juices. In total, these sources are likely to be very small, but they should be kept in mind.

11.7.7 Alkenones

Among the many interesting substances discovered in the ocean, the alkenones are especially important. These are a group of long chain hydrocarbons with a ketone unit near one end (Figure 11.11). The existence of these substances was first learned when they were detected about 35 years ago in extracts from several-million-year-old ocean sediments. They often constitute a significant fraction of the material that can be extracted from ocean sediments with organic solvents. Subsequently it was discovered that the alkenones are manufactured by a few, but very abundant, species of marine algae within the group known as Prymnesiophytes (also called Haptophytes). The most well-known of these are the two species of coccolithophorids known as *Emiliania huxleyi* and *Gephyrocapsa oceanica*; there are also some related but non-carbonate-secreting forms that make alkenones. The most important alkenones are those with 37 carbon atoms and two or three double bonds in the chain; alkenones with 35, 36, 38, and 39 carbon atoms in the chain are also made in smaller amounts, and some with four double bonds are also present. Related long-straight-chain hydrocarbons (alkanes) and long-chain fatty acid methyl esters are also found, as are a few other variants of the alkenone molecules.

The function of the alkenones is not entirely clear. It has been suggested that they might play a role in the structure of membranes and that they may be used as energy stores. It is remarkable that they may constitute as much as 5 to 10% of the total cell carbon.

Many organisms adjust the fatty-acid composition of their lipids in such a way as to maintain the viscosity or consistency of their fat stores or membranes at least somewhat independent of temperature, with saturated lipids in higher proportions at higher temperatures. Organisms that make alkenones also adjust their ratios according to temperature (Figure 11.12). Many field studies and experiments with laboratory cultures have shown that the ratio of the more-saturated to the less-saturated alkenones (expressed as the $U_{37}^{k'}$ ratio; see Figure 11.11) quite closely follows the temperature of growth (Figure 11.12). A compilation of data from surface sediment samples from a wide variety of locations (Figure 11.13) shows that the $U_{37}^{k'}$ratio appears linearly related to the temperature over a broader spatial and temporal scale. A judicious selection of stations where the sediment record is not compromised by lateral advection and the surface-temperature seasonality is low could considerably narrow the spread in these data.

One of the most extraordinary observations about the alkenones is that not only do they survive bacterial and other degrading processes as well or better than the rest of the organic matter during transport from the surface to the sediment and for millions of years in the sediment, but the relative proportions of the C-37:2 and C-37:3 alkenones apparently remain the same, at least in most locations (Herbert 2006). This means that the relation to temperature that has been established in culture or in recent sediments may be used to infer the temperature of the surface water in selected locations during the last several million years. Indeed, an important

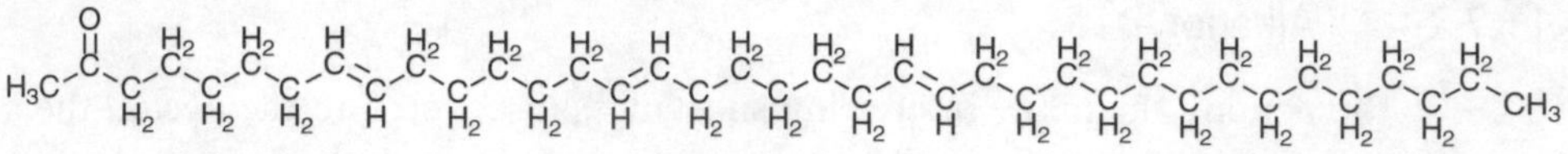

C–37:3

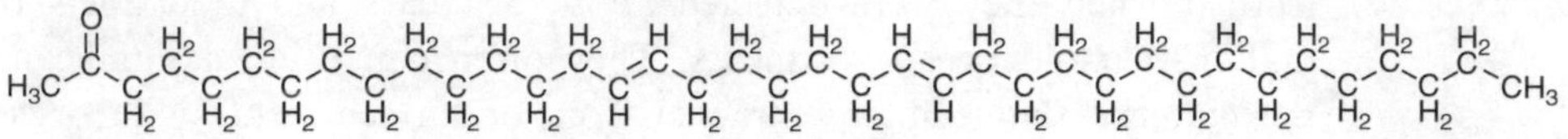

C–37:2

Figure 11.11 Structures of two of the most important alkenones. Note that as with other alkenones of this type there is in each case one ketone unit near the end of the chain. The C-37:2 alkenone has two double bonds while the C-37:3 alkenone has three double bonds. The ratio of these compounds in Prymnesiophyte cells is a function of the temperature at which they grow, and this ratio is apparently maintained in the sediments. The ratio is currently defined as follows (Prahl and Wakeham 1987):

$$U^{k'}_{37} = \frac{C_{37:2}}{C_{37:2} + C_{37:3}}$$

where C is the concentration of each component measured in the sediment. The term $U^{k'}_{37}$ is meant to refer to the unsaturation index of the 37-carbon ketones, and the (′) identifies this as a version modified from that originally proposed.

The double bonds in the alkenones are in the *trans* configuration, which is most unusual for natural lipids.

finding is that surface waters in tropical regions were several degrees cooler during the height of the last glacial age (Herbert 2006).

11.7.8 Humic acids

The term *humic acid* derives from the soil literature. When soils are extracted with alkaline solutions of NaOH, a brown-colored organic material of complex and varied composition is obtained. If this solution is filtered and then acidified, a precipitate designated humic acid is formed. The organic material remaining in solution is called fulvic acid; it is lighter colored, more yellowish, and of lower molecular weight than humic acids. Humic substances (both humic and fulvic acids) have a generally high molecular weight, and contain many functional groups, such as carboxyl, phenolic, hydroxyl, and amine. They contain many aromatic and other ring structures (hence the brown color); all linked in complex branching arrangements. Many of the structural components are recognizable as degradation products of lignin, the three-dimensional matrix that binds cellulose fibers in higher and especially in woody plants. The organic materials that are found in lakes and rivers are similar in many ways to the humic and fulvic acids from soils and are largely derived from these.

The enormous literature on soil and fresh-water humic substances was called upon by chemists investigating the DOM in the ocean, and it was at first thought that these substances could constitute a significant part of this DOM.

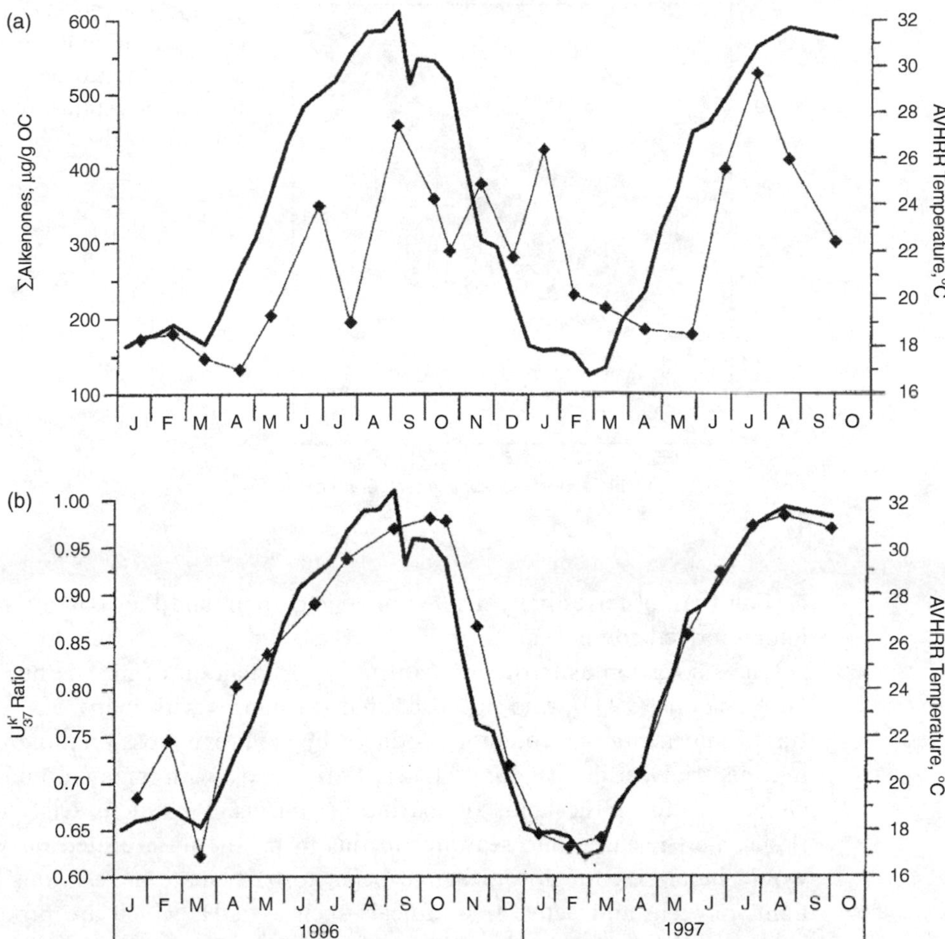

Figure 11.12 Goñi *et al.* (2001) deployed sediment traps at a depth of 500 m in the Guaymas basin near the middle of the Gulf of California. They collected samples every two weeks for more than two years and analyzed the samples for alkenones. They also obtained weekly average temperature data for the area from satellite measurements. The dark lines in the graphs are the temperature data; in this region there is a considerable seasonal swing in temperature. The thin line in (a) shows the concentration of total alkenones collected in the sediment traps. The thin line in (b) shows the ratio of C-37:2 alkenone to the sum of C-37:2 and C-37:3 alkenones.

Humic substance can be extracted from fresh waters by adsorption onto certain non-polar resins (such as XAD-4, -8, -12) and this technique is used to extract some fraction of the DOM in the ocean. The product has therefore been called *marine humics* or *marine humic substance*. Marine humics extracted from the open ocean are chemically different from humics extracted from soils or rivers. They have a lower concentration of aromatic ring structures and are not so strongly colored. The carbon isotope composition is more characteristic of marine-produced organic

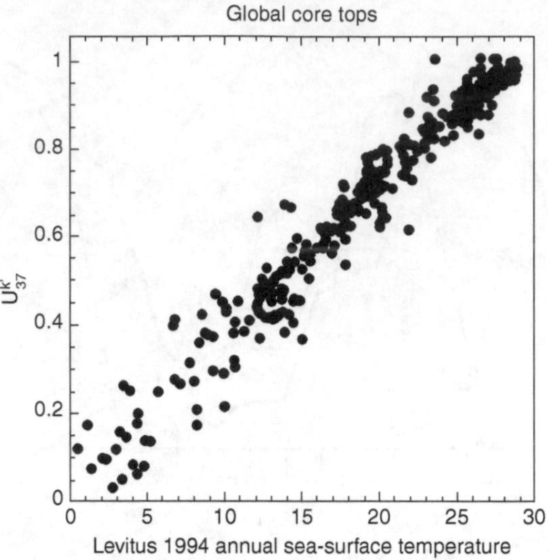

Figure 11.13 This is a compilation (from Herbert 2006) of numerous near-surface sediment measurements of the $U^{k'}_{37}$ index plotted against the long-term average temperature of the surface water above the sediment samples.

carbon than of terrestrially derived organic carbon, and they contain far less of the lignin-derived components.

One suggestion as to the mechanism for production of marine humics is due to Harvey *et al.* (1983), who proposed that substances with many of the characteristics of marine humics could be produced by oxidative cross-linking of unsaturated marine triglycerides (Figure 11.14). Fatty acids with many double bonds are characteristic of the lipids of marine organisms. Present as triglycerides in the tissues and released into seawater in this form, the unsaturated fatty-acid chains would lie close enough together to begin to participate in reactions catalyzed by light, oxygen, and active free radicals such as ·OH, which are produced photochemically in surface water. The reactions are similar to those that convert drying oils such as linseed or tung oil into an insoluble, highly cross-linked material. The product that might be formed by these reactions should have the characteristics of marine humic substances, as far as these are known. For example, they should have a considerable fraction of non-polar components in the molecule but be generally anionic in nature, they should have a small proportion of aromatic rings but many carboxyl and hydroxyl groups, and they should be soluble in seawater. Additional reactions to incorporate nitrogen-containing and other compounds are also at least possible, especially in the presence of sunlight. Chemical structures such as these are likely to be fairly resistant to bacterial attack, in part because they do not have a defined and regular structure and so will not be easily attacked by bacterial metabolic processes geared to dealing with simple fatty acids, sugars, proteins, and related compounds. While there is no direct evidence to support Harvey's suggestion, we lack better models for the production of marine humic substances.

Passing seawater through columns of non-polar resins does not extract all the DOM from the sample. The humic substance extracted from deep water in this way

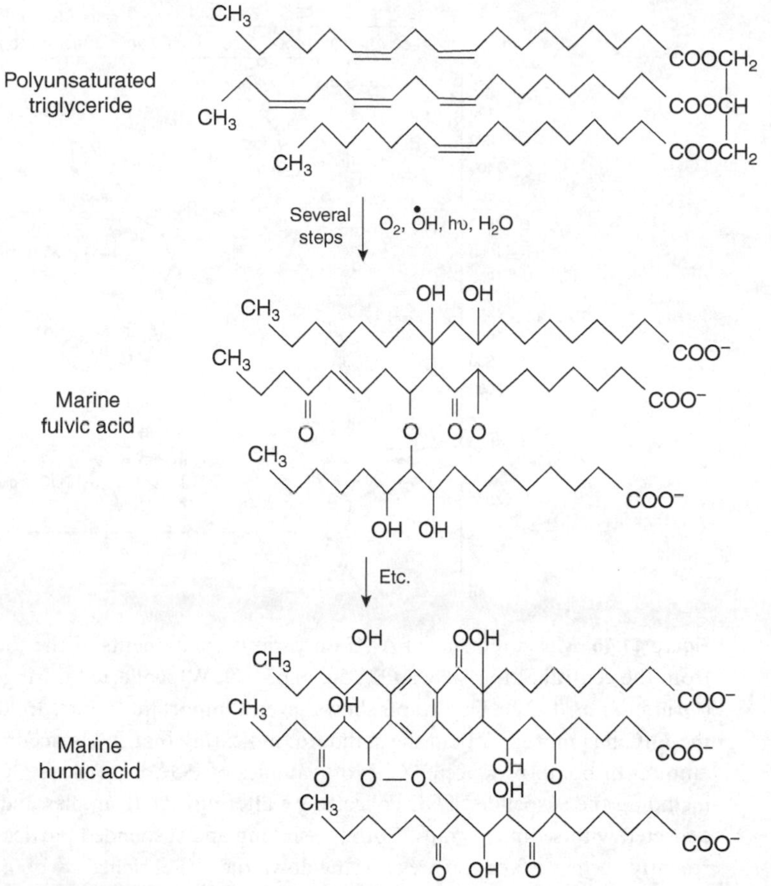

Figure 11.14 Proposed pathway to marine humic substances (Harvey *et al.* 1983). These authors suggested that polyunsaturated lipids are cross-linked by oxidative reactions initiated by hydroxyl radicals or other products activated by ultraviolet light, and possibly catalyzed by transition metals.

may amount to 10 to 20 μ moles of carbon per liter. If the true values for total DOC are around 40 to 50 μM, the extracted humics may account for 25 to 40% of the DOC; it is possible, however, that DOM extracted this way may include some portion of substances not classically defined as humic. Extracted humic materials have been shown by ^{14}C-dating to be older than the bulk of the DOC, however (Figure 11.15). This observation is consistent with the general notion that the humic fraction might be more resistant than other substances comprising the DOM.

11.7.9 Black carbon

The most resistant form of carbon observed in the sea is the so-called "black carbon" Goldberg (1985). This is material mostly derived from biomass burning and can be thought of as little pieces of carbonized vegetable matter, finely divided and carried through the atmosphere, primarily from incomplete combustion during grassland or

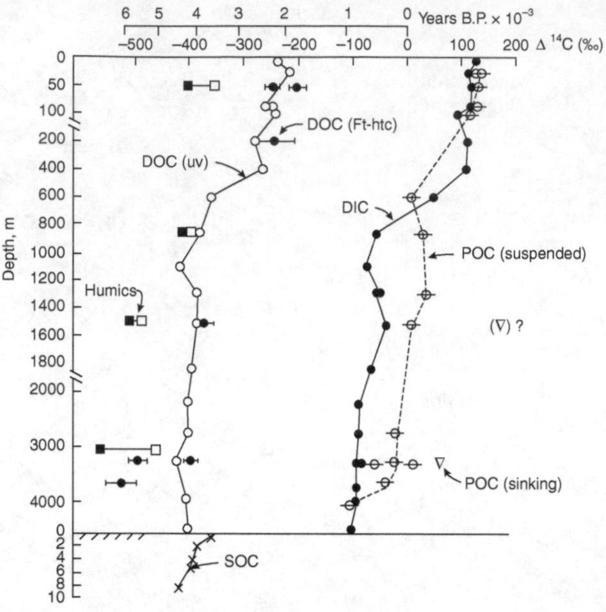

Figure 11.15 Measurements of $\Delta^{14}C$ on various components of the total carbon in seawater from the central Sargasso Sea (32° 50′ N,63° 30′ W), collected during May and June in 1989 (Druffel *et al.* 1992). This graph shows several important points. In all samples above 800 m the DIC has more ^{14}C than the standard, indicating that it has accumulated a considerable amount of bomb-produced ^{14}C. Most samples of POC have more ^{14}C than the standard, including the suspended POC collected by filtering water samples and sinking POC collected with sediment traps. Both the sinking and suspended particles have been made fairly recently at the surface and are carrying down the ^{14}C-enriched carbon. Only two sediment trap samples were measured, and one looks anomalous. Nevertheless, the suggestion is that the sinking, and presumably bigger, particles are the more recently made. Most of the DOC was collected after ultraviolet (uv) oxidation of the samples, but a few were collected by high-temperature combustion (htc) of seawater samples. The latter appear to be slightly younger, but the reason for this remains obscure. The humic fractions (filled squares, isolated with XAD-2 resin; open squares, isolated with XAD-4 resin) are older than the bulk DOC, indicating that some of the most-resistant and long-lived substance is found in these fractions. The top-most sample of sediment organic carbon (SOC), plotted at centimeter depths below the sediment surface, is younger than the DOC in the water column just above. This suggests that it contains a component derived from sinking particles containing fairly young carbon. It is a procedure we have not discussed.

forest fires. An additional source is anthropogenic; this is the incomplete combustion of fossil fuels, especially in diesel engines, and can be thought of as a form of soot. This material, the most resistant form of carbon, falls on the surface of the ocean and eventually ends up in the sediment (Eglington and Repeta 2006). It has been estimated that the flux of black carbon to the sediment may amount to about 10 Tg per year, which corresponds to about 11% of the total burial rate of organic carbon in marine sediments. Measurements of the carbon-14 content of black carbon isolated from marine sediments (Masiello and Druffel 1998) showed that in various places it may be 2500 to 13 000 years older that the rest of the organic carbon in the sediment. This suggests that some fraction of it, at least, has aged somewhere between production and burial. It is most likely that some of this very resistant material has fallen on soils and eventually been carried with other eroded material to the sea. It may be that some of this carbon has survived the sedimentary cycle, and been eroded from sedimentary rocks where its carbon-14 age would appear infinite.

Summary

The concentration of substances containing carbon in organic combination dissolved in seawater is greatest in surface seawater and least in deep water. The concentration in deep water is apparently rather constant at all depths below the top few hundred meters. The absolute amounts present are still a little uncertain, due to analytical difficulties, but in general deep-water values are about 40 μM. The nature of much of this dissolved organic substance is largely unknown. The majority of dissolved organic carbon is old, about 6000 years on average in the deep Pacific. Surface water contains an additional amount of younger organic carbon, so that the average organic matter in surface water is considerably younger.

Many dozens of individual organic compounds have been identified in surface and deep seawater. All are present in picomolar or nannomolar concentrations, so that in total they do not add up to more than a few percent of the total DOC. Even the ill-defined material known as marine humic substance does not constitute more than 20 to 40% of the total. The not yet fully characterized polysaccharide material in surface water may amount to 20% of DOC there, and less in deep water. There still remains mysterious organic substance in the ocean.

The field of marine organic geochemistry is large and active, and there are many fascinating studies of individual compounds.

SUGGESTIONS FOR FURTHER READING

Herbert, T. D. 2006. Alkenone paleotemperature determinations. In *The Oceans and Marine Geochemistry*, ed. H. Elderfield (*Treatise on Geochemistry*, vol. 6, ed. H. D. Holland and K. K. Turekian). Elsevier, New York, pp. 391–432.

This is a thorough and critical review of the potential uses of alkenones in estimating past temperatures in selected, mostly deep-water, regions in the ocean.

Eglington, T. I. and D. J. Repeta. 2006. Organic matter in the contemporary ocean. In *The Oceans and Marine Geochemistry*, ed. H. Elderfield (*Treatise on Geochemistry*, vol. 6, ed. H. D. Holland and K. K. Turekian). Elsevier, New York, pp. 145–180.

An excellent survey of the subject, much broader than the limited treatment possible here.

Thierstein, H. R. and J. R. Young, eds. 2004. *Coccolithophores: From Molecular Processes to Global Impact*. Springer, Berlin.

For those who might be fascinated by the biology and importance of this remarkable group, the book includes surveys of their biochemistry, physiology, cell structure, ecology, and their oceanic and geological importance.

Killops, S. D. and V. J. Killops. 2005. *An Introduction to Organic Geochemistry*, 2nd edn. Blackwell, Malden, MA.

A good general introduction to the general subject.

12 Anoxic marine environments

Absolute stagnation nowhere exists in the ocean, not even at its greatest depths.

DITTMAR 1884

Prior to about 1870, it was thought that the depths of the ocean were devoid of oxygen and devoid of life. In 1873, C. Wyville Thompson published *The Depths of the Sea*, an account of three expeditions in 1868–1870. He and his colleagues dredged and collected water samples for the first time to depths of about 2400 fathoms (more than 4400 m) and found life at all depths. Measurements of gases in water samples all showed the presence of oxygen. Then, during the great *Challenger* expedition, the major ocean basins were surveyed to considerable depths, and all the samples contained oxygen. The statement by Dittmar, above, is still true, because there is always some movement of the water, but there are important areas where oxygen is absent.

Living things exist at all depths. Indeed (with the exception of the extremely hot water feeding the black smokers; see Chapter 13), there is no single liter of seawater anywhere in the ocean that is devoid of life, and all living things metabolize. If oxygen is present it will be used in respiration, and will eventually all be used, if it is not replaced. If all oxygen is gone, life continues, but the types of organisms present are different.

12.1 Rates of oxygen consumption

The first serious attempt to estimate the rate of oxygen consumption by evaluating the distributions of oxygen in the ocean relative to the physical motions of the water was presented by Gordon Riley in a classic paper published in 1951 (see Table 12.1). Another approach is to use the modern sediment-trap data and the equations fitted to these to calculate the oxygen consumption that must be associated with the observed decrease of particle flux as the depth in the water column increases (Figure 12.1). The agreement between the early estimate by Riley and this estimate from sediment-trap data is remarkable. It is, of course, possible or even likely (see the

Table 12.1 An estimate of the rates of oxygen consumption in the subsurface Atlantic Ocean from 54° S to 45° N

Mean depth, m	O_2 consumption, μmol kg^{-1} yr^{-1}
200	9.1
280	3.5
370	2.2
510	2.4
700	1.5
1000	0.6
1250	0.2
1500	0.07
2000	0.06
2500–4000	0.006

From the classic paper by Riley (1951). Riley divided the Atlantic Ocean into 76 segments and used data from the density structure, current velocities, and early estimates of the lateral and vertical coefficients of eddy diffusivity to calculate the probable input of oxygen to each depth level in the Atlantic. He assumed that the observed concentrations were in steady state and estimated the biological consumption required to maintain those concentrations. The biological activity, as expressed by this oxygen consumption, is given above. He estimated that the surface productivity, in terms of carbon, was 21 mol m^{-2} yr^{-1} (with an uncertainty of 12 to 31), and that about 10% of this escaped as particulate carbon below a depth of 200 m to be respired in the depth intervals indicated.

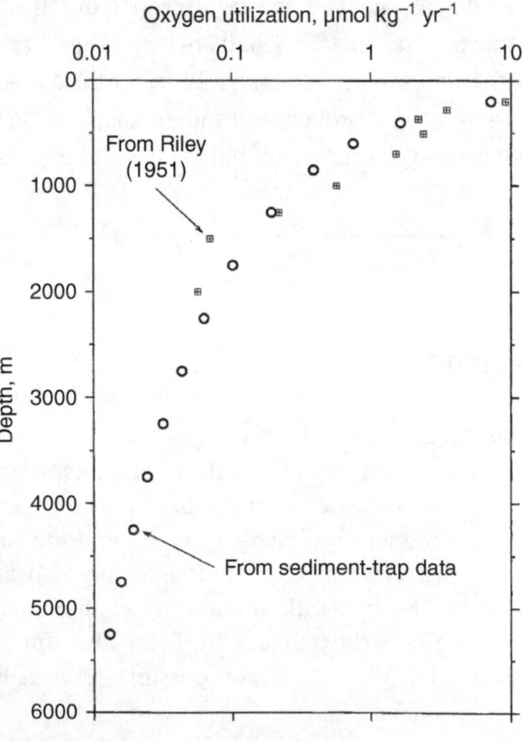

Figure 12.1 A global estimate of the vertical distribution of the consumption of oxygen in the world ocean, based on extrapolations from sediment-trap data. The estimate by Martin *et al.* (1987) that the particulate-carbon flux through the 100-m-depth horizon averages 1.7 mol m^{-2} yr^{-1} was used as a starting point, fluxes at depth were calculated using Eq. (11.4), and the concomitant oxygen consumption calculated over various depth intervals using the O_2/C ratio of 1.43. The estimates of Riley (1951) are also plotted for comparison.

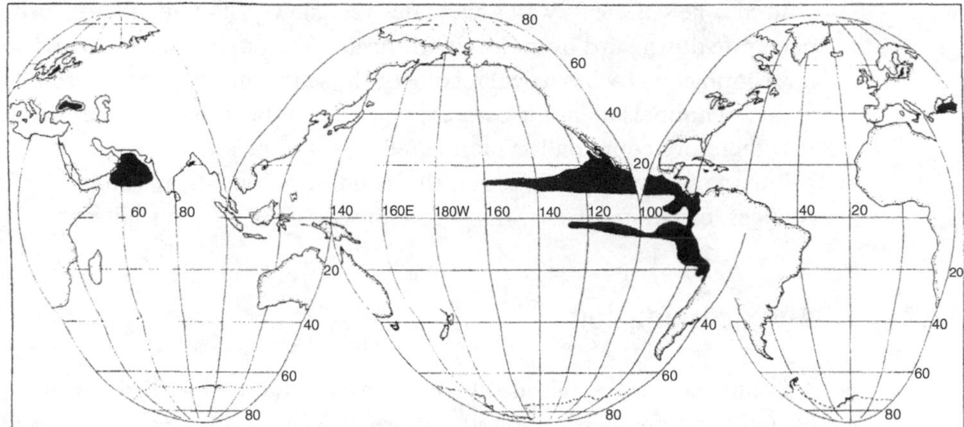

Figure 12.2 Extent of oxygen-deficient ($[O_2] < 10\ \mu mol\ kg^{-1}$, or <3% of saturation values) water in the larger basins of the world ocean. The entire volume of the Black Sea is anoxic below about 100 m. Many other anoxic basins are in continental borderlands or fjord-like environments and, except for the deep anoxic basin in the Baltic Sea, are too small to show on a map at this scale. All other areas that show on this map are low in oxygen or anoxic only at intermediate depths. From Deuser (1975).

discussion in Chapter 11) that both estimates are too low, and in any case the fluxes are certainly quite variable spatially. Nevertheless, these estimates illustrate the point that, given enough time, there would be no oxygen in the deep sea unless it is somehow replaced.

This replacement occurs through the movement of oxygen-rich waters from the surface down to all depths. The sources of such water to the deepest parts of the ocean are the cold polar regions, and there is considerable circulation of water to intermediate depths from intermediate latitudes. Since the deepest parts of all the major ocean basins contain oxygen, it is evident that replacement is sufficiently fast that the oxygen is not exhausted.

If the horizontal movement of water into some basin or other deep region is impeded, however, the metabolic demand for oxygen may equal or exceed the rate of supply, and the concentration can be reduced to zero. At the present time this occurs in a number of isolated basins, and in these basins many biogeochemical processes operate quite differently than in oxygenated seawater. There are also a number of interesting and important regions in the open water column in some parts of the ocean where the input of organic matter exceeds the rate at which oxygen is carried in by moving water, and in these places the oxygen content is reduced essentially to zero. The distribution of anoxic or very low-oxygen regions large enough to show on a small map of the world is shown in Figure 12.2. The obvious correlation with highly productive upwelling zones suggests, correctly, that where the input of sinking organic debris is great enough, the oxygen in the underlying water can be reduced to low or undetectable concentrations. The ancient sedimentary record contains evidence that there were times in the geologic past when these regions were much more extensive than they are today.

In marine sediments water circulates very slowly, if at all, and oxygen must usually penetrate downward by molecular diffusion (although this is often aided in the upper few centimeters by biological mixing of the sediments, called *bioturbation*; and by the action of animals in flushing water through their burrows, a process called *irrigation*). It is therefore common for marine sediments to be anoxic, beginning at some depth below the sediment surface, and this condition dramatically influences the chemical changes, called *diagenesis*, that occur in sediments after burial.

12.2 Anoxic oxidation

All multicellular organisms, nearly all eucaryotic single-celled organisms, and many bacteria and archaea absolutely require oxygen for all or part of their life cycles and are therefore found only where they are able to obtain sufficient oxygen. If usable organic matter or some other oxidizable material is present, however, life does not stop just because all the oxygen is gone.

Many other bacteria and archaea and a few protozoa can live without oxygen; some of these are obligate anaerobes and must live under conditions where no oxygen is present. These organisms obtain energy for life processes by using some of the many energy-yielding reactions that do not require the participation of oxygen. Indeed, it is sometimes said that if any chemical transformation can be carried out to yield biochemical energy, there is generally a bacterium around that can make a living doing it.

A number of substances besides oxygen can act as oxidants (electron acceptors) in biochemical transformations. Besides oxygen, oxidants that can be generally or locally significant in the marine environment include nitrate, nitrite, manganese and iron oxides, and sulfate. Of these, usually sulfate is by far the most abundant and available, so the products of sulfate reduction are noticeably characteristic of marine anoxic environments. The many usual components of organic debris also provide opportunities for a rearrangement of the atoms to yield a small amount of useful energy. This can be thought of as simultaneous oxidation of some components and reduction of others to yield a mixture of reduced and oxidized carbon compounds.

The crude energy yield from a biochemical transformation can be expressed as the released heat (change in enthalpy, ΔH°) during the conversion of reactants to products. Table 12.2 presents equations for some of the overall transformations that are important in the metabolism of organic matter to yield biochemically useful energy. In this table the organic matter is listed with the composition of an average Redfield molecule (from Eq. (8.13)), and the reaction carried out to the stage of complete mineralization, if that is possible. It was assumed that, in the absence of oxygen, the released ammonia is not oxidized further, although in some circumstances, for example in the anammox reaction, it may be. The table gives for each process the total chemical energy released during the transformation from beginning to end (although the practically useful metabolic energy will necessarily in each case be somewhat less than this). In no case can any one organism can carry out the

Table 12.2 Stoichiometry and energy yield ($\Delta H°$) for the biological oxidation of marine organic matter using a variety of natural oxidants

	Reaction	Energy yield
[1]	$(CH_2O)_{104}(CH_2)_{20}(NH_3)_{16} + 166O_2 \rightarrow 124CO_2 + 16NO_3^- + 16H^+ + 140H_2O$	105.6 MJ
[2]	$(CH_2O)_{104}(CH_2)_{20}(NH_3)_{16} + 268MnO_2 + 552H^+ \rightarrow 124CO_2 + 268Mn^{2+} + 16NH_4^+ + 392H_2O$	96.1 MJ
[3]	$(CH_2O)_{104}(CH_2)_{20}(NH_3)_{16} + 107.2NO_3^- + 123.2H^+ \rightarrow 124CO_2 + 53.6N_2 + 16NH_4^+ + 157.6H_2O$	87.1 MJ
[4]	$(CH_2O)_{104}(CH_2)_{20}(NH_3)_{16} + 268Fe_2O_3 + 1088H^+ \rightarrow 124CO_2 + 536Fe^{2+} + 16NH_4^+ + 640H_2O$	49.8 MJ
[5]	$(CH_2O)_{104}(CH_2)_{20}(NH_3)_{16} + 67SO_4^{2-} + 134H^+ \rightarrow 124CO_2 + 16NH_4^+ + 67HS^- + 67H^+ + 124H_2O$	39.8 MJ
[6]	$(CH_2O)_{104}(CH_2)_{20}(NH_3)_{16} \rightarrow 52CO_2 + 52CH_4 + 16NH_4^+ + (C_{20}H_{40})$	19.1 MJ

In each case, the organic matter starts out as 1 mole of Redfield molecules. These estimates were calculated from a starting material with a composition defined by the modified Redfield ratios from Chapter 8 (Eq. (8.13)): $(CH_2O)_{104}(CH_2)_{20}(NH_3)_{16}(H_3PO_4)$. The phosphate ($H_3PO_4$) was not included in the calculations and for simplicity is deleted from both sides of each equation. This procedure follows that of Froelich *et al.* (1979).

Values were calculated using selected standard states of the reactants and products (Haynes 2011 and other sources). An estimate of the enthalpy (heat) of formation of (CH_2O) units was based on that for glucose; for the (CH_2) units the value was taken as $\Delta_f H° = -26$ kJ mol^{-1}, approximately the increment for adding such a unit to a small hydrocarbon chain. The product CO_2 ($\Delta_f H° = -413.26$ kJ mol^{-1}) was actually calculated as HCO_3^- ($\Delta_f H° = 689.93$ kJ mol^{-1}) but is listed in each case as CO_2 in the equations above to make it easier to examine each equation for the effect of the reaction on the alkalinity. For thermodynamic reasons the free-energy yield that can translate into useful work obtainable from a given oxidation is generally somewhat less than the heat of the reaction, and also varies with the concentrations of reactants and the temperature.

It should be understood that the given stoichiometry does not imply much about the details of the reactions, which occur via numerous pathways and intermediate compounds, all within a biological black box. None of the processes summarized in these equations can be carried to completion without the combined activities of several species.

The results provide a roughly quantitative comparison between the several ways in which organic matter can be metabolized. In reaction [6] the residual carbon ($C_{20}H_{40}$), left after all internal oxidations have taken place and no further oxidation is possible, suggests a group of compounds that might contribute to the formation of crude oil.

complete transformation. For example, even in the case of reaction [1], which might be construed as the digestion and metabolism of a phytoplankton cell by a copepod, the copepod would digest most but probably not all of the phytoplankton cell material (e.g. some is excreted as fecal pellets), while the nitrogen would be released largely in the form of ammonia. Only bacteria will complete the oxidation by oxidizing the ammonia to nitrate. Thus, for this reaction, the total energy yield as listed can only be achieved by summing the activities of several species of organisms. While a great deal is known about the efficiency of biochemical oxidation with oxygen, less is known for the other cases about how much of the potentially useful energy is captured by the organisms involved.

The only organisms that are able to make a living carrying out the reactions summarized in reactions [2] through [6] are consortia of bacteria, or in some cases consortia of bacteria and protozoa.[1] The oxides MnO_2 and Fe_2O_3 are highly insoluble, and are found in abundance only in the sediments, but, as we will see later, there are some places where they can play a role higher in the water column. During the metabolism of organic matter in sediments both manganese and iron oxides are reduced, in each case to the divalent state of the metal. It is possible that these metal oxides function by reacting with reducing agents produced by the bacteria and thus enhance the overall effectiveness of the metabolic processes, but some evidence suggests that the bacteria use the oxides directly as electron acceptors. It is an interesting question, how such insoluble materials could be directly utilized. Table 12.2 shows that oxidation of organic matter using nitrate ions and MnO_2 as electron acceptors can potentially yield nearly as much energy per unit of carbon oxidized as oxidation using oxygen as the ultimate electron acceptor. The energy yield using iron oxide is less than that using manganese oxide, but iron oxide is usually more abundant than manganese oxide, so in that sense it is probably a more important oxidant.

The concentration of sulfate in seawater is much greater than that of oxygen; when there is a great deal of organic matter to be oxidized, sulfate may therefore be quantitatively the most important oxidant used. The evidence suggests that in most circumstances the sulfate is not used in the water column, but only in the sediments. It appears that, for the most part, the resulting sulfide in the water column has diffused there from the sediments.

In situations where the sediments are especially rich in organic matter the sulfate can be all used up and only reactions such as are summarized in reaction [6] can be used by bacteria to produce energy for maintenance and growth. The methane produced can diffuse out, be vented as bubbles to the water above, or, depending on temperature and pressure, it can be trapped as *clathrates* (Chapter 2, Figure 2.18). Since these ice-like substances are stable at temperatures considerably above the freezing point of water, at gas pressures greater than that imposed by about 400 m of water, quite considerable quantities of methane exist in this form, trapped mostly in sediments along the continental borderlands under productive waters.

It is generally observed in the sea and in sediments that reactions [1] to [6] are used in the sequence given, with the most energetically unfavorable appearing only when

[1] Strictly speaking, this statement requires careful definition. For example, ruminants such as cattle obtain a great deal of energy from the products of bacterial anaerobic oxidation of their carbohydrate-rich foods. These products include short-chain fatty acids such as acetate that are often intermediates in reaction [6], as well as the bacteria and protozoa themselves that carry out these reactions in the rumen.

the other oxidants are exhausted. Since it is believed that somewhat different consortia of bacteria are involved at different stages of the reduction process, it is not entirely clear why all processes could not go on simultaneously, despite the different efficiencies. Indeed, in regions of overlap it seems that to some extent they do, but it seems likely that the dominant process will generally be the one producing the most energy.

In many circumstances, reduced compounds produced by one process may diffuse into a zone where additional oxidants are present. For example, in sediments the reduced iron and manganese (Fe^{2+} and Mn^{2+}) are soluble (if the concentration of S^{2-} is not too great) and can diffuse upwards into regions where oxygen is present, so they may be oxidized there inorganically or by bacteria that can gain energy by this process.

12.3 The Black Sea

The largest body of water in which the present renewal rate of the deep water is sufficiently low that the oxygen is completely used up is the Black Sea (Figure 12.3). Bounded by Turkey, Bulgaria, Rumania, Ukraine, Russia, and Georgia, the Black Sea is connected to the Mediterranean by narrow and shallow straits. The total of river runoff and precipitation into the basin exceeds the rate of evaporation, so the salinity of the surface waters is reduced to about one half that of normal seawater. This surface water, with an average salinity of about 17.5‰, flows out through the Bosporus to the broad and rather shallow Sea of Marmara, and thence through the Dardanelles into the Mediterranean. During usual weather conditions, and possibly all the time, salty and dense Mediterranean water with an average salinity of 38 to 39‰ flows into the Bosporus towards the Black Sea underneath the fresher surface water. This salty water mixes to some extent en route, and probably enters the Black Sea with a salinity of about 37‰ (Özsoy *et al.* 1993). This water flows along small channels and in a shallow layer across the continental shelf, then over the edge and down the slope. Along the way, it mixes and entrains the fresher but usually colder water above (Figure 12.4). Several lines of evidence suggest that the Mediterranean water moving downslope usually does not make it to the bottom; as it entrains water from the upper layers of the Black Sea it becomes less dense and finally reaches some depth where its density is the same as the surrounding water and it then spreads out in thin layers into the Black Sea basin. These thin layers also carry a small amount of particulate matter resuspended from the sediments over which the water had been flowing. Most of these intrusive layers have been found down to only a few hundred meters below the surface.

The rate of water inflow from the Mediterranean must be quite strongly affected by local winds, because the Dardanelles and the Bosporus are so shallow, so it must on average also vary seasonally, and the average depth of penetration must also vary. The annual rate of input has been difficult to evaluate. One report (Latif *et al.* 1991) suggests an annual average of 312 km^3 per year. At this rate, the total volume of anoxic water (about 465×10^3 km^3) could be replaced in about 1500 years. However,

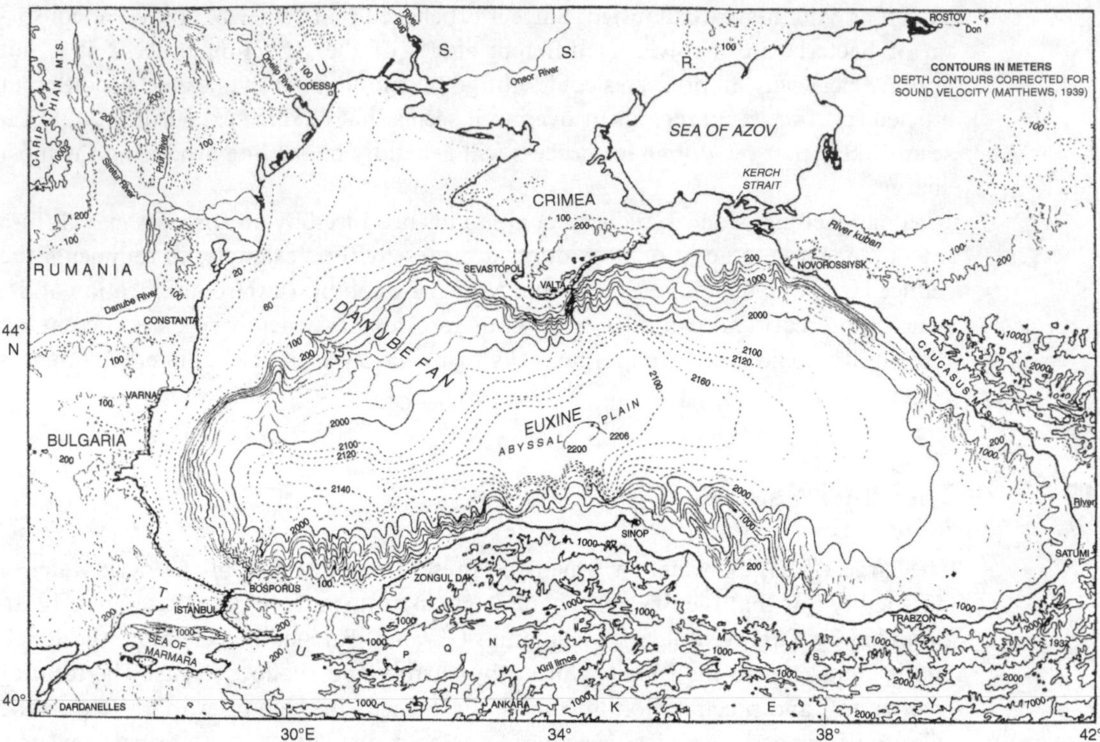

Figure 12.3 Bathymetric chart of the Black Sea (Degens and Ross 1974). The deep anoxic region begins at about the 100-m contour. Seawater from the Mediterranean enters the Black Sea by passing through the narrow straits of the Dardanelles and the Bosporus, the latter being the greater barrier. The Bosporus is about 31 km long, and in some places only 800 m wide, with an average depth of 35.8 m. There is a 32- to 34-m sill a few kilometers from the southern end of the Bosporus, but everywhere else there is some portion of the main channel that is at least 40 m deep.

the deep water in the Black Sea has a salinity of only about 22.3‰, while the salinity of the Mediterranean Sea water entering is about 37‰, so it is evident that some shallow water is entrained into the downflowing input water. From the hydrographic structure and quantitative balances involved it appears that the chief source of the entrained water is likely to be a cold intermediate layer always present in the Black Sea at a depth of about 50 m (and down to about 100 m near the coast), with a salinity of about 18.5‰ and a temperature of about 7 °C. Given these values for the respective salinities, the entrained water might amount to about 1200 km^3 per year, for a total input into the region below about 100 m of 1500 km^3 per year. This reduces the average residence time of the anoxic deep water to about 310 years. As mentioned before, however, the very deepest water may not be not renewed this rapidly, so much of the renewal actually occurs into water at intermediate depths. This situation introduces some uncertainty into models and interpretations of the distributions of substances in the Black Sea. Nonetheless, many vertical profiles appear relatively smooth and much can be learned by examining them.

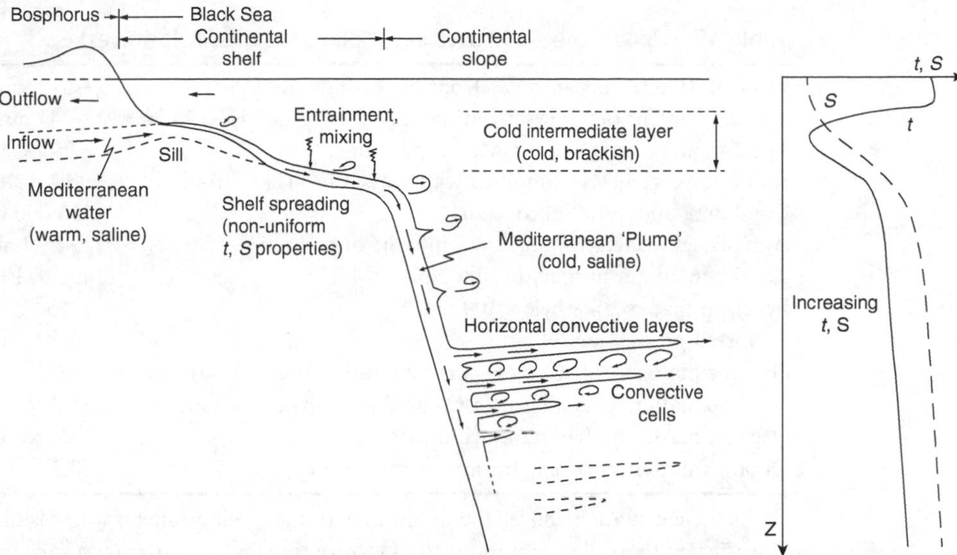

Figure 12.4 Schematic view showing the structure of the flow of Mediterranean water through the Bosporus into the Black Sea (Özsoy *et al.* 1993). The water flows as a thin layer over the shelf and down the slope deeper into the basin, entraining Black Sea water as it goes, and flowing off into thin layers across the basin when the density is right. Note the vertical distribution of temperature in the Black Sea. The cold ($t \approx 7$ °C), brackish ($S \approx 18.5$‰) layer is always present in the major basins, and is renewed annually by cooling over the northern regions in the wintertime. This cold layer, present even in the summer at a depth of 50–75 m, contributes to a cooling of the Mediterranean inflow water as well as reducing its salinity.

How does the input of organic matter into the depths of the Black Sea compare to the input of oxygen that might be consumed in oxidizing this organic matter? There are no really good estimates of the average net productivity of the Black Sea because the numerous measurements that have been made were not well located in space and time to capture both the strong seasonal variability and the spatial patterns (Sorokin 1983), but Sorokin estimated that the overall average may be about 17 mol m^{-2} yr^{-1}, with values in the central gyres of about 14 mol m^{-2} yr^{-1}. The surface area of the Black Sea deeper than the 100 m contour is about 330 000 km^2, and for our purpose here we can assume a surface productivity of 15 mol m^{-2} yr^{-1}. If we further take the common estimate that about 10% of the carbon fixed by surface productivity gets below 100 m each year, then calculation shows that there is much more organic material sinking into the sub-thermocline water than can be oxidized by the oxygen carried down with the input of salty water (Table 12.3). Based on these approximate calculations, it appears that there is only enough oxygen carried by the inflowing Mediterranean water to oxidize about 11% of the organic matter that is likely to sink below 100 m. The amount of oxygen carried by the Black Sea water entrained along with the Mediterranean water is not known; even if this water were completely

Table 12.3 Biogeochemical fluxes in the Black Sea (plausible values)

Flow of Mediterranean water from the Bosporus	~312 $km^3\ yr^{-1}$
Oxygen content (assume saturation at $S = \sim 37‰$, $t = \sim 14$ °C)	~244 mmol m^{-3}
Total input of oxygen from Mediterranean	~7.6 × 10^{10} mol yr^{-1}
Input of oxygen from entrained water (assume 50% saturation)[(a)]	~12 × 10^{10} mol yr^{-1}
Total area above 100-m contour	330 000 km^2
Annual net primary productivity in units of carbon	15 mol $m^{-2}\ yr^{-1}$
Total annual net primary production	495 × 10^{10} mol yr^{-1}
Fraction that escapes below 100 m	10%
Exported production	49.5 × 10^{10} mol yr^{-1}
Organic matter oxidized by O_2 (assume ratio $O_2/C = 1.43$)	–13.7 × 10^{10} mol yr^{-1}
Organic matter remaining after O_2 used (in units of carbon)	35.8 × 10^{10} mol yr^{-1}
Organic matter buried in the sediments[(b)]	–5.5 × 10^{10} mol yr^{-1}
Organic matter remaining for anoxic respiration	30.3 × 10^{10} mol yr^{-1}

[(a)] As the calculation shows, the O_2 input with the Mediterranean water is likely to be sufficient to oxidize only about 10 to 15% of the sinking matter (exported production) so even if the entrained water is 100% saturated, there would not be enough oxygen to oxidize all the remaining organic matter. Much of the entrained water is likely to come from a depth of 50 to 100 m, so it should carry a low concentration of oxygen. Here the calculation is done for entrained water carrying 115 μmol O_2 kg^{-1}, about 50% of saturation.

[(b)] Only a few measurements of the carbon burial rate in the Black Sea exist. The value given here is based on the average of measurements on two cores from 2100 to 2200 m in the deep basin reported by Calvert *et al.* (1991).

oxygenated it could contribute only about 45% of the needed oxidant, but because of its depth it cannot have very much oxygen, so it is impossible for the incoming water to carry enough oxygen to oxidize the sinking organic matter. Approximately another 11% of the organic matter is buried in the sediments, so that the remaining 35 to perhaps 75% of the organic matter must be oxidized by organisms using electron acceptors other than oxygen. Indeed, the deep water of the Black Sea is devoid of oxygen and rich in hydrogen sulfide (Figure 12.5). The exact concentrations of hydrogen sulfide in deep water are a little uncertain because of certain analytical difficulties. The degree of uncertainty may be evaluated by examining Figure 12.6, showing a compilation of data from different sources. It is seems that deep-water concentrations are around 400 μmol L^{-1}, and it also appears that the higher concentrations observed near the bottom in 1969 (Figure 12.5) were not apparent in 1988. It is not known whether this difference is due to changes in the bottom water during the subsequent 20 years, which seems plausible, or to vagaries in the analytical results.

Perhaps the clearest examples of substances with distributions unique to anoxic regions are presented by the vertical distributions of particulate and dissolved manganese (Figure 12.7). In ordinary oxygenated seawater, the concentration of manganese may be about 0.5 nmol kg^{-1} (Table 9.2) or about 0.03 μg kg^{-1}. This is because in

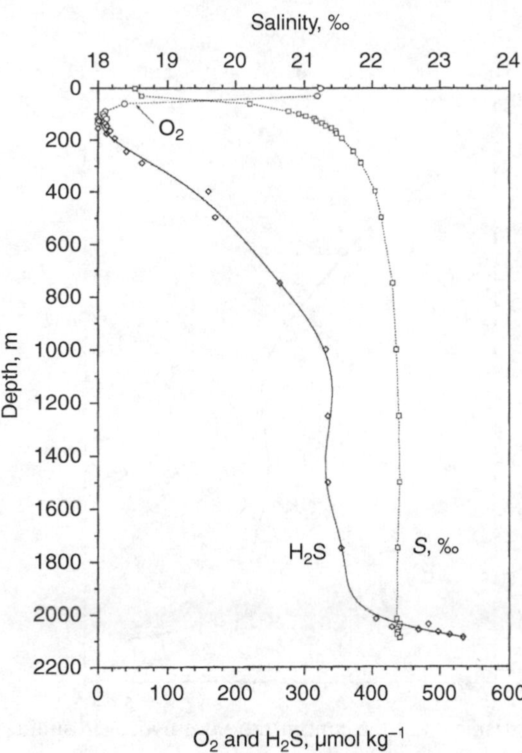

Figure 12.5 Vertical distribution of salinity, oxygen, and hydrogen sulfide in the Black Sea at a station near the middle of the eastern basin (43° 01′ N, 38° 30′ E, taken April 16, 1969). The salinity at the surface was about 18.5‰ and increased to 22.4‰ at the bottom. The oxygen concentration was 324 μmol kg^{-1} at the surface and dropped to 7 at a depth of 99 m, rose slightly to 12.3 μmol kg^{-1} at a depth of 118 m, and was undetectable at 127 m and deeper. At a depth of 127 m, hydrogen sulfide was detectable at 1.1 μmol kg^{-1} and the concentration increased in a complicated way down to over 500 μmol kg^{-1} near the bottom at a depth of about 2100 m. (Data from Brewer 1971.)

the oxidized form manganese (Mn^{4+}, probably present as MnO_2) is very insoluble, and forms particles or attaches to other particles and settles out. In the deep water of the Black Sea, however, manganese is present in the reduced form (Mn^{2+}); concentrations of about 250 μg kg^{-1} are typical and concentrations of up to 450 μg kg^{1} are observed high in the anoxic zone. Indeed, the anoxic portion of the water column of the Black Sea appears to be a trap for dissolved manganese, with concentrations more than 10 000 times greater than those found in ordinary oxygenated seawater.

An explanation for the accumulation of manganese in the deep water becomes apparent from comparing the distributions of dissolved and particulate manganese (Figure 12.7). Any particulate MnO_2 that settles into anoxic water will likely be reduced to soluble Mn^{2+}, either through inorganic reduction by reduced substances such as H_2S, or by bacteria utilizing MnO_2 as an oxidant, so that Mn^{2+} accumulates in the water column. There is enough Mediterranean water along with the entrained Black Sea water entering into the deep basin to raise the water at the depth of the oxygen-zero point upwards by at least 4 to 5 meters per year. This is countered by the downward mixing of the surface water into this water (so that the depth of the oxygen-zero point remains approximately constant year to year). As oxygen is mixed into the water with hydrogen sulfide, the latter is oxidized (both inorganically and by bacteria), along with other reduced substances such as Mn^{2+}. The oxidized manganese forms particles in the water, in the lowest part of the region with oxygen (Figure 12.7), and these particles begin to settle downwards. Some of the settling particles of oxidized manganese are quickly reduced to the

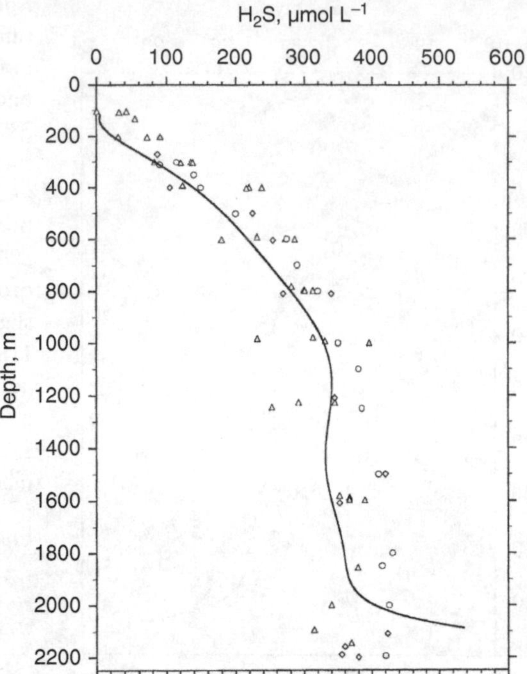

Figure 12.6 Additional measurements of the concentration of hydrogen sulfide in the Black Sea, from a cruise of the RV Knorr in 1988. Concentrations in the upper part of the water column are known to vary somewhat, and the gyral and other currents cause the isopycnals to tilt so that the actual depth at which the concentration of H_2S goes to zero certainly varies by up to 50 m. As far as is currently known, it is reasonable to assume that deep-water values should be fairly uniform horizontally, so that the observed scatter provides some information on the level of agreement between investigators and on our ability to make accurate measurements of this substance in seawater at these concentrations. (The solid line is from the data of Brewer 1971, as plotted in Figure 12.5. The circles are from Millero 1991, the diamonds are from Luther *et al.* 1991, and the triangles are from Kempe *et al.* 1988.)

soluble form again, resulting in a peak in the concentration of dissolved manganese at the top of the anoxic water, and some settles further before being reduced, resulting in the observed profile. This process continues, and the result is that manganese is trapped in the deep anoxic part of the water column and cannot escape, because if carried upward with the rising water it is eventually oxidized, forms particles, and sinks again.

Additional processes in the water column result in the distributions shown in Figure 12.8, where the sulfide and nutrient concentrations are shown for the top 400 m only. The complicated structure in the phosphate profiles is repeated at many stations and is a characteristic feature of the Black Sea. The presence of a nitrate peak and, somewhat deeper, a nitrite peak are also characteristic features.

The peculiar double peak in the concentration of phosphate seems to be due to some process removing dissolved phosphate at the very bottom of the oxygenated

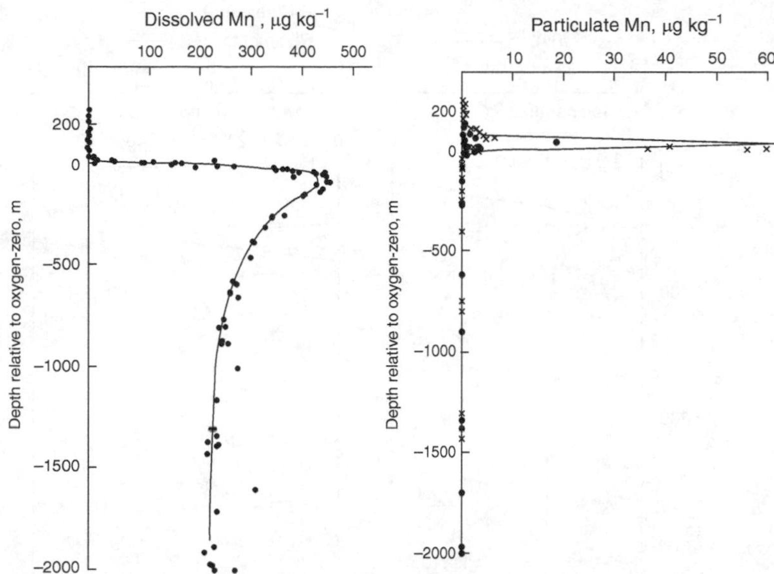

Figure 12.7 Concentrations of dissolved and particulate manganese in the water column of the Black Sea, measured in 1969. The data are from several stations. Since the depth to the point at which oxygen goes to zero, or the depth to any given density surface, varies across the Black Sea from place to place depending on the tilt of the density surfaces due to various currents (being shallowest near the center, and deeper by as much as 100 m near the coasts), the data are plotted relative to the depth at oxygen-zero for each station. From Brewer and Spencer (1974). See also Spencer and Brewer (1971).

zone. The formation of metal oxide particles in the lowest part of the oxygenated zone may explain this. Metal oxide particles, such as MnO_2 and Fe_2O_3 (dissolved and particulate iron exhibit vertical profiles analogous to those of manganese, but the data are somewhat more scattered) are very adsorptive, scavenging phosphate, dissolved organic matter, and other substances from the water column as they settle through. Thus, if manganese and other metal oxide particles are formed at the bottom of the oxygenated zone, they probably adsorb phosphate, which is then carried down, to be released again when the particles dissolve. The upper peak in phosphate apparently represents the beginning of the usual increase in phosphate along with nitrate as organic matter is metabolized. This is truncated, because the adsorptive particles of manganese and iron oxides are formed in this location and the phosphate is adsorbed and then carried down on sinking particles and then released below when these particles dissolve.

The concentrations of many other substances also exhibit striking vertical profiles, and examination of these, has led to interesting insights into the chemical processes involved.

While the Black Sea is by far the biggest and most studied anoxic basin, there are numerous others, and many are good natural laboratories where a considerable variety of natural processes can be studied in detail (e.g. Degens and Ross 1974; Grasshoff 1975; Richards 1965; Skei 1988).

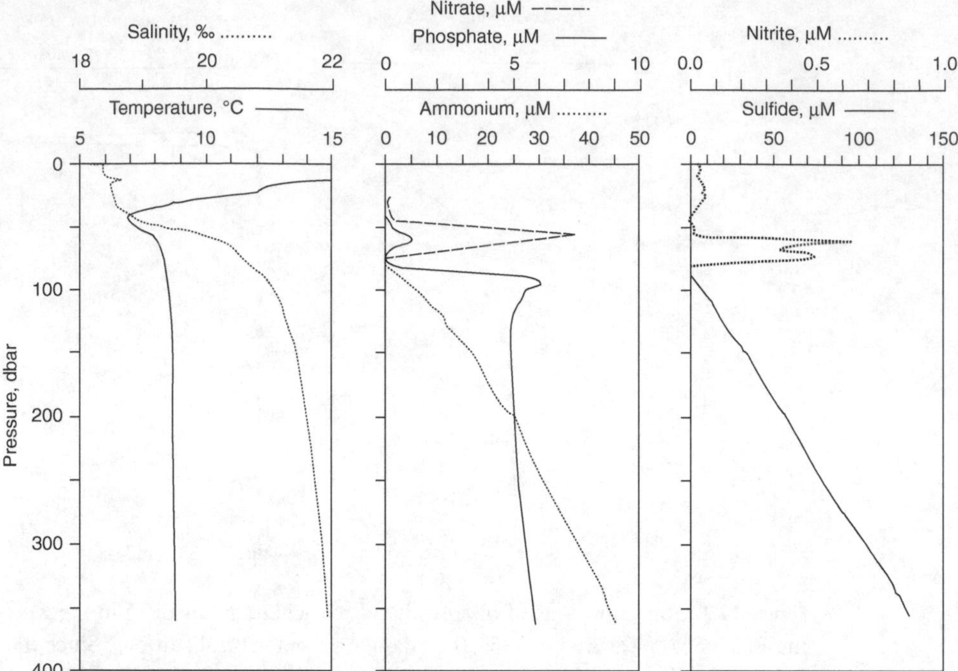

Figure 12.8 Profiles of temperature, salinity, nitrate, phosphate, nitrite, and sulfide from a station in the western gyre of the Black Sea. These profiles were taken in 1988 with a pumping system delivering water to instrumentation making continuous measurements. The concentration of oxygen (not shown on these plots) goes to zero at about 60 to 70 m at this station, just above the depth where sulfide first appears. From Codispoti *et al.* (1991).

Another important feature of anoxic basins derives from the fact that no organisms larger than bacteria (or perhaps a few protozoa) live on and in the marine sediments. The larger organisms present on and in these sediments under oxygenated water are so active in mixing the sediments down to depths of at least several centimeters that the sediment record is usually somewhat smeared over this depth range. In anoxic basins, however, this smearing does not take place, and each annual layer lies just on top of the one below. In favorable circumstances, the seasonally changing character of the sediment deposits makes it possible to distinguish annual *varves*. These varves can be counted and therefore individually dated, much as tree rings can be counted. By careful sectioning of sediment cores it is possible to collect samples of sediment from each year, or small group of years, for analysis. Thus, it is possible to obtain accurate dates for changes that have occurred at some time in the past.

Summary

When the input of oxygen into some region is not sufficient to oxidize the input of organic matter the oxygen will be all used up. With no oxygen, no multicellular organisms are capable of living out their whole life cycle.

Many species of bacteria and archaea specialize in using alternate electron acceptors. These include nitrate (NO_3^-), manganese oxide (MnO_2), iron oxide (Fe_2O_3), and sulfate (SO_4^{2-}). In the marine environment the most abundant electron acceptor is sulfate, and the characteristic odor of H_2S is readily apparent when sulfate has been used.

When all these electron acceptors are absent or used up, there is still energy obtainable from biochemical pathways that result in a series of internal rearrangements of the organic matter; typical end products are methane and carbon dioxide. In organic-rich sediments this process often results in considerable accumulations of methane and methane hydrates.

The largest body of anoxic water is the Black Sea. In this sea, oxygen is absent below about 100 m from the surface, and below that boundary there is an increasing concentration of H_2S, up to about 400 μmol per liter.

Recycling of manganese at the oxic/anoxic boundary results in the trapping of manganese in the basin in its soluble reduced form, where it reaches concentrations up to 10 000 times the concentration on the open ocean. Iron is similarly present in high concentration, limited by the solubility of ferrous sulfide. Presumably other metals with similar solubility and redox behavior may show similar patterns.

Anoxic basins are regions where there is no bioturbation of the sediments, often making them good places for fine-scale studies of the history preserved in the sediments.

SUGGESTIONS FOR FURTHER READING

Caspers, H. 1957. The Black Sea and Sea of Azov. In *Treatise on Marine Ecology and Paleoecology*. vol. 1, ed. J. W. Hedgpeth. Geological Society of America, Washington, DC, pp. 801–889.

This is perhaps the first thorough review of the nature of the Black Sea environment. Now superceded in many technical matters, it is still often cited because of its coverage.

Fenchel, T. and B. J. Finlay. 1994. The evolution of life without oxygen. *Am. Sci.* **82**: 22–29.

A good discussion of the subject of the title.

Izdar, E. and J. W. Murray, eds. 1991. *Black Sea Oceanography*. NATO ASI Series, vol. 351. Kluwer Academic Publishers, Dordrecht,

An entire book devoted to the results of oceanographic studies of the Black Sea.

Murray, J. W. 1991. The Black Sea Oceanographic Expedition: introduction and summary. *Deep-Sea Res.* **38** (Suppl. 2): S655–S661.

This paper introduces an entire supplementary issue of the journal devoted to reports of recent work in the Black Sea.

Oceanography **18**, no. 2, June 2005.

This issue of the journal has nine relatively recent papers on the Black Sea.

Skei, J. M. 1988. Framvaren – environmental setting. *Mar. Chem.* **23**: 209–218.

This article introduces an entire issue of the journal devoted to reports of studies in Framvaren, an anoxic fjord in southern Norway.

13 Exchanges at the boundaries

. . . the ocean may receive supplies of salt from rocks and springs latent in its own bosom, and unseen even by philosophers. ROBERT BOYLE 1673

Our understanding of the processes that affect or control the chemistry of the sea began with the obvious fact that rivers run into the sea, and they carry the washings of the land into the sea. Then came some appreciation of the evaporation of water and the rudiments of the major hydrological cycle. In the twentieth century it become generally appreciated that atmospheric dust is a major contributor to oceanic sediments, and in the latter decades of that century the transport of many substances through the air between land and sea became generally appreciated. In the last half of the century it also became generally well known that throughout the floor of the ocean there is some likelihood of significant chemical exchanges between the sediment and the overlying water, and between seawater and the hot rocks at the spreading centers and other volcanic regions on the sea floor, and that these have important influences on the chemistry of the ocean. None of the fluxes involved is really well known for any substance, however, although the estimates for river inputs of most substances are certainly satisfactory for geochemical calculation, and most of the other fluxes are well enough constrained for first-order estimates and calculations. Much less is known about the factors that might cause changes in the rates of many of the important fluxes.

Deep as it is, the ocean is still a thin layer on the surface of Earth, and during the course of several thousand years nearly every drop of water has some opportunity to be exposed to the air–water boundary at the surface and to the sediments and rocks at the bottom. The chemical exchanges that influence the sea occur at all these boundaries, and in turn the sea profoundly affects the air above and the sediments and rocks below. In the sections that follow, these exchanges are examined further.

13.1 River input

Most of the particulate and dissolved substances that enter the ocean are carried there by the rivers. The present rates include a significant enhancement by the activities of humans. The present sediment discharge of about 20 Gt per year (Table 13.1) is thought to be perhaps two or three times the rate that existed before humans began to greatly increase erosion rates by converting grasslands and forests into pasture and crop-land (Berner and Berner 1987). The present rate at which dissolved solid in the rivers enters the oceans also includes an enhancement by humans by about 10%, due to the discharge of salt and other substances (Tables 13.1 and 13.2). According to Berner and Berner (1987) about 13% of the chloride and 8% of the sodium in the present discharge is *cyclic sea salt*, that component of the riverine salts due to sea spray that is carried inland and falls on the land surfaces. They further estimate that about 2% of the magnesium and sulfate, 1% of the potassium, and 0.1% of the calcium is cyclic sea salt. The remaining dissolved inorganic substance is derived from the chemical leaching and modification of rock minerals or in some cases their complete dissolution. Chloride is a negligible component of silicate minerals; essentially all the chloride in rivers, other than cyclic sea salt and human contributions, comes from deposits of halite (rock salt) and from the leaching of residual sea salt in uplifted marine sedimentary rocks. Some of the sodium comes from the same sources, but a considerable amount is also from the dissolution of silicate minerals, many of which contain sodium in the crystal structure. Most of the calcium is from the dissolution of limestone and gypsum and some is from silicate minerals. Essentially all the silicate comes from the dissolution of silicate minerals.

The 0.36 Gt of total organic carbon (Table 13.1) discharged to the ocean each year from rivers (plus perhaps 0.05 Gt of terrestrial carbon delivered through the atmosphere) amounts to about 0.68% of the 60 Gt of carbon fixed annually on land. The evidence from marine sediments suggests that only a small fraction of terrestrial carbon entering the sea ends up in the sediments, so it must eventually be oxidized in the water column or on the surface of the sediments. As noted in Chapter 11, the burial rate of organic matter in marine sediments amounts to about 0.13 Gt per year, so the difference between delivery of carbon by rivers and atmosphere and the burial in marine sediments of both marine and terrestrial organic matter amounts to 0.28 Gt per year. This causes the world ocean to be heterotrophic; that is, to this small extent the ocean is a net consumer of oxygen and organic carbon, while the land surface is a net producer.

When considering the influence of river-borne substances on the chemistry of the ocean, we cannot assume that all the dissolved material carried by the rivers (Table 13.2) actually enters the ocean in dissolved form. During the formation of soils and the eventual transport of the resulting sediments down rivers, the particles are exposed to the generally slightly acid and low-salt concentrations of fresh-water chemical environments on land and in lakes and rivers. When river water mixes with seawater in estuaries or on the continental shelves, the sediment particles are exposed to an

Table 13.1 Present delivery rates of substances to the world ocean by rivers

Water	
Total flow of all rivers that discharge to the ocean[(a)] ($= 1.185$ Sv $= 37\,400$ km^3 yr^{-1})	37.4×10^{12} m^3 yr^{-1}
Sediments (may be more than double the pre-human rate)	
Suspended sediments carried by rivers[(b)]	20×10^9 t yr^{-1}
Probable total entering the ocean ??	$<20 \times 10^9$ t yr^{-1}
Dissolved inorganic solids	
Suggested pre-human discharge	3.73×10^9 t yr^{-1}
Estimated contribution by humans	0.39×10^9 t yr^{-1}
Dissolved solids delivered to ocean	4.12×10^9 t yr^{-1}
Organic materials	
DOC (dissolved organic carbon)	0.22×10^9 t yr^{-1}
POC (particulate organic carbon)	0.14×10^9 t yr^{-1}
TOC (total organic carbon)	0.36×10^9 t yr^{-1}
TOM (total organic matter; TOM = TOC $\times$ 2.2)	0.79×10^9 t yr^{-1}

River flow from Baumgartner and Reichel 1975; sediments from Milliman and Syvitski 1992; dissolved inorganic solids from Meybeck 1979; and organic substance from Meybeck 1982.

[(a)] Baumgartner and Reichel estimated the flow of glacier ice to the ocean as 2300 km^3 yr^{-1}, giving the flow of both liquid and solid water to the ocean as 39.7×10^{12} m^3 yr^{-1}. These authors suggested that for budgetary purposes the contribution of glacier ice to the transport of dissolved substance is negligible.

[(b)] Milliman and Syvitski were uncertain about how much is trapped by dams (approximately 13% of all fluvial discharge is dammed) and, of that which reaches deltas and large estuarine sinks, how much makes it out to deeper water. The transport of solid particles by glacier discharge is not included; it seems possible that this transport may not be quite negligible.

environment that is somewhat more alkaline and contains much greater concentrations of salt. This causes a number of only partially understood reactions. Clay particles adsorb significant amounts of sodium and potassium, estimated by Berner and Berner (1987) as 16% and 10%, respectively, of the total river transport of these elements. There is also some release of calcium and magnesium. Upon mixing with salt water, some of the organic matter and much of the iron and aluminum carried in solution or in colloidal form in the rivers flocculate and perhaps sink out. It is still an open question whether appreciable amounts of these flocculated materials are trapped in the estuaries or are carried in suspended form out of the estuaries to the ocean. The evidence suggests that they are not retained within the estuaries, but we lack evidence as to whether most of these materials are later deposited in the near-coastal regions or are carried out into the open ocean. In any case, the estuarine and shelf environments are places of considerable chemical processing but are highly variable locally, and it requires careful evaluation for each substance to translate global river transports into estimates of river discharge finally into the global ocean.

Table 13.2 Flux of major river-borne dissolved substances to the ocean

Substance	Concentration,(a) μM Natural	Anthropogenic	Total	Total flux, 10^{12} mol yr^{-1}
Na^+	224	89	313	11.71
K^+	33	2	35	1.31
Ca^{2+}	334	33	367	13.72
Mg^{2+}	138	12	150	5.62
Cl^-	162	71	233	8.70
SO_4^{2-}	68	52	120	4.48
HCO_3^-	852	17	869	32.49
H_4SiO_4	173	0	173	6.47
DOC			479	17.90
Sum of cations			1382	
Sum of anions			1341	
Shortage of anions(b)			41	

Data recalculated from Meybeck 1979, 1982. The concentrations reported are Meybeck's flow-weighted average totals for the world's rivers that drain to the ocean. The estimates of the proportions that are natural and those due to the activities of humans are also from Meybeck, except for sulfate, which is from Berner and Berner 1987. As a common convention, the total carbon dioxide is reported as bicarbonate ion, by far the dominant form. The discharge is calculated on the basis of a total river flow of $37.4 \times 10^{12}\ m^3\ yr^{-1}$ (Table 13.1)

(a) The accuracy of these values is not certain. The inorganic species are almost certainly good to better than 10%, but Meybeck would not claim better than 30% for the DOC. The number of significant figures in the tabulation does not indicate accuracy, and numbers are listed as above to prevent further rounding error in subsequent calculation.

(b) From data in Thurman (1985), it can be estimated that river-borne organic matter may have about 12 mol of anionic carboxylic acid groups ($R–COO^-$) per 100 mol of carbon. The average pK of these groups is about 4.2, which means that ~90% should be ionized at a pH of 5.2. Applying these ratios to the average of 479 μM of organic carbon in rivers results in the estimate that at this pH the riverine organic matter may carry about 52 μM of ionized carboxylic acid groups. Given the uncertainties in all the values for the average concentrations of the various ions and the organic matter, and our lack of knowledge of the average pH of world river water, this near-agreement is remarkable. It suggests that riverine organic matter balances the deficiency of inorganic anions and that the ionic totals are unlikely to be much in error.

13.2 Air–sea exchange

The winds of Earth, in their patterned and random motions, circle the globe every several weeks. In its travels, the air is constantly exchanging substances with the water and land below and carrying these substances some distance before losing them again. The amounts of substance exchanged can be quite significant in worldwide

Table 13.3 Estimates of some fluxes to the ocean via atmospheric transport

	Flux, 10^6 t yr^{-1}		Flux, 10^3 t yr^{-1}
Total mineral aerosol	903	Lead	88
Aluminum	73	Phosphorus	948
Iron	32	Cadmium	3.3
Silica	278	Copper	38
		Nickel	25
		Zinc	164
		Arsenic	6
		Mercury	1.7
		Tim	0.3

Data from Duce *et al.* (1991). Some values are uncertain (by factors ranging from perhaps 2 to 5), but the original article should be consulted for a discussion of these uncertainties.

geochemical fluxes. The movement of water vapor, evaporated from one place and transported to somewhere else where it falls as rain or snow, is of course the greatest part of what is carried by the air. Indeed, some 426×10^3 km^3 of water is evaporated from the ocean each year; most of this falls on the ocean again, with only about 10% falling on land. In addition, however, numerous other gases and surprising amounts of solid substances are transported through the air.

It has proven very difficult, however, to obtain accurate estimates of the amounts of substances transported to and from the sea through the atmosphere. This is due to the high degree of variability in the observed concentrations (often several orders of magnitude from time to time and from place to place), the absence of data from large areas of the ocean, and the difficulty of measuring or calculating the rate at which particulate and gaseous materials fall out, are rained out, or diffuse into or out of the sea. Several direct and indirect lines of evidence can be drawn on to buttress the estimates, however, so most estimates are probably good to within a factor of two or three. Table 13.3 provides some estimates of the transport of solid substances through the atmosphere to the sea.

These fluxes may vary by factors of over 1000 from place to place. For example, mineral aerosol (dust) deposition rates may be less than 10 mg m^{-2} yr^{-1} in the Southern Ocean, while they may exceed 10 000 mg m^{-2} yr^{-1} in the northwest Pacific (due to dust blown from the high northern regions of central Asia, such as the Gobi Desert) and the eastern tropical Atlantic (due to dust blown from the Sahara Desert). Similarly, fluxes of lead may be less than 10 μg m^{-2} yr^{-1} in the Southern Ocean, and may exceed 1000 μg m^{-2} yr^{-1} in the region around Japan and off the east coast of North America. These high values were due to generalized industrial activity and to the lead in gasoline; the latter source has decreased greatly, especially in North America.

It is evident that these inputs may have a very small influence in some regions and a large influence in others. In regions such as the north central Pacific, most of the sediment may be the result of wind-blown dust because the coastlines are so far away that little or no river-borne sediment gets there, biological productivity is low, and the small amount of biological precipitates that do settle tend to dissolve before being stored in the sediment. In the earlier discussion of trace metals in the sea, it was pointed out that the distributions of some trace elements in the ocean may be dominated by wind-borne inputs.

The reverse process, the transport of salt and other substances from sea spray onto land, is most significant near the coasts, but even far inland most of the chloride in rain comes from sea salt. If the estimate of Berner and Berner (1987) is correct, that 13% of the chloride in rivers comes from chloride in rain, the total transport of sea salt to the land must amount to about 74×10^6 t per year. The input of chloride is the most significant of the major components, but the transport of other major ions, approximately in proportion to chloride, contributes a small portion of the amounts that in turn come back to the sea in rivers.

Particles of sea spray, derived from the droplets from bursting bubbles or simply torn from the sea surface by high winds, tend to have a chemical composition somewhat different than that of bulk seawater because the commonly observed surface slick is concentrated on the surfaces of the droplets. This organic material contains, among other things, nitrogen and phosphorus nutrients, which locally may provide a significant input of nutrients to the land along windy coasts with on-shore winds. The curious details of the processes and chemical fractionations that occur when bubbles break and inject minute droplets into the atmosphere were described by MacIntyre (1974a, 1974b).

13.3 Hot rocks

The mid-ocean ridges stretch for some 59 200 km through the major ocean basins (Parsons 1981). These are regions where the great conveyer belts of the ocean floor separate at rates that vary from place to place but average 5.1 cm per year (two-sided). Here, molten rock up-wells from the mantle below and fills the opening spaces as the lithospheric plates move away from each other. The hot lava intrudes as dikes into cracks in the crust, and at the surface of the crust is exposed to seawater and rapidly quenched into the rounded forms of pillow basalts. Seawater finds its way down through cracks in the rocks and cools the rocks below. The cooling opens more cracks so the flow penetrates deeper and deeper and cools the rocks down to possibly a few kilometers below the sea floor (Figure 13.1). The buoyant force on the heated water is enormous; on one cubic meter of water it may amount to more than one-half a ton. The water rises up through cracks in the rock, sometimes concentrated into rapid flows at the famous *black smokers*, where plumes of water heated to temperatures that are often 350 to 400 °C, or more, discharge into the seawater above. In other places, the flow is more diffuse and water is discharged at only a few degrees above ambient temperatures. In many cases, water originally heated to high temperature is mixed with cooler water from channels in the rocks, so that intermediate temperatures are observed. The vents are surrounded by lush

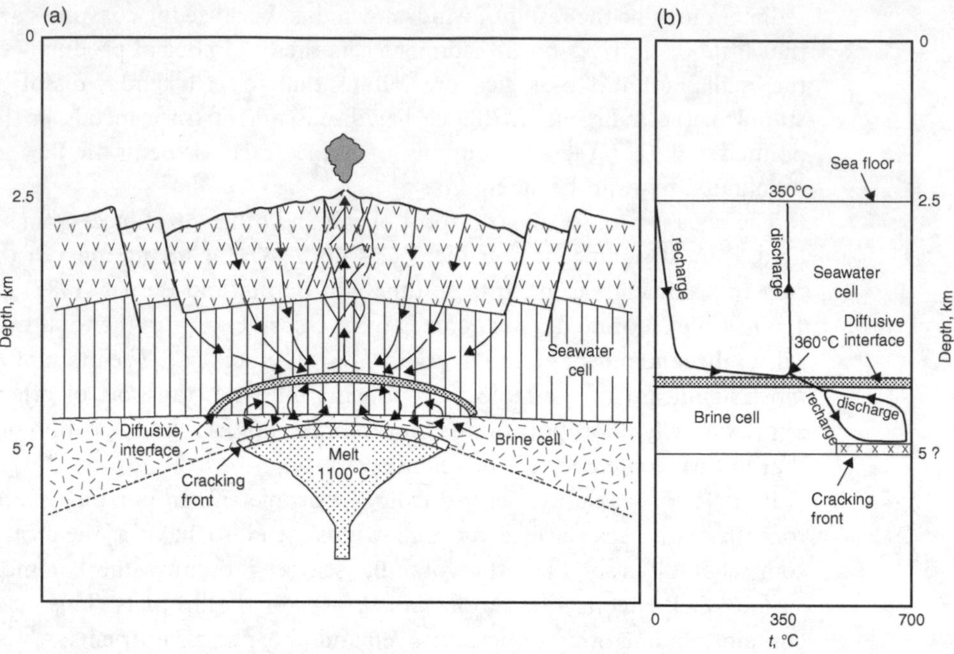

Figure 13.1 Cartoons showing a speculative presentation of the circulation of hydrothermal solutions at the spreading centers (from Bischoff and Rosenbauer 1989). Here, very hot water (~350 °C) exits in the form of black smokers from small orifices; other flows (not shown) exit at lower temperatures as white smokers and at still lower temperatures in more diffuse flows that are harder to find. A double circulation is presented here, where in the upper-cell regular seawater is drawn down through cracks in the sea floor and rises again to exit as black smokers. This is underlain by a circulation cell composed of concentrated brine remaining below after phase separation by boiling of some of the seawater, leaving behind a much denser concentrated brine which sinks deeper and interacts at higher temperatures with rocks below. This presentation helps explain the large salinity variations in the exit water from various locations. Low-salinity exit water is the product of condensation of low-salinity vapor entrained in the rising seawater. High-salinity exit water occurs when the high-salinity water below occasionally becomes entrained in the rising flow. This presentation also helps to explain evidence that the deepest rocks have been chemically modified by very high-salinity water at temperatures (600 to 700 °C) well above the critical temperature of seawater. Bischoff and Rosenbauer (1989) also suggest the possibility of a double diffusive layer separating the two cells, wherein heat, which diffuses rapidly, is transferred to the cell above, while the more slowly diffusing salt is to some extent often left behind.

communities of organisms supported largely or entirely by the bacterial oxidation of reduced substances (primarily hydrogen sulfide and methane) in the vent water (Lutz and Kennish 1993).

Individual black-smoker channels that have been observed are variable but are typically several centimeters in diameter and have flow rates of 1.5 to 2 meters per second, so the water exits at rates of around 1 kilogram per second. Black smokers often occur in groups and in addition there are usually more diffuse flows in the surrounding region, so that an individual vent field might produce several hundred kilograms per second (Converse *et al.* 1984). The plumes rise to 100 to 300 meters above the sea floor before they have entrained enough surrounding seawater to achieve neutral buoyancy, at which point they spread out and move with the general circulation in the region. The vent fields may occur at intervals of some tens of kilometers along the ridge axis, but their global distribution is still a matter of speculation.

In addition to the fairly regular flow (on a timescale of a few months to a few years) from these hot and warm vents, the occurrence of a number of spectacular short eruptions of hot water has been deduced from observations of megaplumes of hydrothermally altered seawater at levels of about 1 kilometer above the Juan de Fuca Ridge (Baker and Massoth 1987; Baker *et al.* 1987), and there is no reason to believe that they do not occasionally occur elsewhere as well. One of these plumes was estimated to have been emitted in little more than a month at a flow rate of 13 000 to 19 000 kilogram per second; it was observable as an oblong layer of water 20 kilometers in diameter and 700 meters thick with a slightly elevated temperature and various chemical anomalies. It seems plausible that this massive release of hot water might have been caused by some cracking of the rocks in the crust, opening up spaces that had earlier been blocked by precipitation of minerals in the channels (Cann and Strens 1989). Such plumes might also be caused by occasional outpourings of lava onto the sea floor.

The chemical composition of lava and the minerals formed when lava first cools to its freezing point (resulting in solid basalt) is such that the minerals are mostly not stable under cool aqueous conditions and are gradually altered upon exposure to sea water. The alterations involve both the uptake of water itself and of mineral substances from the seawater, and the release of various substances into the water. The rate of alteration is dependent upon the temperature, and the types of new minerals formed are also a function of both the temperature at which the alteration takes place and the availability of necessary chemical species in the water. Thus, the resulting minerals and the degree of chemical alteration of the seawater also depend on the ratio of water to rock in the reaction zones.

Because of the considerations noted above, it might be expected that the chemical composition of the water exiting at various places along the mid-ocean ridges would be somewhat variable, and that is indeed what is observed. Only a few places have so far been discovered where it is possible to obtain good samples of heated water coming up from the rocks below, and most efforts have focused on the spectacular black smokers and other quite warm flows.

In order to consider the effect on the oceans caused by chemical modifications of seawater passing over the hot rocks and exiting at the hydrothermal vents, it is necessary to estimate the total flow of this water. This has proved a difficult task, and at present there is no firm consensus. The evaluation is based first on the loss of heat caused by the hydrothermal circulation of seawater that cools the upwelled basalt. As the cooled crust moves away from the spreading center, it gradually

Table 13.4 Data and calculations related to hydrothermal flows

Area of new sea floor formed each year[(a)]		$3.45\ km^2\ yr^{-1}$
(Major ocean basins	$3.00\ km^2\ yr^{-1}$	
Back arc basins	$0.45\ km^2\ yr^{-1}$	
Depth of fresh sea-floor crust		Approx. 5 to 6 km
Volume of formed each year		$1.90 \times 10^{10}\ m^3\ yr^{-1}$
Density of fresh solid crust		$\sim 2800\ kg\ m^{-3}$
Temperature of lava (molten rock)		$\sim 1200\ °C$
Latent heat of crystallization[(b)]		$\sim 6.76 \times 10^5\ J\ kg^{-1}$
Heat capacity[(c)] 1200 °C down to 350 °C		$1100\ J\ kg^{-1}\ K^{-1}$
Heat lost per m^3 while cooling to 350 °C		
$(850\ K \times 1100\ J\ kg^{-1}\ K^{-1} + 6.76 \times 10^5\ J\ kg^{-1}) \times 2800\ kg\ m^{-3}$		$= 4.51 \times 10^9\ J\ m^{-3}$
Total heat loss to form crust:		
$(1.90 \times 10^{10}\ m^3\ yr^{-1})(4.51 \times 10^9\ J\ m^{-3})$		$= 8.57 \times 10^{19}\ J\ yr^{-1}$
Heat-flow deficiency in crust less than 60 million years old[(d)]		$= 1.1 \times 10^{13}\ W$
$(1.1 \times 10^{13}\ J\ s^{-1})\ (3.16\ s\ yr^{-1})$		$= 3.48 \times 10^{20}\ J\ yr^{-1}$
Excess of heat-flow deficiency over amount required to cool crust to 350 °C		
$(3.48 \times 10^{20}\ J\ yr^{-1})–(8.57 \times 10^{19}\ J\ yr^{-1})$		$= 3.48 \times 10^{20}\ J\ yr^{-1}$

[a] Parsons (1982).
[b] Fukuyama (1985).
[c] Stacey (1992).
[d] Stein and Stein (1994).

becomes covered with a layer of sediment. When measurements of the thermal gradient and heat flow through the sediment are made, it is found that the heat flow near the ridges is much less than is predicted based on models of what the heat flow should be from hot basalt, by conduction alone, without prior cooling by permeation with water. If it is assumed that after about 60 million years the thermal gradients and heat flow from the bottom of the crust to the top of the sediment have reached a steady state (and here the models and observations agree), then the deficit in heat flow over the younger crust can be integrated to yield an estimate of the total heat previously lost from the crust by hydrothermal circulation. Stein and Stein (1994) performed this calculation and arrived at a global estimate of the thermal flux to be 1.1×10^{13} watts. Next, this reasonably well-grounded estimate can be compared with the amount of lava up-welled and emplaced as basalt during the production of new seaf loor as the crustal plates pull apart at the spreading centers.

The total area of new sea floor formed each year was estimated by Parsons (1982) to be 3.0 km^2 per year in the major ocean basins, with an additional approximately 0.45 km^2 per year contributed in the less-well-understood marginal basin spreading centers, for a total of about 3.45 km^2 per year. As demonstrated by the calculations shown in Table 13.4, the hydrothermal cooling calculated above exceeds by a factor of 4 the amount of cooling required to reduce the temperature of the newly emplaced lava to that of basaltic crust at 350 °C, so it is clear that cooling by flows of seawater through the hot crust must continue for a considerable time after the sea floor moves away from the spreading centers. This flow will occur at progressively lower temperatures.

For several reasons the calculation in Table 13.4 is too simplistic. But it nevertheless clearly shows that it is impossible for the hot hydrothermal flows to account for all the heat deficit. Probably less than 10% of the heat deficit is carried by this very hot water. While the hydrothermal circulation may start out quite intensely in some locations, the intense flows of very hot water appear to be fairly localized phenomena, accounting for only a small part of the long-term cooling. The hydrothermal circulation must continue at diminishing rates for perhaps millions of years after the initial emplacement of the basalt, so that the total heat loss by hydrothermal circulation includes additional heat input from slow cooling of the crust and diffusion from the underlying mantle (and from internal radioactive decay) during this time.

We can very roughly calculate the minimum likely flow of water through the hydrothermal circulation and out to the ocean above with the following assumptions: 5% of the total heat deficit is carried by very hot water, and the average temperature of this hot water is 373 °C (this gives a temperature rise of 370 °C above that of the ambient water). The corresponding minimum flow of water must be:

$$\text{Flow} = \frac{0.05 \times 3.48 \times 10^{20}\,\text{J yr}^{-1}}{\Delta H}.$$

where ΔH is the enthalpy difference between the in-flowing and the out-flowing water. Lacking data for the heat capacity of seawater over the necessary range of temperature, the value for a 3.2% solution of NaCl (Bischoff and Rosenbauer 1985) is used here for this calculation. For an entering temperature of 3 °C and an exit temperature of 373 °C, the $\Delta H \approx 1.73 \times 10^6$ J per kilogram, and the minimum flow must be 10.1×10^{12} kg per year ($\approx 10.1\ \text{km}^3$ per year or 321 m^3 per second). At this rate the entire volume of the oceans would circulate through the hot basalt and up through black smokers every 133 million years. The remaining heat must be removed by hydrothermal circulation at lower temperatures nearby. With a lesser temperature rise, correspondingly greater volumes of water must circulate to carry away the heat. Since we have no good quantitative estimate of the flow rates at different temperatures, it is not possible to be more specific at this time. As a guess, it is reasonable to suppose that the volume of the oceans circulates through the hot and warm rocks in about 1 to 10 million years, with only a small part of this flow reaching the very high temperatures of 370 °C or even 400 °C, observed in the hottest black smokers.

In general, it must be expected that any region where fresh basalt is being extruded at or near the sea floor will show evidence that seawater circulating through cracks in the basalt has been significantly modified by interaction of the circulating fluid with these hot rocks and, conversely, that the basalt has been chemically modified by exposure to seawater. While the spreading-center ridges are where most undersea volcanism takes place, there are in addition many undersea volcanoes not associated with the ridges. One of these is at the top of the Loihi Seamount (Karl *et al.* 1988), where a new island in the Hawaiian chain is apparently growing under the sea. Here, the exit temperatures of the water that has been sampled are only a few degrees or a

Table 13.5 Characteristics of the hot water at two regions in the mid-ocean ridges

	East Pacific Rise (at 11° and 13° N)	Mid-Atlantic Ridge (at 23° and 26° N)	Seawater (at 35‰)
Temperature, °C	317–380	290–350	~ 2
pH	3.1–3.7	3.7–3.9	~ 7.8
Na (mmol kg^{-1})	290–596	411–849	464
K (mmol kg^{-1})	18.7–29.6	17.0–23.9	9.8
Cl (mmol kg^{-1})	338–760	559–659	541
Ca (mmol kg^{-1})	10.6–55.0	9.9–26.0	10.2
Mg (mmol kg^{-1})	0.0	0.0	52.7
SO_4^{2-} (mmol kg^{-1})	0.0	0.0	27.9
H_2S (mmol kg^{-1})	2.9–12.2	5.9	0.0
SiO_2 (mmol kg^{-1})	14.3–22.0	18.3–22.0	<0.01
Mn (μmol kg^{-1})	925–2532	493–1000	<0.001

Data extracted from a larger summary in Von Damm (1990) of data from sampling the hot water exiting at discrete flows in these two regions. In each case the range of values measured at several locations in each region is presented. A much more extensive tabulation, with many more elements, is provided in German and Von Damm (2006).

few tens of degrees above ambient, and this water also exhibits chemical compositions that differ greatly from normal seawater.

Most studies of the water exiting at the sea floor have been done by collecting samples of fluid from the spectacular black smokers and from other locally concentrated flows at still quite hot temperatures. All these fluids exhibit a chemical composition very different from that of normal seawater (Von Damm 1990). The chemical changes are dramatic and, even after considerable dilution, plumes of changed chemical composition can be detected, often at great distances from the vents. Fluids with the highest temperatures are usually (but not always) the most profoundly changed. In many cases, fluids with lower temperatures show evidence that they have mixed with cooler seawater somewhere in the vent system or during collection, and it is possible to extrapolate back to the composition of the hotter fluid below.

In the fluids coming from the hottest vents, every chemical species measured may differ in concentration from that expected in seawater (Table 13.5). Typically, the concentration of magnesium ion is reduced nearly to zero. This is ascribed to the reaction with basalts, chemically changing these rocks by the absorption of magnesium into new mineral phases. This reaction can extract essentially all the magnesium if the seawater reacts with a sufficient quantity of basalt at temperatures above 150 °C. The evidence for this process is secure enough that the concentration of magnesium can be used to correct the observed concentrations in samples of vent water. It is difficult to collect vent water without some admixture of seawater and, in addition, some ambient seawater may become entrained into the vent flow below the orifice. Any magnesium in water from a hot vent is assumed to come from either or both of these sources, and is therefore subtracted out, along with the proportional amounts of other seawater constituents, to obtain the concentrations of substances in the uncontaminated vent water.

Typically, the concentration of sulfate also is reduced to zero; in this case there are two routes by which the sulfate might be lost. When seawater is heated to a high temperature, anhydrite ($CaSO_4$) will precipitate, and it is known that many vents are coated on the inside with precipitated anhydrite. There is not enough calcium in seawater to precipitate all the sulfate, but additional calcium is released during the interaction of hot basalt with seawater (sometimes nearly as much as the magnesium taken up), enough so that there is often more calcium in the vent water than in the original seawater; therefore a lack of this ion should not limit the precipitation of anhydrite. There is reason to believe that all or most of the anhydrite eventually dissolves again when the flow cools and is not a permanent constituent of the altered basalt pavement on the sea floor. Sulfate is also lost because the hot basalts are reducing in nature, containing reduced iron and manganese. Oxygen and sulfate oxidize these elements; some, or all, of the sulfate is reduced under these conditions, and free H_2S is found in the vent water. In addition, H_2S is present in small concentrations in basalt, so some of the H_2S in the exiting water may have come directly from the basalt. Considerable quantities of sulfide-insoluble metals are also released from the basalt to the very hot seawater, and these are then precipitated in the places where the water exits. These metal sulfides can build up into quite large deposits. Economic deposits of these minerals are found in geologically preserved ancient centers of similar activity.

Concentrations of dissolved silica are dramatically elevated, often several thousand-fold above seawater values, and it seems that the silica may reach solution equilibrium with quartz at temperatures between 300 and 400 °C. Indeed, the observed concentrations, together with the temperature and pressure dependence of the solubility of quartz, are used to guess at both the temperature and the depth of the reaction zones controlling the chemical composition of the vent water in different locations (Von Damm and Bischoff 1987).

The pH of vent water is distinctly acid, with values commonly below 4.0. Because of the acid pH, high temperatures, and reducing conditions, the concentrations of most trace metals are elevated in the black-smoker waters. In particular, manganese and iron are elevated, sometimes to millimolar concentrations, thousands to millions of times their normal concentrations in seawater. It is the cooling and mixture of the vent waters with seawater as they stream out, causing the precipitation of iron and other metal sulfides, that creates the visual appearance of the black smokers.

Concentrations of a number of gases are also elevated, sometimes by large factors. These gases include the hydrogen sulfide already mentioned, methane, carbon dioxide, ^{3}He, ^{4}He, and hydrogen. Concentrations of hydrogen are difficult to evaluate because of possible reactions of the very hot and acid water generating hydrogen by reaction with the walls of the sampling equipment. Concentrations of carbon dioxide are variable but are generally elevated above seawater values, sometimes by almost eight-fold (Merlivat *et al.* 1987), showing that carbon dioxide is out-gassed into the ocean from the mantle at these locations. Normal seawater contains only minute traces of methane, but vent waters always contain rather significant amounts; again, these are variable but range up to 60 μmol kg^{-1} (over 1 $cm^3 \cdot kg^{-1}$). The presence of methane is an indication of the oxidation state of the mantle materials and of the state of some of the carbon stored there. The isotopic compositions of the methane and the carbon dioxide suggest equilibration of these species at temperatures over 600 °C, providing some indication that the hydrothermal circulation might possibly reach temperatures this high in the deepest cracks.

One of the more bizarre features of the hot vent waters is the observed range in concentrations of both sodium and chloride (Table 13.5). Concentrations of both ions may be significantly lower or greater than those in the ambient seawater that feeds the hydrothermal systems. No acceptable suggestions have come forward that can explain these large changes by precipitation or solution of minerals in the vent systems. While it might be possible to explain the range of sodium ion concentrations by release and uptake during the alteration of minerals in basalts, it has not been found possible to explain the chloride results this way, due to the small amounts of chloride in basalts and the rarity of chloride-containing minerals that could be stable under the conditions there. It appears that the temperature conditions deep in the hydrothermal vent systems are such that there can be separation into vapor and gas phases, and that the temperature may often exceed the critical point of seawater (see Chapter 2 and Figure 13.1), so that a concentrated brine phase and a less-concentrated gas phase may be formed. As these two phases make their way up through different channels, with condensation of the gas phase higher up in the vents, and with partial mixing in different channels along the way, the fluids exiting at the sea floor could show the varying concentrations observed. It should be no surprise that changing conditions in the plumbing deep in the reaction regions as paths are occluded or opened, as minerals precipitate or dissolve, and rocks cool or crack, will result in considerable changes in the composition of the fluids coming from each vent. Indeed, this has often been observed (German and Von Damm 2006). The large ranges in concentration found for all elements makes the concept of salinity of the vent fluids essentially meaningless. Since there is not much opportunity for chloride to precipitate or react with the hot rocks it can be thought of as a nearly conservative substance, and it is common to measure the chlorinity of the exit water and to relate the concentrations of other substances to the chlorinity. Even the water itself is not conservative, in the sense that some incorporation of water into the products of seawater–basalt interaction is known to occur, so that some water must also be lost from the circulating seawater. The products of the hydration reactions are thought to contain several percent water, and this water must be to some extent retained within the altered basalt and later carried down into the mantle with subducting slabs of sea-floor material. It is possible that some concentrated brine also is trapped and subducted with the deeper crust into the mantle below.

As mentioned earlier, it is difficult to know the total flow of water through the channels in the hot basalt, but we can still try to make order of magnitude estimates of the possible effects of these flows on the chemistry of the sea. If we take the previously assumed rate of very hot hydrothermal flow (about $18.4 \times 10^9\ m^3\ yr^{-1}$) and, as observed, all the magnesium is removed, the total removal rate is

$$18.4 \times 10^9\ m^3\ yr^{-1} \times 53\ mol\ m^{-3} = 9.8 \times 10^{11}\ mol\ yr^{-1}.$$

This estimate, about 17% of the total transport of magnesium carried by rivers into the sea (Table 13.2), must be a minimum value. Much of the water circulating through the hot rocks must be at a lower temperature but still hot enough to lose magnesium, and water will continue to circulate through partially cooled rocks as long as the channels are open.

The question of what happens to the magnesium is even more complicated and uncertain than appears from the above considerations alone, however, because at cooler temperatures the minerals previously formed during the interaction of basalt with hot seawater may release magnesium again to circulating seawater. Since, at present, it is not possible to obtain quantitative estimates of the exact temperature and water-exposure histories of the average undersea basalt, the full effect of these exchanges remains a matter of speculation. It is plausible to consider that essentially all the magnesium coming down rivers is eventually removed from the ocean by reaction with fresh hot basalt. What can be said for sure, however, is that these exchanges are large and must be a powerful determinant of the steady-state composition of seawater.

It has been thought possible to estimate the total fluxes involved in these exchanges by examining the changed chemical composition of the basalts dredged from the sea floor and obtained by drilling into the altered basaltic crust, and multiplying the changes observed by estimates of the amount of crust formed at various depth intervals. Considerable effort has been expended to attempt this (Staudigel 2004), but so far the results are unsatisfactory because too few cores have been drilled deep enough into the crust, and the recovery of the core material has been poor enough that no convincing conclusions are yet available.

On the basis of the small data set shown in Table 13.5, and more extensive tabulations in German and Von Damm (2006), it seems that, for example, reactions in the hydrothermal systems remove magnesium from seawater at rates that may be enough to balance the river input, that considerable amounts of calcium are introduced into seawater, that while sulfate is largely removed, some of the sulfur is reduced in the systems and comes back into the ocean as H_2S or metallic sulfides, that potassium is added to the ocean, and that large amounts of dissolved silica are added to the ocean.

In addition to those elements listed in Table 13.5, many other elements are either taken up by the basalt or released to seawater during hydrothermal alteration. Elements such as iron and manganese that are insoluble under seawater conditions may be precipitated in the upper parts of the conduits, or may precipitate in the water column above so that the regions around the vents are sometimes mantled with such precipitates. For more complete discussions, see Von Damm (1990), Mottl and Wheat (1994), Elderfield and Schultz (1996), Humphris *et al.* (1995), and German and Von Damm (2006).

13.4 Sediment–water exchange

Nearly the entire bottom of the ocean is covered with sediments; this mantle of generally fine-grained material varies greatly in thickness and in composition from place to place. Over most of the area of the ocean the sediments lie on a pavement of basalt – volcanic material extruded at the spreading centers and moving slowly away, eventually to be subducted into the mantle again. As this ocean crust moves, it receives a slow drizzle of sediment settling from the surface and sometimes also redistributed by ocean currents. The thickness of the sediment depends on both the age of the crust and on the local rate of sedimentation.

The Deep Sea Drilling Project and its successor, the Ocean Drilling Program, have made it possible to obtain long cores of marine sediment from hundreds of locations throughout most of the accessible regions of the oceans. Interstitial pore water from samples of many of these sediment cores has been extracted in order to measure the concentrations of substances dissolved in the pore water, and for most substances a majority of the cores exhibit distinct gradients of concentration with depth in the sediment (Gieskes 1983). When samples of interstitial water from deep in the sediment are examined, concentrations of major ions are usually found to be very different from those in seawater, and they also vary considerably from place to place. Figure 13.2 provides a fairly typical example. Commonly, in most parts of the ocean, the concentration of magnesium decreases greatly in the first few hundred meters below the interface and the concentration of calcium increases. In the case shown here there is a close correspondence between the decrease in magnesium and the increase in calcium, suggesting that the uptake of one may be chemically related to the release of the other. In other places, the relationship may not be nearly so close; also, it is rare to find really smooth curves. Sometimes the bumps and valleys in the curves can be interpreted, sometimes not. As in this example, concentrations of potassium usually decrease with depth below the sediment–water interface, as do those of sulfate. In regions with a greater input of organic matter, the sulfate may be entirely consumed. Other elements, the total alkalinity of the water, and even the isotopic composition of the water itself show marked changes with depth.

These concentration gradients, and their relationships to one another and to the mineralogical composition of the sediments, are important evidence for attempts to understand the chemical diagenesis of the sediments during deep burial over millions of years. The concentration changes are often dramatic, but the great depth over which they occur leads to actual concentration gradients that are quite small. The diffusive flux of substance is therefore usually relatively tiny from deep (several hundred meters) in the sediment up to the overlying water. Consider, for example, the calcium concentration gradient in Figure 13.2 from a depth of 30 m to 450 m. The difference in concentration is 18.8 mM over a distance of 420 m, and for the purpose of a rough calculation can be treated as a linear gradient. The flux can be calculated from Fick's first law, as in Eq. (8.10) and Figure 8.11,

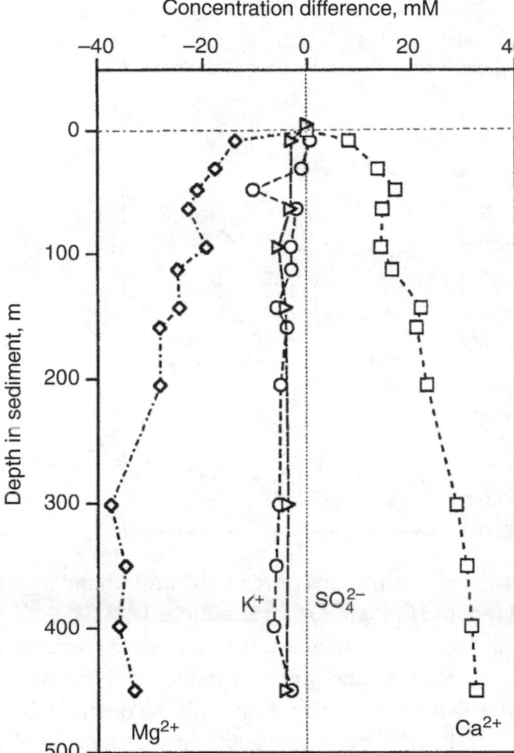

Figure 13.2 Vertical profiles of three abundant cations and sulfate in the interstitial water from marine sediment in a core obtained by the Deep Sea Drilling Project at hole 541 (15° 31′ N, 58° 44′ W; north–north-east of Barbados). The data are plotted as the difference between the concentrations in the pore water and in the seawater above. The uppermost data points are at a depth of 8.5 m; the seawater concentration differences of zero are plotted at the interface. (Data from Gieskes *et al.* 1984.)

$$\text{flux} = \frac{\phi}{\theta^2} D \frac{C_2 - C_1}{Z},$$

where D is the molecular diffusion coefficient in the interstitial water; ϕ, θ are the porosity and tortuosity of the sediment, respectively; C_1 and C_2 are the concentrations at the top and bottom of the gradient interval; and Z is the distance over which the diffusion is taking place.

The effective diffusion coefficient in the interstitial water of marine sediments is not constant, but varies considerably due to variations in the porosity and other properties of the sediments as shown in the above equation (Gieskes 1983). For our purpose here, it will be sufficient to take a typical value for calcium ion in the interstitial water (including the effects of porosity and tortuosity) as about 1×10^{-10} m^2 s^{-1}. Expressing the concentrations per cubic meter, the flux is:

$$\text{flux} = 1 \times 10^{-10}\ \text{m}^2\ \text{s}^{-1}\,\frac{18.8\ \text{mol m}^{-3}}{420\ \text{m}} = 4.5 \times 10^{-12}\ \text{mol m}^{-2}\ \text{s}^{-1}.$$

Suppose this value were to apply over the whole area of the ocean; the yearly flux would be

$$\begin{aligned}\text{flux} &= (4.5 \times 10^{-12}\ \text{mol m}^{-2}\ \text{s}^{-1})(362 \times 10^{12}\ \text{m}^2)(3.16 \times 10^7\ \text{s yr}^{-1})\\ &= 5.12 \times 10^{10}\ \text{mol yr}^{-1}.\end{aligned}$$

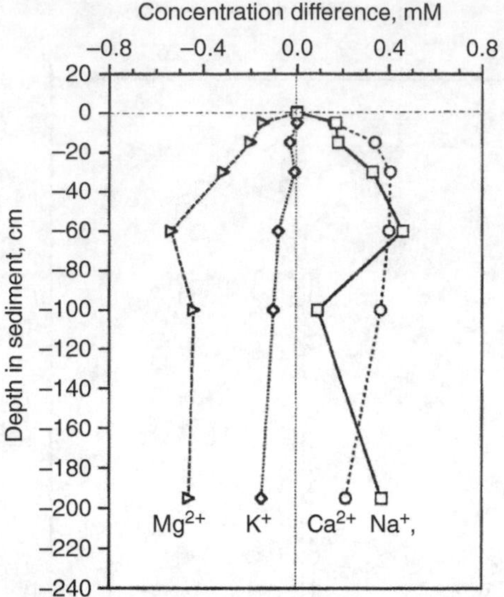

Figure 13.3 Vertical profiles of the concentrations of the four most abundant cations in interstitial water from marine sediments. Data for three stations (AII-78-J, K, N; ~12 to 15° N, 55 to 57° W), situated in the North Atlantic east of Barbados, were averaged and plotted here as the difference between the concentrations in the overlying water and the concentrations in the pore water. Note the differences in both the depth and concentration scales from those in Figure 13.2. The uppermost measurement is at a depth of 5 cm. Concentrations exactly at the sediment surface were not measured, but the values in the water above were plotted at that surface here for the purpose of showing the gradients in the sediment near the surface. Where the difference is positive (to the right of the zero line) the concentrations in the pore water are greater than in the water above, implying a flux of the ion out of the sediments into the overlying water. Conversely, concentrations to the left of the zero line imply a flux into the sediments. (Data from Sayles 1979.)

This potential diffusive flux from deep in the sediments is less than 0.5% of the river flux as listed in Table 13.2. This overly simplistic calculation is nevertheless representative of what could be concluded from a more complete evaluation. The diffusive fluxes deep in the sediments can in most cases be shown to have a negligible effect on the oceanic balance of the elements.

The surface layers of the sediments are another story, however. These are in direct contact with seawater or can engage in diffusive exchange with seawater on a short timescale, and these exchanges are important. In Chapter 8, we saw that the interstitial water in marine sediments is generally characterized by an increasing concentration of dissolved silica with increasing depth in the sediment, and that this vertical gradient leads to a significant diffusive flux of silica from the sediments into the overlying water. Many other substances also exhibit gradients in concentration, so that for each one there is a flux into or out of the sediments. Sayles (1979, 1981) analyzed the interstitial water extracted at intervals from approximately the top two meters of sediment at some 54 stations distributed throughout the North and South Atlantic Ocean. Results for three stations (Figure 13.3) illustrate several common

Table 13.6 Total diffusive flux of some major cations into or out of the sediments of the world ocean

Element	Diffusive flux, 10^{12} mol yr^{-1}	Riverine flux[a], 10^{12} mol yr^{-1}	Sediment flux, % of river flux
Na	4.4	8.38	53
K	−2.0	1.23	163
Ca	6.3	12.5	50
Mg	−5.2	5.16	100

Values are positive for diffusion into the ocean, negative for diffusion from the ocean into the sediments. These estimates are based on an extrapolation of data from stations in the Atlantic Ocean (Sayles 1979); extrapolation to the world ocean serves an illustrative purpose. Many more data are needed to satisfactorily constrain a global budget.

[a] Values derived from Table 13.2, calculated for the estimated natural fluxes only, without the present-day anthropogenic components.

results. The data are not always entirely consistent; even averaging the results from three stations reasonably near each other results in plotted curves that are not always smooth and exhibit a certain amount of "noise." It is the exception rather than the rule to find nice-looking smooth curves that appear easy to interpret. This presentation is, however, typical in that there are clear and usually consistent changes in concentration of all the major cations with depth in the sediment. In each case, the concentrations of both calcium and sodium increase with depth and the concentrations of magnesium and potassium decrease. Relative to the concentration of sodium in the seawater above, the changes observed here are very small, not much above analytical precision, so it may be unwise to rely too strongly on the calculated gradients and fluxes for this element. Data for the other ions are more secure, however, and it is clear that significant exchanges must be taking place. Sayles (1979) fitted equations to the measured concentrations of the various ions as a function of depth to estimate the gradient at the sediment–water interface, then estimated the diffusion coefficients of each type of ion and calculated the fluxes of each. Assuming that the Atlantic values can be extrapolated to the rest of the world ocean, he calculated the worldwide total fluxes (Table 13.6). While we do not have enough information to assess the accuracy of these extrapolations, it is nevertheless evident that fluxes into and out of the sediments must be quite significant to the total oceanic balance of these elements. The flux of sodium is the least accurate of those listed, but, such as it is, it suggests that the fluxes from the sediments and from the rivers are additive. Potassium may be taken up by sediments in amounts that could account for the total river input. The increase of dissolved calcium in the interstitial water comes about in part because of the dissolution of calcium carbonate shell debris in the sediment environment, which is generally somewhat more acid than the overlying seawater. If the ocean is in steady state with regard to the concentration of calcium, the inputs are balanced by the sedimentation and loss from the ocean of that portion of the calcium carbonate that does not dissolve (including the precipitation of calcium carbonate in coral-reef areas).

The sediments have other chemical effects on the water above, but in many cases these exchanges may be considered as a relatively short-term recycling of materials that were not completely processed during settling through the water and before shallow burial in the sediment (e.g. solution of silica, remineralization of other nutrients).

In all but the very lowest sedimentation-rate regimes, however, there is a continuing burial of organic carbon that is very slowly broken down. This leads to a vitally important feature of the sediment environment: Oxygen is biologically consumed in the sediments. Since the only source of oxygen in the sediment is diffusion from the water above, there is a gradient in concentration with depth, and at some point below the surface there is no oxygen left. The sequence of processes previously discussed in examining anoxic waters takes place in the sediments, often nicely separated vertically. Nitrate is produced in the sediment down to the point where oxygen is all gone but is consumed by denitrification at greater depths, and there the product is dinitrogen, N_2. The decrease of $[NO_3^-]$ with depth in the sediment pore waters (Bender *et al.* 1977), and an increase with depth in the concentration of N_2 (Wilson 1978), were measured independently at a half-dozen stations each in the North Atlantic, and provided estimates of the mean denitrification flux of 0.6 and 1.9 mmol N_2 m^{-2} yr^{-1}, respectively. Such measurements have not been repeated in enough places to provide a satisfactory accounting of the global rate of denitrification in the deep-sea sediments. An indication of the magnitude of their possible contribution can be obtained by multiplying the rates above by the area (268×10^6 km^2) of the sea floor covered by pelagic sediments; the result is 4 to 14 Mt of N_2 per year. This is a small but certainly not negligible fraction of the approximately 440 Mt of nitrogen fixed each year (see Table 8.6). Shallow marine sediments in productive areas may denitrify at much greater rates. The estimated average production rate of N_2 in Narragansett Bay (Seitzinger *et al.* 1984) is 260 mmol m^{-2} yr^{-1}, while for all the continental shelves in the North Atlantic the average estimated rate is 126 mmol m^{-2} yr^{-1} (Seitzinger and Giblin 1996). The continental shelves of the North Atlantic alone may be converting fixed nitrogen to N_2 at a rate of about 20 Mt per year. Evidently the major locations for the loss of fixed nitrogen are in the near-shore, shelf, and slope regions. We do not yet have enough data to estimate worldwide rates. Nevertheless, it is clear that throughout the bottom of the ocean the sediments are responsible for much of the global rate of denitrification, and nitrate is consumed perhaps everywhere at rates that are probably controlled largely by the rates of carbon input to the sediments.

Deeper in marine sediments, after nitrate and other oxidants are exhausted, a reduction of sulfate occurs with the production of hydrogen sulfide. The consumption of sulfate causes a generally weak gradient in the concentration of this ion in the pore waters, and consequently there is usually a small flux of sulfate downwards into the sediments. Because at any given constant rate of sedimentation and input of organic carbon the depth at which sulfate reduction occurs is dependent on the concentration of oxygen in the overlying water, the amount of sulfate reduction is on average somewhat inversely proportional to the concentration of oxygen in the overlying water and hence loosely dependent on the concentration in the atmosphere. It is thought that over a very long stretch of geological time this effect helps to buffer the concentration of oxygen in the atmosphere. If the concentration of oxygen rises, more will be consumed in oxidizing sedimentary organic matter. If the concentration of oxygen decreases, a greater proportion of the organic matter will be oxidized by sulfate. This inverse relationship, the oxygen–sulfate buffer, has implications for understanding the long-term controls on the geochemical balances and cycling of sulfur, carbon, and oxygen (Berner and Canfield 1989).

13.5 Warm clay

As the moving plates of oceanic crustal material with their overburden of marine sediments enter subduction zones, a number of things happen (Figure 13.4). In some places, surficial layers of recent land-derived and poorly compacted sediment are scraped off and pile up against the edge of the overriding plate (which is often the edge of a continent). In locations where not too much of this material is present, the largely pelagic oceanic sediments are completely carried down into the mantle with the subducting plate (von Huene and Scholl 1991). In some cases, the leading edge of the overriding plate is also scraped off and subducted as well. Von Huene and Scholl (1991) estimated that sediment and scraped-off materials are subducted at a long-term average rate of about 1.6 km^3 per year. This approximates the long-term rate of land erosion (before rates approximately doubled some tens of millions of years ago because of changes in continental elevation and other geological rearrangements, and before the further approximate doubling due to the influence of humans). Thus, the rate of sediment supply to the ocean is approximately balanced over several hundred million years by the rate at which materials are returned to the mantle (after allowing for a small growth in continental area). At least some of the sediment carried down is melted and appears again incorporated into the lava from volcanoes above the subducted plate. In addition to a great deal of indirect evidence for this, there is direct evidence from the presence of ^{10}Be (half-life = 1.5×10^6 years) in lava from such volcanoes. Background concentrations of this isotope are nearly undetectable in mantle rocks, and the only significant source is cosmic-ray-induced spallation of atmospheric gases. The atmospherically produced ^{10}Be is insoluble; when it reaches the ocean it adsorbs to sediment particles and is concentrated in the sediment. Its appearance in lavas from some volcanoes proves the contribution of sediments to the melt that produced the volcanic lava (Morris *et al.* 1990).

The sediment carried on a subducting plate beneath the edge of an overriding plate is compressed, and sediment piled up against an overriding plate is also compressed, which causes some of the interstitial water to be squeezed out. The water forces its way upward, and the flows are concentrated at faults and other cracks in the overlying sediment. Because this water originates deep beneath an overriding plate or deep within the sediment in front of the overriding plate, it carries upwards some of the heat conducted earlier into the lower layers of sediment from the crustal

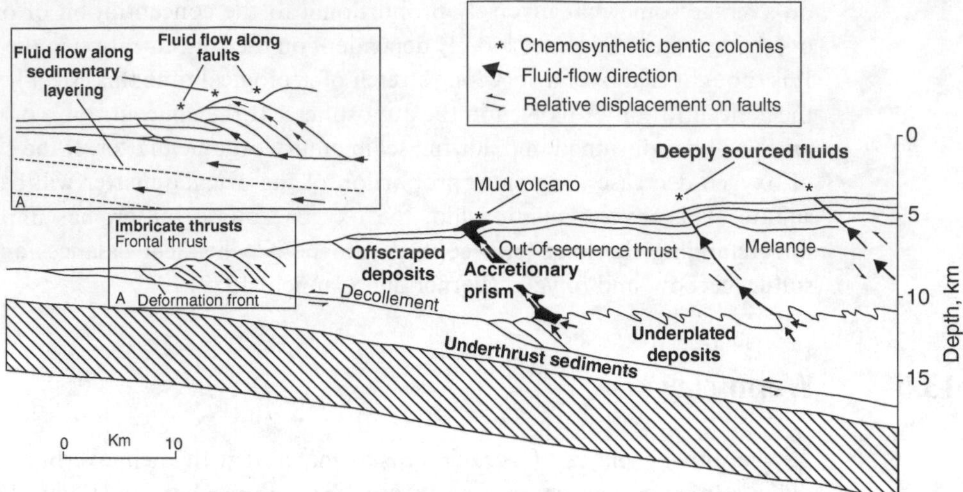

Figure 13.4 Schematic and partially conjectural cross-section of an accretionary prism, showing major features controlling fluid emplacement and expulsion. The bottom layer is oceanic crust, moving towards the right, diving under a continental margin (a common situation) that is stationary or moving to the left. Some portion of the sediments carried on the crustal plate is is scraped off, compressed, and piled up against the accretionary prism of sediment, increasing its bulk. Some of the sediment is carried under the prism and will most likely be subducted partly or totally into the mantle, or at least far enough down to contribute to the volcanoes associated with regions of this sort. Some of the sediment may be plastered as an underplated deposit onto the bottom of the accretionary prism or the overriding plate. The interstitial fluid from the sediment is partially squeezed out by the pressures associated with compression against the accretionary prism and being thrust under a considerable layer of sediment above. The fluid makes its way out along channels in areas of weakness in the sediment, primarily, it is believed, along fault cracks. It appears as diffuse flows at the surface of the sediment, or as mud volcanoes where the flow is especially concentrated. Communities of benthic organisms supported by chemosynthetic bacteria are found associated with regions where the flow is sufficient, and indeed are at present the best way to find such locations. Not shown is a presumed flow of water from the hydrothermally altered crust itself, caused by heating of the crustal slab as it sinks into the mantle and the release of some of the water incorporated into crustal minerals during hydrous alteration. From Moore and Vrolijk (1992).

basement below and is detectably warmer than the bottom seawater. In addition, some water is probably squeezed out from cracks in the fractured crust below the sediment and expelled from hydrated minerals in the crust as it gets warmer. Knowing the amount of oceanic crust produced and subducted each year and the thickness of sediment carried on the subducting plate, it is possible to estimate that the amount of water carried into the subduction and accretion zones each year is about 2.6 km^3 (Table 13.7). This value must approximate the maximum amount of water available that could be expelled from sediments and crust in the subduction zones, allowing that the exact degree of hydration of the crust is poorly known and that the porosity of the sediments has been measured at only a few places.

Table 13.7 Estimates of the volume of water carried by hydrothermally altered crust and overlying sediments into subduction zones and of the amount of water expelled from these zones each year

	Volume transport, $km^3\ yr^{-1}$
Water carried in	
Pore volume in sediment	1.80
Water in hydrous minerals in the sediment	0.43
Pore water and hydrous minerals in the crust[(a)]	0.35
Total	2.6
Water expelled	1.0 to 1.8

Estimates of water carried in from Moore and Vrolijk 1992. Estimates of water expelled from Le Pichon *et al.* 1993.

[(a)] Estimated for a 1-km thickness of hydrothermally altered crust. This is an underestimate by an unknown amount because some hydrothermal alteration is believed to occur considerably deeper. The deepest hole drilled into the oceanic crust is at Site 504 of the Ocean Drilling Program (1° 13.61′ N, 83° 43.82′ W). The total penetration so far (ODP Leg 148; 1993 EOS **74**: 489) is 1837 m into the crust. Some rock samples from even the bottom of the hole have been hydrothermally altered. Other geological evidence suggests that the hydrothermal alteration may occur at least 3 to 4 km below the surface of the crust. However, the degree of exposure of crust to hydrothermal seawater must decrease with depth below the sea floor, and not enough information is available yet to provide an acceptable global estimate.

Water that is not expelled is carried deep into the mantle. A recent evaluation (Parai and Mukhopadhyay 2012) of the long-term net transport into the mantle, based on a considerable literature of disparate results, suggested a value for this flux of about 1 $km^3\ yr^{-1}$; most of this flux is believed to be in the form of hydrous minerals.

Some of the occluded and mineral-bound water is probably carried down all the way into the mantle, some certainly participates in the melting reactions that yield the lava of volcanoes behind the subduction zones, and the rest is forced out by the increased temperature and pressure up through the surficial sediments in these regions. Present estimates of the rate at which water is expelled range from 1.0 to 1.8 km^3 per year or from 40 to 70% of the water potentially available (Table 13.7).

The water that is expelled up through thick layers of sediment appears not to exit in discrete easily sampled flows (at least none have so far been found) but instead in slow flows distributed at intervals where fractures have weakened the sediment, or as *mud volcanoes* (Langseth and Moore 1990). Sampling of this water has so far been confined to the extraction of pore fluids from cores of sediment at various depths in these regions.

The chemical composition of this interstitial water is modified from seawater by processes not yet completely understood. In addition to differences in the ionic composition, it often contains methane and traces of other organic gases. The methane originates both from the bacterial decomposition of organic matter deep in the sediment and apparently also from the thermal cracking of sedimentary organic matter even deeper in the sediment where the temperature is high enough to allow this process to proceed. In places where the water forces its way up through discrete channels to the surface, there often develops a specialized fauna of clams and other organisms based on the oxidation of methane as an energy source (Suess and Whiticar 1989; Le Pichon *et al.* 1992), and in fact one of the best ways to find such seeps is to search for this fauna. When examined in accessible regions near the sediment surface these fluids often appear to be supersaturated with calcium carbonate; this mineral is precipitated in the interstitial spaces of the sediment just below the surface and even in chimneys of calcite rising above the sediment surface (Han and Suess 1989). The latter authors suggest that the precipitation of calcium in these places may be of sufficient magnitude on a global scale to be an important entry in mass-balance calculations of the exchange of calcium between various of the geochemical reservoirs of this element.

One of the most remarkable observations concerning the chemistry of the interstitial water sampled from cores of sediment taken in these regions is that the salinity is commonly less than that of seawater. Specifically, the chloride concentration is reduced below that of seawater, sometimes reaching values as little as one-half the 545 mmol kg^{-1} present in typical seawater (Kastner *et al.* 1991). As chloride is not known to be sequested in any minerals formed under the conditions in these regions, it seems that the seawater from the pore spaces in the sediments has been diluted with fresher water. Two sources of this fresher water have been proposed. One suggestion is that meteoric water (rain water that has percolated down through the continental crust) is somehow entrained into the flows of water squeezed out of the sediments and crust below. The other, much more plausible, suggestion is that the hydrated minerals in the sediments and in the hydrothermally altered crust below lose considerable amounts of water upon being heated and compressed as the subducting plate penetrates the mantle below the overriding plate, and this water makes its way along faults and cracks and dilutes the more seawater-like fluids from the pore spaces of the sediments. As can be imagined from the occasional finding of quite low chloride concentrations, there must be a significant amount of water available, enough to effect a considerable dilution of the pore water.

At the present time, there is not yet enough information to make possible a convincing calculation of the various chemical fluxes associated with the sediment and crustal dewatering process noted above. In comparison with a river flow of more than 37 000 km^3 per year or the flows of certainly hundreds of cubic kilometers per year from cooling of the crust at and near the spreading centers, these flows of not more than 1 to 2 km^3 per year seem small. It would seem reasonable to conclude that while the flow of interstitial water squeezed out in the subducting plate margins is certainly of local importance to the sediment chemistry and to the biota colonizing the regions where this water escapes from the sediment, it is unlikely in the short term to have a major impact on the chemistry of the oceans as a whole.

The examination of the warm water expelled up through the sediments is, however, of considerable interest when addressing questions of the amount of water and other substances carried down into the mantle, and thus questions of long-term controls on the amount of water in the ocean and in the mantle, and with various geochemical fluxes between the mantle and the surface of Earth.

13.6 Residence times

The concept of residence time was introduced by Barth (1952) as follows:

$$\tau = \frac{\text{amount in the sea}}{\text{supply per year}}.$$

Because the idea of residence time can sometimes be confused with the average lifetime of a substance, the simple concept defined above is often called the *replacement time*, which may be less ambiguous. For our purpose here, however, I still use the term residence time.

As noted in Chapter 4, the concept of residence time, in this simple form, requires the assumption that the reservoir be well mixed. In turn, this requires that the residence time be long with respect to the mixing time of the reservoir. If the residence time of some substance entering the ocean is similar to the mixing time, the ocean cannot have a uniform concentration. This is so because it is likely that the substance will enter in some regions and be removed in some other regions. The concentrations must therefore vary, and the lifetime of an individual molecule will not be uniform throughout the reservoir because it depends on where it enters and where it leaves. The calculation and presentation of short residence times is still useful, however, because any substance whose calculated residence time is short must be chemically quite reactive, while a substance with a very long residence time must be chemically unreactive in the ocean (at least with respect to removal processes). An insight into the chemical behavior of a substance is therefore quickly acquired from this calculation.

Values for the residence times of most of the elements have been calculated and tabulations prepared (e.g. Goldberg 1963; and various authors since). It is common in discussing the geochemistry of any element to present a calculation of the residence time.

Even for substances that have a long residence time, however, the calculation requires careful definition. For example, the residence time of sodium in the ocean can be calculated from the mass of the ocean (1.4×10^{21} kg; Appendix J, Table J.1), the average salinity (34.72‰; Table 1.3), the concentration of sodium at 35‰ (467 mmol kg^{-1}; Table 4.1), and the river flux (11.7×10^{12} mol yr^{-1}; Table 13.2) as follows:

$$\frac{(1.4 \times 10^{21}\text{kg})(469 \times 10^{-3}\text{ mol kg}^{-1})(34.72/35.00)}{11.7 \times 10^{12}\text{ mol yr}^{-1}} = 55.7 \times 10^{6}\text{yr}.$$

This is a very useful result, because it tells us that the ocean responds only slowly to changes in the input of sodium. The ocean mixes on the order of each one or two thousand years, so that (with the exception of salinity) it will have been fairly well homogenized perhaps 25 000 times while one ocean mass of sodium is introduced. Therefore, sodium will be quite uniformly distributed within the ocean. Indeed, this is the reason that sodium is a conservative element in seawater, bearing a constant ratio to the other major elements with similarly long residence times.

It is not wise, however, to put too much faith in the exact numerical value of a residence time without consideration of the various fluxes involved. Take the case of sodium, calculated above. That residence time was based on the recently measured total river flux of 11.71×10^{12} mol yr^{-1}. Of this total, Berner and Berner (1987) estimated that 8% is from cyclic sea salt (that is, salt that is blown inland from the ocean and returns via the rivers). If we subtract this short-term recycling from the input flux, the residence time becomes 60×10^6 years. Then it was estimated by Meybeck (see Table 13.2) that 3.33×10^{12} mol yr^{-1} is due to anthropogenic influences of various kinds. If we subtract this value and calculate a pre-anthropogenic flux of sodium, the residence time becomes 87×10^6 years. The suggestion by Berner and Berner (1987) that about 16% of the sodium coming down rivers is removed by adsorption onto clay particles during the mixing of river water with seawater further reduces the total flux of sodium into the ocean and raises the calculated residence time to 104×10^6 years. The estimate that perhaps 4.4×10^{12} mol of sodium diffuses out of marine sediments each year (Table 13.6) poses a further question. Is this a flux into the ocean or only another short-term recycling? If it is treated as a flux into the ocean, it reduces the residence time; otherwise it has no effect. As mentioned earlier, this sediment flux is a very uncertain value, and in any case it is probably best to treat it as a relatively short-term recycling process. It is unclear whether chemical exchanges during hydrothermal circulation through the hot rocks at the spreading centers result in a gain or loss of sodium. The rocks are probably a sink for sodium, but if there is a net release, then this flux must be included in the input flux to the ocean, reducing the residence time. What should be done about the sodium carried up with the water expelled from the subduction zones? Suppose the flow amounts to 1 km^3 per year (a conservative estimate; see Table 13.7) and suppose that this carries a sodium concentration of 400 mmol kg^{-1}; then the flux is 0.4×10^{12} mol yr^{-1}. This relatively small input reduces the last calculated residence time from 104 to 97.5×10^6 years. The flow could be double this estimate, however, but whether it should be included is uncertain and may depend on the purpose of the calculation. The water expelled comes from crust and sediments that are generally several tens of millions of years old and could be considered either as recycled ocean water or as a long-term input.

The previous discussion shows that the calculation of residence times is often not straightforward. It is a matter of judgment which fluxes should be included. In many cases, there is not enough information to make anything more than an approximate calculation. As mentioned earlier, however, this calculation is nevertheless worth doing, because it provides general information about the substance of concern, and the importance of the various fluxes becomes evident.

Summary

The composition of the ocean is largely controlled by the balance between the exchanges of substances across its boundaries.

For most substances input by rivers is of predominant importance. These inputs are the best and most accurately known fluxes.

The atmosphere carries significant amounts of sea salt derived from spray entrainment over land surfaces and this returns to the sea again in rivers. The atmosphere carries important loads of dust that falls into the ocean. This can be locally of some importance.

The exchanges between seawater heated to sometimes over 400 °C where it contacts freshly emplaced hot rocks at the undersea spreading centers must have important effects on the chemistry of the ocean. The much larger flows of water at lower temperatures are probably of greater importance, but their quantitative and qualitative effects are even less well known.

A further exchange takes place when slabs of subducting plates lose water and much additional substance squeezed out early in the subduction process.

Throughout the ocean there is a continuing exchange of material with the sediments, driven in part by continuing chemical and biochemical changes within the sediments, and these exchanges are large enough to have a major role in controlling the chemistry of the overlying water.

None of these processes are quantitatively well enough understood to make possible a complete budget of any substance.

SUGGESTIONS FOR FURTHER READING

Burdige, D. 2006. *Geochemistry of Marine Sediments*. Princeton University Press, Princeton, NJ.

Exchanges with marine sediments have a large role in affecting the composition of the ocean, and the extensive information on the chemical processes involved is very well covered and explained in David Burdige's recent book.

Emerson, S. and J. Hedges. 2006. Sediment diagenesis and benthic flux. In *The Oceans and Marine Geochemistry*, ed. H. Elderfield (*Treatise on Geochemistry*, vol. 6, ed. H. D. Holland and K. K. Turekian), Elsevier, New York, pp. 293–319.

Another excellent review of sediment diagenesis and associated fluxes.

German, C. R. and K. L. Von Damm. 2006. Hydrothermal processes. In *The Oceans and Marine Geochemistry*, ed. H. Elderfield (*Treatise on Geochemistry*, vol. 6, ed. H. D. Holland and K. K. Turekian). Elsevier, New York, pp. 181–222.

These authors provide a fine summary of the nature of hydrothermal circulation at the oceanic spreading centers, and of the possible impact of this circulation on the chemistry of the ocean.

In addition, the InterRidge Global Database of Active Submarine Hydrothermal Vent Fields (the "InterRidge Vents Database") is a frequently updated data base of undersea hot vent fields; see http://www.interridge.org/irvents/.

14 Chemical extraction of useful substances from the sea

The oceans are a storehouse of some 5×10^{16} tons of mineral matter.

JOHN L. MERO 1965

The ocean is vast, and all (or nearly all) of the chemical elements are present dissolved in the water, so the ocean must contain huge quantities of important and valuable minerals. This idea has led to many schemes to extract one or another of the substances needed or wanted by mankind. In fact there are only a few rather mundane substances that can be extracted at a sufferable cost. The extraction of common salt has been of crucial importance for some peoples in the past, and continues to be economically favorable today. The extraction of water makes it possible for many to live where it would otherwise be impossible.

14.1 Salt

Consider the Maya. The Classical Maya civilization flourished some 1200 years ago in Central America in what is now Guatemala and Belize, and the adjacent parts of Mexico and Honduras. The evocative remains of their great cities such as Copán and Tikal, their extraordinary stone carvings and other objects, their elaborate and bizarre system of writing, and their detailed and accurate calendar with its implied care in astronomical observations have caught the imagination of many people since the first reports by John Stevens, who discovered Copán and other cities in 1841. This remarkable civilization existed in a hot and humid tropical environment, a situation that imposes certain constraints. With an estimated population of 5 million and little wild animal food available, the Mayans turned much of what now looks like unbroken tropical forest into agricultural fields. With no draft animals, they must have worked hard to tend these fields. The population of Tikal (near the middle of the region) is estimated to have been at least 45 000, a population not reached by the city of London, England, until centuries later.

People living and working outdoors, in a climate characteristic of the region around Tikal, require much evaporative cooling and therefore have a relatively high need for salt. The average salt requirement for a population in this environment is about 8 g per person per day, so the people in Tikal alone required about 130 tons per year, and the Mayan population of 5 million must have consumed in total about 15 000 tons per year.

Vegetable foods alone do not provide the necessary salt, and these people had only minimal supplies of meat. There are no deposits of rock salt in central Yucatán and no other source of salt. There are a number of salt springs in the southern highlands, but these are quite inadequate to supply the necessary tonnage.

This information alone is an entirely sufficient base from which to draw certain conclusions about the technology, travel, and commerce of that time. That civilization could not have existed without an adequate supply of salt. The only sufficient source of salt is the ocean, and the Maya must have had at least a primitive knowledge of how to harvest it. Lacking a knowledge of chemistry, they most likely worked according to a number of empirical rules which led to the harvest of adequately pure sodium chloride and the exclusion of most of the other substances in seawater. The city of Tikal, far from the ocean, may have received about 100 tons of salt per year, carried at least part of the way there on human backs. The transport of possibly more than 10 000 tons of salt annually throughout the region implies well-developed trade routes and the exchange of other materials as well. Furthermore, the absolute necessity of getting salt into the interior suggests that political power probably accrued to those who controlled the salt works or the trade routes. Even if there were no direct evidence for the extraction of salt from the sea, we could be sure that those people knew and practiced the technology to harvest salt from the sea and that it must have taken place on a large scale. In fact, the remains of many ancient salt works, some still in use today, have been found along the coast of Yucatán (Andrews 1983, McKillop 2002).

Today, we take the availability of salt for granted, but the necessity of salt in the diet has influenced human settlement and political structures, initiated patterns of travel and commerce, created great wealth, and caused wars (Bloch 1963). People living in cool climates with an abundance of meat may get by with little salt, but those living in a hot climate whose diet contains mostly vegetable matter must have a significant supply. The importance of salt in the ancient world is reflected by our use of the word "salary," from the Latin *salarium argentum*, which means "silver salt" (Roman soldiers were sometimes paid part of their salary in silver and part in salt, so valuable was this product). Perhaps the first record of the importance of salt in trade and commerce is a directive issued 4400 years ago that Shandong Province on the coast of China was to provide salt to the court of Emperor Yu, presumably from the sea (Braitsch 1971).

Evidently, the technology to get salt from salt deposits, salt lakes, and from the sea has been widely practiced for a very long time. Many deposits of rock salt are found throughout the world, and their exploitation by mining the salt can yield a product that is directly usable, but getting salt for human consumption from the sea requires some technological expertise, which was evidently widespread. Throughout history, the most important chemical substance extracted from the sea has been common salt.

14.2 Evaporation of seawater

When seawater becomes isolated, either in a lagoon or a laboratory beaker, and is allowed to evaporate, nothing much happens until about three-quarters of the water is lost (raising the salinity to about 140‰). At this point or earlier, depending on conditions, a small precipitate of calcium carbonate may be observed. As evaporation continues another larger precipitate begins to form, and grows until the volume of liquid remaining is about 10 or 12% of the original volume (Figure 14.1). This second precipitate consists of nearly pure gypsum, $CaSO_4 \cdot 2H_2O$.

As the solution continues to evaporate, a different type of crystal appears and settles on top of the gypsum. These crystals are sodium chloride NaCl, also known as *halite*. The mass of crystals grows rapidly with further evaporation, and this layer is nearly all NaCl with only small contributions of additional gypsum and some other salts. Sodium chloride continues to precipitate while the solution shrinks, until, when the remaining solution has a volume of only 30 or 40 mL from an original liter of solution, about 21 g of NaCl has crystallized on top of about 0.1 g of $CaCO_3$ and 1.7 g of gypsum. Along with the last crystals of NaCl are traces of $MgSO_4$, $MgCl_2$, and NaBr. The remaining approximately 30 mL is a quite concentrated solution containing ions of magnesium, sodium, potassium, sulfate, chloride, and bromide. The high concentration of magnesium ion causes the solution to taste quite bitter, and the common name for this residual liquid is *bitterns*. As can be inferred from the legend to Figure 14.1, the literature does not contain much information on the early stages of the evaporation process; most investigative effort has been devoted to the later stages of crystallization (e.g. Harvie *et al.* 1980).

The behavior of salts during the evaporation of seawater lends itself to the rather convenient separation of NaCl, the major salt component. If the solution is allowed to evaporate until nearly all the gypsum is crystallized and settled out, then transferred to another container and allowed to evaporate until most of NaCl is crystallized, and then the liquid drained away, the crystals in the second container will be almost pure sodium chloride. After a quick single washing with water to remove the surface film of brine and a little crystallized magnesium salts, the remaining NaCl is about 99% pure. The procedure (or some less-well-controlled variation of it) has been used in one way or another for thousands of years to make some of the salt used by humans.

Nowadays, the practice is to pump seawater (or allow it to flow in at high tide) into large flat and shallow ponds up to several hectares in size and allow it to evaporate. After the initial deposit of gypsum, the water is allowed to flow or is pumped into the next pan where the NaCl crystallizes out. Most commonly the bitterns (as the remaining brine is called) are drained off to the sea, but in some cases this solution is further processed for other minerals. From the air, the evaporation ponds, or *salterns*, as they are called, present an interesting patchwork of colors. The ponds where the first evaporation takes place may be quite green, while ponds with a higher concentration of brine turn a sort of rusty red. The green ponds are places where

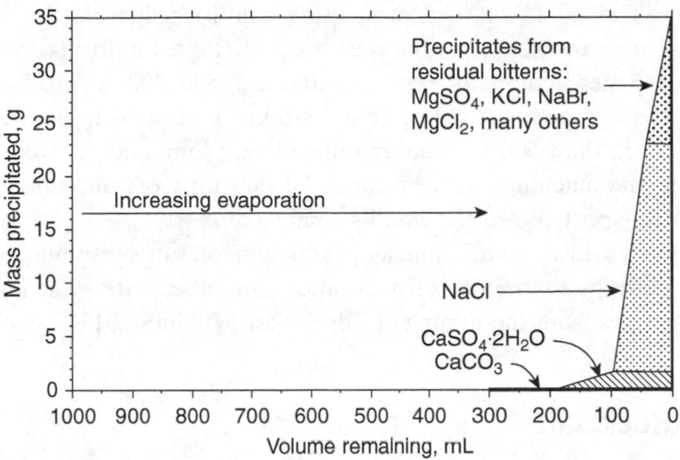

Figure 14.1 Sketch of the sequence of events as 1 L of seawater is allowed to evaporate and to precipitate a series of solid crystalline deposits, of which only the first three are explicitly shown. The classic series of experiments to demonstrate what happens was carried out by Usiglio (1849), who evaporated 5-L batches of Mediterranean water, and analyzed the supernatant and crystalline deposits at intervals. These results suffer from inherent analytical problems of that time, and also somewhat from the evaporation procedures used, but appear never to have been repeated. Herrmann *et al.* (1973) analyzed the supernatant water and some of the crystalline deposits from a series of brine evaporation pans at a commercial salt works in what is now Slovenia, and obtained results that differed in some details from those of Usiglio. The latter observations suffered from analytical imprecision and from a lack of control on the history of each sample. The solubilities of many of the salts are quite temperature-sensitive; Usiglio did his experiments at 40 °C under conditions of unknown but probably very low partial pressures of CO_2, while the pans in Slovenia differed from one another in temperature. The sketch above derives from both sets of observations but is certainly oversimplified. In particular, the point where $CaCO_3$ begins to precipitate is uncertain, as is the point where gypsum ($CaSO_4 \cdot 2H_2O$) deposition begins. Also, the boundaries between gypsum and halite (NaCl) and between halite and the later salts are drawn horizontal, but in reality they must have some slope because there is certainly a little concurrent deposition.

An additional point to consider is that there may be back-reactions between the early deposits and the later more concentrated brines, and the details must depend on how the process is carried out. For example, in a commercial salt works, the brine may be transferred to the next pan after the initial deposition of gypsum is completed, to be replaced with less concentrated brine, so that the deposited gypsum is never in contact with a much more concentrated brine. In a natural evaporation basin, the gypsum may be covered by the deposited halite, again not coming in contact with a more concentrated brine. However, if the circumstances do allow this contact, the gypsum may lose water to the very concentrated brine, being transformed into a new crystalline form called anhydrite ($CaSO_4$). In general, the later stages of the precipitation reactions are exceedingly complex (Borchert 1965; Braitsch 1971; Eugster *et al.* 1980). ■

there is a considerable growth of green, single-celled algae, while in the ponds with a still higher concentration of salt some of the green algae (*Dunaliella salina*) turn red, although most of the red color in the most saline ponds is due to large populations of halophilic archaea and bacteria (Oren and Rodríguez-Valera 2001).

The world's largest production area for salt from the sea is adjacent to Laguna Guerrero Negro on the west coast of Baja California, Mexico. Here, a Mexican–Japanese company produces about 7 500 000 tons of salt annually in immense shallow ponds (totaling about 330 km^2) into which seawater is pumped at a rate of more than 500 tons per minute all year long and allowed to evaporate. Every day, giant machines harvest about 20 000 tons of salt; this is loaded onto ships and transported into the world's commerce.

Worldwide, the annual production of salt is about 290×10^6 tons, of which perhaps 20 to 30% is from evaporation of seawater. The rest of the salt in commerce comes from the mining of salt deposits (USGS 2012).

14.3 Rock salt

The history of the salt deposits in the geological record has been a focus of considerable geological interest. How were they formed? Consider Figure 14.2a. If a lagoon or even a sizeable branch of the ocean is partially isolated by the presence of a bar or sill, and if it lies in a geographic region where the rate of evaporation from the water surface exceeds the rate of rainfall and river input, the water inside will get progressively saltier with time. If the sill is not too shallow, some interchange continues between the lagoon and the ocean. The salty dense water flows out over the sill and lighter, fresher water flows in over top of the outflow. This is precisely the situation for the Mediterranean and Red seas today. The balance of inflow, outflow, and evaporation control the salinity of the lagoon or seas. Both the Mediterranean and Red seas are saltier than the ocean by 2 to 4 salinity units. One can easily imagine that this difference could be much greater for only a small change in the physical arrangements. Consider Figure 14.2b. Here the sill is shallow, no return flow occurs, and the water level of the enclosed area drops. If the sill is closed completely, the seawater will just evaporate and deposit a series of salts, as described earlier. If seawater continues to run in at some rate, the precipitation sequence could be quite complex, depending on details and variations in rates of evaporation and input of fresh and salt water. In Figure 14.2c an arrangement is shown in which it even seems possible that the water could deposit almost pure gypsum in the first basin, NaCl in the second, and the remainder in the third. In addition, there is the possibility that saline brines may percolate down into the sediments, with several possible consequences. The salt deposits known in the geological record are sufficiently varied and complex that perhaps every conceivable variation in circumstance is recorded somewhere.

Many of the known salt deposits are spectacularly thick, giving evidence of repeated inundation and evaporation, and some cover enormous areas. One of the classic deposits, in the sense that it has been commercially important for many centuries, and has been extensively studied, is at Strassfurt in Germany. The deposit is late Permian in age ($\sim 225 \times 10^6$ years old) and is sometimes called the Zechstein deposit, after the European name for part of the late Permian. The total thickness may amount to 1200 m of salt, but in any one area the thickness is not more than 600 m. Even 600 m of salt would require that about 35 000 m of seawater be evaporated. Nowhere is the

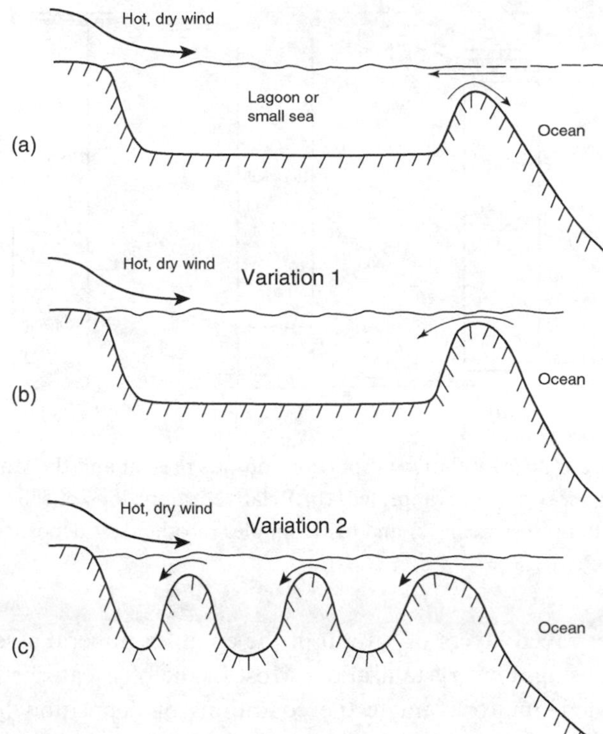

Figure 14.2 Sketches of various situations leading to evaporative concentration of seawater. (a) The sill is not shallow enough to prevent salty water from leaving. The resulting salinity depends on the rate of net evaporative water loss and the degree of interchange allowed. This situation applies to the Mediterranean Sea and the Red Sea.
(b) The sill is too shallow to allow exchange; the water inside is trapped and any entering water only flows in one way, continuously or intermittently (at high tide or because of storm surges) and continues to evaporate so that eventually salt is precipitated on the bottom. Many salt deposits must have formed in situations similar to this. (c) Different degrees of exchange between small basins can be imagined, leading to considerable variation in the nature of the salt deposits. Many other complications can easily be imagined. ■

sea that deep. Therefore it is clear that the deposit must have resulted from repeated flooding and evaporation of a shallow sea. That sea must have covered considerable territory, for extensions of the deposit are present from Poland to England.

Indeed, one of the most famous salt mines in Europe is the Wieliczka mine near Krakow in Poland. This mine was started in the eleventh century; the salt was traded locally and by ship at least as far away as England. The trade yielded such wealth that the king at the time endowed one of the first universities in Europe, the Jagelonian University in Krakow, at which Copernicus later taught. It is a remarkable salt mine, with passages, rooms, and even a whole church carved into the layers of salt, some 200 m or more underground.

Many of the layers of rock salt in the Strassfurt deposit display annual varving, and hundreds of annual layers may be recognized. In some places there are

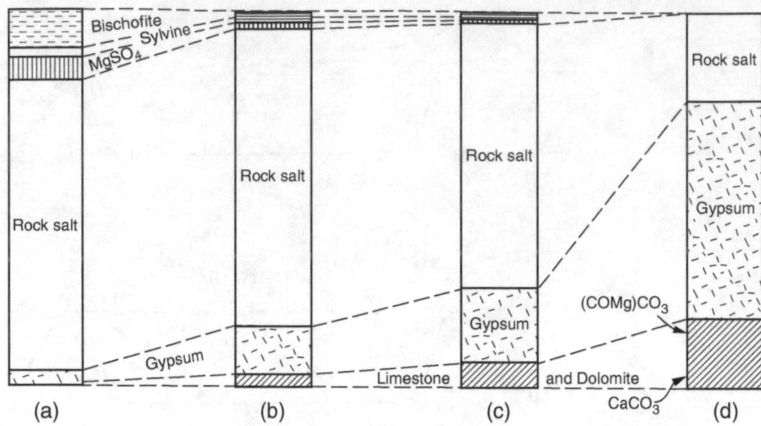

Figure 14.3 (a) Relative proportions of salts present and the sequence in which they precipitate when seawater is evaporated. (b) Relative proportions of salts in an idealized section of the German Zechstein. (c) and (d) Examples of other salt deposits with a marine origin. From Borchert (1965).

interleaved layers of several of the complex mineral assemblages resulting from the later stages of crystallization. Most of the physical–chemical studies that have been carried out to evaluate the conditions of deposition have focused on these later stages, where many minerals of magnesium, potassium, sulfate, and chloride precipitate along with sodium- and bromide-containing minerals. Many dozens of such minerals have been identified (Borchert 1965; Braitsch 1971), their varying occurrence depending on local conditions of deposition and also on post-depositional alteration due to the combined effects of temperature and movement of groundwater around and through the deposits. Various other salt deposits also show great variation in the relative amounts of even the major minerals (Figure 14.3). In those cases where there is more gypsum than halite in a deposit, it is clear that the shallow sea or lagoon where this formed must have had intermittent open communication with the ocean, so that concentrated brine from which not all the NaCl had precipitated could drain away or be diluted and flushed out with new seawater.

Despite the variety of mineral phases in different places, certain major features of large salt deposits remain well established. In every case where a favorable preservation exists, the deepest (first-deposited) layers contain $CaCO_3$, the next above is gypsum, and above that is halite (NaCl). This sequence, observed since the Cambrian, provides evidence that the ratios of the major ions in seawater have been roughly the same throughout the last 600 million years.

14.4 Magnesium

Another substance obtained from seawater is magnesium. The world production of magnesium (as the metal or in various chemical combinations) in 2011 was close to 7 million tons, of which only about 5 to 6% was obtained from seawater (USGS 2012).

The proportion produced from seawater used to be much greater, but it turns out that now it is often cheaper to produce magnesium from various terrestrial brines or from certain minerals.

The process for extracting magnesium from seawater is as follows (Shigley 1951).

1. Oyster shell, or some other relatively clean source of $CaCO_3$, is heated (*calcined*) in large (90-m long) rotary kilns to make CaO, or *quicklime*:

$$CaCO_3 + heat(\sim 1300°C) \rightarrow CaO + CO_2\uparrow. \quad (14.1)$$

2. The calcium oxide is mixed with water to make calcium hydroxide, or *slaked lime*, which is the cheapest alkali available:

$$CaO + H_2O \rightarrow Ca(OH)_2. \quad (14.2)$$

3. The slaked lime is mixed with seawater in large vats, raising the pH to about 11. This causes magnesium hydroxide to precipitate:

$$Ca(OH)_2 + Mg^{2+} \rightarrow Mg(OH)_2 \downarrow + Ca^{2+}. \quad (14.3)$$

4. The white precipitate of magnesium hydroxide, called *milk of magnesia*, is allowed to settle, then filtered and washed. Magnesium hydroxide may be used commercially in this form.
5. Most magnesium hydroxide is calcined or heated to convert it to magnesium oxide:

$$Mg(OH)_2 + heat \rightarrow MgO + H_2O\uparrow \quad (14.4)$$

   Large amounts of magnesium oxide are used in the steel industry. There are many other industrial uses for magnesium oxide.
6. In order to make magnesium metal, the $Mg(OH)_2$ is dissolved with HCl to make a solution of $MgCl_2$. The $MgCl_2$ is then concentrated to dryness and introduced into electrolytic cells where the magnesium metal is obtained by electrolysis. Magnesium metal is an important component in light-weight alloys.

There are still several companies in the USA, China, Brazil, Ireland, and other countries that each year produce several hundred thousand tons of magnesium compounds from seawater.

14.5 Bromine

The total global production of bromine is about 660 000 tons per year (USGS 2012). Two decades ago, most of the U S production came from a Dow Chemical plant in Freeport, Texas, where bromine was extracted from seawater. There is about 67 mg of bromine in each kilogram of seawater. This can be extracted as follows:

1. Seawater is made acid with sulfuric acid, and chlorine gas is bubbled in. Under acid conditions the following reaction takes place:

$$2HBr + Cl_2 \rightarrow 2HCl + Br_2. \quad (14.5)$$

The seawater solution containing the volatile bromine gas is allowed to cascade down through big columns with baffles while air is pumped upwards through the columns, stripping the bromine from the water.

2. In turn, the bromine in the air can be captured by passing it through a solution of SO_2, which reduces the bromine to bromide:

$$Br_2 + 2H_2O + SO_2 \rightarrow 2HBr + H_2SO_4. \quad (14.6)$$

The bromine may enter commerce as Br_2 or HBr. It is used in production of flame-retardant plastics, pesticides, high-density drilling fluids, and other chemicals.

At one time, the production of bromine in Texas required pumping more than one cubic kilometer of seawater from the Gulf of Mexico each year. The acid effluent from the bromine extraction and the alkaline effluent from magnesium extraction, also done at that plant, were combined and discharged back into the Gulf of Mexico. This facility closed in 1998. In the USA, it has been found more economical to obtain bromine from certain salt deposits in Arkansas that are rich in the residual products from the ancient crystallization of salt from seawater. Wells are drilled, water is pumped down, and the brine that is formed, which contains a high concentration of bromine (about 5000 mg kg^{-1}) as well as other salts, is pumped up and the bromine extracted as described above. In other parts of the world, some of the bromine produced comes from extracting the bromide-rich bitterns left after the production of common salt from seawater, or potassium salts from Dead Sea water.

14.6 Gold

The lure of gold, that has so dramatically affected human history generally, did not leave marine scientists untouched. Much excitement was generated in 1872 when it was announced (Sonstadt 1872) that the waters of the English Channel contained 65 mg of gold in each ton of water. The immense popular interest at the time was seldom tempered by the realization that the value of the gold in one ton of water was, at prevailing prices, only about 6.5 cents.

Later, Svante Arrhenius (in about 1900) reported that the concentration was only 6 mg per ton, then worth about 0.6 cents (Riley 1965). While not much in a ton, there were speculations of possible immense wealth in the sea, because at 6 mg per ton the amount of gold in the world ocean was estimated to be enough to make every living person on Earth a millionaire, with nearly enough left over to retroactively make every person who had ever lived in the past a millionaire.

During the first four decades of the twentieth century, there were at least 50 patents issued in various countries for methods to extract gold from seawater. Fraudulent schemes abounded, and stock companies were organized to extract money from the pockets of credulous investors (Riley 1965). In 1920, Fritz Haber (who had won the Nobel prize for his process to fix atmospheric nitrogen) planned to help pay off the German World War I debt with gold extracted from the sea. He developed newly accurate analytical procedures and set off to find the best location, where the concentration was the greatest. This interest was one stimulus for the great German *Meteor* expedition, focused mostly on physical oceanography, which set a new standard for competence and thoroughness in marine investigations. Haber's analyses showed that gold was present in seawater at a concentration of only 0.004 mg per ton (worth then about 0.004 cents). At this point, no gold had ever been extracted and it did not seem economical to attempt it.

During the 1950s, Dow Chemical Company, which then processed immense quantities of seawater in order to extract magnesium and bromine, also investigated the extraction of gold. They processed 15 tons of seawater and collected in pure form exactly 0.09 mg of gold, which was then carefully shaped into the form of a small cross for display under a magnifying glass. At that time, the value of this gold was about one-hundredth of a cent. To date, this is the bulk of all the gold ever extracted from seawater (Mero 1965), and at the time it was estimated that the effort had cost Dow Chemical Co. about $50 000.

Since Haber, a number of investigators have published a variety of estimates of the concentration of gold in seawater, but up until 30 years ago (e.g. Bruland 1983) no estimate was thought better than Haber's value of 0.004 mg per ton ($\approx$ 20 pmol per kilogram). Then, Kelly Falkner, a student of John Edmond at MIT, devoted a Ph.D. thesis to the study of gold in seawater. Falkner and Edmond (1990) showed that the true concentration of gold in the ocean is about 50 fmol kg^{-1} in both the Atlantic and Pacific, with values sometimes found up to three times greater, and in the Mediterranean the concentrations may be up to 200 fmol kg^{-1}. The average value of 50 fmol kg^{-1} translates to about 10^{-5} mg per ton, two orders of magnitude less than was accepted only a few years earlier. This works out to about 14 000 tons in the global ocean. In 2010, the total gold held by all the world's central banks was about 35 000 tons.

The gold extracted by Dow Chemical Company was about 600 times more than should have been there (assuming the near-shore water they used was typical of open-ocean water with respect to its gold content). It is probable that the reagents and containers used provided the bulk of the gold collected.

14.7 Water

Most of the mass of seawater is water, and water comprises most of the substance extracted from seawater. Thousands of plants worldwide produce in total about 60×10^6 m^3 of water every day (60×10^6 t per day). More plants are being constructed, and the rate is increasing. The cost of this water varies with the source of energy; in the USA, fresh water is produced from seawater for about $0.60 to $0.80 per cubic meter.

The first procedure for obtaining fresh water from the sea involved simply distilling the water and condensing the resulting vapor. Modern plants are the result of considerable engineering development, involving multistage evaporators and condensers that minimize the amount of fuel used. Originally all plants used distillation processes.

The development of semipermeable membranes and the associated support structures, pumps, and controls made it possible to use reverse osmosis to force the fresh water out of seawater. We learned earlier (Chapter 4) that a one molar solution can (neglecting activity coefficient effects) exhibit an osmotic pressure of 24.6 atmospheres, or 2493 kPa. Seawater at a salinity of 35‰ has about 1.15 moles of ions in each liter so it should exhibit an osmotic pressure (again neglecting activity coefficients) of about 2867 kPa, or about 28 atmospheres. If the seawater is contained inside a chamber with walls made of a semipermeable membrane that allows water but not salt to pass through, and a pressure greater than 2867 kPa is applied, the water will be forced out through the membrane, leaving the salt behind. Here the energy requirement is supplied by the electricity to run the pumps. It turns out that this procedure uses somewhat less energy than the distilling processes. The majority of new plants to make fresh water from seawater now use reverse osmosis. Nearly all such plants in the USA use reverse osmosis, and more than half worldwide (NRC 2008).

Summary

Despite the enormous amounts of many substances dissolved in the ocean, it is not at present a very good or economic source for any of them except salt. The largest solid product, by weight, extracted from sea water is common salt, NaCl. In the past, significant tonnages of magnesium and bromine were obtained, and there are still a few places where relatively small amounts of these are still extracted. It appears uneconomic to harvest any other of the many dissolved components. In places with an inadequate supply of fresh water, the production of fresh water from seawater is now a large-scale industrial process, of growing importance.

SUGGESTIONS FOR FURTHER READING

I have not found any single source that covers most of the material in this chapter. All the information is drawn from the references listed, which should be consulted for additional perspective.

The production of fresh water from seawater is covered very well up to the time of publication in the following volume, which, despite its title, provides good information on the global situation:

National Research Council. 2008. *Desalination, a National Perspective*. National Academies Press, Washington, DC.

15 Geochemical history of the oceans

But the whole vital processes of the earth take place so gradually and in periods of time which are so immense compared with the length of our life, that these changes are not observed, and before their course can be recorded from beginning to end whole nations perish and are destroyed.

ARISTOTLE ~320 BCE

There is much evidence that the age of Earth is $\sim 4.6 \times 10^9$ years. Our knowledge of events during the first half-billion years of Earth's history is fragmentary at best. No rocks have yet been found that are older than 4.0×10^9 years. One important piece of evidence is that the concentration of volatile elements in the atmosphere and Earth's crust is very low compared to expectation from Solar System abundances (Holland 1984). These elements appear to have been lost because of some catastrophe, or series of catastrophes, that occurred during the later stages of the formation of Earth. The nature of such catastrophes is not known, but there are several candidates. The massive collision that appears to have led to the formation of the Moon could have stripped most of the atmosphere from early Earth. Beyond evidence from the compositions of the atmosphere, the solid Earth, the Moon, and many meteorites, the history of the first half-billion years is largely a matter of speculation, model building, and calculation. After that, increasingly abundant sedimentary remains bear evidence, although often unclear, about conditions when they were deposited. In the following discussion, I have assumed that Earth had reached essentially its present size by 4.0×10^9 years ago, and the calculations take this time as a beginning. A few sedimentary rocks not much younger than that have survived, giving us a faint glimmer of the conditions prevalent at that time; those rocks suggest the presence of both liquid water and a sub-aerial source for the sediments (Holland 2006).

15.1 Illustrative rates

Some calculations may be helpful to set in perspective the later history of the oceans and to obtain a simple appreciation of the magnitudes of various processes involved.

Flux of water required to fill the ocean once. The constant flux of water required to fill the present ocean basin once during Earth's 4×10^9 year history is

$$\frac{1.35 \times 10^{18}\ \text{m}^3}{4 \times 10^9\ \text{yr}} = 338 \times 10^6\ \text{m}^3\ \text{yr}^{-1}.$$

This impressive rate, amounting to about 11 tons per second throughout all geological history, provides some sense of the vastness of the ocean using units that are accessible on a human scale. The source of this water is still unclear and is discussed later.

River input. Given the total volume of the ocean at 1.35×10^{18} m^3 and the annual discharge of present rivers at about 3.74×10^{13} m^3 yr^{-1} (Baumgartner and Reichel 1975), the present rivers discharge the present volume of the oceans in

$$\frac{1.35 \times 10^{18}\ \text{m}^3}{3.74 \times 10^{13}\ \text{m}^3\ \text{yr}^{-1}} = 36\ 000\ \text{yr}.$$

Assuming present rates, during the last 4×10^9 years the total volume of the ocean water has been recycled through rivers about

$$\frac{4 \times 10^9\ \text{yr}}{36 \times 10^3\ \text{yr}} = 111\,000\ \text{times}.$$

Transport of sediment to the ocean. Milliman and Syvitski (1992) estimated that the particulate matter transported to the oceans by rivers each year amounts to 20×10^9 t. If we assign a density of 2.5 t m^{-3} in sediment made up of this material, the volume transported each year is

$$\frac{20 \times 10^9\ \text{t}}{2.5\ \text{t m}^{-3}} = 8 \times 10^9\ \text{m}^3\ \text{yr}^{-1}.$$

This sediment would fill up the total volume of the ocean basins in

$$\frac{1.35 \times 10^{18}\ \text{m}^3}{8 \times 10^9\ \text{m}^3\ \text{yr}^{-1}} = 1.69 \times 10^8\ \text{yr}$$

or 169 million years. At present rates, the total volume of the ocean basins could have been filled with the sediments eroded off the continents some

$$\frac{4 \times 10^9}{169 \times 10^6} = 24\ \text{times}$$

during its history. Present rates of sediment transport to the ocean may be as much as twice the estimated rates before humans began modifying Earth's environment (Berner and Berner 1987). The pre-human rates of sediment transport might therefore have filled the present volume of the ocean basins only some 12 times during Earth's history.

Transport of solubles to the ocean. The average concentration of dissolved solids in the world's rivers is about 100 g m^{-3} (Berner and Berner 1987). At a flow of 3.74×10^{13} m^3 yr^{-1} the total transport of dissolved materials is 3.74×10^9 t yr^{-1}. Taking a density for the evaporated solids of 2.2 t m^{-3}, the volume of dissolved substance transported annually is 1.7×10^9 m^3, which would fill the ocean basins in

$$\frac{1.35 \times 10^{18}\ \text{m}^3}{1.7 \times 10^9\ \text{m}^3\ \text{yr}^{-1}} \approx 8 \times 10^8\ \text{yr}.$$

Before the middle of the last century, it was not possible to formulate an adequate explanation for the recycling of substances entering the sea and textbooks generally avoided such calculations. Beginning in the 1950s and 1960s, the remarkable and revolutionary development of our knowledge about the dynamic behavior of Earth's crust has given us the necessary general understanding of those active processes that now renew the ocean basins, and these processes provide an essential framework within which to consider the history of seawater. Since the 1960s it has been recognized that Earth's crust consists of large plates that move relative to one another, exposing fresh basaltic material at regions where they separate, while downwelling one beneath the other and also throwing up mountain ranges at regions where they come together. Only since that revolution in geological understanding have we had plausible mechanisms to explain the recycling of both sediments and water. This comprehension of *plate tectonics* does not, however, provide us with answers to a number of questions we can ask in considering the geochemical history of the oceans.

- Has the volume of the oceans changed over time, and in what direction?
- Has the total mass of salt changed over time?
- What is the history of the composition of seawater?

These questions are discussed in the following sections.

15.2 Early history of the ocean volume

This subject cannot be discussed properly without reference to the way in which Earth condensed out of the primordial gas and dust cloud surrounding the Sun. Early theories of the condensation process supported either of two alternatives. On the one hand was the theory of a hot beginning: Earth condensed out of an extremely hot gas cloud as a ball of molten rock and subsequently cooled. On the other hand the theory of a cold beginning was advanced: Earth formed from cold lumps of material condensed in the frigid temperatures of interplanetary space and subsequently heated by the release of gravitational energy, radioactive decay, and thermal radiation from a newly-forming Sun. Neither simple theory alone accounted for all the known facts.

The history of the ocean would be very different in the two cases outlined above. We may imagine that in the case of a hot beginning, the volatiles, including water, would all be in an atmosphere surrounding the ball of molten rock. As the rocks cooled, the oceans would condense as liquid water. Subsequent weathering of the cooled rocks would absorb water, not only by the formation of crystal hydrates more stable at low temperatures, but through the occlusion of water in sediments. Thus, the volume of the oceans would tend to decrease through time. In the case of a cold beginning, much of the water would be trapped in the interior of Earth, to be released as part of the out-gassed volatiles as Earth heated up. Thus the volume of water in the oceans would increase with time and might well be increasing still.

As information from meteorites and Moon rocks has provided new insights into the history of these bodies, and as ever more detailed theory has developed concerning the nature of the space environment and the interactions of clouds of objects in various gravitational orbits, there has grown a more assured understanding of Earth's early history. Nevertheless, there remain significant uncertainties about the details of this history. Some of these uncertainties will perhaps always remain, if only because of the occasionally dominating importance of singular events, such as the collision of Earth with large planetesimals during and subsequent to the initial accretion process. It is thought that this initial accretion, during the initial condensation within the solar nebula, took perhaps 100 million years, followed by the slower growth to roughly its present size. The earliest rocks known from Earth's surface were formed about 4.0×10^9 years ago; these mark the beginning of the earliest (Archaean) eon in Earth's history (Figure 15.1).

Any theory of the early history of Earth must take into account numerous kinds of information, among which the following are noted.

- Meteorites come in a considerable range of compositions, and some bear evidence of complex diagenetic histories before impact with Earth. Evidence of considerable differences in isotopic composition suggest either that the solar nebula was inhomogeneous with respect to the origin of the elements, and perhaps that the substance of the meteorites, or particles within the meteorites, or even the meteorites themselves, did not all come into the Solar System at the same time. The Earth appears to have been formed from a diverse assemblage of particles, of which not all types are currently present among the known meteorites.
- The deeper convecting layers of Earth's mantle are even now not completely homogeneous in composition.
- The discovery of high levels of ^{3}He in the deep Pacific (Clark *et al.* 1969) proved that Earth is still out-gassing. This is because there is no quantitatively significant source of ^{3}He on Earth except from the primordial material forming the early Solar System.
- The shortage of certain heavy inert gases in Earth's atmosphere, compared to their abundance in the Sun, is evidence that at some early stage Earth lost most or all of its atmosphere.
- The very existence of the Moon, its dynamical relationship to Earth, and its composition all provide important evidence. This evidence is best explained by the hypothesis that the Moon formed from debris thrown off during a massive collision with a protoplanet of about the size of Mars, after the Earth had reached close to its present size. Most of the Moon substance came from the protoplanet (Canup and Asphaug 2001).

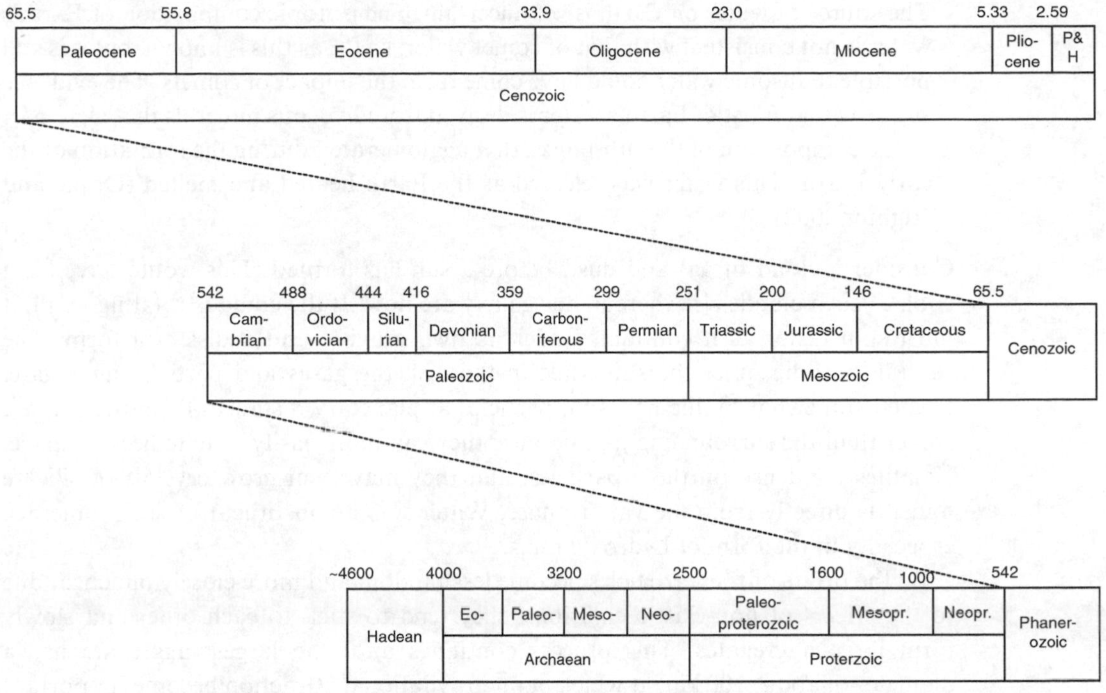

Figure 15.1 Simplified representation of the history of Earth, showing the major named divisions of the geologic record, along with the ages (in millions of years ago) of their boundaries. The bottom bar shows the whole time span, beginning at about 4600 million years ago when it is believed the Earth began forming by the agglomeration of solid materials in orbit around the condensing Sun. Subsequent bars are each expanded by a factor of 10. The earliest rocks so far found are about 4000 million years old. All the early boundaries are arbitrary time divisions, until the beginning of the Phanerozoic which marks the earliest appearance of certain fossils visible to the unaided eye (other visible fossils do exist in the late Proterozoic). Subsequent boundaries are all based on the fossil record. The last unit on the top bar began 2.59 million years ago and is compossed of the Pleistocene epoch and the Holocene epoch. The latter time period, the Holocene, began about 12 000 years ago. The numerical values and boundaries are derived from the International Stratigraphic Chart (ICS 2009).

- Calculations of the orbital dynamics and statistics of accreting bodies in a rotating disc around a protosun suggest that the initial body would be cold, but that it would later become hot on the surface, as its gravitational mass grew by accretion and the energy of impacts correspondingly increased. The subsequent warming and reworking of the interior provide interesting physical problems (Stevenson 1983).
- It is widely agreed that the initial accretion of Earth and its bombardment by debris at rates sufficient to significantly increase its mass must have come to an end some time before 4×10^9 years ago. Nevertheless, some bombardment continued at a decreasing rate and continues today; the annual increment of mass is still about 50 000 tons per year, but most of this is dust (Love and Brownlee 1993). The evidence clearly shows (Shoemaker 1983) that there have been numerous impacts of large asteroid- or comet-sized objects during the last 4×10^9 years.

- The source of water on Earth is yet uncertain. The isotopic composition of Earth's water is not consistent with that of comet water, so far as this is known, but it is still possible that some water could have come from the impact of comets. The evidence suggests that most of Earth's water is derived from hydrous minerals that were part of the composition of the substances that agglomerated during the formation of the early Earth. This water was released as the Earth heated and melted (Drake and Righter 2002).

Consider a cloud of gas and dust, before a sun has formed. This would have been cool, as such clouds elsewhere in our galaxy are now. If the cloud is rotating at all, it must spin faster as it contracts under its own gravity, and a disk will form. The central mass becomes the sun while in the disk the gases and particles aggregate. Heated somewhat by the new sun, particles at planetary distances from the sun are colder than the surrounding gas, because they can more easily radiate heat to space. Volatiles condense on these particles, and they may even grow crystals of silicate minerals directly from the vapor phase. Water was a constituent of such minerals, especially in the form of hydroxyl ion.

As the orbits of these particles become less random and more closely bunched, due to the effects of non-elastic collisions, they tend to stick to each other and slowly form large aggregates. This process continues until the largest aggregate has a diameter of about 100 km, at which point gravitational attraction becomes important and more particles are rapidly attracted. A rapid increase in size will then occur until all particles in the immediate neighborhood of the protoplanet are swept up. Because of the small size of the protoplanet, little gravitational energy is released on impact during this initial phase, and the object might reach the size of the present Moon while still remaining fairly cold with volatiles trapped inside. It is estimated that this process might take about 30 000 000 years, after which growth continues at a much slower rate as particles in orbits some distance away gradually diffuse into Earth's orbit and are swept up.

Analysis of orbital dynamics suggests that additional particles and aggregates would become available slowly enough that the total accumulation of essentially the present mass of Earth took perhaps 300×10^6 to 400×10^6 years. The surface temperature had time to cool down after each impact, and the volatiles accumulated. As the mass grew, the gravitational attraction served to hold an increasing fraction of the volatiles. These volatiles were released from the incoming material during the heating and melting that occurred at the initial impact. This impact would release enough energy to melt a significant proportion of the incoming mass. The water in this released material probably entered the atmosphere and condensed shortly thereafter (Arrhenius and Alfven 1974).

The very low atmospheric abundance of several of the inert gases, such as neon, relative to the proportions present in the original solar nebula (judging from the abundance present in the Sun), however, is evidence that at some point that cannot have been too early in the condensation process Earth's atmosphere was swept largely free of volatile matter (Walker 1977). This loss was once thought to have occurred because of a flare-up of heat from the Sun, but a massive collision, such as

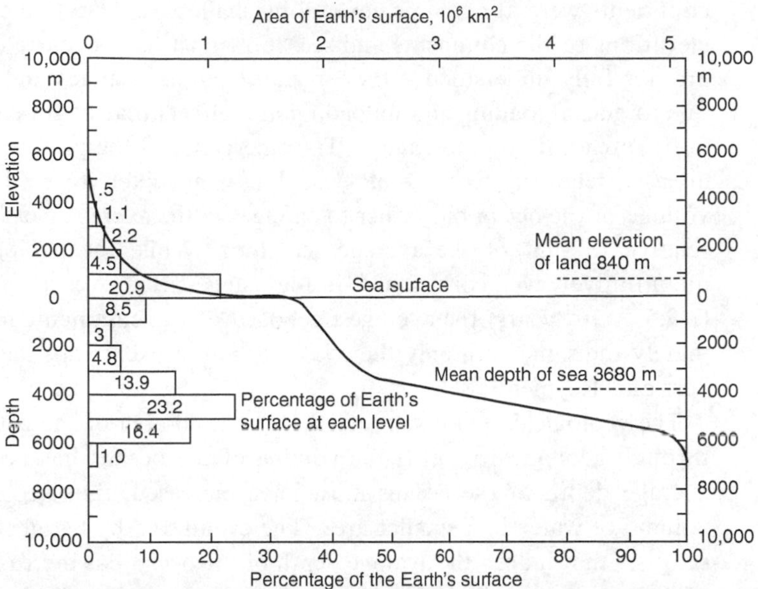

Figure 15.2 Hypsographic curve showing the area of Earth's solid surface above any given level of elevation or depth. At the left of the figure is a frequency histogram giving the distribution of elevations and depths for 1000-m intervals. (Redrawn from Kennett 1982.) Mean depth of the sea from Charette and Smith (2010).

that which led to the formation of the Moon, now seems more likely to have been the major cause. In any case, the present atmospheric components must have then subsequently accumulated by some combination of out-gassing from Earth's interior and input by meteorites and comets. Water is one of these components.

It seems likely, but by no means certain, that a large part of the present ocean volume had accumulated at the end of this last build-up period, about 4×10^9 years ago. Some out-gassing of volatile materials such as water may be continuing today because ^{3}He is certainly out-gassing, but the flux of water is likely to be small, and we have no certain means of detecting it. Neither do we have good estimates of the rate at which water is carried into the mantle by subducting plates, aside from the speculative calculations presented earlier. This water must be carried down both as hydrated mineral phases formed by the reaction of hot basalt with seawater, and as interstitial liquid in that portion of marine sediments that is dragged down by the descending plate. Most subducted water must be returned to the surface through volcanoes resulting from magma generated above the subducting plate. It is unlikely that the volume of water delivered to the deep mantle exceeds 0.1 km^3 per year (Parai and Mukhopadhyay 2012).

There is no direct geological evidence on the question of changes in the volume of the ocean over very long time periods. The hypsographic curve of the present Earth (Figure 15.2) suggests that rather large areas of land would be exposed or inundated by small changes in the volume of the ocean. Indeed, large areas of the present continents were at one time covered by shallow seas, so that changes in the relative elevations of the continents and sea floor must have occurred, but the mechanisms are not fully understood – they include volume changes and isostatic adjustments due to glacial loading and unloading and effects that might ensue if the rate of sea-floor spreading were to change. The emergence of the present low-lying continental areas in relatively recent geological time is not likely to be due to a change in the volume of the ocean but rather to changes in the average elevation of the continents relative to that of the average sea floor. While the geological evidence is not quantitatively well constrained, it does suggest that over a very long period of time ($\sim 2.5 \times 10^9$ years) the average freeboard of the continents has remained approximately the same with only the relatively minor excursions mentioned above (Schubert and Reymer 1985).

The geological record suggests that the total area of the continents has increased through geological time. If the volume of the oceans has been constant, then the average depth of the oceans must have increased, thus accommodating the same volume of water in a smaller area. The evaluation by Schubert and Reymer (1985) suggests that indeed the average depth of the ocean has increased (due to a lessening of the total heat flow) by an amount just sufficient to balance the increase in continental area. If this is true, then the total volume of the ocean may have been approximately constant during much of Earth's history.

Since processes operating during the last several billion years (such as cometary input, out-gassing from the mantle, and subduction into the mantle) appear sufficient to exhaust or double the ocean, while such evidence as there is suggests a nearly constant volume, it is useful to ask if this constant volume is the result of chance compensation, or whether there is some feedback control that helps to maintain a constant volume. One such mechanism was suggested by Kasting and Holm (1992) and Holm (1996). They point out that the changes in density and heat capacity of water as it approaches the critical point at high temperature and pressure in the hydrothermal circulation cells near the fast spreading centers provide a possible feedback mechanism. The cooling efficiency of water in these circulation cells increases dramatically and reaches a maximum as it approaches the critical point. Suppose the ocean volume were less than it is now; the water depth (hydrostatic pressure) over a spreading center would be less that it is now. The water involved in the hydrothermal circulation would be less efficient at extracting heat, the rocks would not be so well cooled, and fracturing of the rock to allow deeper penetration would be less effective. Consequently, the crust would contain less water when eventually it was subducted. Mantle degassing would continue to discharge water into the ocean, and the volume would increase. Eventually, the increased hydrostatic pressure would cause more efficient extraction of heat, the crust would be more completely hydrated, and the subduction of water into the mantle would increase to equal the mantle-degassing rate and any other inputs of water to the system. This mechanism is highly speculative, but it suggests that it is indeed possible that feedback in the system could keep the volume of the ocean approximately constant over time, which appears to be in agreement with the geological evidence.

15.3 Glacially caused changes in ocean volume

At present, the volume of ice present in the Greenland and Antarctic ice caps is estimated to be 28×10^6 km^3 or 2.8×10^{16} m^3. This volume is equal to a layer on the ocean about 78 m thick. The amount of additional glacier ice formed during the ice ages is estimated at about 4.3×10^{16} m^3 (equivalent to a layer on the ocean about 120 m thick). Therefore, during maximum glaciation the total volume of ice would have been about 7.1×10^{16} m^3. This value is about 5.3% of the present volume of the oceans. Since the ice caps and glaciers are essentially fresh water, the average salinity of the ocean would be expected to undergo an excursion of 1.9‰ between complete melting of ice caps and the height of a glacial age. If the above values for the volumes of glacial ice are correct, the salinity of the ocean during maximum glaciation must have increased from its present average value of 34.72 to about 35.8‰, and if all the present ice melted the salinity of the ocean would average about 34.0‰. Glacially induced variations such as these change only the mass of water in the ocean, leaving the mass of salt unchanged.

15.4 Mass of salt in the ocean

The very large opportunity for complete geochemical reprocessing of seawater that is evident from the introductory calculations might suggest that the ocean is more or less in a steady state with respect to its total content of salt. There is some reason to believe, however, that the total mass of salt in the sea could have changed somewhat, even during geologically short periods of time.

Suppose that the Strait of Gibraltar were to become shallower as the result of some rather minor tectonic movements. If the barrier were then only a few meters deep, the shallow surface flow into the basin could continue, but the deep return flow of dense salty water out of the Mediterranean would be blocked. Since the Mediterranean is a region of greater evaporation than precipitation, seawater from the Atlantic would continue to flow into the Mediterranean. The water would leave as a vapor and the salt would remain behind. Nothing except major climate change or further tectonic movement would stop this process: salt would accumulate in the Mediterranean until the entire basin was filled with a solid mass of crystallized salt having a volume of 3.77×10^6 km^3. At a density of 2.2 t m^{-3}, the mass of salt precipitated in the Mediterranean basin would be

$$(3.77 \times 10^6\ \text{km}^3)(10^9\ \text{m}^3\ \text{km}^{-3})(2.2\ \text{t m}^{-3}) = 8.29 \times 10^{15}\ \text{t, or } 8.29 \times 10^{18}\ \text{kg.}$$

The total mass of all salt in the ocean is about

$$(1.35 \times 10^{18}\ \text{m}^3)(35.7\ \text{kg m}^{-3}) = 4.82 \times 10^{19}\ \text{kg, or } 4.82 \times 10^{16}\ \text{t.}$$

Therefore, the Mediterranean, if completely filled with salt, could hold about

$$\frac{8.29 \times 10^{18}}{4.82 \times 10^{19}} \times 100 = 17.2\%\ \text{of all the salt in the ocean.}$$

It is by no means incredible that the equivalent of filling the Mediterranean with dry salt could have happened in the past, since it could so easily happen now. In fact, it can happen remarkably fast. If the net evaporation over precipitation in the Mediterranean is 1 m per year, the entire basin could be solid salt in 95 000 years – barely a blink in geological time. (In reality, such an event would cause additional effects; for example, the extra mass of salt would cause downwarping of the crust and even more salt could be stored there.)

In fact, the Mediterranean has experienced a remarkable event only a little less dramatic than the scenario sketched above (Hsü *et al.* 1973; Ryan 1973). For about 2 million years, centered about 6 million years ago and ending at the very end of the Miocene epoch, the Strait of Gibraltar was apparently closed intermittently, and the Mediterranean Sea dried down to a salt lake that is thought to have been 2000 m or more below sea level. The Strait may have opened several times and the sea poured in like a gigantic waterfall, only to evaporate again. Salt deposits under the present sea floor of the basin reach thicknesses of up to 1800 m and locally contain shallow-water deposits along the sides, now at the bottom of the sea. The total volume of salt precipitated there is somewhat uncertain, but estimates range from about 1×10^6 to about 1.5×10^6 km^3. The conservative estimate, one million cubic kilometers, amounts to 2.2×10^{15} t or about 4.6% of all the salt in the ocean. If it is indeed true, as Ryan (1973) and others believe, that the basin of the Mediterranean was largely empty of water during the maximum drying periods, then the sea level in the rest of the world ocean must have been about 10 m higher that it would have been if the Mediterranean had been full of water.

It has been estimated (Holland 1972) that perhaps one-half of all the NaCl in the crust and oceans resides in various sedimentary deposits of salt. These, while locally sometimes very large, never approach the volume of the entire Mediterranean, but there are a great many such deposits. Simply on statistical grounds, it seems unlikely that all the salt-bearing sediments on Earth could be eroded and the salt released without some precipitation of salt occurring somewhere else. The deposits are sufficiently scattered throughout the world that, on a long-term average, they are probably eroded as fast as they are formed, and the salt content of the oceans has probably remained fairly constant at least over the last several hundred million years. The maximum reasonable excursion of salinity due to trapping of sea salt in or release from evaporite basins does not seem likely to exceed 10 or 20% of its present value.

It is worth noting, however, that the present time does seem to be unusual. Nowhere on Earth today do we see examples of the precipitation of sea salt on the scale known from the past. One of the largest areas now precipitating salt is the Gulf of Kara-Bogaz-Gol, an arm of the Caspian Sea with an area of about 13 000 km^2. The gulf is some 3 meters lower than the Caspian Sea (itself 28 m below sea level) and water flows continuously, at a rate of several cubic kilometers per year, through a narrow passage and over a waterfall into the Gulf, where it evaporates and deposits salt (but this is mostly Na_2SO_4, rather than NaCl). No equivalent situation connected to the ocean is active today, and it appears that the Mediterranean evaporite was the last large deposit to have been formed. Many of the ancient salt deposits continue to be eroded, however, so that there must be a corresponding net increment of salt to the ocean and an increase in salinity. From estimates in Berner and Berner (1987), it can be calculated that to balance the present (pre-industrial) input of chloride from rivers, the ocean would someplace have to be precipitating NaCl at the rate of 274×10^6 t per year. Since this is not happening today, we can assume that the oceanic salinity is currently increasing, but at a rate only fast enough to double the salinity in 160×10^6 years.

Over the much longer span of 4 billion years we do not know very much about whether the oceans might be getting saltier slowly or not, or when the present salinity was established. Weathering reactions continue, and sodium, for example, continues to be released as rocks dissolve or change in mineral character. Ions such as sodium continue to associate with clay minerals and are carried into the sediments in other ways. Chloride, however, is the major ion in seawater, and chloride is not a major constituent of many minerals. It must have been largely expelled (as HCl) from the mantle early in Earth history. It is carried down into the mantle with subducting plates, and returns in volcanic discharges. Whether there is still any net release from the mantle is not known. We do not know whether the global balance reached today is close to that reached during the first billion years of Earth's history. We do not have evidence to show that it is not.

15.5 Composition of sea salt

Some of the same considerations applied previously in evaluating the likelihood that the total salt content of the oceans has remained roughly constant for a very long time also apply to the composition of the salt in the sea. The whole volume of the oceans has been exposed numerous times to intimate contact with both hot and cold crustal rocks, even including at least one passage through a complete cycle of sedimentation, diagenesis of sediments, and re-erosion. From this it seems plausible that the salt would long ago have reached a steady state at approximately its present composition, as least with respect to the broad general features. We are still far from understanding the principal reactions that control the composition of seawater, however, and therefore we are on shaky ground if we try to speculate about changes that might still be occurring, or about geochemical or geophysical processes that could cause changes in the ratios of the major ions.

There are, however, certain limits that can be set on the possible variations in the ionic ratios that could have occurred since the late Proterozoic. As pointed out by Holland (1972, p. 642): "No changes in composition are needed to explain the mineralogy of marine evaporites from late Precambrian to the present time." Conversely, this mineralogy provides information that sets boundaries on the possible variations in the past. Consider, for example, the typical (although not always seen) evaporite sequence:

$NaCl$
$CaSO_4 \cdot 2H_2O$
$CaCO_3$

When marine evaporite is found deposited on some underlying formation, the first (bottom) layer is often $CaCO_3$, the next is gypsum, and above that is halite (NaCl). This sequence corresponds to what happens when seawater is evaporated today: the first crystals to settle out of the concentrating brine are composed of $CaCO_3$, the next of $CaSO_4{\cdot}2H_2O$, and only after a significant amount of these have settled out and the solution is further concentrated does NaCl begin to crystallize and settle out on top of the gypsum.

If we suppose that the concentration of sulfate has stayed constant, then the solubility of gypsum sets an upper limit on the concentration of calcium. A three-fold increase in calcium concentration would cause present normal seawater to exceed the solubility of gypsum. If either the concentration of calcium or of sulfate, or their product, were three-fold greater in the past, then gypsum would be found much more commonly and interspersed in all halite deposits. This is not found in typical marine evaporite sequences, examples of which are found from all geological epochs since the late Proterozoic.

At the lower end of the concentration range, a limit is imposed because gypsum is commonly found precipitated before NaCl. This observation requires that calcium always be more than 3% of its present concentration and probably much closer than that to its present concentration relative to NaCl.

A reduction in calcium concentration to less than one-tenth of its present value would result in the exhaustion of all calcium as a $CaCO_3$ precipitate before all the carbonate was exhausted. Then no gypsum would be formed. Again, since gypsum *is* found, the calcium must have been not less than one-tenth of its present value, if the carbonate concentration had its present value.

Similar arguments can be used in the case of magnesium. The presence of dolomite, $(Ca,Mg)CO_3$, throughout the geological record, associated with calcium carbonates, suggests that the concentration of magnesium must have been near its present value throughout the last 3.5×10^9 years. We do not know precisely the physical and chemical conditions required for the formation of dolomite. It is generally thought to be a recrystallization product of calcium carbonate in contact with seawater. Possibly certain deep-lying carbonates in reefs are slowly dolomitizing now. It is unlikely this process could proceed if the magnesium concentration were much lower than it is now.

If we assume that the concentration of magnesium has not changed much since at least the late Proterozoic, then an upper limit can be set on the pH in seawater. Above pH 9.5, $Mg(OH)_2$ precipitates in a mineral form known as brucite. Since no brucite is found in the record of shallow marine sedimentary deposits, the pH could not commonly have reached as high as pH 9.5 over any extensive area, and the average surface-water pH must have been lower (probably no higher than pH 9.0) throughout this geological time.

A further limit on the typical values of surface-water pH is set by the presence of marine limestones throughout the geologic record. If the average pH in surface water were much below 7.5, it would have been difficult to form the extensive deposits of $CaCO_3$ that are so dramatically important in the sedimentary record.

A number of additional constraints can be applied to the ancient composition of seawater (Holland 1972), and in every case the present composition of seawater is about in the middle of the limits established. This lends additional support to the view that the general concentration and composition of sea salt has remained approximately constant throughout at least the last 600 to 1000 $\times$ 10^6 years and possibly for quite a lot longer. Nevertheless, the limits are still quite broad and certainly allow for a considerable variation on the ratios of major and minor constituents.

One geological mechanism that might change the oceanic ratios of major ions involves changes in the rate of sea-floor spreading and consequently the amount of seawater passing through the hot rocks at the spreading centers. Larson (1991) has presented evidence that from 120 to 80 million years ago (during part of the Cretaceous period) oceanic spreading rates may have been greater by as much as 50% compared to the time before or to the present time. Since magnesium is removed quantitatively from seawater passing through the hot rocks and is replaced by calcium, this process plays a major role in determining the composition of seawater. With respect to these ions and several others, the ionic ratios in the ocean represent a balance between the input from rivers and the losses or inputs at the mid-ocean ridges. If the spreading rate changed by as much as 50%, the effect would be potentially quite large. Spencer and Hardie (1990) have calculated that an increase in the hydrothermal flux by 25% would change the magnesium-to-calcium ratio from the present value of about 5.2 to about 0.9. Many other ionic ratios would also change; for example, the increased calcium input would lead to additional precipitation of calcium carbonate, and according to their calculations the dissolved bicarbonate in the ocean would decrease from the present value of 2.2 mM to about 0.3 mM. It seems likely, therefore, that at various times in the past the ionic composition of seawater may have been significantly different, although probably still within the broad limits suggested by Holland.

Hardie (1996) used proxy evidence (such as estimated sea-level changes) relating to the rate of sea-floor spreading to extend models of the magnesium-to-calcium ratio (and other ion ratios) in seawater back in time some 550 million years to the beginning of the Cambrian. He calculated quite significant changes in this ratio (for example, from 5.2 today to about 1.0 in the mid Cretaceous) and showed that the mineral assemblages in salt deposits, precipitated during the late stages of crystallization after the halite is deposited, were entirely consistent with expectation if the ratios in seawater were changed as suggested, as were changes in the mineralogy of marine limestones. Thus, we have strong suggestive evidence that the major ion composition of seawater has indeed undergone considerable changes, driven by changes in the rates of geological processes.

Further exploration of the rate of hydrothermal exchange as a function of the spreading rate, and the consequent effects on the composition of seawater, would have to take into account and budget all the processes that affect the major ion composition of the ocean, including sea-floor exchanges and river-borne input. These latter fluxes are even more difficult to estimate for times in the distant past than are the hydrothermal exchanges. This is still a fruitful area for speculation.

15.6 Oxygen

The history of the oxygen content of the atmosphere and therefore of the surface ocean is a subject great importance because our understanding of the influence of life on the geochemistry of Earth depends in part on understanding the processes that control the concentration of oxygen. The presence of the enormous banded iron formations in the geological record from the late Archaean and early Proterozoic on several continents is interpreted as evidence that the atmosphere, while having some free oxygen during those times, had only very low concentrations of this gas. Earlier, it is assumed, no oxygen was present, and the atmosphere was reducing. It is also commonly assumed that the oxygenation of the atmosphere was largely due to the activities of photosynthetic organisms, but other processes need consideration.

Hydrogen is light enough to escape at a certain rate from Earth's gravity and be lost to space. This process goes on today, but at a low rate because not much hydrogen is present in the upper layers of the atmosphere. A reducing atmosphere containing ammonia, methane, and water vapor, however, would provide a hydrogen-rich composition in the upper layers of the atmosphere. There the ultraviolet light from the Sun can dissociate the molecules, and enough hydrogen would always be present, to allow the total loss rate to be much greater than it is today (Hunten and Donahue 1976). This situation is likely to have been the case during Earth's earliest history. The progressive loss of hydrogen would leave carbon, nitrogen, and oxygen behind, so the world would become progressively oxygenated. If this process was great enough in early times, it might result in an oxygen atmosphere independently of the process of photosynthesis and burial of reduced carbon.

The huge mass of organic carbon buried in sediments and in rocks derived from sediments, however, must have produced more than thirty times the mass of oxygen present in the atmosphere today, and presumably much more in the past. The continuing processes of upwelling reduced mantle substances at the mid-ocean ridges, hot spots, and volcanoes, and the erosion of many parts of the land mass, results in a continuing consumption of oxygen today, and this must have consumed the excess. It is an open question whether it may have consumed more than has been biologically produced.

The geological and fossil record can be interpreted as being consistent with the idea that the oxygen content of the atmosphere had come close to its present value by Cambrian times and has not changed greatly since then. On the other hand, the volume of water in the major ocean basins, which has at various times become anoxic, has varied greatly. The reasons for this are still speculative.

The question of the history of the oxygen content of the atmosphere and oceans has been a subject of considerable work and speculation, as has, of course, the closely related question of when life began and the nature of such life (Schopf 1983, Holland 2006). The subject is best treated in a dedicated book of its own.

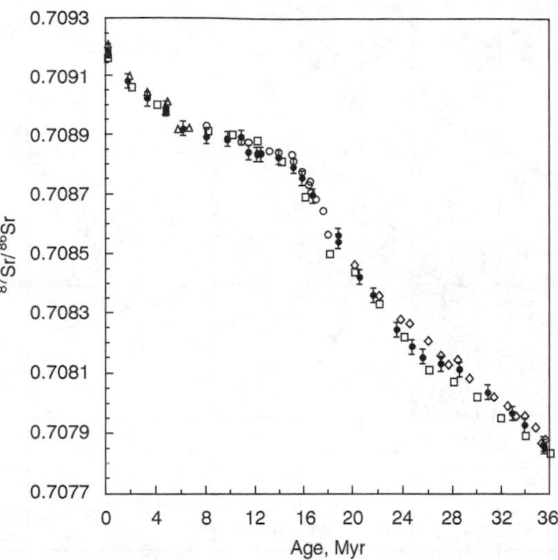

Figure 15.3 Strontium isotope composition in seawater over the last 36 Myr as reflected in the isotopic composition of strontium present as a trace component in marine carbonates (mostly planktonic foraminifera) and marine barites. Data are from five groups of investigators. Modified from Paytan *et al.* (1993).

15.7 Strontium isotopes

Unequivocal evidence for at least some change in the composition of the ocean over time resides in the ratios of two isotopes of strontium (Figures 15.3 and 15.4). Strontium is always present as a trace constituent in carbonates and barites ($BaSO_4$) formed in a seawater environment, and the isotopic ratio $^{87}Sr/^{86}Sr$ of the strontium present in these substances reflects the ratio in seawater. This ratio can be measured with high precision.

The ratio $^{87}Sr/^{86}Sr$ in seawater today is 0.70916, but 36 Myr ago, at the beginning of the Oligocene epoch, the value was about 0.70785. It was as low as 0.7068 during the Permian, 250 Myr ago, and was about the same as today at the end of the Cambrian. What could be responsible for these changes?

Both ^{86}Sr and ^{87}Sr are stable, but ^{87}Sr is the daughter product from decay of ^{87}Rb. Rubidium is a member of the alkali-metal group of elements (lithium, sodium, potassium, and rubidium) and its occurrence in rocks is different from that of strontium. In particular, it is more concentrated in granite-type rocks and is less concentrated in basalts. Rubidium has only two naturally occurring isotopes; ^{85}Rb and ^{87}Rb, the latter being 27.8% of the total. Only ^{87}Rb is radioactive, and it has the exceptionally long half-life of 48.8×10^9 years. Thus, rocks that contain sufficiently high concentrations of Rb gradually accumulate ^{87}Sr, and in suitable circumstances this can be used to date their age of formation. In general, granites will have higher ratios of $^{87}Sr/^{86}Sr$ than basalts of the same age. Limestones, which have very little rubidium and relatively large concentrations of strontium, will have $^{87}Sr/^{86}Sr$ ratios that do not change with time and reflect instead the ratio present in seawater when they were first formed.

The concentration of strontium and the ratios of the isotopes in river water are, for each river, a reflection of the types of the rocks in the drainage basin. If the global average type of geological terrain eroded by rivers has changed over time, then it is possible that the isotopic ratio in average river water has also changed. The biggest such geological

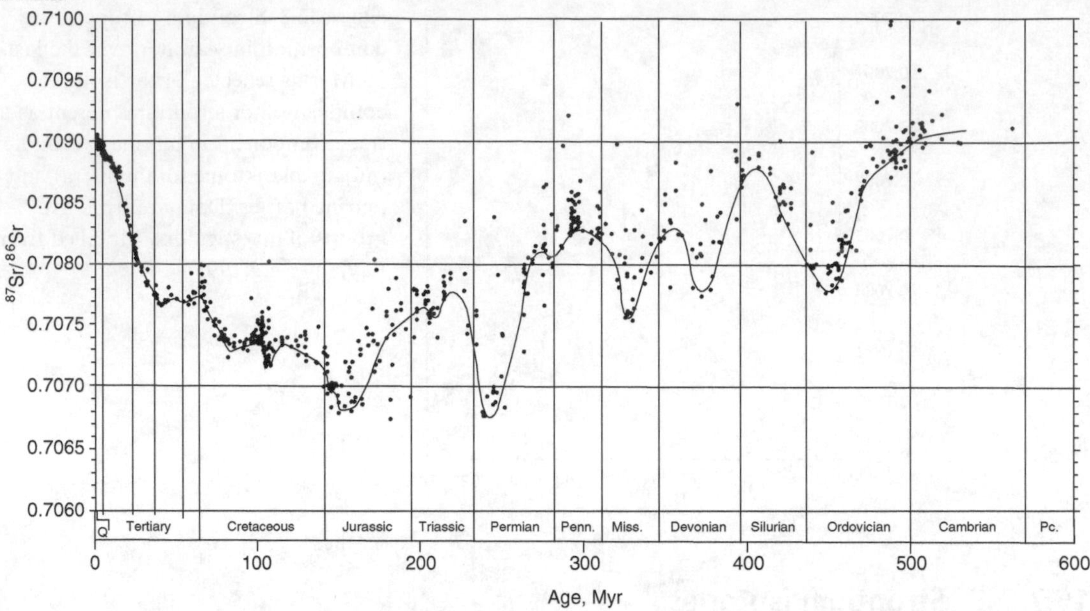

Figure 15.4 Strontium isotope composition of seawater since the early Cambrian, approximately 540 Myr (Burke *et al.* 1982). In this graph a line similar to that in Figure 15.3 is extended far back in time. Evidently the data derived from the strontium in carbonate sediments becomes progressively more noisy the older the sediments, presumably due to varied diagenetic changes and uncertainties in dating, and the assignment of a best estimate of the composition of seawater becomes increasingly a matter of judgment with the older material. Nevertheless, it is abundantly clear that dramatic changes occurred in the distant past, just as they have during the most recent 36 Myr.

change evident during the last several tens of millions of years is due to the movement of the plate bearing India north into Asia, causing the elevation of the Himalayas and the Tibetan plateau. The rivers draining this region, especially the Ganges and Brahmaputra, have the right combination of high flow rates, high concentrations of strontium, and high ratios of $^{87}Sr/^{86}Sr$ to have a significant effect on the ocean (Palmer and Edmond 1989). It seems likely that the increasing oceanic ratio is due to this effect. The average ratio in river water seems to have been raised by the elevation of these mountains, which raised a sufficient amount of granite-type rocks and exposed them to very fast rates of erosion.

There are other fluxes to take account of, however. The average isotopic ratio in most of the rest of the world's rivers that have been sampled is greater than in the ocean. The other source of strontium in the ocean is dissolution from the hot rocks at the spreading centers. The average $^{87}Sr/^{86}Sr$ ratio in the few hydrothermal fluids sampled is 0.7035 (Palmer and Edmond 1989), so strontium released from the mid-ocean-ridge basalts carries a lower ratio than that in the ocean and tends to lower the oceanic ratio. A full analysis of the budget requires a quantitative evaluation of all the fluxes and their possible changes over time. At present it seems that the ocean is not quite in balance, with a slow rise in the ratios caused by the continuing input of high-ratio strontium from rivers draining the mountainous regions north of India. It

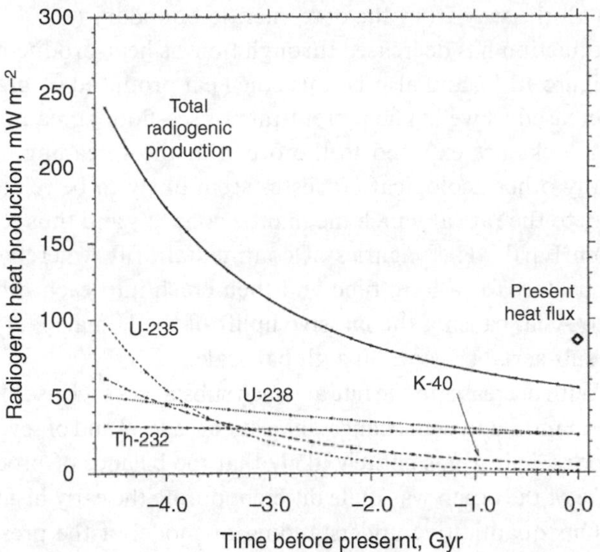

Figure 15.5 Heat production by radioactive decay since Earth was formed. The present average heat flux at the surface of the solid Earth, 86.6 mW m^{-2}, (or 44.2 TW for the world as a whole; Pollack *et al.* 1993) is greater than the estimated present production rate of 54.9 mW m^{-2} (= 28 TW in total; Stacey 1992) by radioactive decay. There are several reasons for this. Heat from past production is still making its way to the surface. The production rate is uncertain because of the difficulty in estimating the concentrations of the radioactive elements in the inaccessible mantle by extrapolation from known concentrations in rocks available at the surface. Stacey (1992) estimated that latent heat released as the core solidifies and gravitational energy released by evolution of the core, differentiation of the mantle, and thermal contraction of Earth amount to perhaps 3.9 TW. This leaves a flux of perhaps 12.3 TW to be accounted for by the cooling of Earth as it gradually loses heat stored from all these processes in the past. Whether the total heat loss in the past maintained the same proportionality to or excess over the instantaneous radiogenic production is a matter of speculation.

might also be suggested that the global rate of continental erosion is higher now than in earlier geological epochs due to the unusually high rates from those regions.

It is a matter of speculation whether the effects of these relatively recent (from the geological perspective) high rates of erosion and/or changes in the average of the types of rocks eroded might also be reflected in changes in the relative concentrations of major or other minor components of the salts in seawater. Such changes are likely to be small, but their possibility should be kept in mind when thinking about the history of the composition of seawater.

15.8 The churning of the Earth

Heat produced deep within Earth is transported, largely by the convection of the mantle, to the surface where it is lost. The shapes and sizes of the convective structures is a matter of current research; it appears that there is a widespread small-scale convection interspersed with occasional plumes of varying sizes that come up all the way from the core–mantle boundary (e.g. Larson 1991). The rate of heat production has decreased through time as heat-producing radioactive elements decay (Figure 15.5), and also because the heat produced by gravitational settling must have decreased as well. The present rate of sea-floor spreading (and thus the rate at which hot rocks are exposed to the ocean at the spreading centers) and also the rates of many other geological processes seem likely to be related in some not entirely clear way to the rate at which the mantle convects and thus to the rate at which heat is lost from Earth. The vagaries of continental drift, whereby some continents may move separately for a long time and then crash into each other (as India is now crashing into Asia, causing the massive uplift of the Himalayas) can change the total amount of sub-aerial erosion on a global scale.

With decreases in the rate at which substances such as chlorine and carbon out-gas from the mantle, and with changes in relative sea level and of several rates of chemical-exchange processes, it seems entirely likely that the balance of processes maintaining the present state of the ocean was quite different during the early history of Earth.

Our quantitative understanding or model of the present ocean is not sufficient to explain everything that we see today. Therefore, we are on shaky ground if we claim any quantitative certainty about our guesses as to the state of the ocean early in Earth's history when Earth was churning faster. We can, however, be fairly sure that there have been significant changes over time, and using what we do know about present conditions it is always fun to speculate about the conditions in the distant past. It is becoming increasingly possible to seem realistic about these speculations and to test the results.

Stay tuned.

Summary

A series of simple calculations from the current known rates and fluxes of substances shows that everything in the ocean, and the ocean itself, has been recycled a few times in some cases and many times in other cases. The composition of the ocean is therefore a result of many competing processes, and some rates at least may have changed with time, so the composition has surely also changed over time.

Nevertheless, the evidence from ancient salt deposits suggests a broad similarity in at least the major ion composition for at least the last 3500 million years.

Liquid water has been abundant on the surface of the Earth for at least 4000 million years. It is uncertain whether its volume is increasing or decreasing at the present time. The usual assumption is that the total mass of water on the surface of Earth has not changed much since early evidence of water at about 4000 million years ago.

The total mass of salt (mostly NaCl) resides about half in the ocean and about half in salt deposits found in many terrestrial locations. This ratio seems likely to have varied over time. At the present time there is no significant formation of salt deposits, while some are being eroded, so that the mass of salt and consequently the salinity is increasing, though at a very slow rate.

The gross composition of sea salt appears not to have changed much (within rather broad limits) since the late Proterozoic.

The oxygen concentration reached about its present value in the late Proterozoic, so the surface layers of the sea have been oxygenated since then, but there have been episodes of anoxia in the deeper water.

As demonstrated by changes in the ratio of the strontium isotopes, during the last 600 million years there have been changes in the proportions of different rock types eroded from the continents. It is a matter of speculation whether this led to significant changes in other aspects of seawater chemistry.

Heat production by radioactive decay within the Earth has decreased greatly since the formation of the Earth, and thus probably the rate of sea-floor spreading and corresponding rates of reactions between seawater and the hot basalt will have decreased with time.

SUGGESTIONS FOR FURTHER READING

Bekker, A. and H. D. Holland. 2012. Oxygen overshoot and recovery during the early Paleoproterozoic. *Earth Planet. Sci. Lett.* **317–318**: 295–304.

A recent paper that provides a detailed evaluation of the evidence for changes in the oxygen content of the atmosphere during the Paleoproterozoic and an entrée into the recent literature on early oxygen.

Holland, H. D. 1984. *The Chemical Evolution of the Atmosphere and Oceans*. Princeton University Press, Princeton, NJ.

This book by Heinrich Holland is a superb summary of the subject up until the date of publication.

Holland, H. D. 2006. The geologic history of seawater. In *The Oceans and Marine Geochemistry*, ed. H. Elderfield (*Treatise on Geochemistry*, vol. 6, ed. H. D. Holland and K. K. Turekian). Elsevier, New York, pp. 583–625.

The subject is brought up date in this summary review.

APPENDIX

A The chemical elements

The periodic table of the elements is a comforting document. Within a few lines it manages to tame the seemingly chaotic world of matter into a reliable regularity. Each atom has its own pigeonhole, its place in the order of things. And the beauty of the periodic table is that the way it groups atoms allows some prediction of how they should behave. The noble gases such as helium remain aloof and unreactive,while the alkali metals can be expected to be silvery and reactive. THE ECONOMIST **374**: JAN 15, 2005, P. 75.

According to the present evidence from cosmological research, our universe began about 13.7×10^9 ($\pm 0.2 \times 10^9$) years ago with the *big bang*. Calculation also suggests that only the three lightest elements were formed under the conditions at that time. Elements from about number 4 upwards could only be formed later, under the conditions in stars. Ordinary stars burn hydrogen and then successively heavier elements, and apparently can synthesize all the elements up to iron (number 26) by processes of thermonuclear fusion. All elements above iron must be made by several slow and rapid processes of neutron and proton addition as stars age and sometimes explode. The contents of stars are redistributed into space by gas streaming and under the remarkable conditions when they explode as novae or supernovae. The present mixture of elements on Earth (and in the Solar System generally) seems to require multiple processes for synthesis and thus seems to be the product of more than one generation or episode of star formation and explosion. Information from meteorites also suggests heterogeneity in source material.

The number of elements that are known to exist on Earth depends somewhat on arbitrary definition. The conventional periodic table of the elements contained 92 entries. In fact, at least 95 elements are known to occur naturally somewhere. All the elements from numbers 1 to 94 occur naturally on Earth, except for technetium (Tc; the half-life of its longest-lived isotope = 2.6×10^6 yr) and promethium (Pm; the half-life of its longest-lived isotope = 17.1 yr). These two elements are present in at least some stars and are therefore made there, but their half-lives are too short for them to survive and be found on Earth, and there is no significant natural production of them on Earth. Some of the naturally occurring elements are not very abundant. For example, francium (Fr) has no stable isotopes and its longest-lived isotope has a half-life of 22 min. It occurs only as a minor product of radioactive decay. It is practically undetectable on Earth; it is estimated that the entire crust of Earth contains about 30 g. Elements above uranium (U) are called *transuranics*. Two transuranics, neptunium

(Np) and plutonium (Pu) are naturally present on Earth because they are formed, though in extremely small amounts, in ore bodies that contain high concentrations of radioactive minerals. At least two transuranics, plutonium (Pu) and americium (Am), are sufficiently abundant products of human nuclear activities that they are now detectable nearly everywhere on the surface of Earth and in seawater. In total, humans have made at least 19 transuranic elements; all are radioactive but most have very short half-lives. Thus, 91 or 92 elements, two mostly anthropogenic, can be found throughout the surface of Earth.

A modern form of the periodic table and an alphabetical list of the elements with their names, conventional symbols, and atomic numbers are provided at the front of the book. The list of elements is abbreviated; no element above number 98 is included, uncertainties in the last significant figure are not given, and where the atomic weights are known to more than five or six significant figures this is also not listed. For this information see Wieser (2006) and updates in the IUPAC website.

Extensive listings of the properties of the individual elements, their isotopes, and their compounds may be found in Haynes (2011). The geochemical occurrences and behavior of most elements are reviewed in great detail in Wedepohl (1969).

APPENDIX

B Symbols, units, and nomenclature

The difficulties and confusion caused by the great variety of units, symbols, and conventions widely used in science during the last several centuries were not entirely eliminated by the introduction and spread of the metric system after the French Revolution, nor by the near universal adoption (except in the United States) of the modern metric system, the International System of Units (Système International d'Unités, abbreviated as SI). In 1960, the 11th General Conference on Weights and Measures (CGPM) adopted the abbreviation SI and specified the rules for primary and supplementary units and nomenclature; and a number of improvements and clarifications of detail have subsequently been added (Goldman and Bell 1981).

In 1960, the differing bases for the calculation of chemical masses were reconciled by definition; the mass 12 isotope of carbon was assigned the mass of precisely 12.000. The masses of all other elements or isotopes are, therefore, relative atomic masses based on the mass of carbon-12. With regard to chemical calculations and nomenclature, an important decision, formalized in 1971 (14th CGPM), was to define the mole on the basis of the previous assignment of mass 12 to the isotope 12 of carbon, and to generalize its usage. (See also Petley 1996.)

- *The mole is the amount of substance of a system that contains as many elementary entities as there are atoms in 0.012 kilogram of carbon-12.*
- *When the mole is used, the elementary entities must be specified and may be atoms, molecules, ions, electrons, other particles, or specified groups of such particles.*
- *In the definition of the mole, it is understood that unbound atoms of carbon-12, at rest and in their ground state, are referred to.*

One effect of adopting these rules was to make obsolete such terms as the gram-atom, gram-molecule, gram-ion, and so on. This greatly simplified chemical nomenclature in some usages and even increased the aesthetic quality of some chemical writing.

According to the principles behind the definitions above, the number of elementary particles in the mole (known as Avogadro's number) is not a matter of definition but of measurement. Many measurements have been made; the most recent recommended value (Haynes 2011) is 6.022143×10^{23}.

The special problems of nomenclature in oceanography led to the establishment of a working group under the International Association for the Physical Sciences of the Ocean (IAPSO) to address these issues and make recommendations, published in the SUN Report (Menaché *et al.* 1979). The SUN Report was expanded, revised, and re-issued in 1985 (UNESCO 1985). The 1985 report forms the basis for the

Table B.1 SI base units

Quantity	Name	Symbol
Length[a]	meter	m
Mass[b]	kilogram	kg
Time[c]	second	s
Electric current[d]	ampere	A
Thermodynamic temperature[e]	kelvin	K
Amount of substance[f]	mole	mol
Luminous intensity[g]	candela	cd

All other units should derive in some way from these units, or be defined in terms of these units.

[a] Defined in terms of the speed of light and the second. It is the length traveled by light in a vacuum during 1/299 792 458 of a second. The relation is exact, by definition.

[b] Defined as the mass of the international prototype kilogram maintained in Paris.

[c] Defined as the duration of 9 192 631 770 periods of the radiation corresponding to the transition between the two hyperfine levels of the ground state of cesium-133 atoms. This preserves the original relationship, defined as 1/86 400 of a mean solar day, but the new definition is both absolute and more precise.

[d] Defined as the current that will produce, between two parallel thin conducting wires one meter apart in a vacuum, a force equal to 2×10^{-7} newton per meter of length.

[e] Defined by reference to absolute zero and the triple point of water. The latter is defined as 273.16 K, and one kelvin is therefore 1/273.16 of the thermodynamic temperature at that point. This preserves equivalence between the kelvin and the degree celsius, which had of course provided the initial definition of the size of a kelvin. Under normal atmospheric pressure the freezing point of water is 0.00 °C or 273.15 K, and the triple point is at 0.01 °C. As far as I know, the isotopic composition of the water referred to is not defined, but I suppose that eventually it will be referred to that of SMOW.

[f] For chemists, this is the crucial definition.

[g] An awkward unit that we need not discuss here.

recommended units given here. Tables B.1 to B.4 give units selected as useful in the present context, or are those needed to construct them. For a complete listing and discussion, the sources should be consulted.

Table B.2 SI derived units of importance in marine chemical studies

Quantity	Name	Other symbol	SI base units	Units
Frequency	hertz	Hz		s^{-1}
Force	newton[(a)]	N		$m\ kg\ s^{-2}$
Pressure, stress	pascal[(b)]	Pa	$N\ m^{-2}$	$m^{-1}\ kg\ s^{-2}$
Energy, work, quantity of heat	joule[(c)]	J	N m	$m^2\ kg\ s^{-2}$
Power, radiant flux	watt	W	$J\ s^{-1}$	$m^2\ kg\ s^{-3}$
Quantity of electricity	coulomb	C		s A
Electric potential	volt	V	$W\ A^{-1}$	$m^2\ kg\ s^{-3}\ A^{-1}$
Temperature, celsius	degree celsius[(d)]	°C		K
Activity (of a radionuclide)	becquerel[(e)]	Bq		s^{-1}
Absorbed dose	gray	Gy	$J\ kg^{-1}$	$m^2\ s^{-2}$

[(a)] The newton is the force that will accelerate 1 kilogram at a rate of 1 m s^{-2}.

[(b)] The pascal is the force of 1 newton applied over 1 square meter. For much oceanographic use, this is an inconveniently sized unit. Standard atmospheric pressure is 101 325 Pa. A much more convenient unit is the bar (Table B.3). 1 bar = 10^5 Pa. Standard atmospheric pressure is 1.01325 bar. Meteorologists commonly use millibars (mbar). In the ocean, pressure increases roughly 1 bar per 10 m, or 1 decibar per meter, so oceanographers often use decibars. While some authorities recommend against use of the bar, it will not soon disappear from atmospheric and oceanographic usage.

[(c)] The most commonly used of the several calories, the 15 °C calorie, = 4.1855 J. However, for thermochemical calculations, the thermochemical calorie (= 4.184 J) is used. Additional common energy units, not in the SI system, are:
the erg: 1 J = 10^7 ergs
the electron volt: 1 J = 6.242×10^{18} eV.

[(d)] The scale introduced by Celsius ran from 0 degrees at the boiling point of water to 100 degrees at the freezing point. It was Christin who inverted it to give us the scale we use today.

[(e)] 1 Bq = one radioactive nuclear transformation per second.

Table B.3 Other recommended units and symbols

Name	Symbol	Definition	
Minute	min	1 min	= 60 s
Hour	h	1 h	= 3600 s
Day[(a)]	d	1 d	= 86,400 s
Liter	L	1 L	= 10^{-3} m^3
Metric ton[(b)]	t	1 t	= 10^3 kg
Nautical mile	—		= 1852 m
Knot	—		= 1852/3600 = 0.5144 m s^{-1}
Ångstrom	Å	1 Å	= 10^{-10} m
Bar	bar	1 bar	= 10^5 Pa
Curie	Ci	1 Ci	= 3.7×10^{10} Bq
Electron volt	eV	1 eV	= 1.602×10^{-19} J

These are not in the SI family, but are allowed, at least temporarily, by agreement of the representatives to CGPM.

(a) The week, month, and year have no precise definition, and their use is discouraged, which demonstrates a certain lack of realism among those who do this sort of thing. By general agreement, however, the symbol "a" (for anno) is recommended as the abbreviation for "year." Nevertheless, many people in the scientific community still use "yr" for this purpose, as I do in this book. At the present time, on average, 1 a or 1 yr = 3.1557×10^7 s.

(b) The ton (European: tonne) is a convenient unit of mass in oceanography; it is approximately the mass of 1 m^3 of water or seawater, and some geochemical units of transport are commonly given in tons (e.g. the release of fossil fuel CO_2 in Gt of C). In this text I use the word "ton" to refer only to the "metric ton" = 1000 kg.

Table B.4 Prefixes for factors, as currently approved

Factor	Prefix	Symbol	Factor	Prefix	Symbol
10^1	deka	da	10^{-1}	deci	d
10^2	hecto	h	10^{-2}	centi	c
10^3	kilo	k	10^{-3}	milli	m
10^6	mega	M	10^{-6}	micro	μ
10^9	giga	G	10^{-9}	nano	n
10^{12}	tera	T	10^{-12}	pico	p
10^{15}	peta	P	10^{-15}	femto	f
10^{18}	exa	E	10^{-18}	atto	a
10^{21}	zetta	Z	10^{-21}	zepto	z
10^{24}	yotta	Y	10^{-24}	yocto	y

Most authorities recommend not using prefixes that refer to factors raised to powers that are not multiples of three; their use is much less common in science than in commerce.

The original intent was that the factors above 10^0 would be represented by words derived from Greek and by upper-case letters, while words for the smaller factors would be Latin-derived and represented by lower-case letters. This move has evidently not been entirely successful; for example, femto and atto derive from Danish.

APPENDIX

C Physical properties of seawater

In this section, I have provided brief tables summarizing the density, freezing point, viscosity, vapor pressure, and heat capacity of seawater, calculated using algorithms from UNESCO (1981b) and from compilations by Fofonoff and Millard in UNESCO (1983b). The density algorithms are also given in Boxes 3.3 and 3.4 in Chapter 3.

C.1 Density of seawater

As discussed in Chapter 3, the absolute values of the density are based on estimates of the density of pure water with the isotopic composition of SMOW. Recent re-evaluations provide evidence that the absolute density of pure SMOW differs from the values given here by less than 1 gram per cubic meter at all temperatures below 25 °C and the difference does not exceed five grams per cubic meter up to 40 °C. Relative values can be measured to even better precision, however, and physical oceanographers sometimes use calculations carried out to an additional significant figure.

It is awkward, repetitive, and unnecessary to use such big numbers as are given in Table C.1.1. In physical oceanography the *density anomaly* is used:

$$\sigma = \rho - 1000$$

so that, for example, at $S = 35‰$ and $t = 20$ °C,

$$\sigma_{(35,20)} = 1024.763 - 1000 = 24.763 \text{kg m}^{-3}.$$

This allows greater sensitivity in the computation and is otherwise easier to use. Other transformations, such as the comparably calculated specific volume anomaly, are in common use.

In examining the physics of ocean circulation, it is essential to take into account the temperature change (which can exceed 1 °C) caused by the pressure change as water sinks to the depths or rises to the surface. Thus, the *potential temperature*, θ, is the temperature that some water mass at some depth will have if it is brought adiabatically (without exchanging heat with the surroundings) to some other specified depth or to the surface. For a full explanation of these and other physical properties of seawater, a textbook in physical oceanography should be consulted.

Tables C.1.1 and C.1.2 give the density (kg m^{-3}) over a range of temperature and salinity that encompasses or even exceeds what is ever seen in the ocean.

Table C.1.1 Density of water and seawater (kg m^{-3}) under a total pressure of 1atmosphere, as it varies with temperature and salinity

	Temperature, °C									
S, ‰	−2	0	5	10	15	20	25	30	35	40
0	999.670	999.843	999.967	999.702	999.102	998.206	997.048	995.651	994.036	992.220
5	1003.780	1003.913	1003.949	1003.612	1002.952	1002.008	1000.809	999.380	997.740	995.906
10	1007.860	1007.955	1007.907	1007.501	1006.784	1005.793	1004.556	1003.095	1001.429	999.575
15	1011.928	1011.986	1011.858	1011.385	1010.613	1009.576	1008.301	1006.809	1005.118	1003.244
20	1015.992	1016.014	1015.807	1015.269	1014.443	1013.362	1012.050	1010.527	1008.810	1006.915
25	1020.054	1020.041	1019.758	1019.157	1018.279	1017.154	1015.806	1014.252	1012.509	1010.593
30	1024.119	1024.072	1023.714	1023.051	1022.122	1020.954	1019.570	1017.986	1016.217	1014.278
32	1025.746	1025.685	1025.298	1024.611	1023.661	1022.477	1021.078	1019.482	1017.702	1015.755
33	1026.559	1026.492	1026.090	1025.391	1024.431	1023.238	1021.833	1020.230	1018.446	1016.494
34	1027.373	1027.299	1026.883	1026.172	1025.202	1024.001	1022.588	1020.979	1019.190	1017.234
35	1028.187	1028.107	1027.677	1026.953	1025.973	1024.763	1023.343	1021.729	1019.934	1017.973
36	1029.001	1028.914	1028.469	1027.734	1026.744	1025.526	1024.099	1022.479	1020.679	1018.714
38	1030.631	1030.530	1030.056	1029.298	1028.288	1027.054	1025.613	1023.980	1022.170	1020.196
40	1032.261	1032.148	1031.645	1030.863	1029.834	1028.583	1027.128	1025.483	1023.662	1021.679

Values calculated from the algorithm given in Box 3.4. The density at $S = 0$ ‰ is that provisionally adopted for the water in SMOW (UNESCO 1976). The salinity *S* is based on the PSS 1978. For salinities less than about 35 the densities at −2 °C refer to supercooled water. For a more detailed evaluation of the density, for the various transformations of density into density anomaly, specific volume anomaly, etc., and for the calculation of other physical properties, see UNESCO (1983b).

Table C.1.2 The density of pure water and seawater (kg m^{-3}) under pressures up to 1000 bar

	Temperature, °C						
S, ‰	0	2	5	10	15	20	25
----- 0 bar -----							
0	999.843	999.943	999.967	999.702	999.102	998.206	997.048
30	1024.072	1023.968	1023.714	1023.051	1022.122	1020.954	1019.570
32	1025.685	1025.569	1025.298	1024.611	1023.661	1022.477	1021.078
33	1026.492	1026.370	1026.090	1025.391	1024.431	1023.238	1021.833
34	1027.299	1027.171	1026.883	1026.172	1025.202	1024.001	1022.588
35	1028.107	1027.972	1027.676	1026.953	1025.973	1024.763	1023.343
36	1028.914	1028.774	1028.469	1027.734	1026.744	1025.526	1024.099
38	1030.530	1030.378	1030.056	1029.298	1028.288	1027.054	1025.613
----- 100 bar -----							
0	1004.873	1004.902	1004.830	1004.430	1003.721	1002.739	1001.512
30	1028.826	1028.666	1028.335	1027.563	1026.545	1025.307	1023.866
32	1030.422	1030.250	1029.903	1029.109	1028.072	1026.818	1025.364
33	1031.221	1031.043	1030.688	1029.883	1028.837	1027.574	1026.113
34	1032.020	1031.836	1031.473	1030.657	1029.601	1028.330	1026.863
35	1032.819	1032.629	1032.259	1031.431	1030.366	1029.087	1027.613
36	1033.618	1033.423	1033.045	1032.205	1031.131	1029.845	1028.364
38	1035.218	1035.011	1034.617	1033.756	1032.663	1031.361	1029.867
----- 200 bar -----							
0	1009.789	1009.750	1009.586	1009.058	1008.244	1007.178	1005.884
30	1033.477	1033.263	1032.858	1031.982	1030.879	1029.572	1028.076
32	1035.057	1034.832	1034.412	1033.515	1032.394	1031.072	1029.564
33	1035.848	1035.617	1035.190	1034.282	1033.152	1031.823	1030.308
34	1036.638	1036.402	1035.968	1035.049	1033.911	1032.574	1031.053
35	1037.429	1037.188	1036.746	1035.817	1034.670	1033.325	1031.798
36	1038.221	1037.974	1037.525	1036.585	1035.429	1034.077	1032.544
38	1039.804	1039.547	1039.083	1038.123	1036.949	1035.582	1034.037
----- 300 bar -----							
0	1014.595	1014.491	1014.240	1013.587	1012.674	1011.526	1010.167
30	1038.029	1037.762	1037.287	1036.311	1035.126	1033.753	1032.204
32	1039.593	1039.316	1038.827	1037.831	1036.630	1035.242	1033.681
33	1040.375	1040.093	1039.598	1038.592	1037.382	1035.987	1034.421
34	1041.158	1040.871	1040.369	1039.353	1038.135	1036.733	1035.161
35	1041.941	1041.649	1041.140	1040.114	1038.888	1037.479	1035.901
36	1042.725	1042.428	1041.912	1040.876	1039.642	1038.226	1036.642
38	1044.293	1043.986	1043.456	1042.401	1041.150	1039.721	1038.125
----- 400 bar -----							
0	1019.293	1019.128	1018.793	1018.022	1017.012	1015.787	1014.365
30	1042.482	1042.166	1041.624	1040.552	1039.289	1037.852	1036.252
32	1044.031	1043.705	1043.150	1042.060	1040.781	1039.330	1037.719
33	1044.806	1044.476	1043.914	1042.815	1041.528	1040.071	1038.454
34	1045.581	1045.246	1044.678	1043.569	1042.275	1040.811	1039.189

Table C.1.2 (*cont.*)

	Temperature, °C						
S, ‰	0	2	5	10	15	20	25
35	1046.356	1046.017	1045.443	1044.325	1043.023	1041.552	1039.925
36	1047.133	1046.789	1046.208	1045.081	1043.771	1042.294	1040.661
38	1048.686	1048.333	1047.739	1046.594	1045.268	1043.778	1042.134
----- 500 bar -----							
0	1023.885	1023.661	1023.247	1022.364	1021.262	1019.962	1018.479
30	1046.840	1046.477	1045.871	1044.709	1043.370	1041.871	1040.222
32	1048.374	1048.003	1047.385	1046.204	1044.852	1043.340	1041.680
33	1049.142	1048.766	1048.142	1046.953	1045.593	1044.075	1042.410
34	1049.910	1049.530	1048.900	1047.702	1046.335	1044.811	1043.140
35	1050.678	1050.294	1049.658	1048.451	1047.077	1045.547	1043.871
36	1051.447	1051.058	1050.416	1049.201	1047.820	1046.284	1044.603
38	1052.986	1052.589	1051.935	1050.703	1049.307	1047.758	1046.067
----- 600 bar -----							
0	1028.373	1028.094	1027.606	1026.616	1025.427	1024.055	1022.513
30	1051.105	1050.698	1050.032	1048.782	1047.372	1045.814	1044.117
32	1052.625	1052.210	1051.533	1050.267	1048.843	1047.273	1045.566
33	1053.386	1052.967	1052.284	1051.009	1049.579	1048.004	1046.292
34	1054.147	1053.724	1053.035	1051.753	1050.316	1048.735	1047.018
35	1054.908	1054.481	1053.787	1052.497	1051.053	1049.466	1047.744
36	1055.670	1055.239	1054.539	1053.241	1051.790	1050.198	1048.471
38	1057.195	1056.756	1056.045	1054.731	1053.267	1051.663	1049.927
----- 700 bar -----							
0	1032.760	1032.430	1031.871	1030.781	1029.508	1028.068	1026.470
30	1055.279	1054.831	1054.107	1052.776	1051.298	1049.683	1047.940
32	1056.786	1056.330	1055.596	1054.249	1052.758	1051.133	1049.380
33	1057.540	1057.080	1056.341	1054.986	1053.489	1051.858	1050.101
34	1058.294	1057.831	1057.086	1055.724	1054.221	1052.585	1050.823
35	1059.049	1058.582	1057.832	1056.463	1054.953	1053.311	1051.545
36	1059.804	1059.333	1058.579	1057.201	1055.685	1054.038	1052.268
38	1061.317	1060.838	1060.073	1058.681	1057.152	1055.494	1053.714
----- 800 bar -----							
0	1037.048	1036.670	1036.045	1034.860	1033.509	1032.003	1030.353
30	1059.365	1058.878	1058.101	1056.691	1055.148	1053.480	1051.692
32	1060.859	1060.364	1059.577	1058.154	1056.599	1054.921	1053.124
33	1061.607	1061.108	1060.317	1058.886	1057.325	1055.642	1053.841
34	1062.355	1061.853	1061.056	1059.618	1058.052	1056.363	1054.558
35	1063.103	1062.598	1061.796	1060.352	1058.779	1057.085	1055.276
36	1063.852	1063.343	1062.537	1061.085	1059.507	1057.808	1055.995
38	1065.352	1064.836	1064.020	1062.554	1060.963	1059.255	1057.433
----- 900 bar -----							
0	1041.240	1040.816	1040.130	1038.856	1037.431	1035.864	1034.162
30	1063.365	1062.841	1062.013	1060.531	1058.927	1057.207	1055.376

Table C.1.2 (*cont.*)

	Temperature, °C						
S, ‰	0	2	5	10	15	20	25
32	1064.846	1064.316	1063.479	1061.983	1060.368	1058.639	1056.800
33	1065.588	1065.054	1064.213	1062.710	1061.089	1059.356	1057.513
34	1066.330	1065.792	1064.947	1063.438	1061.811	1060.073	1058.226
35	1067.072	1066.532	1065.681	1064.166	1062.534	1060.791	1058.940
36	1067.815	1067.271	1066.417	1064.894	1063.257	1061.509	1059.654
38	1069.303	1068.752	1067.888	1066.353	1064.704	1062.947	1061.085
----- 1000 bar -----							
0	1045.337	1044.872	1044.128	1042.772	1041.277	1039.652	1037.902
30	1067.280	1066.722	1065.848	1064.297	1062.635	1060.866	1058.995
32	1068.750	1068.186	1067.303	1065.740	1064.067	1062.290	1060.411
33	1069.486	1068.918	1068.031	1066.462	1064.784	1063.002	1061.119
34	1070.222	1069.651	1068.760	1067.184	1065.501	1063.715	1061.829
35	1070.959	1070.385	1069.489	1067.907	1066.219	1064.429	1062.539
36	1071.696	1071.119	1070.219	1068.631	1066.938	1065.143	1063.249
38	1073.172	1072.589	1071.681	1070.080	1068.376	1066.573	1064.671

The temperature is the value *in situ*. Density values were calculated using the algorithm given in Box 3.3. See comments in text and in Table C.1.1.

Everything in the first table can be found somewhere, but it is obvious that listings in Table C.1.2 for water at zero salinity and a pressure of 1000 bar (reflecting a depth of approximately 10 000 m) represents water that is not seen in the ocean. Similarly, water at temperatures of 5 to 25 °C is not found at a pressure of 1000 bar on the open water column, but could be found several hundred meters down in the sediments.

Figure C.1.1 shows the relationship between temperature, salinity, and density. This is plotted so that the density increases downward and denser water is found below less-dense water. A important point here is the notable uniformity of deep water. The little ellipse labeled "BW" encompasses essentially all the water filling the bottom kilometer of the world ocean, with few exceptions. Two exceptions are the Mediterranean "M" and Red "R" seas. All their deep waters are encompassed well within the two respective ellipses. It is easy to see that the deep water in the Mediterranean is denser that anything in the world ocean outside. This suggests that it is plausible that the deep ocean could be filled with water much warmer than is the case today if the controlling geological structures were rearranged, even if the climate were roughly similar to that of today.

Figure C.1.2 shows the changing relative importance of salinity and temperature in affecting the density of seawater. Because the change in density with unit change in temperature is so small at cold temperatures, the importance of changes in salinity becomes relatively greater.

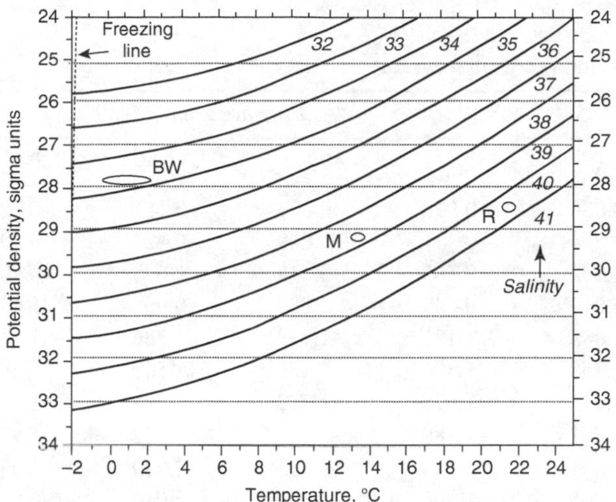

Figure C.1.1 Density (ρ in kg m^{-3}) of seawater at 1 atmosphere pressure, plotted in sigma units where $\sigma = \rho - 1000$, for salinities from 32 to 41‰. The dashed line is the freezing point of seawater. The curves are lines of constant salinity, identified at each upper right-hand end. Note that the density scale increases downwards, so that the densest water is at the bottom of the graph. At temperatures near 0 °C the density becomes much less sensitive to changes in temperature and very slightly more sensitive to changes in salinity, compared to the situation at higher temperatures. The field labeled BW identifies characteristics of much of the densest water lying on the bottom of the Pacific, Indian, and Atlantic oceans. Properties characterizing the deep water in the basins of the Mediterranean (M) and Red (R) Seas are all found well within their respective ellipses.

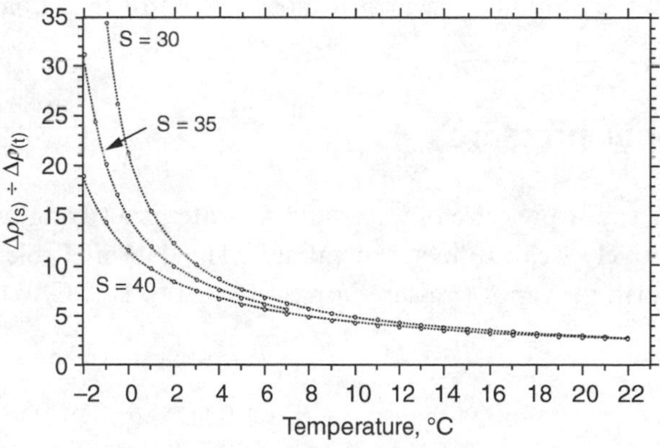

Figure C.1.2 Ratio of the change in density of seawater due to a salinity change (0.5‰) to the change in density due to a temperature change (0.5 °C), calculated for salinities of 30, 35, and 40‰.

C.2 Freezing point of seawater

The formulation currently used to calculate the freezing point of seawater is based on measurements published by Doherty and Kester (1974). These data were fitted to an algorithm appearing in UNESCO (1983b) as follows:

Table C.2 Freezing point of seawater

	Pressure (dbar)						
S, ‰	0	50	100	200	300	400	500
0	0.000	−0.038	−0.075	−0.151	−0.226	−0.301	−0.377
5	−0.274	−0.311	−0.349	−0.424	−0.500	−0.575	−0.650
10	−0.542	−0.580	−0.618	−0.693	−0.768	−0.844	−0.919
15	−0.812	−0.849	−0.887	−0.962	−1.038	−1.113	−1.188
20	−1.083	−1.121	−1.159	−1.234	−1.309	−1.384	−1.460
25	−1.358	−1.396	−1.434	−1.509	−1.584	−1.660	−1.735
30	−1.638	−1.676	−1.713	−1.788	−1.864	−1.939	−2.014
32	−1.751	−1.789	−1.826	−1.902	−1.977	−2.052	−2.128
33	−1.808	−1.846	−1.883	−1.959	−2.034	−2.109	−2.184
34	−1.865	−1.903	−1.940	−2.016	−2.091	−2.166	−2.242
35	−1.922	−1.960	−1.998	−2.073	−2.148	−2.224	−2.299
36	−1.980	−2.017	−2.055	−2.130	−2.206	−2.281	−2.356
38	−2.095	−2.133	−2.471	−2.246	−2.321	−2.397	−2.472
40	−2.212	−2.250	−2.287	−2.363	−2.438	−2.513	−2.589

$$t_f = -0.0575S + 1.710523 \times 10^{-3} S^{1.5} - 2.154996 \times 10^{-4} S^2 - 7.53 \times 10^{-4} p, \quad \text{(C.2.1)}$$

where t_f is the freezing temperature (°C), S is salinity (‰) and p is pressure (dbar). Conventionally, as with most pressure measurements in an oceanographic context, zero pressure here means 1 atmosphere or 10.1325 dbar.

C.3 Vapor pressure

The vapor pressure of water and seawater is a strong function of temperature, and a relatively weak function of salinity. The data in Table C.3 were calculated with the algorithm: vapor pressure in pascals = A × B × C × 100, where

$$A = 6.1094 \times e^{\left(\frac{17.625t}{243.04+t}\right)}$$

$$B = 1.00071 \times e^{0.0000045p}$$

$$C = 1 - (5.096 \times 10^{-4})S - (7.231 \times 10^{-7})S^2 \quad \text{(C.3.1)}$$

where t is temperature in °C, p is atmospheric pressure in millibars, S is salinity in ‰.

Factor A is from Alduchov and Eskridge (1996), and accounts for the vapor pressure of pure water in the absence of air. Factor B, also from Alduchov and Eskridge (1996), accounts for the increased water vapor in the presence of air, the so-called enhancement factor. The data below were calculated at the standard atmospheric pressure of 101 325 Pa. Ordinary changes in atmospheric pressure at sea level would not have much effect on this small correction.

Table C.3 Vapor pressure, in pascals, of pure water and the water in seawater

	Temperature, °C									
S, ‰	0	5	10	15	18	20	25	30	35	40
0	614.2	876.2	1232	1711	2071	2346	3178	4259	5647	7414
5	612.6	873.9	1229	1707	2065	2340	3170	4248	5633	7395
10	611.0	871.6	1226	1702	2060	2334	3162	4237	5618	7375
15	609.4	869.3	1223	1698	2054	2327	3154	4226	5603	7356
20	607.8	867.0	1220	1693	2049	2321	3145	4214	5588	7336
25	606.1	864.6	1216	1688	2043	2315	3137	4203	5573	7316
30	604.4	862.2	1213	1684	2038	2308	3128	4191	5557	7296
35	602.7	859.8	1209	1679	2032	2302	3119	4179	5542	7275
40	600.9	857.3	1206	1674	2026	2295	3110	4167	5526	7254

Factor C accounts for the effect of salt in depressing the vapor pressure; it derives from the measurements of the effect of sea salt on the vapor pressure of seawater over a salinity range from 17 to 38 and at 25 °C made by Robinson (1954). Calculations of the effect of sea salts on the osmotic coefficient of seawater and thus on the vapor pressure at intervals over a salinity range of 5 to 40 ‰ presented by Millero and Leung (1976) do not significantly differ from the values in Table C.3, after allowing for the different vapor pressures assigned to pure water.

C.4 Viscosity of seawater

The viscosity of pure water from 8 to 70 °C and at atmospheric pressure was measured by Korson *et al.* (1969). They fitted their data to an algorithm, as follows:

$$\log = \frac{\eta_t}{\eta_{20}} = \frac{1.1709(20 - t) - 0.001827(t - 20)^2}{t + 89.93} \tag{C.4.1}$$

where t is in Celsius, and η_{20}, the viscosity of water at 20 °C, used as a standard point of reference, is 1.0020 centipoises. This equation was used to generate the values for viscosity at $S = 0$ (Table C.4).

The values for η below 8 °C are beyond the range of measurement, but are included by extrapolation to –1 °C to make possible an approximate calculation of the viscosity of seawater near the freezing point. The centipoise is numerically equivalent to the mPa s (millipascal second). The pascal second has units of kg m^{-1} s^{-1}; this is the SI unit.

The effect of salt on the viscosity of water, and thus the viscosity of seawater, was discussed by Millero (1974). The viscosity of seawater can be represented by the following equation (derived in the same way that Riley and Skirrow 1975, p. 576, did, by assuming a linear extrapolation and interpolation between the relationships, at 5 °C and 25 °C, reported by Millero 1974):

Table C.4 Viscosity of pure water and seawater

	Temperature °C									
S, ‰	**–1**	**0**	**5**	**10**	**15**	**18**	**20**	**25**	**30**	**35**
A. Dynamic viscosity (η); (units: centipoises, = 10^{-3} kg m^{-1} s^{-1})										
0	–	1.791	1.519	1.307	1.138	1.053	1.002	0.890	0.797	0.719
5	–	1.804	1.532	1.319	1.150	1.064	1.013	0.901	0.808	0.729
10	–	1.818	1.544	1.330	1.160	1.074	1.023	0.910	0.816	0.738
15	–	1.831	1.556	1.341	1.171	1.084	1.033	0.919	0.825	0.746
20	1.909	1.844	1.568	1.352	1.181	1.094	1.042	0.929	0.834	0.754
25	1.922	1.857	1.580	1.364	1.192	1.105	1.052	0.938	0.843	0.762
30	1.936	1.871	1.592	1.375	1.202	1.115	1.062	0.947	0.851	0.771
35	1.950	1.884	1.605	1.386	1.213	1.125	1.072	0.956	0.860	0.779
40	1.964	1.898	1.617	1.398	1.223	1.135	1.082	0.965	0.869	0.787
B. Kinematic viscosity (ν); (units: centistokes, = 10^{-6} m^2 s^{-1})										
0	–	1.791	1.519	1.307	1.139	1.055	1.004	0.893	0.801	0.724
5	–	1.797	1.526	1.314	1.146	1.062	1.011	0.900	0.808	0.731
10	–	1.803	1.532	1.320	1.152	1.068	1.017	0.906	0.814	0.737
15	–	1.809	1.538	1.326	1.158	1.074	1.023	0.912	0.820	0.742
20	1.878	1.815	1.544	1.332	1.164	1.080	1.029	0.917	0.825	0.748
25	1.884	1.821	1.550	1.338	1.170	1.085	1.034	0.923	0.831	0.753
30	1.890	1.827	1.556	1.344	1.176	1.091	1.040	0.929	0.836	0.758
35	1.896	1.833	1.562	1.350	1.182	1.097	1.046	0.934	0.842	0.764
40	1.902	1.839	1.568	1.356	1.188	1.103	1.052	0.940	0.847	0.769

$$\eta_s = \eta_w\left(1 + \left(\frac{0.1067 + 0.0519t}{1000}\right)\left(\frac{S \times \rho}{1806.55}\right)^{0.5} + \left(\frac{2.591 + 0.033t}{1000}\right)\left(\frac{S \times \rho}{1806.55}\right)\right), \quad \text{(C.4.2)}$$

where S is the salinity of the seawater, ρ is the density (kg m^{-3}) of the seawater at salinity S and temperature t, t is the temperature in °C, ηs is the viscosity of the seawater at salinity S and temperature t, and ηw is the viscosity of pure water at temperature t (=ηt above).

Some error will be present if the relationship departs measurably from linearity. Probably, the last significant figures given in Table C.4 are uncertain. These data are also in reasonable agreement with the earlier data of Miyake and Koizumi (1948).

The kinematic viscosity, v, is the so-called dynamic viscosity divided by the density of the solution. The SI unit is m^2 s^{-1}. Values in Table C.4 are given in centistokes, 1 cSt, = 10^{-6} m^2 s^{-1}.

C.5 Specific heat of seawater

The amount of energy required to raise the temperature of 1 kilogram of seawater by 1 °C is a little less than that required to raise the temperature of fresh water by the same amount. The difference is in part because there is about 3.5% less water in a kilogram of seawater. Measurements of the specific heat of seawater have been fitted to the following algorithm (Millero *et al.* 1973, UNESCO 1983b):

$$C_p(S,t) = C_p(0,t) + (a_0 + a_1 t + a_2 t^2)S + (b_0 + b_1 t + b_2 t^2)S^{1.5}. \qquad \text{(C.5.1)}$$

where $C_p(0,t)$ is the specific heat of pure water: $C_p(0,t) = C_0 + C_1 t + C_2 t^2 + C_3 t^3 + C_4 t^4$. Here C_p is the amount of energy (joules per kilogram) that increases the temperature by 1 °C when the water is at temperature t and salinity S. The constants are:

$C_0 = +4217.4$	$a_0 = -7.643575$	$b_0 = +0.1770383$
$C_1 = -3.720283$	$a_1 = +0.1072763$	$b_1 = -4.07718 \times 10^{-3}$
$C_2 = +0.1412855$	$a_2 = -1.38385 \times 10^{-3}$	$b_2 = +5.148 \times 10^{-5}$
$C_3 = -2.654387 \times 10^{-3}$		
$C_4 = +2.093236 \times 10^{-5}$.		

The specific heat of seawater decreases by several percent under pressures up to 1000 bar. For an algorithm to calculate the effect of pressure on the specific heat see UNESCO (1983b).

Table C.5 Specific heat of pure water and seawater, in units of J kg^{-1}

	Temperature, °C						
S, ‰	**0**	**5**	**10**	**15**	**20**	**25**	**30**
0	4217.4	4202.0	4191.9	4185.5	4181.6	4179.4	4178.2
5	4181.2	4168.1	4159.9	4155.2	4152.7	4151.5	4151.0
10	4146.6	4135.6	4129.3	4126.1	4124.8	4124.5	4124.7
20	4080.4	4073.3	4070.3	4070.0	4071.0	4072.5	4073.9
30	4017.2	4013.7	4013.8	4016.1	4019.1	4022.2	4024.7
31	4011.0	4007.9	4008.3	4010.8	4014.1	4017.3	4019.9
32	4004.9	4002.1	4002.8	4005.5	4009.0	4012.4	4015.1
33	3998.7	3996.3	3997.3	4000.2	4003.9	4007.5	4010.3
34	3992.6	3990.5	3991.8	3995.0	3998.9	4002.6	4005.5
35	3986.5	3984.8	3986.3	3989.8	3993.9	3997.7	4000.7
36	3980.5	3979.0	3980.9	3984.6	3988.8	3992.8	3995.9
37	3974.4	3973.3	3975.5	3979.4	3983.8	3988.0	3991.2
38	3968.4	3967.6	3970.1	3974.2	3978.9	3983.1	3986.4
39	3962.4	3961.9	3964.7	3969.0	3973.9	3978.3	3981.7
40	3956.4	3956.3	3959.3	3963.9	3968.9	3973.5	3977.0

APPENDIX

D Gases

Three relationships are of basic importance when working with gases.

[1] **Law of partial pressures**. In a mixture of gases, the total pressure is the sum of the individual partial pressures of each gas. For all the gases listed in Table D.1 the partial pressures of each gas are very nearly proportional to their concentrations by volume. Any deviations from this relationship are less than1%, under atmospheric conditions.

[2] **The ideal gas law**:

$$PV = nRT. \tag{D.1}$$

Here P is the pressure, V is the volume, n is the number of moles of gas, R is the gas constant, and T is the temperature in kelvins.

The value of R depends on the units of the other parameters:

For P in atmospheres and V in liters,	$R = 0.08205746$	L atm K^{-1} mol^{-1}
For P in bars and V in liters,	$R = 0.08314462$	L bar K^{-1} mol^{-1}
For P in pascals and V in cubic meters	$R = 8.314462$	m^3 Pa K^{-1} mol^{-1}

From these values it follows that 1 mole of gas at 0 °C (273.15 K) will occupy 22.414 L when under a pressure of 1 atmosphere, and 22.711 L if under a pressure of 1 bar.

No gas is an ideal gas; all deviate to a greater or lesser extent depending on conditions. For precise work it is often necessary to make an adjustment for non-ideality through a calculation of the *fugacity*, the thermodynamically correct term analogous to the partial pressure. This is done through the use of an additional equation, such as the so-called "virial equation":

$$PV = RT(1 + \frac{B_T}{V} + \ldots) \tag{D.2}$$

where B_T is called the "second virial coefficient." Subsequent coefficients may exist that are needed for circumstances more extreme than are found in the atmosphere, but we are not concerned with them here. The calculation (Eq. (D.2)) gives a value that applies to the pure gas. In gas mixtures, such as CO_2 in air, the effect is less, but difficult to calculate. Carbon dioxide is a gas where the non-ideality effect is just measurable; the numerical value for the fugacity is about 0.3 to 0.5% less than the partial pressure calculated from [1] above (Weiss 1974). See Section E.3 for more comment on fugacity.

Table D.1 Composition of the dry atmosphere (mean molecular weight = 28.97)

	Concentration parts per million by volume (ppmv)	Approximate uncertainty
N_2	780 840	± 40
O_2[(a)]	209 460	± 20
Ar	9340	± 10
CO_2[(b)]	390	± 1
Ne	18.18	± 0.04
He	5.24	± 0.02
Kr	1.14	± 0.01
Xe	0.087	± 0.001
H_2	0.53	± 0.04
CH_4[(c)]	1.79	± 0.02
N_2O[(d)]	0.323	± 0.002
Misc. gases containing halides	~2	
Hydrocarbons (non-methane)[(e)]	~1.5	

Values are calculated for dry air. For CO_2, N_2O, and CH_4 the concentration values are estimated for the year 2010. Data derived from Haynes (2011), Graedel (1978), Oliver *et al.* (1984), the National Oceanic and Atmospheric Administration (NOAA) Earth System Research Laboratory (www.esrl.noaa.gov) and the Carbon Dioxide Information Analysis Center (CDIAC) – see http://cdiac.esd.ornl.gov/home.html)

From the data above, the mean molecular weight of the dry atmosphere is 28.97.

The data for N_2, O_2, and the noble gases, are the conventional values (including uncertainties) found in standard reference works. They are given here for consistency. It should be noted that these values are based on measurements taken between 1903 and 1950; nitrogen was not measured, but calculated by difference. The recent decreases in O_2 and increases in CO_2, and uptake of CO_2 by the ocean, lead to the following approximate present values, when derived from the numbers in the table above: $N_2 = 780\ 902$, $O_2 = 209\ 319$ ppmv.

The situation with regard to major and conservative components needs to be re-evaluated.

[(a)] Haynes (2011) gives 20.9476% O_2, but this must be a typo, originally present in the source of Haynes' data. The properly cited ultimate source is the review by Glueckauf (1951), who suggests the value 20.946%. That value is also quite problematic, since, without explanation, it is based on three sets of data that do not average to the value selected. In any case, the concentration of oxygen must be decreasing at a rate of about 3.0 ppmv per year.

[(b)] The value for the year 2010. Carbon dioxide is currently increasing in the atmosphere at about 2.0 ppmv per year.

[(c)] Methane is increasing at an uneven rate.

[(d)] Nitrous oxide is increasing in the atmosphere at a rate of about 0.74 ppbv per year.

[(e)] Well over 200 non-methane hydrocarbons have been identified in the atmosphere, mostly from samples collected in anthropogenically affected regions. These hydrocarbons include both straight- and branched-chain alkanes, alkenes, and alkynes, as well as many terpenes and aromatic compounds. All are variable in concentration but a surprising number of observations of individual compounds are in the range of several parts per billion (ppbv). In total, they may amount to 1 or 2 ppmv.

Tabulated values of B_T for various gases as functions of temperature may be found in the *Handbook of Chemistry and Physics* (Haynes 2011).[a]

In most of the discussion that follows, we will consider that the partial pressure is the operative parameter, and this is simply calculated by the law of partial pressures.

[3] **Henry's law**:

$$[\mathrm{G}] = H_\mathrm{G} \times pp_\mathrm{G}. \tag{D.4}$$

The concentration of each gas in a solution in equilibrium with a gas phase is linearly proportional to the partial pressure, *pp*, (or, more precisely, the fugacity) of each gas. The constant of proportionality is the Henry's law constant, H_G. Deviations from linearity are not apparent at the concentrations of gas generally found in the ocean.

There have been many ways to express the concentrations of gases in solution, and many ways to express the solubility. At present, when dealing with solutions in seawater, the recommended procedure (Kester 1975) is to report the concentration, [G], in mol kg^{-1}. That is, [G] refers to the number of moles of a gas, G, dissolved in 1 kilogram of seawater.

Within the range of concentrations and pressures at the surface of the sea, most gases of interest may be considered as showing ideal behavior in the gas phase, and as obeying Henry's law in solution, so that the concentration of each gas in a solution in equilibrium with a gas phase is linearly proportional to the partial pressure, *pp*, of each gas.

For perhaps historical reasons, and certainly for convenience in dealing with at least the conservative atmospheric gases, it is not usual to see tabulations of the Henry's law constant itself. The common approach is to list the solubility of a gas as the concentration in solution under some defined conditions of pressure and temperature.

Pressures have been expressed in many kinds of units (e.g. feet of water, lb in^2, mm mercury, kg cm^{-2}, atmospheres, pascals, bars, and various derivatives of these). International authorities recommend only the pascal, but the bar (and mbar and dbar) is commonly used in meteorology and oceanography, while the atmosphere is still in fairly common use:[b] 1 atm = 101 325 Pa or 1013.25 mbar.

(a) Another formulation that has often been used is the Van der Waals Equation:

$$\left(p + \frac{n^2 a}{V^2}\right)(V - nb) = RT \tag{D.3}$$

where the constants a and b were introduced to account for the mutual attractions of the molecules and the space they occupy. Values of the Van der Waals constants are listed in Haynes (2001). However, the virial equation is now generally preferred.

(b) Following a recommendation in 1979 by the International Union of Pure and Applied Chemistry, the National Institute of Standards and Technology (NIST) publishes tables of thermodynamic data based on a standard pressure of 100 000 Pa (= 1 bar), rather than standard atmospheric pressure, but this move is still propagating into many fields of chemistry. For some time to come it will be necessary to watch carefully when using tabulated values for boiling points, the gas constant, and thermodynamic values.

Once the pressure units have been established, however, the gas pressure may be defined in one of four ways, depending on the purpose. For example, concentrations in solution may be referred to the situation where:

a. The partial pressure of the pure gas is 101.325 kPa (= standard atmospheric pressure). This gives the so-called "dry-gas solubility." Alternatively, the partial pressure may be specified as 100 000 Pa (=1 bar), a convention that is becoming more common.
b. The partial pressure of the pure gas plus the vapor pressure of water is 101.325 kPa. In this case the gas phase is saturated with water vapor at the temperature of measurement. This yields the "wet-gas solubility."
c. The gas phase is dry air, and the gases in solution are in equilibrium with air at the normal atmospheric concentrations, and the total gas pressure is 101.325 kPa. This gives the "dry-air solubility," sometimes called the "normal atmosphere equilibrium concentration," or NAEC.
d. The gas phase is wet air, the gases in solution are in equilibrium with air at the normal atmospheric concentrations, and the total gas pressure is 101.325 kPa. The total pressure therefore includes the vapor pressure of water at the temperature of measurement. This gives the "wet-air solubility." This convention is adopted here for the tables in Appendix D that refer to gases generally considered to be conservative in the atmosphere. This is the usual convention in oceanographic procedure.

A logical SI-related basis, in the general form whereby the Henry's law constant itself is tabulated in appropriate units, e.g. mol kg^{-1} Pa^{-1}, referring to the partial pressure of the pure gas, is not in common use at the present time.

If only the conservative gases (i.e. those that are found in the same relative proportions everywhere in the atmosphere) are to be dealt with, then for many practical problems in oceanography the most generally useful formulation is "d" above, giving the wet-air solubility. With tabulations made on this basis the effect of the vapor pressure of water has been taken into account, so that it is only necessary to know the temperature and salinity to obtain a good approximate value for the concentration of the gas at equilibrium in seawater. For precise work a correction for barometric pressure must also be made.

Even within the metric system, there has been a remarkably large number of conventions and ways in which gas concentrations and gas solubilities have been expressed. For example, concentrations have been expressed as volume per volume (mL L^{-1}, or mL mL^{-1}), as mass per volume (mg L^{-1}), in the molar scale (M = moles per liter, also mM, μM, nM, etc., as well as the obsolete term, "gm-atoms" per liter), and in the mole per mass scale (mol kg^{-1}, mmol kg^{-1}, μmol kg^{-1}, etc.). Only the last one (also called the molinity scale) is currently favored, based on its conservative nature, in that concentrations do not change with temperature and pressure, and on the ease of calculating stoichiometric relationships.

Certain expressions for the solubilities of gases may be found in the literature, identified by name:

$$\text{Bunsen absorption coefficient}: = \frac{\text{cm}^3 \text{ gas at STP}}{\text{cm}^3 \text{ solvent}[t]}$$

$$\text{Kuenen coefficient}: = \frac{\text{mL gas at STP}}{\text{g solvent}}$$

$$\text{Ostwald absorption coefficient}: = \frac{\text{vol. gas at}[t]}{\text{unit vol. solvent}[t]}$$

where "gas" means the gas dissolved in solution, [*t*] refers to the temperature of measurement and (STP) refers to standard temperature and the old standard of pressure: 0 °C and 101.325 kPa. In each of these cases the coefficient applies to a situation where the solution is in equilibrium with the gas phase such that the partial pressure of the gas in question is 1 atm. The Bunsen and Ostwald coefficients suffer from the additional complication that the volume of the solution is greater than the volume of the solvent to the extent that the dissolved gas increases the volume. The use of these old-fashioned units is discouraged.

It should also be noted that various authors have redefined several of these coefficients, as well as the Henry's law constant, in various ways. For example, the Bunsen coefficient has sometimes been defined (e.g. by Weiss 1974) as the number of moles of dissolved gas per unit volume of solution per unit partial pressure of the gas above the solution. The Greek symbols α and β have both been used by various authors to refer to either the Bunsen coefficient or the Ostwald coefficient.

It has been found that the formulation due to Weiss (1970) is generally useful in fitting the original data to get satisfactory algorithms for calculating solubilities:

$$\begin{aligned}\ln[\text{G}] = A_1 + A_2(100/T) + A_3\ln(T/100) + A_4(T/100) \\ +S[B_1 + B_2(T/100) + B_3(T/100)^2]\end{aligned} \tag{D.5}$$

where [G] is the concentration of the gas, in μmol kg^{-1}, T is the temperature in kelvins (= °C + 273.15), S is the salinity in ‰, and the As and Bs are constants determined for each gas and the units used.

For some gases a modified equation (D.6) is used because it provides increased precision and accuracy. This equation was originally introduced by García and Gordon (1992) for use with oxygen, a special case because of its importance. Subsequently, Eq. (D.6) was used by Hamme and Emerson (2004) to fit their data for nitrogen, argon, and neon.

$$\begin{aligned}\ln[G] = A_0 + A_1T_s + A_2T_s^2 + A_3T_s^3 + A_4T_s^4 + A_5T_s^5 + \\ S(B_0 + B_1T_s + B_2T_s^2 + B_3T_s^3) + C_0S^2\end{aligned} \tag{D.6}$$

Where [G] is the concentration of the gas, in μmol kg^{-1}, T_s is a special scaled temperature: = ln[(298.15 − t)/(273.15 + t)], t is the temperature in °C, S is the salinity in ‰ (with the other constants given in Table D.2b).

Table D.2a Constants for Eq. (D.5)

	He	Kr	CH_4	CO	H_2	N_2O
A_1	−156.5120	−101.9782	−68.8862	−47.6148	−47.8948	−55.6567
A_2	216·3442	153.5817	101.4956	69.5068	65.0368	100.2520
A_3	139·2032	74.4690	28.7314	18.7397	20.1709	25.2049
A_4	−22·6202	−10.0189	0	0	0	0
B_1	−0·044781	−0.011213	−0.076146	0.045657	−0.082225	−0.062544
B_2	0·023541	−0.001844	0.043970	−0.040721	0.049564	0.035337
B_3	−0·0034266	0.0011201	−0.0068672	0.0079700	0.0078689	−0.0054699

For He and Kr these constants yield the natural log of the concentration of each gas in seawater (nmol kg^{-1}) relative to air at a total air pressure of 101.325 kPa (= standard atmospheric pressure) when the air is saturated with water vapor (100% relative humidity). Modified from tabulations in Weiss (1970, 1971) and Weiss and Kyser (1978).

For CH_4, CO and H_2 the constants are from Wiesenberg and Guinasso (1979), and yield the natural log of the Bunsen absorption coefficient (mL of gas at STP contained in 1 mL of solution under a gas partial pressure of 1 atm at the temperature of measurement). Additional calculation is required to arrive at the methane and carbon monoxide data shown in Tables D.7 and D.8.

The constants for N_2O yield the natural log of the concentration in nmol kg^{-1} under a partial pressure of 1 Pa. These constants were modified from Weiss and Price (1980).

Table D.2b Constants for Eq. (D.6)

	O_2	N_2	Ne	Ar
A_0	5.80871	6.42931	2.18156	2.79150
A_1	3.20291	2.92704	1.29108	3.17609
A_2	4.17887	4.32531	2.12504	4.13116
A_3	5.10006	4.69149	–	4.90379
A_4	-9.86643×10^{-2}	–	–	–
A_5	3.80369	–	–	–
B_0	-7.01577×10^{-3}	-7.44129×10^{-3}	-5.94737×10^{-3}	-6.96233×10^{-3}
B_1	-7.70028×10^{-3}	-8.02566×10^{-3}	-5.13896×10^{-3}	-7.66670×10^{-3}
B_2	-1.13864×10^{-2}	-1.46775×10^{-2}	–	-1.16888×10^{-2}
B_3	-9.51519×10^{-3}	–	–	–
C_0	-2.75915×10^{-7}	–	–	–

Constants for Eq. (D.6) which yield the solubility of these gases in seawater, relative to air at a total gas pressure of 101.325 kPa (= standard atmospheric pressure) when the air is saturated with water vapor (100% relative humidity). With these constants the solubility is expressed in μmol kg^{-1}. García and Gordon (1992) also provide additional constants to express the solubility of oxygen in volume units.

Sources: O_2, García and Gordon (1992); N_2, Ne, and Ar, Hamme and Emerson (2004).

Table D.3 Concentration of oxygen in fresh water and seawater (μmol kg^{-1}) in equilibrium with air

	S, ‰													
t, °C	0	5	10	15	20	25	30	31	32	33	34	35	36	38	40
−2	–	–	–	–	–	–	–	–	–	–	–	–	363.9	358.2	352.6
−1	–	–	–	–	401.9	386.4	371.5	368.6	365.7	362.8	360.0	357.2	354.4	348.8	343.4
0	457.0	439.6	422.8	406.6	391.1	376.1	361.7	358.9	356.1	353.4	350.6	347.9	345.2	339.9	334.6
1	444.3	427.5	411.3	395.7	380.7	366.3	352.4	349.7	347.0	344.3	341.7	339.1	336.4	331.3	326.2
2	432.2	416.0	400.4	385.3	370.9	356.9	343.5	340.9	338.3	335.7	333.1	330.6	328.1	323.1	318.2
3	420.7	405.0	389.9	375.4	361.4	348.0	335.0	332.5	330.0	327.5	325.0	322.5	320.1	315.2	310.5
4	409.6	394.5	380.0	365.9	352.4	339.4	326.9	324.4	322.0	319.6	317.2	314.8	312.4	307.7	303.1
5	399.1	384.5	370.4	356.9	343.8	331.2	319.1	316.7	314.3	312.0	309.7	307.4	305.1	300.5	296.1
6	389.0	374.9	361.3	348.2	335.5	323.3	311.6	309.3	307.0	304.7	302.5	300.3	298.0	293.7	289.3
7	379.4	365.7	352.6	339.9	327.6	315.8	304.4	302.2	300.0	297.8	295.6	293.4	291.3	287.0	282.9
8	370.2	356.9	344.2	331.9	320.0	308.6	297.6	295.4	293.2	291.1	289.0	286.9	284.8	280.7	276.6
9	361.3	348.5	336.2	324.3	312.8	301.7	291.0	288.9	286.8	284.7	282.7	280.6	278.6	274.6	270.7
10	352.9	340.5	328.5	316.9	305.8	295.0	284.6	282.6	280.6	278.6	276.6	274.6	272.6	268.8	264.9
12	337.0	325.3	314.0	303.2	292.7	282.5	272.7	270.8	268.9	267.0	265.1	263.3	261.4	257.7	254.1
14	322.3	311.3	300.7	290.4	280.5	271.0	261.7	259.9	258.1	256.3	254.5	252.8	251.0	247.5	244.1
16	308.8	298.4	288.4	278.7	269.3	260.2	251.5	249.8	248.1	246.4	244.7	243.0	241.4	238.1	234.8
18	296.3	286.5	277.0	267.8	258.9	250.3	242.0	240.4	238.8	237.1	235.6	234.0	232.4	229.3	226.2
20	284.7	275.4	266.4	257.6	249.2	241.1	233.2	231.6	230.1	228.6	227.0	225.5	224.0	221.1	218.1
22	273.9	265.0	256.5	248.2	240.2	232.4	224.9	223.5	222.0	220.5	219.1	217.7	216.2	213.4	210.6
24	263.8	255.4	247.3	239.4	231.8	224.4	217.2	215.8	214.4	213.0	211.7	210.3	208.9	206.2	203.6
26	254.4	246.4	238.6	231.1	223.9	216.8	210.0	208.7	207.3	206.0	204.7	203.4	202.1	199.5	197.0
28	245.5	237.9	230.5	223.4	216.4	209.7	203.2	201.9	200.6	199.4	198.1	196.9	195.6	193.2	190.8
30	237.3	230.0	222.9	216.1	209.5	203.0	196.8	195.6	194.3	193.1	191.9	190.7	189.6	187.2	184.9
32	229.4	222.5	215.7	209.2	202.9	196.7	190.7	189.6	188.4	187.2	186.1	184.9	183.8	181.5	179.3
34	222.0	215.4	208.9	202.7	196.6	190.7	185.0	183.8	182.7	181.6	180.5	179.4	178.3	176.2	174.0
36	215.0	208.7	202.5	196.5	190.6	185.0	179.5	178.4	177.3	176.3	175.2	174.2	173.1	171.0	169.0
38	208.3	202.2	196.3	190.6	185.0	179.5	174.3	173.2	172.2	171.2	170.1	169.1	168.1	166.1	164.2
40	202.0	196.1	190.4	184.9	179.5	174.3	169.2	168.2	167.2	166.3	165.3	164.3	163.3	161.4	159.5

Values are calculated from Eq. (D.6), giving results for the situation where the air is saturated with water vapor under the temperature of measurement and the total pressure is 101.325 kPa (one standard atmosphere). The units are [O2] in μmol kg^{-1}.

Table D.4 Concentration of nitrogen in fresh water and seawater (μmol kg^{-1}) in equilibrium with air

	S, ‰														
t, °C	0	5	10	15	20	25	30	31	32	33	34	35	36	38	40
−2	–	–	–	–	–	–	–	–	–	–	–	653.3	647.8	637.0	626.4
−1	–	–	–	–	722.1	692.7	664.4	658.9	653.4	648.0	642.6	637.3	632.0	621.6	611.3
0	830.5	796.9	764.6	733.7	704.0	675.5	648.2	642.9	637.6	632.4	627.2	622.0	616.9	606.8	596.9
1	808.9	776.4	745.3	715.4	686.8	659.2	632.8	627.6	622.5	617.5	612.4	607.4	602.5	592.7	583.1
2	788.3	757.0	726.9	698.0	670.2	643.5	618.0	613.0	608.1	603.2	598.3	593.5	588.7	579.2	569.9
3	768.6	738.3	709.2	681.3	654.5	628.7	603.9	599.1	594.3	589.5	584.8	580.1	575.5	566.3	557.2
4	749.9	720.6	692.4	665.3	639.4	614.4	590.4	585.7	581.0	576.4	571.8	567.3	562.8	553.9	545.1
5	731.9	703.6	676.3	650.1	624.9	600.7	577.4	572.9	568.4	563.9	559.4	555.0	550.7	542.0	533.5
6	714.8	687.3	660.9	635.5	611.1	587.6	565.0	560.6	556.2	551.9	547.6	543.3	539.1	530.7	522.4
7	698.4	671.7	646.1	621.5	597.8	575.0	553.1	548.8	544.6	540.4	536.2	532.0	527.9	519.8	511.8
8	682.6	656.8	632.0	608.1	585.1	563.0	541.7	537.5	533.4	529.3	525.3	521.2	517.2	509.3	501.5
9	667.6	642.6	618.5	595.3	572.9	551.4	530.8	526.7	522.7	518.7	514.8	510.8	507.0	499.3	491.7
10	653.2	628.9	605.5	583.0	561.3	540.4	520.2	516.3	512.4	508.5	504.7	500.9	497.1	489.6	482.2
12	626.2	603.2	581.1	559.8	539.3	519.5	500.5	496.7	493.0	489.4	485.7	482.1	478.5	471.4	464.4
14	601.3	579.5	558.6	538.4	519.0	500.2	482.2	478.6	475.1	471.6	468.2	464.7	461.3	454.6	448.0
16	578.3	557.7	537.8	518.7	500.2	482.4	465.2	461.9	458.5	455.2	451.9	448.6	445.4	439.0	432.7
18	557.1	537.5	518.6	500.4	482.8	465.9	449.5	446.3	443.1	439.9	436.8	433.7	430.6	424.5	418.5
20	537.4	518.8	500.8	483.4	466.7	450.5	434.9	431.8	428.8	425.7	422.7	419.8	416.8	411.0	405.2
22	519.2	501.4	484.2	467.7	451.6	436.2	421.2	418.3	415.4	412.5	409.6	406.8	404.0	398.4	392.8
24	502.3	485.3	468.8	452.9	437.6	422.8	408.5	405.7	402.9	400.1	397.4	394.6	391.9	386.6	381.3
26	486.5	470.2	454.4	439.2	424.5	410.3	396.5	393.8	391.2	388.5	385.9	383.3	380.7	375.5	370.4
28	471.7	456.1	441.0	426.3	412.2	398.5	385.3	382.7	380.2	377.6	375.1	372.6	370.0	365.1	360.2
30	457.9	442.9	428.3	414.3	400.7	387.5	374.8	372.3	369.8	367.3	364.9	362.5	360.1	355.3	350.6
32	444.9	430.5	416.4	402.9	389.8	377.1	364.8	362.4	360.0	357.6	355.3	352.9	350.6	346.0	341.5
34	432.7	418.8	405.2	392.2	379.5	367.2	355.4	353.1	350.8	348.5	346.2	343.9	341.7	337.2	332.8
36	421.2	407.7	394.7	382.0	369.8	357.9	346.4	344.2	341.9	339.7	337.5	335.3	333.1	328.8	324.6
38	410.4	397.3	384.6	372.4	360.5	349.0	337.9	335.7	333.6	331.4	329.3	327.1	325.0	320.8	316.7
40	400.0	387.4	375.1	363.2	351.7	340.5	329.8	327.6	325.5	323.5	321.4	319.3	317.3	313.2	309.2

Values are calculated from Eq. (D.6), giving results for the situation where the air is saturated with water vapor under the temperature of measurement and the total pressure is 101.325 kPa (one standard atmosphere). Data derive from measurements by Hamme and Emerson (2004).

Table D.5 Concentrations of helium, neon, and argon in water and seawater in equilibrium with air

	Temperature, °C								
S, ‰	0	5	10	15	18	20	25	30	35
				[He], nmol kg^{-1}					
0	2.188	2.124	2.072	2.031	2.010	1.997	1.969	1.944	1.921
10	2.059	2.004	1.960	1.924	1.906	1.896	1.872	1.851	1.831
20	1.939	1.891	1.853	1.823	1.808	1.799	1.779	1.762	1.746
30	1.825	1.784	1.752	1.727	1.715	1.708	1.692	1.678	1.665
33	1.792	1.753	1.723	1.700	1.688	1.681	1.666	1.653	1.641
34	1.781	1.743	1.713	1.690	1.679	1.673	1.658	1.645	1.633
35	1.770	1.733	1.704	1.681	1.670	1.664	1.649	1.637	1.625
36	1.760	1.723	1.694	1.672	1.662	1.655	1.641	1.629	1.618
34	1.781	1.743	1.713	1.690	1.679	1.673	1.658	1.645	1.633
40	1.718	1.683	1.657	1.637	1.627	1.621	1.608	1.597	1.587
				[Ne], nmol kg^{-1}					
0	10.084	9.538	9.069	8.668	8.457	8.328	8.043	7.808	7.619
10	9.459	8.963	8.537	8.175	7.984	7.868	7.613	7.404	7.238
20	8.873	8.422	8.037	7.710	7.538	7.434	7.205	7.020	6.876
30	8.323	7.915	7.566	7.271	7.117	7.024	6.820	6.657	6.531
33	8.165	7.768	7.430	7.144	6.995	6.905	6.709	6.551	6.432
34	8.113	7.720	7.386	7.103	6.955	6.866	6.672	6.517	6.399
35	8.061	7.672	7.341	7.061	6.915	6.827	6.635	6.482	6.366
36	8.009	7.625	7.297	7.020	6.876	6.788	6.599	6.448	6.333
37	7.958	7.578	7.253	6.979	6.836	6.750	6.563	6.414	6.301
40	7.807	7.438	7.123	6.857	6.719	6.636	6.455	6.312	6.204
				[Ar], μmol kg^{-1}					
0	22.30	19.50	17.26	15.44	14.51	13.95	12.70	11.65	10.74
10	20.64	18.11	16.08	14.42	13.57	13.06	11.92	10.95	10.11
20	19.11	16.82	14.98	13.47	12.70	12.23	11.18	10.29	9.515
30	17.69	15.62	13.95	12.58	11.88	11.45	10.49	9.670	8.957
33	17.28	15.28	13.65	12.33	11.64	11.22	10.29	9.492	8.796
34	17.15	15.16	13.56	12.24	11.56	11.15	10.22	9.433	8.743
35	17.02	15.05	13.46	12.16	11.49	11.07	10.16	9.375	8.691
36	16.89	14.94	13.37	12.08	11.41	11.00	10.10	9.317	8.638
37	16.76	14.83	13.27	11.99	11.33	10.93	10.03	9.260	8.586
40	16.37	14.51	12.99	11.75	11.11	10.72	9.841	9.089	8.432

Helium data were calculated with Eq. (D.5) and the constants of Weiss (1971). Neon and argon data were calculated with Eq. (D.6), and the constants evaluated by Hamme and Emerson (2004). In each case the last significant figure is probably uncertain.

Table D.6 Concentrations of krypton and xenon in water and seawater in equilibrium with air

	Temperature, °C								
S, ‰	0	5	10	15	18	20	25	30	35
				[Kr], nmol kg^{-1}					
0	5.54	4.71	4.06	3.54	3.27	3.11	2.76	2.46	2.22
10	5.12	4.37	3.77	3.29	3.05	2.90	2.58	2.31	2.08
20	4.73	4.04	3.50	3.06	2.84	2.70	2.41	2.16	1.96
30	4.37	3.75	3.25	2.85	2.64	2.52	2.25	2.03	1.84
33	4.27	3.66	3.18	2.79	2.59	2.47	2.21	1.99	1.81
34	4.23	3.63	3.15	2.77	2.57	2.45	2.19	1.97	1.79
35	4.20	3.60	3.13	2.75	2.55	2.43	2.18	1.96	1.78
36	4.17	3.58	3.11	2.73	2.53	2.42	2.16	1.95	1.77
37	4.14	3.55	3.08	2.71	2.51	2.40	2.15	1.94	1.76
40	4.04	3.47	3.01	2.65	2.46	2.35	2.10	1.90	1.73
				[Xe], nmol kg^{-1}					
0	0.90	0.74	0.61	0.51	–.–	0.44	0.37	0.32	–.–
10	0.83	0.68	0.56	0.47	–.–	0.40	0.35	0.30	–.–
20	0.76	0.63	0.52	0.44	–.–	0.37	0.32	0.28	–.–
30	0.69	0.58	0.48	0.40	–.–	0.35	0.30	0.26	–.–
33	0.67	0.57	0.47	0.39	–.–	0.34	0.29	0.26	–.–
35	0.66	0.56	0.46	0.39	–.–	0.33	0.29	0.25	–.–
37	0.65	0.55	0.45	0.38	–.–	0.33	0.28	0.25	–.–
40	0.63	0.53	0.44	0.37	–.–	0.32	0.28	0.25	–.–

Krypton data were calculated with Eq. (D.5) and the constants of Weiss and Kyser (1978) in Table (D.2a). The xenon data are from Kester (1975), who stated that the values were provisional and in need of re-determination.

Coefficients for Eqs. (D.5) and (D.6), appropriate for several gases, are listed in Tables D.2a and D.2b. Wet-air solubilities of several gases in seawater, calculated using Eqs. (D.5) and (D.6), are given in the subsequent tables.

The solubilities of methane, carbon monoxide, and nitrous oxide, non-conservative in the atmosphere, are recalculated from Eq. (D.5) and coefficients in Table D.2a to give the concentrations under a stated partial pressure of the gas. In the case of methane (Table D.7), the partial pressure is 1 bar, convenient when considering the solubility under conditions deep in the sediment where methane hydrates form. Carbon monoxide and nitrous oxide data are presented as concentrations under a partial pressure of 1 Pa.

Mercury is an interesting and important special case, but the solubility is still not well known. The solubility of the liquid metal in water increases with temperature, because the vapor pressure increases more then the solubility of the vapor decreases with rising temperature. Some measurements of the solubility of the vapor have been made using the liquid metal and calculating the vapor pressure according the relationship illustrated in Figure D.1. Others have been made by equilibrating the vapor

Table D.7 Solubility of methane in water and seawater

	Temperature, °C												
S, ‰	–2.0	0.0	2.0	4.0	6.0	8.0	10.0	15.0	18.0	20.0	25.0	30.0	35.0
	$[CH_4]$, mmol kg^{-1} bar^{-1}												
0.0	–.—	2.527	2.379	2.244	2.123	2.012	1.912	1.698	1.592	1.528	1.392	1.282	1.193
10.0	–.—	2.331	2.197	2.075	1.965	1.865	1.774	1.580	1.482	1.424	1.299	1.199	1.117
20.0	–.—	2.150	2.029	1.919	1.820	1.729	1.646	1.470	1.381	1.328	1.214	1.121	1.046
25.0	–.—	2.065	1.950	1.846	1.751	1.665	1.586	1.418	1.333	1.282	1.173	1.084	1.012
30.0	–.—	1.983	1.874	1.775	1.685	1.603	1.528	1.367	1.286	1.238	1.133	1.048	0.979
33.0	–.—	1.936	1.830	1.734	1.647	1.567	1.494	1.338	1.259	1.212	1.110	1.027	0.960
35.0	2.019	1.905	1.801	1.707	1.621	1.543	1.472	1.319	1.242	1.195	1.095	1.014	0.947
37.0	1.986	1.874	1.773	1.681	1.597	1.520	1.450	1.300	1.224	1.179	1.080	1.000	0.935
40.0	1.938	1.830	1.731	1.642	1.560	1.486	1.418	1.272	1.198	1.154	1.059	0.980	0.916

The original data of Yamamoto *et al.* (1976) were refit by Wiesenberg and Guinasso (1979) to an equation of the type used by Weiss (1970) to yield values of β, the Bunsen absorption coefficient (Table D.2a). For this Table D.7, the β values were then converted to concentrations in (mmol kg^{-1}) at equilibrium under a partial pressure of 1 bar.

in air or nitrogen gas with samples of water. These measurements are difficult and authors differ. Most investigators use the data of Sanemasa (1975), because he is the only person who investigated both water and seawater, and because his data appear internally consistent (Figure D.2). For pure water, Sanemasa's data show a considerably steeper slope with temperature than those of other investigators. If the others are correct (and Clever *et al.* 1985, who have reviewed previous work, do not prefer Sanemasa's data) then at low temperatures the relationship given in Figure D.2 will underestimate the solubility. The difference is large, more than a factor of 2 at 5 °C. Models of mercury cycling in the ocean and atmosphere may be significantly compromised. Most workers agree on the solubility at about 25 °C while Sanemasa would overestimate the solubility at higher temperatures. The procedure to achieve equilibration of mercury vapor with the water involved bubbling the carrier gas (usually air) through the liquid; because the pressure inside the bubbles was undoubtedly slightly higher than atmospheric, the values calculated by Sanemasa may all be a few parts in a hundred too high. The salinity of the seawater used by Sanemasa was not reported, but presumably was about 34 ‰, typical of surface seawater near Japan. In a recent paper Andersson *et al.* (2004) report that their measurements of solubility in pure water agree almost exactly with those of Sanemasa; however, they did not find any difference between the solubility in seawater and fresh water. If the latter observation is correct, mercury vapor is uniquely the only gas whose solubility is not affected by the presence of salt.

Here I retain the solubility data of Sanemasa because they include seawater data with a plausible offset from pure water. It is clear, however, that the question of the solubility of mercury vapor needs to be revisited.

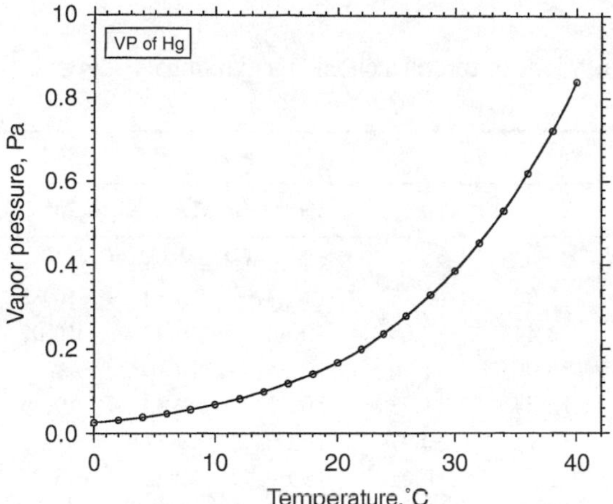

Figure D.1 Vapor pressure of metallic mercury, calculated from relationships in Ambrose and Sprake (1972). Over this range, at least, the curve may be linearized and data calculated by the equation: $\log_{10}$ VP = 10.204 – 3218.7/T, where VP is in Pa, and T is in kelvin. At a room temperature of 20 °C the pressure of 0.169 Pa is sufficient to produce at saturation an atmospheric concentration of about 15 mg m^{-3}, far above the level at which this element is toxic under conditions of chronic exposure. This is why industrial exposure used to cause mercury poisoning, and why it is undesirable to allow any mercury spilled from broken thermometers or laboratory apparatus to remain uncollected.

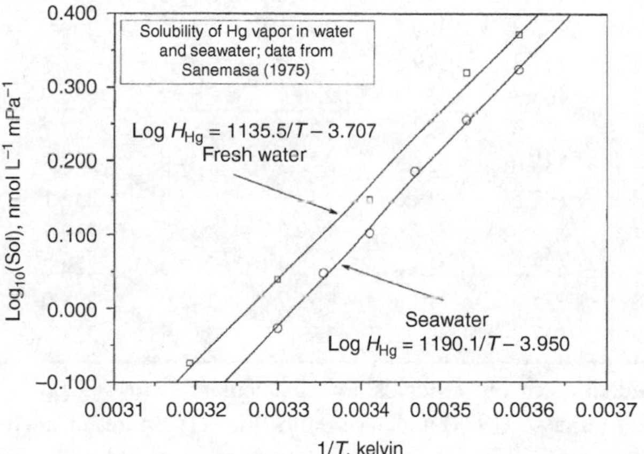

Figure D.2 The data on solubility of mercury vapor in fresh water and in seawater obtained by Sanemasa (1975) were converted to SI units and plotted here as the log of the Henry's law constant (where [Hg] = $H_{Hg} \times pp$Hg). The equations of the straight regression lines fitted to the data points were used to calculate the solubilities given in Table D.9. The units used for H_{Hg}, nmol L^{-1} mPa^{-1}, appear a little awkward, but are used here to emphasize the fact that vapor pressures of mercury cannot get as high as 1 Pa without condensing liquid mercury, unless the temperature is above 50 °C (see Figure D.1).

Table D.8 Concentrations of carbon monoxide and nitrous oxide, each under a partial pressure of 1 Pa, in water and seawater

	Temperature, °C								
S, ‰	0	5	10	15	18	20	25	30	35
				[CO], nmol kg^{-1} Pa^{-1}					
0	15.64	13.90	12.49	11.33	10.74	10.38	9.58	8.92	8.37
5	15.11	13.44	12.09	10.98	10.42	10.07	9.32	8.69	8.17
10	14.60	13.00	11.70	10.64	10.10	9.77	9.06	8.46	7.97
15	14.10	12.57	11.33	10.32	9.80	9.49	8.80	8.24	7.78
20	13.62	12.15	10.97	10.00	9.51	9.21	8.56	8.03	7.59
25	13.16	11.75	10.61	9.69	9.22	8.94	8.32	7.82	7.41
30	12.71	11.36	10.27	9.39	8.94	8.67	8.09	7.62	7.23
33	12.45	11.14	10.08	9.22	8.78	8.52	7.95	7.50	7.13
34	12.37	11.06	10.01	9.16	8.73	8.47	7.91	7.46	7.09
35	12.28	10.99	9.95	9.10	8.68	8.42	7.86	7.42	7.06
36	12.20	10.91	9.88	9.05	8.62	8.37	7.82	7.38	7.02
37	12.11	10.84	9.82	8.99	8.57	8.32	7.78	7.34	6.99
40	11.87	10.62	9.63	8.82	8.42	8.17	7.65	7.22	6.89
				[N_2O], nmol kg^{-1} Pa^{-1}					
0	585.6	478.2	396.5	333.5	302.5	284.2	245.3	214.2	189.2
5	565.9	462.8	384.1	323.4	293.5	275.9	238.3	208.2	184.0
10	546.9	447.8	372.1	313.6	284.8	267.8	231.5	202.4	178.9
15	528.6	433.3	360.5	304.1	276.4	259.9	224.8	196.7	174.0
20	510.8	419.3	349.3	294.9	268.1	252.3	218.4	191.2	169.2
25	493.7	405.8	338.4	286.0	260.2	244.9	212.1	185.8	164.5
30	477.1	392.6	327.8	277.4	252.5	237.7	206.1	180.6	160.0
33	467.4	385.0	321.6	272.3	247.9	233.5	202.5	177.6	157.3
34	464.2	382.5	319.6	270.7	246.4	232.1	201.3	176.6	156.4
35	461.1	379.9	317.6	269.0	245.0	230.7	200.2	175.6	155.6
36	457.9	377.5	315.6	267.4	243.5	229.3	199.0	174.6	154.7
37	454.8	375.0	313.6	265.7	242.0	228.0	197.8	173.6	153.8
40	445.6	367.7	307.7	260.9	237.7	223.9	194.4	170.6	151.3

Carbon monoxide and nitrous oxide data were calculated with Eq. (D.5), with constants from Wiesenberg and Guinasso (1979) and Weiss and Price (1980), respectively. In the case of CO, the Bunsen coefficients from Table D.2a were recalculated to give the solubility in the units above.

The rate at which gases exchange between the atmosphere and the sea is a difficult subject, and evaluations require the Schmidt number (Table D.11), and that in turn requires an estimate of the diffusion coefficients of the individual gases Table D.10).

Table D.9 Solubility of mercury vapor in distilled water and in seawater

	nmol L^{-1} mPa^{-1}			nmol L^{-1} mPa^{-1}	
t, °C	DI water	Seawater	*t*, °C	DI water	Seawater
0	2.82	2.55	22	1.38	1.21
2	2.63	2.37	24	1.30	1.14
4	2.46	2.21	26	1.23	1.07
6	2.30	2.06	28	1.16	1.00
8	2.15	1.92	30	1.09	0.946
10	2.01	1.79	32	1.03	0.891
12	1.88	1.67	34	0.977	0.841
14	1.77	1.57	36	0.925	0.794
16	1.66	1.47	38	0.876	0.750
18	1.56	1.37	40	0.830	0.709
20	1.47	1.29			

Values were converted to SI units and recalculated from Sanemasa (1975) according to the relationships in Figures D.1 and D.2, and are expressed here as H_{Hg}, the Henry's law constant, as in ($[Hg] = H_{Hg} \times ppHg$), in units of nmol L^{-1} mPa^{-1}. The last significant figure, at least, should be considered uncertain; see text.

Table D.10 Diffusion coefficients (10^{-9} m^2 s^{-1}) of some gases in water

		Temperature, °C								
Gas	*S*, ‰	0	5	10	15	18	20	25	30	35
He	0	4.74	5.20	5.68	6.19	6.51	6.73	7.29	7.89	8.50
	35	4.46	4.90	5.37	5.87	6.18	6.39	6.94	7.52	8.13
Ne	0	2.34	2.63	2.94	3.28	3.5	3.65	4.04	4.46	4.91
Ar	0	1.20	1.42	1.66	1.93	2.11	2.24	2.58	2.96	3.38
Kr	0	0.88	1.03	1.20	1.39	1.52	1.61	1.85	2.11	2.41
Xe	0	0.66	0.79	0.93	1.09	1.20	1.27	1.47	1.70	1.96
Rn	0	0.57	0.68	0.81	0.96	1.07	1.14	1.34	1.56	1.81
H_2	0	2.83	3.22	3.64	4.10	4.39	4.59	5.13	5.71	6.33
	35	2.77	3.11	3.49	3.89	4.15	4.33	4.80	5.30	5.84
CH_4	0	0.94	1.09	1.25	1.43	1.55	1.63	1.85	2.09	2.35
CO_2	0	0.93	1.09	1.26	1.46	1.59	1.68	1.92	2.18	2.47
O_2	0	1.29	1.49	1.72	1.96	2.13	2.24	2.54	2.87	3.23
N_2	0	1.40	1.63	1.90	2.19	2.39	2.52	2.89	3.29	3.73

Fitting parameters

	A	*E*		*A*	*E*		*A*	*E*
He (0)	818	11 700	Xe (0)	9007	21 610	CO_2 (0)	5019	19 510
He (35)	886	12 020	Rn (0)	15 877	23 260	O_2 (0)	4200	18 370
Ne (0)	1608	14 840	H_2 (0)	3338	16 060	N_2 (0)	7900	19 620
Ar (0)	10 600	20 630	H_2 (35)	1981	14 930			
Kr (0)	6393	20 200	CH_4 (0)	3047	18 360			

Values for O_2, N_2, and Ar were calculated from the relationships in Wise and Houghton (1966) and the rest from Jähne *et al.* (1987a), who measured these coefficients (except for Rn, fitted to literature values) at 5, 10, 15, 25, and 35 °C and fitted an equation of the form:

$$D = Ae^{\left(\frac{-E}{RT}\right)}$$

where D is in units of 10^{-9} m^2 s^{-1}, E is the energy of activation of diffusion in J mol^{-1}, R is the gas constant (8.3145 J mol^{-1} K^{-1}), T is temperature in kelvin, and A is a fitting parameter. All coefficients were measured in distilled water, and those for He and H_2 were also measured in solutions of NaCl to approximate the effects of salt in seawater ($S \approx 35$ ‰). Values below 5 °C in this table are extrapolated outside the range of measurement, to suggest approximately the diffusion constants to be expected at these lower temperatures. These measurements are difficult to make, but Jähne *et al.* suggest that the values here are likely to depart from true values by less than 5%.

Table D.11 Estimates of the Schmidt number in seawater for a variety of gases

	Temperature, °C									
Gas	**–1**	**0**	**5**	**10**	**15**	**18**	**20**	**25**	**30**	**35**
He	431	410	320	252	202	178	165	136	111	86
Ne	903	855	653	503	394	344	315	255	203	150
Ar	2038	1909	1375	999	745	636	576	455	346	211
Kr	2345	2205	1619	1196	899	768	694	544	415	270
Rn	3644	3413	2451	1766	1297	1093	981	756	559	328
O_2	2085	1953	1407	1022	762	651	589	466	354	216
N_2	2356	2206	1588	1155	863	738	670	532	409	256
CH_4	2163	2039	1518	1138	868	746	678	538	417	285
CO_2	2202	2073	1530	1136	859	735	666	525	402	267
N_2O	2457	2301	1657	1,04	900	769	698	555	426	267
SF_6	3770	3532	2544	1849	1379	1177	1066	842	640	391

All values calculated for a salinity of 35‰.
The Schmidt number is defined as:

$$Sc = \frac{v}{D}$$

where Sc is the Schmidt number, v is the kinematic viscosity ($m^2\ s^{-1}$), and D is the diffusion coefficient in SI units ($m^2\ s^{-1}$). Wanninkhof (1992) took measured diffusion coefficients from Jähne *et al.* (1987a), added theoretically derived diffusion coefficients for gases not measured by Jähne *et al.*, calculated a table of Sc, then fitted the resulting values to third-degree polynomials and published the polynomial coefficients for each gas. The values in the table were calculated from these algorithms of Wanninkhof.

APPENDIX

E Carbon dioxide

This section contains information on the relationships of CO_2 in solution, including the solubility of the gas, constants describing the association of carbonate species with protons, and the degree of isotopic fractionation between the various species.

E.1 Acid–base buffering

A complete evaluation of the acid–base buffer in seawater requires knowledge of the appropriate physical constants for all significant substances that react with protons. The dominant buffer in seawater is the carbon dioxide system, and the necessary constants for that system are included in this section.

Several different pH scales have been used in marine work, and the definitions and numerical values of most of the important constants are correspondingly affected. For many decades the common approach has been to use measurements with a glass electrode standardized with NBS buffers. Because of small uncertainties associated with the liquid junction potential of glass electrodes, and the greater ionic strength of seawater, this approach leaves some slight uncertainty in the final values. (It seems possible that the spectrophotometric method of measurement will come to be preferred for precise work.) Of the several scales suggested, the pH_T scale is currently favored for the most precise work in the open ocean, calibrated with buffers made up to the ionic strength of (usually) seawater with $S = 35‰$. There is no commercial source for the buffers needed to calibrate electrodes on the pH_F, pH_T, and pH_{SWS} scales, and these must be made over a range of salinities in the laboratory, a process requiring considerable care and chemical skill (Dickson *et al.* 2007). See Appendix F for further comments on pH issues. Extensive discussions of pH measurements in seawater are presented in Dickson *et al.* (2007) and in Riebesell *et al.* (2010).

In this book, I have retained the traditional approach in which the hydrogen ion activity is defined relative to the NBS buffers and measurement is by glass electrode. There are several reasons for this. There is a large body of work in this scale, the buffers are readily available, physiological work is carried out with this scale, and it is more practical for work in the coastal zone and in estuaries where salinities may vary from zero to that of full-strength seawater (Cai and Wang 1998, Frankignoulle and Borges 2001).

Several investigators have evaluated the dissociation constants of the carbonate species. Currently, the best constants appear to be those determined (on the pH_{NBS} scale) by Mehrbach *et al.* (1973), and these are used throughout the oceanic community. (For use with the pH_T scale they have been adjusted to this scale by use of the f_H

Table E.1 Algorithms for calculating the carbon dioxide constants in seawater

In the following formulations:

T	is the temperature (kelvin: K = t (°C) + 273.15)
S	is the salinity (‰) (PSS 1978)
$ppCO_2$	is the partial pressure of CO_2, usually in atmospheres (1 atm = 101.325 kPa)
$[CO_2aq]$	is the concentration of $[CO_2(f)] + [H_2CO_3]$ (mol kg^{-1} of seawater)
H_{CO2}	is the Henry's law constant
pK_1^*	is negative log of the first apparent dissociation constant of carbonic acid
pK_2^*	is negative log of the second apparent dissociation constant of carbonic acid

(1) Solubility of CO_2 (form slightly modified from Weiss 1974):
$\ln H_{CO2} = 9345.17/T - 167.8108 + 23.3585 \ln T + (0.023517 - 2.3656 \times 10^{-4}\, T + 4.7036 \times 10^{-7}\, T^2)\, S.$

(2) Concentration of CO_2aq in solution:
$[CO_2aq] = H_{CO2} \times ppCO_2$.
For a given partial pressure of CO_2 the concentration of CO_2aq is always calculable from the Henry's law constant. If the solution is made sufficiently acid that the dissociation of carbonic acid is completely suppressed, then no other forms are present and CO_2aq constitutes the total CO_2 in the solution.

(3) Dissociation constants of carbonic acid (from Mehrbach *et al.* 1973 as modified by Plath *et al.* 1980); for use in seawater with $S > 20$‰:
$pK_1^* = 17.788 - 0.073104\, T - 0.0051087\, S + 1.1463 \times 10^{-4}\, (T^2)$,
$pK_2^* = 20.919 - 0.064209\, T - 0.011887\, S + 8.7313 \times 10^{-5}\, (T^2)$.

(4) Dissociation constants of carbonic acid (from Millero 2010) on the pH_{SWS} scale for use in seawater with $S < 20$‰:
$pK_1^* = 6320.813/T - 126.34048 + 19.568224 \ln T + 13.4038 S^{0.5} + 0.03206 S - 0.00005242 S^2 + (-530.659 S^{0.5} - 5.8210 S)/T - 2.0664 S^{0.5} \ln T.$
$pK_2^* = 5143.692/T - 90.18333 + 14.613358 \ln T + 21.3728 S^{0.5} + 0.1218 S - 0.0003688 S^2 + (-788.289 S^{0.5} - 19.189 S)/T - 3.374 S^{0.5} \ln T.$
Each resulting K value is multiplied by the f_H factor calculated as in Table F.2 to yield K_1^* and K_2^* values on the pH_{NBS} scale, as in Table E.4.

factor.) Throughout this book, all calculations relative to the carbon dioxide system have been made using the solubility of the gas from the measurements of Weiss (1974) and the dissociation constants measured by Mehrbach *et al.* (1973) as fitted to improved algorithms by Plath *et al.* (1980). These dissociation constants were measured using the pH_{NBS} scale and the moles per kilogram of solution concentration scale. Algorithms for calculating the constants associated with the carbon dioxide system in seawater are given in Table E.1. Selected values for the solubility of CO_2 gas, and for the dissociation constants of carbonic acid, are given in Tables E.2, E.3a, E.3b, and E.4.

The measurements by Mehrbach *et al.* (1973) were carried out over the salinity range of only $S = 20$ to $S = 40$‰, and their constants do not extrapolate to known and well-established values at zero salinity. Accordingly, the numerical values tabulated in Tables E.3a and E.3b are listed only for the salinity range of 20 to 40‰. Millero (2010) has combined results from two sets of measurements made over the full range of salinity relative to the pH_{SWS} scale with those of Mehrbach *et al.* (1973),

Table E.2 Solubility of CO_2 in fresh water and seawater

	Temperature, °C												
S, ‰	–2	0	2	4	6	8	10	15	18	20	25	30	35
	mmol kg^{-1}												
0	–	77.57	71.73	66.49	61.77	57.51	53.67	45.56	41.55	39.16	34.06	29.95	26.62
5	–	75.28	69.63	64.55	59.99	55.87	52.15	44.30	40.42	38.12	33.19	29.22	26.00
10	–	73.05	67.58	62.67	58.25	54.27	50.67	43.08	39.33	37.10	32.33	28.50	25.39
15	–	70.89	65.60	60.84	56.57	52.71	49.23	41.89	38.26	36.11	31.5	27.80	24.80
20	–	68.80	63.67	59.07	54.93	51.20	47.84	40.74	37.23	35.15	30.7	27.12	24.22
25	–	66.76	61.80	57.35	53.35	49.74	46.48	39.61	36.22	34.21	29.91	26.45	23.66
30	–	64.79	59.99	55.68	51.80	48.31	45.16	38.52	35.24	33.30	29.14	25.80	23.11
31	–	64.40	59.63	55.35	51.50	48.03	44.90	38.31	35.05	33.12	28.99	25.68	23.00
32	–	64.01	59.27	55.02	51.20	47.75	44.64	38.09	34.86	32.94	28.84	25.55	22.89
33	–	63.63	58.92	54.70	50.90	47.48	44.39	37.88	34.67	32.76	28.69	25.42	22.78
34	–	63.25	58.57	54.38	50.60	47.20	44.13	37.67	34.48	32.58	28.54	25.30	22.67
35	–	62.87	58.22	54.05	50.31	46.93	43.88	37.46	34.29	32.41	28.39	25.17	22.57
36	67.65	62.49	57.88	53.74	50.01	46.66	43.63	37.25	34.10	32.23	28.24	25.05	22.46
38	66.84	61.75	57.19	53.10	49.43	46.12	43.13	36.83	33.73	31.89	27.95	24.80	22.25
40	66.03	61.01	56.51	52.48	48.85	45.58	42.63	36.42	33.36	31.54	27.66	24.55	22.04

These data refer to the concentration of CO_2aq, where CO_2aq is $[CO_2(f)] + [H_2CO_3]$. The values are calculated for a situation where the water is in equilibrium with a gas phase in which the partial pressure (the fugacity) of CO_2 is 101 325 Pa (= 1 atm.). They are calculated using the algorithm (Table E.1) from Weiss (1974). The values are equivalent to a Henry's law constant expressed in units of mmol kg^{-1} atm^{-1}, so the values above are in mmol kg^{-1}.

adjusted to this scale, and presented algorithms that enable the calculation of constants over the range of 0 to 40‰ appropriate for the pH_F, pH_T, and pH_{SWS} scales. Here I have adjusted the algorithm (shown in Table E.1) to yield values on the pH_{NBS} scale, and tabulated values are listed in Table E.4 for salinity values below 20. Thus, the full range is available through Tables E.3a, E.3b, and E.4

E.2 Air–sea exchange of carbon dioxide

Carbon dioxide is an example of a gas that exhibits non-ideal behavior (Weiss 1974). At normal atmospheric temperatures CO_2 is close enough to its condensation temperature that the chemical activity, or *fugacity*, of the gas departs from the simple law of partial pressures to an extent that is detectable with precise measurements. The intermolecular attractions of the molecules reduce the effective chemical activity of the gas. The solubility data given in Table E.2 are based on the fugacity calculated according to the virial equation for gases (Weiss 1974).

All the equations in this book dealing with CO_2 in solution that yield the partial pressure of CO_2 yield the fugacity, expressed in units of pressure (e.g. Pa, μatm). The

Table E.3a First apparent dissociation constant, K_1^*, of carbonic acid, based on Mehrbach *et al.* (1973)

	Temperature, °C															
	−2	**0**	**2**	**4**	**6**	**8**	**10**	**12**	**14**	**16**	**18**	**20**	**25**	**30**	**35**	**40**
S, ‰	multiply all values by 10^{-7}															
20	–	5.369	5.628	5.888	6.147	6.403	6.656	6.905	7.148	7.383	7.610	7.828	8.323	8.733	9.043	9.241
25	–	5.694	5.969	6.245	6.519	6.791	7.060	7.323	7.581	7.830	8.072	8.302	8.827	9.262	9.591	9.801
30	–	6.039	6.331	6.623	6.914	7.203	7.487	7.767	8.040	8.305	8.560	8.805	9.362	9.823	10.172	10.395
31	–	6.111	6.406	6.701	6.996	7.288	7.576	7.859	8.135	8.403	8.662	8.910	9.473	9.939	10.292	10.518
32	–	6.183	6.482	6.781	7.079	7.374	7.665	7.952	8.231	8.503	8.764	9.015	9.585	10.057	10.414	10.642
33	–	6.256	6.558	6.861	7.162	7.461	7.756	8.046	8.329	8.603	8.868	9.122	9.698	10.176	10.537	10.768
34	–	6.330	6.636	6.942	7.247	7.550	7.848	8.141	8.427	8.705	8.973	9.230	9.813	10.296	10.662	10.896
35	–	6.405	6.715	7.024	7.333	7.639	7.941	8.237	8.527	8.808	9.079	9.339	9.929	10.418	10.788	11.025
36	6.169	6.481	6.794	7.107	7.420	7.729	8.035	8.335	8.628	8.912	9.186	9.449	10.047	10.542	10.916	11.155
38	6.316	6.635	6.956	7.277	7.596	7.913	8.226	8.533	8.833	9.124	9.405	9.674	10.286	10.792	11.176	11.421
40	6.466	6.793	7.121	7.450	7.777	8.102	8.422	8.736	9.043	9.341	9.629	9.905	10.531	11.049	11.442	11.693

Values from $S = 20$ to $S = 40$‰ calculated from the data of Mehrbach *et al.* (1973) as fitted by the algorithm presented by Plath *et al.* (1980), in Table E.1.

Table E.3b Second apparent dissociation constant, K_2^*, of carbonic acid, based on Mehrbach *et al.* (1973).

	Temperature °C															
	−2	**0**	**2**	**4**	**6**	**8**	**10**	**12**	**14**	**16**	**18**	**20**	**25**	**30**	**35**	**40**
S‰	Multiply all values by 10^{-10}															
20	–	2.203	2.375	2.556	2.747	2.947	3.157	3.377	3.605	3.844	4.091	4.347	5.025	5.750	6.513	7.305
25	–	2.526	2.723	2.931	3.150	3.379	3.620	3.872	4.134	4.407	4.691	4.985	5.762	6.593	7.469	8.376
30	–	2.896	3.122	3.361	3.612	3.875	4.151	4.440	4.740	5.054	5.379	5.716	6.607	7.560	8.564	9.605
31	–	2.976	3.209	3.454	3.712	3.983	4.266	4.563	4.872	5.194	5.528	5.874	6.790	7.770	8.802	9.871
32	–	3.059	3.298	3.550	3.815	4.093	4.385	4.689	5.007	5.338	5.681	6.037	6.978	7.985	9.046	10.145
33	–	3.144	3.389	3.648	3.921	4.207	4.506	4.819	5.146	5.486	5.839	6.205	7.172	8.207	9.297	10.427
34	–	3.231	3.483	3.750	4.030	4.323	4.631	4.953	5.289	5.638	6.001	6.377	7.371	8.434	9.555	10.716
35	3.075	3.321	3.580	3.854	4.141	4.443	4.760	5.091	5.436	5.795	6.168	6.554	7.575	8.669	9.820	11.014
36	3.160	3.413	3.679	3.961	4.256	4.567	4.892	5.232	5.587	5.956	6.339	6.736	7.786	8.909	10.093	11.319
38	3.338	3.605	3.887	4.183	4.496	4.824	5.167	5.526	5.901	6.291	6.695	7.115	8.224	9.410	10.661	11.956
40	3.526	3.808	4.105	4.419	4.749	5.095	5.458	5.837	6.233	6.645	7.072	7.515	8.686	9.940	11.260	12.629

Values from $S = 20$ to $S = 40$‰ calculated from the data of Mehrbach *et al.* (1973) as fitted by the algorithm presented by Plath *et al.* (1980), in Table E.1.

Table E.4 First and second apparent dissociation constants, K_1^* and K_2^*, of carbonic acid, on the pH_{NBS} scale, based on Millero (2010), for $S = 0$ to 15‰

	Temperature, °C														
	0	**2**	**4**	**6**	**8**	**10**	**12**	**14**	**16**	**18**	**20**	**25**	**30**	**35**	**40**
S, ‰	K_1^*, multiply all values by 10^{-7}														
0	1.999	2.120	2.241	2.361	2.481	2.599	2.715	2.827	2.937	3.043	3.144	3.375	3.570	3.725	3.836
5	3.460	3.676	3.894	4.114	4.335	4.558	4.779	5.000	5.219	5.436	5.649	6.162	6.637	7.066	7.440
10	4.079	4.327	4.579	4.834	5.091	5.349	5.608	5.867	6.125	6.381	6.634	7.249	7.831	8.369	8.852
15	4.557	4.826	5.098	5.375	5.653	5.933	6.215	6.496	6.777	7.056	7.333	8.009	8.655	9.257	9.807
	K_2^* multiply all values by 10^{-10}														
0	0.178	0.191	0.204	0.217	0.231	0.245	0.260	0.274	0.289	0.304	0.318	0.355	0.391	0.425	0.457
5	0.964	1.046	1.132	1.224	1.320	1.420	1.526	1.636	1.752	1.872	1.997	2.331	2.694	3.084	3.501
10	1.453	1.577	1.709	1.848	1.996	2.152	2.316	2.489	2.671	2.862	3.063	3.606	4.210	4.877	5.608
15	1.847	2.001	2.165	2.340	2.525	2.721	2.928	3.147	3.378	3.621	3.877	4.577	5.363	6.244	7.222

Values calculated from the algorithm for pH_{SWS} of Millero (2010), then adjusted to the pH_{NBS} scale by multiplying by the f_H factor from Appendix F.

Table E.5 Water-vapor and fugacity adjustments for CO_2 in air when the concentration in the atmosphere is 400 ppmv

	Temperature, °C										
	–2	0	5	10	15	18	20	25	30	35	40
	Partial pressure after subtracting water vapor, μatm										
Atm. press.											
98 000	384.8	384.5	383.5	382.1	380.2	378.9	377.8	374.6	370.4	365.0	358.2
100 000	392.7	392.4	391.4	390.0	388.1	386.7	385.7	382.5	378.3	372.9	366.1
101 325	397.9	397.6	396.6	395.2	393.4	**392.0**	390.9	387.7	383.5	378.1	371.3
102 000	400.6	400.3	399.3	397.9	396.0	394.6	393.6	390.4	386.2	380.8	373.9
105 000	412.5	412.1	411.1	409.7	407.9	406.5	405.4	402.2	398.0	392.6	385.8
	Fugacity after making fugacity adjustment, μatm										
98 000	383.1	382.9	381.9	380.7	378.9	377.6	376.5	373.4	369.3	364.0	357.2
100 000	391.0	390.7	389.8	388.5	386.8	385.4	384.4	381.2	377.1	371.8	365.1
101 325	396.1	395.9	395.0	393.7	391.9	**390.6**	389.6	386.4	382.3	377.0	370.3
102 000	398.8	398.5	397.6	396.3	394.6	393.3	392.2	389.1	385.0	379.7	372.9
105 000	410.5	410.2	409.4	408.1	406.3	405.0	404.0	400.9	396.8	391.5	384.7

Atmospheric pressure in pascals. Standard atm. pressure is 101 325 Pa, shown in bold. The global average sea surface temperature is 18 °C.
Calculated for salinity = 35‰. Water-vapor correction calculated from Appendix C.3. Fugacity correction calculated as in Dickson *et al.* (2007).

fugacity differs from the partial pressure calculated from the ideal law of partial pressures by about 0.35%, though this varies with temperature (Table E.5).

Concentrations of CO_2 in the atmosphere are most commonly reported in units of parts per million by volume (ppmv) or as the mole fraction, expressed relative to dry air. For rough calculation of air–sea exchange one could take the partial pressure of CO_2 in air that has, for example, 400 ppmv, to be equal to 400 μatm. However, three adjustments are needed to obtain an estimate of the exact partial pressure of the CO_2 near a water surface in the real atmosphere.

The first adjustment is because atmospheric pressure varies, often by several percent, so it is only on average about 1 atm or 101 325 Pa. In situations where the pressure is known, this can be taken into account. Otherwise, one usually just assumes that the atmosphere is average, which introduces some uncertainty (Table E.5).

The second is due to the specification that the concentration is based on dry air, while the atmosphere right at the sea surface is generally assumed to be saturated with water vapor. This dilutes the dry air according to the vapor pressure of the water. The effect is considerable. The water-vapor effect reduces the partial pressure from 0.6 to over 7% through the temperature range of 0 to 40 °C.

The third adjustment is due to the measurably non-ideal behavior of CO_2, where the activity of the molecules is detectably reduced. The resulting diminished partial pressure, the thermodynamically correct activity of the CO_2 molecules, is called fugacity, and the solubility given in Table E.2 is based on the fugacity, expressed in pressure units. The departure from ideality is a function of temperature and also depends on the concentration of CO_2 in the atmosphere (Weiss 1974). Under current conditions the effect amounts to between 0.25 to 0.44%. That is, at 400 ppmv in dry air the fugacity ranges from 398.3 to 399.0 μatm over the temperature range of 0 to 40 °C.

The following equation, derived from Dickson *et al.* (2007, see Weiss 1974), gives the fugacity, *f*, of CO_2 expressed in micro-atmospheres (μatm) calculated from the concentration in dry air expressed as the mole fraction:

$$f(CO_2) = X_{CO2}\frac{A}{101325}e^{\frac{[B+2(1-X_{CO2})\times(57.7-0.118T)]\times A}{RT}}, \quad (E.1)$$

where X_{CO2} = ppmv × 10^{-6}, A = atmospheric pressure in Pa (if not known, use 101 325 Pa), B = (–1636.75 + 12.0408T – 3.27957 × 10^{-2} T^2 + 3.16528 × 10^{-5} T^3), R = the gas constant, 8.31446; and T = temperature in kelvins.

Table E.5 provides examples of the size of the effect of the adjustments from the concentration in dry air, including first the effect of water vapor, and also the effect of changes in total barometric pressure, expressed in pascals, followed by the adjustment for non-ideality to arrive at the fugacity, expressed in this case in units of μatm.

Fluxes of CO_2 across the air–sea boundary are not calculated using the concentration of total CO_2 (TCO_2) in the water, but rather by using the concentration of dissolved CO_2 gas (for which the value of CO_2aq is generally used though it overstates the concentration of dissolved gas by about 0.1%), as this is the species that directly crosses the boundary. Because of this, the relationship between the TCO_2 in the water and the gradient between the water and the atmosphere depends on all those factors that affect the proportions of the various constituents of the carbonate system. The relationship depends on the temperature, salinity, and pH of the water; the latter is of course influenced by changes in the concentration of TCO_2.

Imagine a situation where the surface mixed layer is exactly in equilibrium with the atmosphere. Then a plankton bloom occurs. The O_2 concentration will increase, and the CO_2 concentration will correspondingly decrease. As shown in Chapter 8, the ratio of the changes in the two gases during biological activity varies somewhat, and is frequently a little uncertain, but it is convenient as an example for this evaluation to use an O_2 increase of 100 μmol kg^{-1} and a CO_2 decrease of 100 μmol kg^{-1}. Both gases will then tend to re-equilibrate with the atmosphere, O_2 decreasing and CO_2 increasing. The rate at which CO_2 equilibrates will be much slower than that of O_2 because only a small proportion of the total CO_2 is in the form of dissolved CO_2. The difference in the fluxes is quite considerable and also is dependent on conditions, of which temperature is especially important. The fraction of the TCO_2 that is present as CO_2aq depends on the pH; when CO_2 is removed by photosynthesis the pH increases, and when it is produced by respiration the pH decreases. Accordingly, the rate at which CO_2 in the water equilibrates with the atmosphere will be different for the case where the departure from equilibrium results from respiration or from photosynthesis (Figure E.1). The ratio of the two rates is also a function of salinity, increasing at lower salinity.

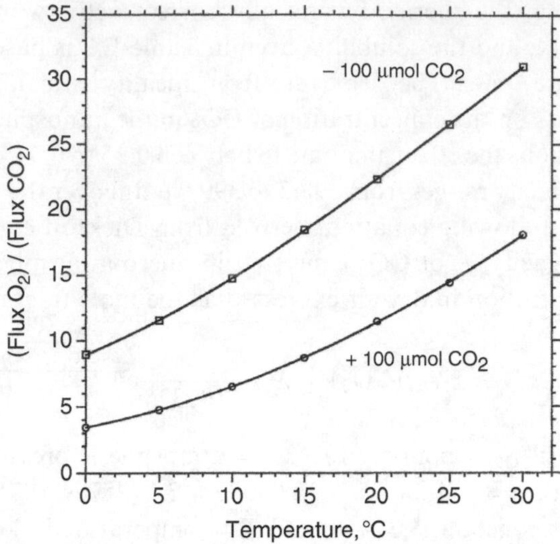

Figure E.1 The ratio of the air–sea flux of oxygen to that of carbon dioxide at equivalent departures from the equilibrium concentrations. For this example the salinity was set at 35‰ and the total alkalinity was set at 2320 μmol kg^{-1}. At each temperature the calculation was done as follows. The concentration of O_2 was set at equilibrium with the atmosphere; the concentration of TCO_2 was set to yield a partial pressure of 390 μatm. The CO_2 system was then recalculated to yield a TCO_2 concentration increased by 100 μmol kg^{-1} and the oxygen concentration was decreased by 100 μmol kg^{-1} (this could occur if there is a net respiration amounting to that magnitude in the water column). Then the initial rates of exchange were calculated according to Eq. (5.10), and their ratio plotted above. (In this simple formulation the result is independent of wind speed; I used 4 m s^{-1}.) The process was then repeated for the situation driven by a net photosynthesis, where the O_2 concentration was increased by 100 μmol kg^{-1} and the TCO_2 concentration was decreased by 100 μmol kg^{-1}.

E.3 Isotopic relationships in solution

The isotopic composition of carbon-containing compounds, organic or inorganic, is often of great importance to biological or geochemical investigations. The use of the radioactive isotope carbon-14 is of course well known, and is discussed elsewhere. There are two stable isotopes, carbon-12 and carbon-13. Because the masses are slightly different, the rates of reaction of the two isotopes are slightly different. This leads to differences in the ratio of carbon-13 to carbon-12 (the usual convention) between different phases at equilibrium and between reactants and products during reaction processes. The dissolved species that are the sources for synthesis of organic matter and for the formation of carbonate minerals differ quite measurably in their isotopic ratios, and this can be taken into account when considering the mechanisms that yield the isotopic ratios in the various products.

It is not at present possible to make highly accurate and precise measurements of the absolute ratio of $^{13}C/^{12}C$ in any sample, so it is necessary to reference all

Table E.6 Carbon isotope notation, and fractionation among dissolved species

$R = \frac{C^{13}}{C^{12}}$	$\delta(^0/_{00})\left(\frac{R_x}{R_{std}}-1\right)1000$	$\alpha^a_b = \alpha_{a-b} = \alpha_{ab} = \frac{R_a}{R_b}$
$\varepsilon = (\alpha - 1)1000$	$\alpha = \frac{\varepsilon}{1000} + 1$	
VPDB standard:	R_{std}* = 0.011180: ^{12}C = 98.8944%, ^{13}C = 1.1056%	

Atmospheric CO_2: $\delta^{13}C$ is decreasing, due to dilution by fossil fuel; currently ~ –8‰.

Following are relationships from Zhang *et al.* (1995), who should be consulted for some evaluation of uncertainties, where g is gaseous CO_2, d is CO_2aq, b is HCO_3^-, c is CO_3^{2-}, DIC is total CO_2, and t is in °C:

$\varepsilon_{dg} = +\ 0.0049t - 1.31‰$	$\varepsilon_{bg} = -\ 0.1141t + 10.78‰$
$\varepsilon_{cg} = -\ 0.052t + 7.22‰$	$\varepsilon_{DICg} = -\ 0.105t + 10.51‰$

* It is extremely difficult to make measurements of absolute ratios. Indeed, the value originally reported by Craig (1957) would yield the delta value $\delta^{13}C$ = +5.5‰ relative to the value given here for the standard. Several different measurements have been reported. The value here was adopted by Hayes (2002) and by Fry (2006).

measurements to some standard. The standard originally selected to represent zero on the delta scale was a ground-up sample of an ancient belemnite fossil (a Cretaceous shelled mollusk) found in the "Pee Dee" Formation in North Carolina, designated "PDB." This sample having been exhausted, a working group of the International Atomic Agency in Vienna recommended that the zero point on the delta scale should be a value as nearly identical as could be achieved to the original sample, and this standard should be designated VPDB. No actual sample of VPDB exists. The most important physical reference standard is a powdered marble limestone provided by the US National Institute of Standards and Technology (NIST), designated "NBS 19" (also known as RM 8544). In turn NBS 19 is defined as having $\delta^{13}C$ = +1.95 exactly, relative to VPDB. Thus the primary standard is NBS 19, but results are adjusted to refer to VPDB.

The notations and relationships used to describe isotopic ratios take some getting used to, and are summarized in Table E.6. Included there is the absolute ratio in the ultimate standard used as a reference. The oceans are not in isotopic steady state with respect to the atmosphere, because the CO_2 from fossil fuels is much lighter (has a lower concentration of carbon-13) than atmospheric CO_2, so the burning of fossil fuels is reducing the proportion of carbon-13 in the atmosphere. At the present time the average isotopic composition of CO_2 in the atmosphere, in standard delta notation, is $\delta^{13}C \approx -8.0$.

The notations used and some of the relationships among the dissolved species are listed Table E.6, and estimates of the fractionation factors and resulting isotopic ratios in the dissolved components are in Tables E.7 and E.8.

To pursue further the use of stable isotopes of carbon, it is useful to begin with Fry (2006) and Hayes (2002).

Table E.7 Epsilon values in the carbonate system in seawater as a function of temperature, calculated from relationships in Zhang *et al*. (1995)

t, °C	ε_{dg}	ε_{bg}	ε_{cg}	ε_{DICg}	ε_{bd}
0	−1.310	10.780	7.220	10.510	12.106
5	−1.286	10.210	6.960	9.985	11.510
10	−1.261	9.639	6.700	9.460	10.914
15	−1.237	9.069	6.440	8.935	10.318
18	−1.222	8.726	6.284	8.620	9.960
20	−1.212	8.498	6.180	8.410	9.722
22	−1.202	8.270	6.076	8.200	9.483
25	−1.188	7.928	5.920	7.885	9.126
30	−1.163	7.357	5.660	7.360	8.530

Table E.8 Delta values in the carbonate system, assuming equilibrium with atmospheric CO_2 at $\delta^{13}C$ = −8.0 ‰, calculated from data in Table E.6

t, °C	CO_2aq	HCO_3^-	CO_3^{2-}	Total CO_2
0	−9.296	+2.697	−0.834	+2.430
5	−9.272	+2.131	−1.092	+1.909
10	−9.247	+1.566	−1.350	+1.388
15	−9.223	+1.000	−1.608	+0.867
18	−9.208	+0.660	−1.763	+0.555
20	−9.199	+0.434	−1.866	+0.346
22	−9.189	+0.207	−1.969	+0.138
25	−9.174	−0.132	−2.124	−0.174
30	−9.150	−0.698	−2.382	−0.695

APPENDIX

F Dissociation constants and pH scales

> The choice of a suitable pH scale for the measurement of oceanic pH remains for many people a confused and mysterious topic.
>
> DICKSON 1993

One of the more confusing aspects of the recent literature of marine chemistry concerns the development and elaboration of the necessary dissociation constants for the many substances in seawater that associate with or release a hydrogen ion (carbonate, sulfate, phosphate, borate, etc.). Investigators in this field have used three different concentration scales for the reactants: mole per kilogram of water scale (*molality*); the mole per liter scale (*molarity*) identified by the symbol M; or a mole per kilogram of solution scale (if the solution is seawater, it is the *seawater scale*, sometimes called *molinity*). This last concentration scale is now generally universal for measurements in open ocean waters.

There are several different pH scales proposed or in use for describing the concentration or activity of the hydrogen ion in solution. Authors have not always been entirely explicit about the conventions used, and different symbols or descriptors have sometimes been used for the same convention. Caution is advised when reading the literature.

In the recent literature, the hydrogen ion activity or concentration has been defined in the following ways.

(a) $pH = \log\{H^+\}$, as measured with a glass electrode, calibrated with NBS buffers. The symbol $\{H^+\}$ stands for the activity of hydrogen ion. This scale is usually identified as pH_{NBS}, where the subscript NBS stands for National Bureau of Standards (now NIST), and NBS buffers are used. This scale, and its ease and simplicity of use, has led to near-universal adoption, except for recent work in the open ocean. Its main disadvantages are that it does not measure exactly the activity of hydrogen ion (which is in principle not measurable anyway), and it is subject to apparently unavoidable variation due to instrumental artifacts. The liquid junction potential is variable between electrodes and the difference in ionic strength between NBS buffers and seawater introduces additional uncertainties (Whitfield *et al.* 1985).

(b) $pH_F = -\log [H^+]_F$, where $[H^+]_F$ is the concentration of H^+ in solution. This is termed the *free hydrogen ion scale*. This scale is conceptually the simplest, and perhaps the soundest, but for work in seawater it has not been adopted, though it is unclear why not.

Table F.1 Apparent dissociation constant, K_w^*, of water in seawater

	Temperature, °C									
S, ‰	0	5	10	15	18	20	25	30	35	40
	Each value must be multiplied by the factor 10^{-14}									
0	0.087	0.14	0.222	0.343	0.441	0.518	0.768	1.12	1.59	2.28
5	0.172	0.285	0.460	0.725	0.943	1.12	1.69	2.51	3.65	5.21
10	0.218	0.364	0.591	0.938	1.23	1.46	2.22	3.30	4.83	6.95
15	0.259	0.432	0.706	1.12	1.47	1.75	2.68	4.00	5.88	8.48
20	0.296	0.497	0.813	1.30	1.70	2.03	3.11	4.66	6.85	9.90
25	0.333	0.560	0.918	1.47	1.93	2.30	3.53	5.29	7.80	11.28
30	0.369	0.622	1.02	1.64	2.15	2.57	3.94	5.92	8.73	12.64
35	0.406	0.685	1.13	1.81	2.38	2.84	4.37	6.57	9.68	14.01
40	0.444	0.750	1.24	1.99	2.61	3.12	4.80	7.23	10.66	15.43

The algorithm given by Millero (1995):

$\ln K_w^* = 148.9802 - 13847.26/T - 23.6521 \ln T + (-5.977 + 118.67/T + 1.0495 \ln T)S^{0.5} - 0.01615S$,

where T is in kelvin, and S is the salinity, gives $K_w^* = [H^+] \times [OH^-]$, where the scale of concentration is in moles per kilogram of solution for both ions. This gave the dissociation constant in the pH_T scale. Each value was then multiplied by the appropriate f_H factor from Table F.2, to give the values above which can be used with measurements made on the pH_{NBS} scale. (For conversion back to the pH_T scale, divide by the f_H factors.)

(c) $pH_{SW} = -\log [H^+]_T$, where $[H^+]_T$ is the concentration of H^+, plus the concentration of HSO_4^-, plus the concentration of HF in solution. That is:

$$[H^+]_{SW} = [H^+]_F + [HSO_4^-] + [HF]$$

This is called the seawater scale, also identified as pH_{SWS}. The difference between (c) and (d) amounts to about 0.001 of the $[H^+]$. Despite its name, this scale is not generally in current use.

(d) $pH_T = -\log [H^+]_T$, where $[H^+]_T$ is the concentration of H^+, plus concentration of HSO_4^- in solution. That is:

$$[H^+]_T = [H^+]_F + [HSO_4^-]$$

This is now generally called the *total hydrogen ion scale*. This scale is the basis for all current measurements of high precision made in the open ocean where the salinity does not differ much from the common value of 35‰.

Inter-conversions between concentration scales are often necessary. The following conversion between molality (mol kg^{-1} water) and seawater (mol kg^{-1} seawater) scales is appropriate:

$$C_{(seawater)} = \left(\frac{1000 - 1.0057\ S}{1000}\right) \times C_{(molality)} \qquad (F.1)$$

or

$$C_{(\text{seawater})} = (1 - 0.00100575\ S) \times C_{(\text{molality})},$$

where C is the concentration of some substance and S is the salinity in ‰.

Some authors have measured the dissociation constants as a function of the ionic strength. This seems undesirable for use in seawater, as the ionic strength of seawater is not easy to determine exactly because of the complexity of seawater solutions. Furthermore, the complex ionic reactions between seawater constituents are not captured in simpler solutions made up to the ionic strength of seawater. When it is necessary to convert from an ionic strength of seawater to the equivalent salinity, the following algorithm (Millero and Leung 1976) may be used,

$$I = \frac{19.92S}{1000 - 1.00488S} \tag{F.2}$$

where I is the ionic strength and S is the salinity in ‰.

It must be appreciated that a conversion between the pH_{NBS} scale and the free hydrogen scale involves the conversion from an apparent activity of hydrogen ion to its concentration. This involves a factor for both the activity coefficient of the hydrogen ion and the liquid junction potential of the electrode system used. Both are functions of temperature and salinity and include non-thermodynamically rigorous approximations; the conversion is an empirically derived approximation.

A conversion from the pH_{NBS} scale to the pH_T scale involves in addition a factor for the amount of H^+ associated with sulfate as the HSO_4^- ion and sometimes with fluoride as the HF molecule. The sum of all these factors is given the symbol f_H. Values of f_H generally are in the range of 0.67 to 0.80. An algorithm derived from empirical data (Millero *et al.* 1988) was used to derive useful empirical values given in Table F.2.

The f_H factor is applied as follows.

$$[H^+]_T = \frac{\{H^+\}}{f_H} \tag{F.3}$$

$$pH_T = pH_{NBS} + \log(f_H) \tag{F.4}$$

$$pK_T = pK_{NBS} + \log(f_H) \tag{F.5}$$

$$K_{NBS} = K_T \times f_H \tag{F.6}$$

Since, as noted above, there has been some variation in the definitions of these constants, attention to the definitions is needed before making the conversions.

The current practice is to use the total hydrogen scale, (item (d) above) for high-precision work in water near $S = 35‰$ (Dickson *et al.* 2007, Riebesell *et al.* 2010). This scale has been used on all recent expeditions during the evaluation of the carbonate systems in oceanic regions.

While nearly all recent pH measurements have been made using glass electrodes, the method of choice for the most precise measurements of the hydrogen ion concentration on any scale seems to be the spectrophotometric method (Byrne and Breland 1989; Clayton and Byrne 1993); this requires the use of a high-quality spectrophotometer.

Both potentiometric and spectrophotometric techniques should be calibrated with buffers designed for use in seawater (e.g. Tris, a commonly used organic material for making buffers); such buffers are not commercially available at certified pH values but must be made up in the laboratory and calibrated there. Further discussion may be found in Dickson (1984), Dickson (1993), Millero *et al.* (1993), Millero (1995), Grasshoff *et al.* (1999) Dickson *et al.* (2007), and Riebesell *et al.* (2010).

Dissociation constants for several acids that can release or take up protons under natural conditions in seawater or during acid titration down to about pH 4 are given in this section (except for carbonic acid itself, which is in Appendix E). There is a long history of measurements of these constants. Often measurements have been made in simpler salt solutions (NaCl or $KClO_4$, for example); these have intrinsic value, but have been shown to be unsuitable for use in real seawater. Measurements of various constants in real seawater, difficult as these may be, yield data that are much better supported by the totality of other evidence. There may yet be unknown reasons for this, but some are known. For example, boric acid has been shown to have a special reaction with some carbonate species, which to some small extent changes their titration behavior (McElligott and Byrne 1998). Borate also forms complexes with carbohydrates and other compounds with hydroxyl groups. To the extent that these substances are present they may have an effect on the apparent constants measured for boric acid. For this reason, measurements of the carbonate dissociation constants in borate-free artificial seawater, or of borate constants in carbonate-free artificial seawater, will yield results that are not exactly applicable to real seawater. Likewise, fluoride ion forms an ion pair with magnesium ion, so that the free fluoride available to react with a proton is reduced by a significant factor (Perez and Fraga 1987). If one is interested in really precise data, it is always important to examine exactly how the measurements were made.

In the tables presented in this section, and the tables for the K_1^* and K_2^* for the dissociation of carbonic acid in Section E, the constants are on the pH_{NBS} scale unless explicitly stated otherwise, but may be converted to the pH_T scale by use of Eqns. (F.3), (F.4), (F.5), or (F.6), as appropriate.

Table F.2 Apparent activity coefficients f_H of the hydrated proton or hydrogen ion

	Temperature, °C									
S, ‰	0	5	10	15	18	20	25	30	35	40
0	0.757	0.757	0.757	0.758	0.758	0.758	0.758	0.759	0.759	0.759
5	0.752	0.750	0.747	0.744	0.743	0.742	0.739	0.737	0.734	0.731
10	0.752	0.746	0.741	0.735	0.732	0.730	0.724	0.719	0.713	0.708
15	0.756	0.747	0.739	0.730	0.725	0.722	0.713	0.705	0.696	0.688
20	0.763	0.752	0.740	0.729	0.722	0.718	0.706	0.695	0.684	0.672
25	0.775	0.760	0.746	0.732	0.723	0.718	0.703	0.689	0.675	0.660
30	0.790	0.773	0.756	0.739	0.728	0.721	0.704	0.687	0.670	0.652
35	0.810	0.790	0.770	0.749	0.737	0.729	0.709	0.689	0.669	0.649
40	0.833	0.810	0.787	0.764	0.750	0.741	0.718	0.695	0.672	0.649

Values calculated with the following algorithm, from Millero *et al.* (1988),

$$f_H = 0.739 + 0.0307S + 0.0000794S^2 + 0.00006443T - 0.000117ST,$$

where S is the salinity and T is the temperature in kelvins.

The f_H factor includes a number of relationships, such as the true activity coefficient of the ion, the fraction bound by two proton-accepting groups (SO_4^- and F^-) and the liquid junction potential of the electrode used. Since the latter component varies somewhat with the electrode used, these values are not absolute, but provide a working approximation that can be used for inter-conversion among constants obtained while working with different pH scales.

Table F.3 Apparent dissociation constant, K_B^*, of boric acid

	Temperature, °C														
S, ‰	−2	0	2	4	6	8	10	12	14	16	18	20	25	30	35	40
	All values must be multiplied by the factor 10^{-10}															
0	–	2.38	2.53	2.68	2.83	2.99	3.15	3.31	3.48	3.65	3.81	3.98	4.40	4.81	5.22	5.61
5	–	4.64	4.93	5.23	5.53	5.85	6.17	6.50	6.84	7.18	7.54	7.90	8.83	9.80	10.81	11.87
10	–	5.69	6.04	6.40	6.77	7.15	7.54	7.94	8.36	8.78	9.21	9.66	10.82	12.04	13.35	14.72
15	–	6.58	6.98	7.39	7.80	8.24	8.68	9.14	9.61	10.09	10.59	11.09	12.42	13.85	15.36	16.99
20	–	7.41	7.85	8.30	8.77	9.24	9.73	10.24	10.75	11.29	11.83	12.40	13.87	15.45	17.15	18.98
25	–	8.23	8.71	9.20	9.70	10.22	10.75	11.30	11.86	12.44	13.03	13.64	15.24	16.96	18.81	20.81
30	–	9.06	9.57	10.10	10.64	11.20	11.77	12.36	12.96	13.58	14.22	14.87	16.58	18.42	20.40	22.55
31	–	9.22	9.74	10.28	10.83	11.40	11.98	12.57	13.18	13.81	14.45	15.11	16.85	18.71	20.72	22.89
32	–	9.39	9.92	10.46	11.02	11.59	12.18	12.78	13.40	14.04	14.69	15.36	17.11	19.00	21.03	23.23
33	–	9.56	10.09	10.64	11.21	11.79	12.39	13.00	13.62	14.27	14.93	15.60	17.38	19.29	21.35	23.58
34	–	9.72	10.27	10.83	11.40	11.99	12.59	13.21	13.84	14.50	15.16	15.85	17.65	19.58	21.66	23.91
35	9.36	9.89	10.44	11.01	11.59	12.19	12.80	13.42	14.07	14.73	15.40	16.10	17.91	19.87	21.97	24.25
36	9.52	10.06	10.62	11.20	11.78	12.39	13.00	13.64	14.29	14.96	15.64	16.34	18.18	20.16	22.28	24.59
37	9.68	10.23	10.80	11.38	11.98	12.59	13.21	13.85	14.51	15.19	15.88	16.59	18.45	20.45	22.60	24.93
38	9.84	10.40	10.98	11.57	12.17	12.79	13.42	14.07	14.74	15.42	16.12	16.84	18.72	20.73	22.91	25.26
40	10.17	10.74	11.33	11.94	12.56	13.19	13.84	14.50	15.19	15.88	16.60	17.33	19.26	21.31	23.53	25.94

Calculated from the algorithm presented by Dickson 1990b,

$\ln K_B^* = (-8966.90 - 2890.53S^{0.5} - 77.942S + 1.728S^{1.5} - 0.0996S^2)/T + (148.0248 + 137.1942S^{0.5} + 1.62142S) + (-24.4344 - 25.085S^{0.5} - 0.2474S) \ln T + (0.053105S^{0.5})T$, and then adjusted by use of the f_H factor to be appropriate for use with the pH_{NBS} scale.

Table F.4 Apparent dissociation constant, K^*_{HSO4}, of HSO_4^- ion in seawater

	Temperature, °C														
S, ‰	−2	0	2	4	8	10	12	14	16	18	20	25	30	35	40
	Each number must be multiplied by the factor 10^{-2}														
0	1.65	1.56	1.48	1.40	1.25	1.19	1.12	1.06	1.00	0.94	0.89	0.77	0.67	0.57	0.49
5	6.74	6.36	6.00	5.65	5.01	4.71	4.43	4.16	3.91	3.67	3.44	2.93	2.49	2.11	1.79
10	10.37	9.70	9.08	8.50	7.44	6.95	6.50	6.08	5.68	5.31	4.96	4.18	3.53	2.97	2.50
15	13.38	12.43	11.54	10.73	9.26	8.61	8.01	7.45	6.93	6.45	6.00	5.02	4.20	3.52	2.96
20	16.04	14.78	13.62	12.56	10.71	9.90	9.15	8.47	7.84	7.26	6.73	5.59	4.65	3.88	3.25
25	18.70	17.08	15.62	14.30	12.02	11.04	10.15	9.35	8.61	7.95	7.34	6.03	4.99	4.15	3.47
30	21.81	19.74	17.90	16.25	13.47	12.30	11.24	10.29	9.44	8.67	7.97	6.50	5.34	4.43	3.69
31	22.53	20.35	18.42	16.70	13.81	12.59	11.49	10.51	9.63	8.83	8.11	6.61	5.43	4.49	3.75
32	23.29	21.00	18.98	17.18	14.16	12.89	11.76	10.74	9.83	9.01	8.27	6.72	5.51	4.56	3.80
33	24.11	21.70	19.57	17.69	14.53	13.21	12.04	10.98	10.04	9.19	8.43	6.84	5.60	4.63	3.86
34	24.98	22.44	20.20	18.23	14.94	13.56	12.34	11.24	10.27	9.39	8.61	6.97	5.70	4.71	3.92
35	25.92	23.24	20.88	18.81	15.37	13.93	12.66	11.52	10.51	9.61	8.80	7.11	5.81	4.79	3.99
36	26.92	24.09	21.61	19.44	15.83	14.33	13.01	11.83	10.78	9.84	9.00	7.26	5.93	4.88	4.07
37	28.01	25.02	22.40	20.11	16.33	14.76	13.38	12.15	11.06	10.09	9.22	7.43	6.05	4.98	4.15
38	29.18	26.02	23.25	20.84	16.87	15.23	13.78	12.50	11.37	10.36	9.46	7.61	6.19	5.09	4.24
40	31.84	28.27	25.18	22.49	18.08	16.28	14.70	13.30	12.07	10.97	10.00	8.01	6.50	5.34	4.44

The data were calculated from the algorithm given in Dickson 1990a as follows:

$\ln K_T = -4276.1/T + 141.328 - 23.093 \ln T + (-13856/T + 324.57 - 47.986 \ln T)I^{0.5} + (35474/T - 771.54 + 114.723 \ln T)I + (-2698/T)I^{1.5} + (1776/T)I^2$,

where T is in kelvin, I is the ionic strength, and the resulting values are on the pH_T hydrogen scale and on the molal concentration scale. Salinity was converted to ionic strength by use of Eq. (F.2), and then values were converted to the scale of mol kg^{-1} of seawater by Eq. (F.1). Finally, the constants were multiplied by the f_H factor from Table F.2 so that the constants above are appropriate to use with measurements of $\{H^+\}$ with electrodes calibrated on a pH_{NBS} scale.

Table F.5 Apparent dissociation constant, K_F^*, of HF

S, ‰	Temperature, °C								
	5	**10**	**15**	**18**	**20**	**25**	**30**	**35**	**40**
	Each value is to be multiplied by the factor 10^{-3}								
5	1.86	1.76	1.66	1.61	1.58	1.50	1.43	1.37	1.31
10	2.06	1.95	1.84	1.79	1.75	1.67	1.59	1.51	1.45
15	2.23	2.11	2.00	1.93	1.89	1.80	1.72	1.64	1.57
20	2.38	2.25	2.13	2.07	2.03	1.93	1.84	1.75	1.67
25	2.52	2.39	2.26	2.19	2.15	2.04	1.95	1.86	1.77
30	2.66	2.52	2.38	2.31	2.26	2.15	2.05	1.96	1.87
35	2.79	2.64	2.50	2.43	2.38	2.26	2.15	2.06	1.96
40	2.92	2.76	2.62	2.54	2.49	2.37	2.25	2.15	2.06

These values were calculated by the algorithm given by Perez and Fraga (1987):

$\ln \beta_{HF} = 9.68 - 874/T - 0.111S^{0.5}$,

where T is temperature in kelvins, S is salinity in ‰, and β is the association constant, so that $K_F^* = 1/\beta$. The values were converted to K_F^* values. The data used to derive this algorithm spanned only the ranges 9 to 33 °C and 11 to 35‰ in salinity. Measurements were made in the pH_{NBS} scale and concentrations were on the molinity scale.

Table F.6 Apparent first dissociation constant, K_{1S}^*, of H_2S

S, ‰	Temperature, °C									
	0	**5**	**10**	**15**	**18**	**20**	**25**	**30**	**35**	**40**
	All values are to be multiplied by the factor 10^{-7}									
0	0.279	0.357	0.448	0.551	0.619	0.667	0.793	0.930	1.074	1.224
5	0.524	0.667	0.833	1.021	1.144	1.230	1.458	1.702	1.959	2.223
10	0.628	0.797	0.991	1.211	1.353	1.453	1.715	1.993	2.284	2.582
15	0.704	0.889	1.101	1.339	1.493	1.601	1.882	2.178	2.485	2.796
20	0.762	0.959	1.184	1.434	1.595	1.706	1.997	2.302	2.615	2.929
25	0.809	1.015	1.248	1.506	1.671	1.785	2.081	2.388	2.700	3.010
30	0.848	1.061	1.299	1.562	1.729	1.844	2.142	2.447	2.755	3.057
35	0.882	1.099	1.342	1.608	1.776	1.891	2.188	2.490	2.791	3.083
40	0.911	1.132	1.378	1.646	1.814	1.929	2.224	2.521	2.814	3.096

These values were calculated according to the algorithm of Millero *et al.* (1988):

$pK_1^{TH} = -98.080 + 5765.4/T + 15.0455 \ln T - 0.1498S^{0.5} + 0.0119S$,

where pK_1^{TH} is the negative log of the dissociation constant, expressed on the total hydrogen (pH_T) scale and on the mol kg^{-1} concentration scale, S is the salinity in ‰ and T is the temperature in kelvins. The values given by this algorithm were multiplied by the f_H factor so the resultant K^* values above may be used with the pH_{NBS} scale.

Table F.7 Apparent dissociation constants of phosphoric acid, H_3PO_4

	Temperature, °C									
S, ‰	0	5	10	15	18	20	25	30	35	40
	K_{P1}* of H_3PO_4 (all values are to be multiplied by 10^{-2})									
5	1.16	1.14	1.11	1.08	1.05	1.04	1.00	0.96	0.91	0.87
10	1.39	1.36	1.33	1.29	1.27	1.25	1.20	1.15	1.10	1.05
15	1.55	1.53	1.49	1.45	1.42	1.40	1.35	1.30	1.24	1.18
20	1.69	1.67	1.63	1.58	1.55	1.53	1.48	1.41	1.35	1.28
25	1.82	1.79	1.75	1.70	1.67	1.64	1.58	1.51	1.44	1.37
30	1.93	1.90	1.85	1.80	1.77	1.74	1.67	1.60	1.52	1.44
35	2.03	2.00	1.95	1.90	1.86	1.83	1.76	1.68	1.60	1.51
40	2.13	2.10	2.05	1.99	1.95	1.92	1.84	1.76	1.66	1.57
	K_{P2}* of H_3PO_4 (all values are to be multiplied by 10^{-7})									
5	1.54	1.69	1.84	1.97	2.04	2.09	2.19	2.28	2.35	2.41
10	2.37	2.62	2.86	3.08	3.20	3.28	3.46	3.61	3.74	3.84
15	3.11	3.44	3.77	4.07	4.24	4.35	4.60	4.81	4.99	5.12
20	3.75	4.17	4.57	4.95	5.16	5.30	5.60	5.87	6.09	6.25
25	4.32	4.80	5.27	5.72	5.96	6.12	6.48	6.79	7.04	7.23
30	4.79	5.34	5.87	6.37	6.65	6.83	7.23	7.57	7.85	8.06
35	5.19	5.80	6.38	6.92	7.23	7.42	7.86	8.23	8.53	8.74
40	5.52	6.17	6.79	7.38	7.71	7.91	8.38	8.77	9.08	9.30
	K_{P3}* of H_3PO_4 (all values are to be multiplied by 10^{-10})									
5	0.219	0.270	0.331	0.402	0.451	0.486	0.584	0.697	0.826	0.975
10	0.835	1.041	1.287	1.58	1.78	1.93	2.33	2.80	3.35	3.97
15	1.72	2.17	2.71	3.36	3.80	4.13	5.04	6.10	7.34	8.77
20	2.60	3.31	4.17	5.21	5.93	6.46	7.94	9.69	11.74	14.13
25	3.20	4.12	5.25	6.62	7.58	8.28	10.27	12.63	15.41	18.66
30	3.43	4.47	5.75	7.32	8.42	9.23	11.54	14.31	17.58	21.43
35	3.31	4.35	5.66	7.28	8.43	9.27	11.69	14.60	18.08	22.19
40	2.94	3.91	5.13	6.67	7.77	8.58	10.91	13.74	17.14	21.19

The first, second, and third dissociation constants of phosphoric acid were calculated from the algorithms in Millero (1995), who derived these from earlier work:

$\ln K_{P1}^* = 115.54 - 4576.752/T - 18.453 \ln T + (0.69171 - 106.736/T)S^{0.5} + (-0.01844 - 0.65643/T)S;$

$\ln K_{P2}^* = 172.1033 - 8814.715/T - 27.927 \ln T + (1.3566 - 160.340/T)S^{0.5} + (-0.05778 + 0.37335/T)S;$

$\ln K_{P3}^* = -18.126 - 3070.75/T + (2.81197 + 17.27039/T)S^{0.5} + (-0.09984 - 44.99486/T)S.$

where *T* is the temperature in kelvins, and *S* is the salinity in ‰, and the *K*s are on the total hydrogen pH_T scale. The values were then converted to the pH_{NBS} scale by multiplication with the f_H factor from Table F.2.

APPENDIX

G Solubility of calcium carbonate

It has been difficult to measure the solubility of calcium carbonate in seawater, due to:

(a) the necessity of first knowing the constants of the carbonate system in the dissolved phase;
(b) the presence in seawater of organic matter, which coats and makes passive the surfaces of particles;
(c) the presence of impurities in artificial seawater that may affect the rate of approach to equilibrium;
(d) the existence of two mineral phases (calcite and aragonite) that must be dealt with; and
(e) the presence of magnesium in seawater and the possible formation of solid phases containing some magnesium substituting for calcium in calcite.

In addition, the calcite precipitated by marine organisms may contain varying amounts of magnesium, and it is this calcite that is often of most interest in evaluating the conditions in the sea.

The solubility of calcium carbonate has been measured by several investigators. The solubility product constants given in Table G.1 are from Mucci (1983), and the subject was summarized in UNESCO (1987).

Solubilities are expressed as the solubility product constant at equilibrium.

The conventional definition of the thermodynamic solubility product constant, K^0_{sp}, is:

$$K^0_{sp} = \{Ca^{2+}\}\{CO_3^{2-}\} \tag{G.1}$$

where {} denote the activities of the ions concerned. For work in seawater, the stoichiometric solubility product constant is defined as follows:

$$K^*_{sp} = [Ca^{2+}][CO_3^{2-}] \tag{G.2}$$

where the concentrations may be expressed either on a molal basis or as per kilogram of seawater, although this must be specified as the constants are different. Here the values are based on the scale of per kilogram of seawater.

The effect of temperature on K^0_{sp} is given by Plummer and Busenberg (1982) for calcite, and Mucci (1983) for aragonite, as follows.

$$\log K^0_{sp(C)} = -171.9065 - 0.077993T + \frac{2839.319}{T} + 71.595 \log T, \tag{G.3}$$

Table G.1 Fitting coefficients for the calcite and aragonite solubility algorithms

Constants	b_0	b_1	b_2	c_0	d_0
Calcite	–0.77712	0.0028426	178.34	–0.07711	0.0041249
Aragonite	–0.068393	0.0017276	88.135	–0.10018	0.0059415

These values are from Mucci (1983).

$$\log K^0_{sp(A)} = -171.945 - 0.077993T + \frac{2903.293}{T} + 71.595 \log T, \qquad (G.4)$$

where $K^0_{sp(C)}$ is the thermodynamic solubility product constant for calcite, $K^0_{sp(A)}$ is the thermodynamic solubility product constant for aragonite, and T is the temperature on the kelvin scale.

Values of K^*_{sp} at several different temperatures and salinities were determined by Mucci (1983) and the data fitted to an equation of the following form:

$$\log K^*_{sp} = \log K^0_{sp} + (b_0 + b_1 T + \frac{b_2}{T})S^{0.5} + c_0 S + d_0 S^{1.5} \qquad (G.5)$$

Values for the various coefficients are listed in Table G.1.

These measurements are difficult, and Mucci (1983) suggested that the present precision of the measurements is such that the K^*_{sp} constants are known to perhaps ±5%.

Tabulated values of $K^*_{sp(C)}$ and $K^*_{sp(A)}$ calculated according to Eq. (G.5) are given in Table G.2.

The degree of saturation of a sample of seawater is usually expressed as:

$$\Omega = \frac{IP}{K^*_{sp}}$$

where the ion product IP = K^*_{sp} = $[Ca^{2+}][CO_3^{2-}]$, represents the product of observed concentrations in a sample.

Table G.2 Solubility-product constants for calcite and aragonite

	Temperature, °C									
***S*, ‰**	**0**	**5**	**10**	**15**	**18**	**20**	**25**	**30**	**35**	**40**
					Calcite					
0	0.416	0.403	0.389	0.371	0.360	0.352	0.331	0.309	0.286	0.263
5	5.465	5.374	5.255	5.107	5.006	4.934	4.739	4.524	4.292	4.047
10	10.98	10.85	10.68	10.45	10.29	10.17	9.855	9.495	9.099	8.671
15	16.89	16.77	16.57	16.31	16.12	15.98	15.58	15.11	14.59	14.02
20	23.03	22.93	22.76	22.50	22.30	22.15	21.72	21.20	20.60	19.93
25	29.35	29.32	29.20	28.98	28.80	28.66	28.24	27.71	27.08	26.36
30	35.94	36.00	35.96	35.82	35.68	35.57	35.20	34.70	34.09	33.36
33	40.06	40.19	40.23	40.15	40.05	39.95	39.63	39.18	38.61	37.90
34	41.47	41.63	41.69	41.64	41.55	41.46	41.16	40.73	40.17	39.47
35	42.91	43.09	43.17	43.15	43.07	43.00	42.72	42.31	41.76	41.08
36	44.36	44.57	44.69	44.69	44.63	44.57	44.31	43.92	43.39	42.72
40	50.45	50.78	51.02	51.14	51.16	51.14	51.01	50.72	50.28	49.69
					Aragonite					
0	0.652	0.627	0.598	0.567	0.547	0.533	0.497	0.460	0.423	0.386
5	10.08	9.838	9.536	9.183	8.949	8.785	8.348	7.880	7.389	6.883
10	19.80	19.44	18.97	18.39	18.00	17.72	16.97	16.15	15.27	14.35
15	29.46	29.05	28.49	27.77	27.27	26.91	25.92	24.82	23.62	22.34
20	38.87	38.48	37.89	37.11	36.54	36.13	34.97	33.66	32.20	30.63
25	48.23	47.91	47.35	46.55	45.96	45.51	44.25	42.78	41.13	39.32
30	57.88	57.68	57.20	56.43	55.83	55.38	54.06	52.48	50.67	48.66
33	64.00	63.89	63.48	62.75	62.16	61.71	60.37	58.74	56.86	54.73
34	66.12	66.04	65.65	64.94	64.36	63.91	62.57	60.93	59.02	56.86
35	68.28	68.24	67.88	67.19	66.61	66.16	64.82	63.17	61.23	59.04
36	70.49	70.50	70.16	69.49	68.92	68.47	67.13	65.47	63.51	61.28
40	79.95	80.13	79.93	79.35	78.82	78.39	77.06	75.36	73.33	70.97

All values are to be multiplied by 10^{-8}.

These values were calculated with Eq. (G.5). They give the value of $K^*_{sp} = [Ca^{2+}][CO_3^{2-}]$, at saturation when the concentrations are expressed in mol kg^{-1} of seawater.

APPENDIX

H Effects of pressure

When uncharged molecules in solution dissociate into ions, there is an associated change in the volume of the solution, often amounting to some tens of milliliters per mole of ion. This is because the charged ions attract and orient the polar water molecules around them. The effect of this is always to decrease the volume occupied by the ions and oriented water molecules, relative to the volume of the unionized molecule and nearby water molecules. Because the direction of change is to a lesser volume upon ionization, the effect of applied pressure is to shift each equilibrium towards increased ionization.

Millero (1983) gives a general equation, which expresses the effect of pressure on a dissociation constant,

$$\ln \frac{K_{\mathrm{p}}^*}{K_0^*} = -\frac{\Delta V^*}{RT}(P) + \frac{0.5\Delta K^*}{RT}(P^2) \tag{H.1}$$

Where K_0^* is the stoichiometric dissociation constant at zero pressure (the present oceanographic convention is to define 1 atmosphere (= 1.01325 bar) as zero water pressure, this being zero depth at the top of the water column), K_{p}^* is the stoichiometric dissociation constant under pressure, $-\Delta V^*$ is the partial molal volume change for the reaction considered, $-\Delta K^*$ is the change in partial molal compressibility associated with the reaction considered, P is the pressure (bars), T is the temperature (K), and R is the gas constant (83.144 $\mathrm{cm^3\ bar\ mol^{-1}\ K^{-1}}$ in units appropriate for these calculations).

The numerical values of ΔV^* and ΔK^* are largely functions of temperature only. The effect of varying the salinity on the pressure coefficients is not well known but appears to be small. Salinity changes very little at the depths in the ocean where pressure is a significant variable, so our lack of information on salinity effects is not important. Values for ΔV^* and ΔK^* (Millero 1979, 1995) can be calculated from the following equations:

$$\Delta V^* = a_0 + a_1 t + a_2 t^2 \tag{H.2}$$

$$\Delta K^* = b_0 + b_1 t \tag{H.3}$$

where t is in °C.

The factors in Table H.1 for these equations are from Zeebe and Wolf-Gladrow (2001), who provide a convenient tabulation of corrected values.

An equation of the same form, Eq. (H.1), may be used to estimate the effect of pressure on the solubility of various solids, such as $CaCO_3$, which, in general, become more soluble under pressure; factors for $CaCO_3$ are also listed in Table H.1.

Table H.1 Coefficients for the calculation of ΔV^* and ΔK^* in Eqs. (H.2) and (H.3)

Substance	Constant	a_0	a_1	$10^3 a_2$	$10^3 b_0$	$10^3 b_1$
H_2CO_3	K_1*	–25.50	0.1271	0.0	–3.08	0.0877
HCO_3^-	K_2*	–15.82	–0.0219	0.0	1.13	–0.1475
$B(OH)_3$	K_B*	–29.48	0.1622	2.6080	–2.84	0.0
H_2O	K_w*	–25.60	0.2324	–3.6246	–5.13	0.0794
HSO_4^-	K_S*	–18.03	0.0466	0.3160	–4.53	0.0900
HF	K_F*	–9.78	–0.0090	–0.942	–3.91	0.054
Calcite	$K_{sp(C)}$*	–48.76	0.5304	0.0	–11.76	0.3692
Aragonite	$K_{sp(A)}$*	–46.00	0.5304	0.0	–11.76	0.3692

From Zeebe and Wolf-Gladrow (2011).

Table H.2 Effect of pressure on the first and second dissociation constants of carbonic acid

	Temperature, °C							
	0	5	10	15	0	5	10	15
P, bar	$K^*_{1(P)}/K^*_{1(0)}$				$K^*_{2(P)}/K^*_{2(0)}$			
100	1.12	1.11	1.11	1.10	1.07	1.07	1.07	1.07
200	1.25	1.24	1.23	1.22	1.15	1.15	1.15	1.14
300	1.39	1.37	1.36	1.34	1.24	1.23	1.23	1.22
400	1.55	1.52	1.50	1.47	1.33	1.32	1.31	1.30
500	1.72	1.69	1.65	1.62	1.43	1.41	1.40	1.39
600	1.91	1.87	1.82	1.78	1.53	1.52	1.50	1.49
700	2.12	2.06	2.01	1.96	1.65	1.63	1.61	1.59
800	2.35	2.28	2.21	2.15	1.77	1.74	1.72	1.69
900	2.60	2.51	2.43	2.35	1.91	1.87	1.84	1.80
1000	2.87	2.77	2.67	2.58	2.06	2.01	1.96	1.92

Listed here are the factors by which the dissociation constant is increased under pressure. They are expressed as $K^*_{(P)}/K^*_{(0)}$, where the K^*_s are the apparent dissociation constants under high pressure and at zero (surface, ≈ 1 bar) pressure, respectively. The pressure is given in bars (= 10^5 Pa). Calculated using Eqs. (H.1), (H.2), and (H.3).

Caution should be used in evaluating the results of calculations using the various constants derived from these formulations. It is not easy to obtain high accuracy and precision in the experiments from which the constants for the fitting equations are derived. Mucci (1983) estimated that the precision with which the K_{sp} constants for aragonite and calcite at atmospheric pressure are determined is about ±5% of the reported values. Errors associated with measurements under pressure must be greater. However, the direction and approximate magnitude of the effects may be considered well established. Table H.2 gives values for the ratio of the dissociation constants of the carbonate species under pressure to the values at 1 bar. Table H.3

Table H.3 Effect of pressure on the dissociation constants of boric acid and water

	Temperature, °C							
	0	**5**	**10**	**15**	**0**	**5**	**10**	**15**
P, bar		$B(OH)_3$					H_2O	
100	1.14	1.13	1.12	1.12	1.12	1.11	1.10	1.10
200	1.29	1.28	1.26	1.24	1.25	1.23	1.22	1.21
300	1.47	1.44	1.41	1.39	1.39	1.36	1.34	1.33
400	1.66	1.62	1.58	1.54	1.54	1.51	1.48	1.45
500	1.88	1.83	1.77	1.71	1.71	1.66	1.62	1.59
600	2.13	2.05	1.98	1.90	1.89	1.83	1.78	1.74
700	2.41	2.31	2.21	2.10	2.08	2.01	1.95	1.90
800	2.71	2.59	2.46	2.33	2.29	2.20	2.13	2.07
900	3.06	2.90	2.74	2.58	2.52	2.41	2.32	2.26
1000	3.44	3.24	3.04	2.84	2.76	2.63	2.53	2.46

Listed here are the factors by which the dissociation constant is increased under pressure. They are expressed as $K^*_{(P)}/K^*_{(0)}$, where the K^*_s are the apparent dissociation constants under high pressure and at zero (surface, ≈ 1 bar) pressure, respectively. The pressure is given in bars (= 10^5 Pa). Calculated using Eqs. (H.1), (H.2), and (H.3).

Table H.4 Effect of pressure on the solubility of calcite and aragonite

	Temperature, °C							
	0	**5**	**10**	**15**	**0**	**5**	**10**	**15**
P, bar		Calcite					Aragonite	
100	1.24	1.22	1.20	1.18	1.22	1.20	1.19	1.17
200	1.52	1.48	1.44	1.40	1.48	1.44	1.40	1.37
300	1.86	1.78	1.71	1.65	1.79	1.72	1.65	1.59
400	2.26	2.15	2.04	1.94	2.16	2.05	1.94	1.85
500	2.74	2.57	2.41	2.27	2.58	2.42	2.27	2.14
600	3.30	3.06	2.85	2.65	3.07	2.85	2.65	2.47
700	3.96	3.63	3.35	3.09	3.64	3.34	3.08	2.85
800	4.72	4.30	3.92	3.59	4.28	3.91	3.57	3.28
900	5.60	5.06	4.58	4.17	5.02	4.54	4.12	3.76
1000	6.61	5.93	5.34	4.82	5.85	5.26	4.75	4.30

Listed here are the factors by which the solubility is increased under pressure. This is expressed as $K_{sp}{}^*{}_{(p)}/K_{sp}{}^*{}_{(0)}$, where $K_{sp}{}^*$ is the stoichiometric solubility product constant under pressure and at zero (surface, ≈ 1 bar) pressure, respectively. The pressure is given in bars (= 10^5 Pa). Calculated with Eqs. (H.1), (H.2), and (H.3).

gives values for the ratio of the dissociation constants of boric acid and water under pressure to the values at 1 bar, calculated from the formulations given above. These serve to show the rather dramatic effects of pressure on the equilibrium constants and provide a basis for estimating conditions present at depth in the ocean. Table H.4 gives estimates of the effect of pressure on the solubility of calcium carbonate.

APPENDIX

I Radioactive decay

As in many fields, the units and their definitions have changed with time. Originally, the primary unit was selected to be the amount of radiation produced by 1 gram of radium, and it was named the "curie," after Marie Sklodowska Curie. A remarkable scientist, she is the only woman to have won two Nobel prizes, and the only person to have won two Nobel prizes in different fields of science. The first was the physics prize (1903), shared with Henrie Becquerel who had discovered radioactivity, and with her husband and co-worker, Pierre Curie, for discovering polonium and radium. The second was the chemistry prize (1911), for her work with the properties of radium.

The original definition of the curie was unsuitable for several reasons; notably there are three isotopes of radium, with different properties, and the measurements are difficult to make. The curie is still in general use because of its history and familiarity, but it is now defined as being equal to 3.7×10^{10} Bq.

Table I.1 lists the common units used to express the amount of radioactivity and to express the energy carried by various particles and X-ray and γ-ray photons. The evaluation of biological effects of radiation is complex, and some special units that are used for this purpose are also listed.

To set the magnitude of some of these units in perspective it is perhaps helpful to note that the natural radiation dose absorbed from decays internal to the human body amounts to 0.2 mGy per year. Since this is largely from ^{40}K and ^{14}C yielding γ-rays and weak β-particles, it is also 0.2 mSv per year.

Figures I.1, I.2 and I.3 display the three longest and most important of the decay schemes of importance in marine and other geochemical investigations. The data for these figures come from Firestone and Shirley (1996).

Table I.1 Units for expressing the amount of radioactivity

Name	Symbol	Units	Description
Becquerel	Bq	s^{-1}	One radioactive nuclear transformation per second
dpm	dpm	min^{-1}	One radioactive nuclear disintegration per minute
Curie	Ci	s^{-1}	3.7×10^{10} Bq
Roentgen	R		Quantity of X- or γ-radiation that produces in a mass of dry air at STP an amount of ions $= 2.58 \times 10^{-4}$ C kg^{-1}
Electron volt	eV		Energy acquired by any charged particle of unit charge falling through a potential difference of 1 volt, 1 eV $= 1.602 \times 10^{-19}$ J
	Special units to describe biological doses		
Gray	Gy	J kg^{-1}	Absorbed dose, energy per unit mass
Sievert	Sv	J kg^{-1}	Equivalent absorbed dose, $(Gy \times Q)^{(a)}$
rad	rad		The rad was originally defined as = 100 erg g^{-1} 10^{-2} Gy
rem	rem		10^{-2} Sv (= 1 cSv)

(a) For each kind of radiation, Q is a quality factor that corrects for the variable amount of damage caused by different types of radiation at the same total energy deposited in the tissue. Typical factors are:

- 1 for γ-rays, X-rays, and weak β-particles
- 1.7 for strong β-particles
- 3–10 for neutrons
- 10 for protons and α-particles

Other factors may also have to be accounted for.

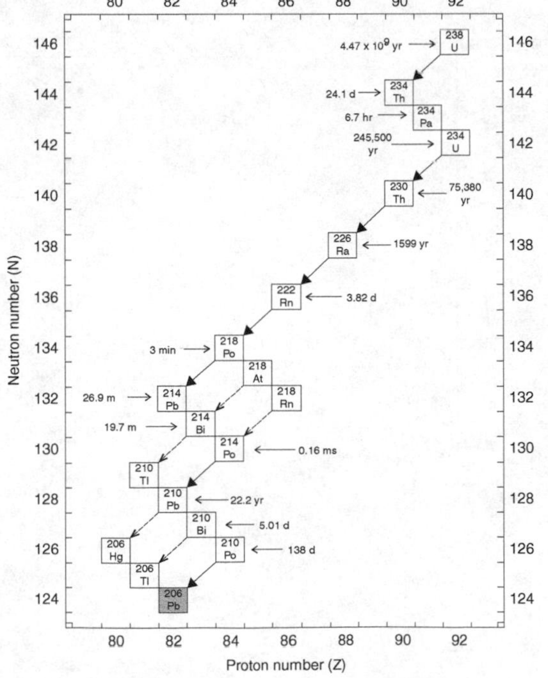

Figure I.1 Decay scheme of uranium-238. Solid arrows identify the main routes of decay; dashed lines identify minor branches. Half-lives are listed for major branches only. The shaded box (^{206}Pb, a stable isotope of lead) identifies the stable end product.

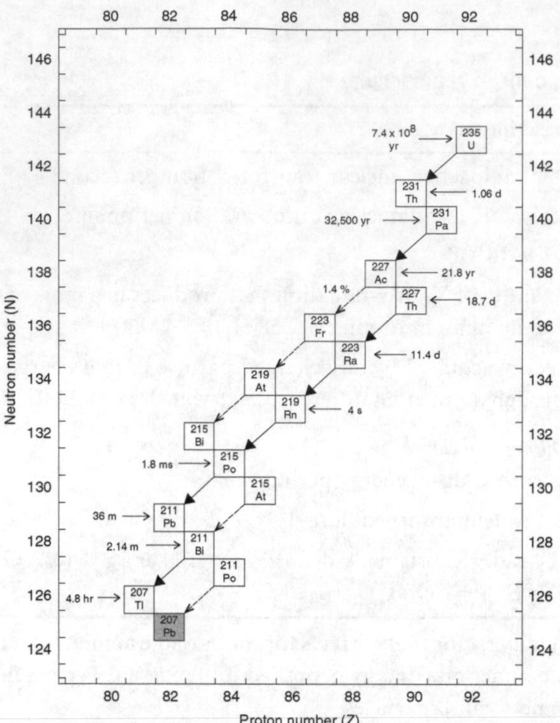

Figure I.2 Decay scheme of uranium-235. Solid arrows identify the main routes of decay; dashed lines identify minor branches. Half-lives are listed for major branches only. The shaded box (^{207}Pb, a stable isotope of lead) identifies the stable end product.

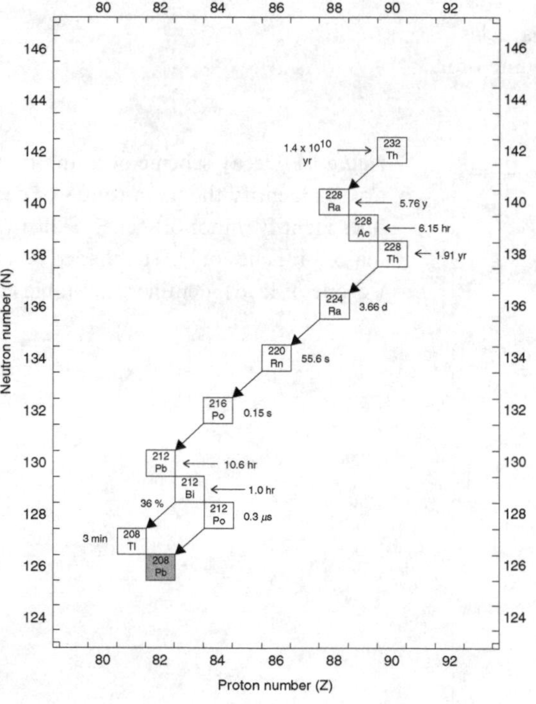

Figure I.3 Decay scheme of thorium-232. Solid arrows identify the main routes of decay. The shaded box (^{208}Pb, a stable isotope of lead) identifies the stable end product.

APPENDIX

J Geochemical reservoirs, and some rates

Table J.1 Estimates of the total masses of various geochemical reservoirs on Earth, and of the areas of land and sea

	Mass, 10^{18} tons
Earth[a]	5974
Core[a]	1927
Mantle[a]	4019
Crust[b]	28.50
Granites	9.81
Basalts	16.44
Sediments	2.25
77% shales	
15% sandstones and greywackes	
8% limestones	
Oceans[c]	1.40
Atmosphere[d]	Mass, 10^{15} tons
Dry mass	5.1316
Water vapor	0.0125
Total mass	5.1441
Areas[a]	10^6 km^2
World	510
Land	148
Sea	362
Shallow margins	~52

[a] Adapted from Stacey (1992.)

[b] Ronov and Yaroshevsky (1969). Their *volcanic–sedimentary* is included in basalts. As given here, the division of the igneous rock types into only basalts and granites is of course an oversimplification.

[c] Calculated from the volume of 1.3499×10^{18} m^3 (Menard and Smith 1966) and the volume-weighted mean density, estimated here at about 1.0373 t m^{-3}, from data in Levitus and Boyer (1994a,b) and Levitus *et al.* (1994).

[d] Trenberth and Guillemot (1994). The values listed correspond to a global average atmospheric pressure at sea level of about 1011 mbar. The total mass is better known than the dry mass. The uncertainty is mainly with respect to water: $\pm 0.005 \times 10^{15}$ t. The world total mass of water vapor varies over a total range of $\sim 0.002 \times 10^{15}$ t seasonally. This water vapor corresponds to ~2.5 cm if all precipitated. In the 20 years since the estimate was made the increased temperature of the ocean and atmosphere and the increase in CO_2 will have in each case increased the last significant figure.

Table J.2 Average composition of the continental crust of Earth, including only the major components

Component	Mass (10^{18} t)	Weight %	Component	(mol t^{-1})	(mol %)
SiO_2	13.17	61.81	O	29 501	62.37
Al_2O_3	3.21	15.09	Si	10 254	21.68
Fe_2O_3	1.32	6.20	Al	2950	6.24
CaO	1.15	5.40	Na	1027	2.17
MgO	0.780	3.66	Ca	961	2.03
Na_2O	0.726	3.41	Mg	905	1.91
K_2O	0.551	2.59	Fe	774	1.64
CO_2[a]	0.156	0.732	K	547	1.16
TiO_2	0.141	0.663	C	166	0.350
P_2O_5	0.0371	0.174	Ti	83.8	0.177
MnO	0.0198	0.093	F	28.0	0.0584
BaO	0.0125	0.0586	P	24.4	0.0517
SrO	0.0084	0.0395	S	21.7	0.0460
ZrO	0.0059	0.0275	Cl	13.3	0.0281
Cr_2O_3	0.0039	0.0185	Mn	13.0	0.0276
V_2O_3	0.0031	0.0145	N	7.1	0.0150
ZnO	0.0021	0.0097	Ba	4.25	0.0090
Li_2O	0.0008	0.0039	Sr	3.80	0.0080
B_2O_3	0.0007	0.0033	Li	2.59	0.0055
F	0.0006	0.0028	Cr	2.42	0.0051
S[a]	0.0005	0.0022	Zr	2.23	0.0047
Cl	0.0003	0.0013	V	1.924	0.0041
N[a]	0.0001	0.0006	B	1.018	0.0022
			Zn	0.994	0.0021
Total crust[a]	21.3	100.0		47,295	100.0

Calculated from Wedepohl (1995) except N from Goldblatt *et al.* (2009). The mass of the components listed here totals 99.9% of the total water-free crust[a] and was set at 100% for these calculations. The mass of the continental crust is taken as 21.3×10^{18} t. The review by Rudnick and Gao (2004) should also be consulted.

[a] The values presented by Wedepohl were based on a water-free crust. In fact, according to Wedepohl (1995), the continental crust contains an average of about 2% water, mostly in the upper layers which have a large component of sedimentary rocks. If that water were included, all percentages would change slightly, and hydrogen, not appearing above, would be the fourth most abundant element, approximately 2220 mol of H per ton. Note that C, S, and N, at least, are also special cases and are arbitrarily specified. Some C is not in an oxidized form, being present as graphite or hydrocarbons, etc. Sulfur is present in both oxidized and reduced states, while N is probably mostly present as ammonia. All the other 67 elements not listed total less than 0.1% of the mass of the continental crust.

Table J.3 Earth's inventory of water

	Volume (km^3)	%
Oceans	1 336 030 000	97.39
Polar ice caps and glaciers	27 820 000	2.03
Ground water, soil moisture	8 062 000	0.59
Lakes and rivers	225 000	0.016
Atmosphere	12 500	0.001
Total	1 372 150 000	100.0
	Fresh water only	
Polar ice caps, glaciers, icebergs	27 820 000	77.02
Ground water to 800 m depth	3 561 000	9.86
Ground water 800 to 4000 m	4 461 000	12.35
Soil moisture	61 400	0.17
Lakes	126 000	0.35
Rivers	1100	0.003
Hydrated earth minerals	360	0.001
Plants, animals	1100	0.003
Atmosphere	12 500	0.04
Total fresh water	36 120 000	~ 100.0

Other facts:

Atmospheric water ≈ 2.5 cm, if all condensed.

Average life-time of water molecules in atmosphere ≈ 9.5 d.

World average evaporation: ocean, 117.6 cm yr^{-1} ; land, 48.0 cm yr^{-1}; world, 97.3 cm yr^{-1}.

World average precipitation: ocean, 106.6 cm yr^{-1}; land, 74.6 cm yr^{-1}; world, 97.3 cm yr^{-1}.

Total discharge from land: 39.7×10^3 km^3 yr^{-1} (about 1.26 Sv). If river-borne substance is to be calculated, then the discharge of ice from Antarctica and Greenland should be subtracted, leaving river runoff at $\sim 37.4 \times 10^3$ km^3 yr^{-1} (~1.19 Sv).

These estimates of the volumes of water present in the major reservoirs in the upper part of the crust, on the surface, and in the atmosphere, in solid, liquid, and gaseous forms, are from Baumgartner and Reichel (1975), except ocean volume from Table J.6 and water vapor from Trenberth and Guillemot (1994). All these estimates have varying levels of uncertainty, and in any case some are moving targets. Probably several thousand cubic kilometers of water have departed from glaciers and ice caps since 1975, and entered the ocean.

Table J.4 Areas and volumes of the major ocean basins

Region	Area, 10^6 km^2	Volume, 10^6 km^3
Pacific	166.241	696.189
Asiatic Mediterranean	9.082	11.366
Bering Sea	2.261	3.373
Sea of Okhosk	1.392	1.354
Yellow and East China Seas	1.202	0.327
Sea of Japan	1.013	1.690
Gulf of California	0.153	0.111
Pacific and adjacent seas, total	181.344	714.410
Atlantic	86.557	323.369
American Mediterranean	4.357	9.427
Mediterranean	2.510	3.771
Black Sea[(a)]	0.461	0.534
Baltic Sea	0.382	0.038
Atlantic and adjacent seas, total	94.267	337.139
Indian	73.427	284.340
Red Sea	0.453	0.244
Persian–Arabian Gulf	0.238	0.024
Indian and adjacent seas, total	74.118	284.608
Arctic	9.485	12.615
Arctic Mediterranean	2.772	1.087
Arctic and adjacent seas, total	12.257	13.702
World ocean, total[(b)]	361.986	1349.859

Data are from Menard and Smith (1966).

[(a)] Menard and Smith give the area and volume of the Black Sea as 0.508 and 0.605 in the above units, respectively, but simple planimetry shows these values to be too high. Here I adopt the value of 0.423×10^3 km^2 for the area of the Black Sea proper from Ross *et al.* (1972), and 0.038×10^3 km2 for the Sea of Azov from Caspers (1957), for a total of 0.461×10^3 km^2. The volume of the Black Sea from Ross *et al.* (1972) is 0.534×10^3 km^3. The Sea of Azov, with an average depth of only about 10 m, does not add a unit to the volume at this level of precision. The totals are adjusted accordingly.

[(b)] For further comments on the volumes, and possibly better estimates of the total volumes of the major basins, see Table J.6.

Table J.5 Areas of sea floor within various depth intervals (10^6 km^2)

	Depth interval (km)											
Ocean region	**0 to 0.2**	**0.2 to 1**	**1 to 2**	**2 to 3**	**3 to 4**	**4 to 5**	**5 to 6**	**6 to 7**	**7 to 8**	**8 to 9**	**9 to 10**	**10 to 11**	**Totals**
Pacific	2.712	4.294	5.403	11.397	36.233	58.162	44.691	2.896	0.313	0.105	0.032	0.002	166.241
Asiatic Mediterranean	4.715	0.841	0.948	1.104	0.608	0.707	0.149	0.007	0.005	0	0	0	9.082
Bering Sea	1.050	0.135	0.172	0.234	0.670	0	0	0	0	0	0	0	2.261
Sea of Okhosk	0.368	0.549	0.311	0.047	0.115	0	0	0	0	0	0	0	1.392
Yellow and East China seas	0.977	0.137	0.072	0.015	0.001	0	0	0	0	0	0	0	1.202
Sea of Japan	0.238	0.154	0.199	0.204	0.218	0	0	0	0	0	0	0	1.013
Gulf of Calif.	0.071	0.032	0.040	0.010	0	0	0	0	0	0	0	0	0.153
Totals	10.131	6.142	7.145	13.011	37.845	58.869	44.840	2.903	0.318	0.105	0.032	0.002	181.344
Atlantic Ocean	6.080	4.474	3.718	7.436	16.729	28.090	19.324	0.639	0.058	0.010	0	0	86.557
American Mediterranean	1.021	0.465	0.589	0.667	0.906	0.586	0.112	0.008	0.002	0	0	0	4.357
Mediterranean	0.513	0.564	0.437	0.766	0.224	0.006	0	0	0	0	0	0	2.510
Black Sea	0.148	0.037	0.140	0.134	0	0	0	0	0	0	0	0	0.459
Baltic Sea	0.381	0.001	0	0	0	0	0	0	0	0	0	0	0.382
Totals	8.143	5.541	4.884	9.003	17.859	28.682	19.436	0.647	0.060	0.010		0	94.265
Indian Ocean	2.622	1.971	2.628	7.364	18.547	26.906	12.476	0.911	0.001	0	0	0	73.427
Red Sea	0.188	0.195	0.068	0.003	0	0	0	0	0	0	0	0	0.453
Persian/ Arabian Gulf	0.238	0	0	0	0	0	0	0	0	0	0	0	0.238
Totals	3.048	2.166	2.696	7.367	18.547	26.906	12.476	0.911	0.001	0	0	0	74.118
Arctic Ocean	3.858	1.569	0.968	1.249	1.573	0.269	0	0	0	0	0	0	9.485
Arctic Mediterranean	1.913	0.567	0.174	0.118	0	0	0	0	0	0	0	0	2.772
Total	5.771	2.136	1.142	1.367	1.573	0.269	0	0	0	0	0	0	12.257
World ocean totals	27.093	15.985	15.867	30.747	75.824	114.725	76.753	4.461	0.380	0.115	0.032	0.002	361.984

Data from Menard and Smith 1966, for all values except those of the Black Sea, which was remeasured giving slightly different values. The slightly revised values are entered here and the totals adjusted accordingly. The interval between 0 and 200 m is a very rough approximation to the area of the continental shelves.

Table J.6 Volumes of the major ocean basins and the world ocean, by depth interval

Interval (km)	Pacific (10^6 km^3)	Atlantic (10^6 km^3)	Indian (10^6 km^3)	Arctic (10^6 km^3)	Totals (10^6 km^3)
0–0.2	35.25	18.03	14.52	1.844	69.65
0.2–1	134.51	66.67	55.99	4.307	261.47
0–1	169.76	84.70	70.51	6.153	331.11
1–2	161.48	78.13	67.55	3.766	310.93
2–3	151.37	71.15	62.49	2.494	287.50
3–4	125.52	57.53	49.28	0.938	233.26
4–5	77.70	34.45	25.64	0.090	134.88
5–6	25.43	8.224	5.931	0	35.58
6–7	1.685	0.337	0.314	0	2.337
7–8	0.283	0.035	0.001	0	0.318
8–9	0.081	0.003	0	0	0.084
9–10	0.015	0	0	0	0.015
10–11	0.001	0	0	0	0.001
Totals	707.32	333.56	281.70	13.439	1336.03

Calculated from the data in Table J.5. The volume of the upper layer, to a depth of 200 m, gives an approximate estimate of the volume of the ocean that may interact with the atmosphere during the course of 1 year. Menard and Smith (1966) provide a total world ocean volume of 1349.929 in the above units. It makes a difference how the volumes are calculated. They assumed a linear relationship between the areas at each depth and calculated the volumes as one would the area of a trapezoid on a plane figure. For a three-dimensional figure, still assuming a linear interpolation of the areas, the formula for a frustum of a cone is:

$$\mathrm{Vol} = (A_1 + A_2 + \sqrt{A_1 \times A_2}) \times \frac{h}{3}.$$

This formula, applied to the data in Table J.5, gives a volume of the ocean smaller by 14 million cubic kilometers. Further corrections for non-linearity of the areas between the depths measured would make further changes. One may conclude that the volumes, presented by Menard and Smith (1966) to six significant figures, and given above to five significant figures, are in fact not known for certainty in the third figure. Accumulations of new data and calculation from digitized depth data will surely result in some small future adjustments.

Table J.7 Concentrations and fluxes of dissolved substances in river water

	Concentration (mmol m^{-3})		Total flux (10^{12} mol yr^{-1})		
Substance	Actual	Natural	Actual	Natural	Anthropogenic flux
Na^+	313	224	11.7	8.38	3.33
K^+	36	33	1.35	1.23	0.11
Ca^{2+}	367	334	13.7	12.5	1.23
Mg^{2+}	150	138	5.61	5.16	0.45
Cl^-	233	162	8.71	6.06	2.66
SO_4^{2-}	120	86	4.49	3.22	1.27
HCO_3^-	869	852	32.5	31.9	0.6
SiO_2	173	173	6.47	6.47	0.0

Data from Meybeck 1979.
The values above are presented as the weighted mean values in the world rivers that flow to the ocean, and the total fluxes to the ocean. Meybeck provides an estimate of the enhancement of the flux due to anthropogenic activities (these estimates must pertain to some span of years ending in the 1970s). All concentration values in mmol m^{-3}, and fluxes in Tmol yr^{-1}, based on the estimated total flow of liquid water to the ocean given in Table J.3.

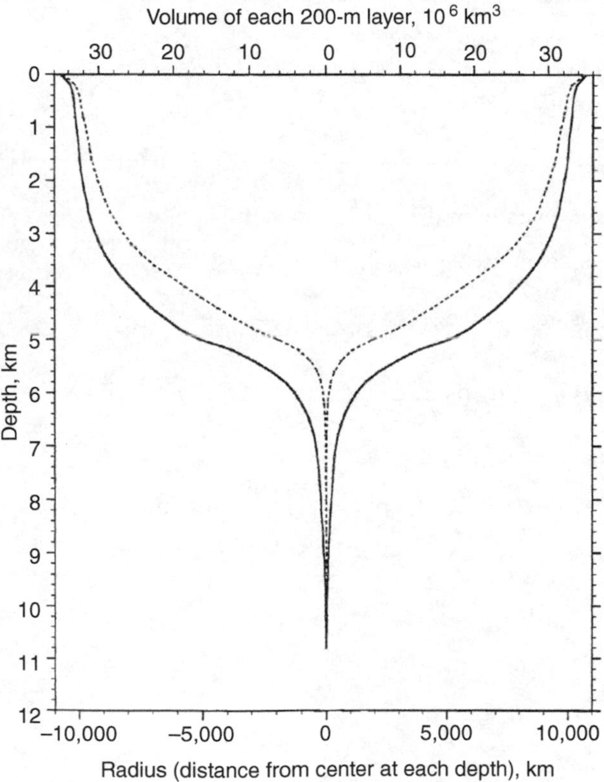

Figure J.1 The wine-glass hypsometric curve. Imagine the whole world ocean converted into a circular basin, maintaining the average depth, area, and volume properties of the present ocean. The bottom curve then gives a cross section of this basin across the middle, with the distances given plus and minus from the center. The upper curve shows how the volume varies with depth, expressed as the volume of each half of the basin at each 200-m interval of depth (i.e., 0 to 200 m, 200 to 400 m, and so on). To make this graph, the data presented by Menard and Smith (1966) were graphically interpolated at intervals of 0.2 km and plotted here with an interpolated line between the points. Note that the greatest depth of the ocean is about 11 km, and that the area and volume at each depth continue to decrease roughly exponentially from a depth of 6 km down to the deepest point. This decrease cannot be seen on a graph at this linear scale. It is evident that regions with depths below 6 km, while intrinsically very interesting, nevertheless contain a vanishingly small volume of the whole ocean.

APPENDIX

K Sound absorption

The remarkably large effect that several dissolved substances have in reducing the transmission of sound through seawater has been difficult to study, in part because of the chemical complexity of the medium. The molecular mechanisms are not at all well understood; it might be more accurate to say that some are still mysterious.

Empirical observations from both laboratory and field have been fitted to empirical algorithms that have been used for several decades to calculate the effects. A recent formulation (Ainslie and McColm 1998) is presented here. The variables involved are:

f, the frequency of the sound waves, in kHz;
f_1, the relaxation frequency of the magnesium sulfate interaction:
$Mg^{2+} + SO_4^{2-} \leftrightarrows MgSO_4^0$;
f_2, the relaxation frequency of the borate species (whatever they are);
t, temperature in °C;
S, salinity in ‰;
pH, important for the borate interaction; and
z depth, a proxy for pressure, in km.

For magnesium sulfate, $f_1 = 42\ e^{\left(\frac{t}{17}\right)}$.
For borate, $f_2 = 0.78\sqrt{\frac{S}{35}}\ e^{\left(\frac{t}{26}\right)}$.

The absorbance is expressed as α, usually in dB km^{-1}.

The absorbance of pure water is given as: $\alpha_{H_2O} = 0.00049\, f^2 e^{-\left(\frac{t}{27}+\frac{z}{17}\right)}$.

The absorbance due to $MgSO_4$ is: $\alpha_{Mg} = 0.52\left(1+\frac{t}{43}\right)\left(\frac{S}{35}\right)\frac{f_1 f^2}{f_1^2+f^2}e^{-\left(\frac{z}{6}\right)}$.

The absorbance due to borate is: $\alpha_B = 0.106\frac{f_2 f^2}{f_2^2+f^2}e^{-\left(\frac{pH-8}{0.56}\right)}$.

The total absorbance is $\alpha_T = \alpha_{H_2O} + \alpha_{Mg} + \alpha_B$.

A test value for $f = 3$ kHz, $S = 37$‰, $t = 10$ °C, pH $= 8.1$, $z = 0.1$ km is: $\alpha_T = 0.2115$ dB km^{-1}.

EPILOGUE

There is a pleasure in the pathless woods,
There is a rapture in the lonely shore,
There is society where none intrudes,
By the deep sea, and music in its roar:
I love not Man the less, but nature more,
From these our interviews, in which I steal
From all that I may be or have been before,
To mingle with the universe, and feel
What I can ne'er express, yet cannot all conceal.

Roll on, thou deep and dark blue Ocean – roll!
Ten thousand fleets sweep over thee in vain;
Man marks the earth with ruin – his control
Stops with the shore; – upon the watery plain
The wrecks are all thy deed, nor doth remain
A shadow of man's ravage, save his own,
When, for a moment, like a drop of rain,
He sinks into thy depths with bubbling groan,
Without a grave, unknelled, uncoffined, and unknown.

His steps are not upon thy paths – thy fields
Are not a spoil for him – thou dost arise
And shake him from thee; the vile strength he wields
For earth's destruction thou dost all despise,
Spurning him from thy bosom to the skies,
And send'st him shivering in thy playful spray,
And howling, to his Gods, where haply lies
His petty hope in some near port or bay,
And dashest him again to earth – there let him lay.

The armaments which thunderstrike the walls
Of rock-built cities, bidding nations quake,
And monarchs tremble in their capitals,
The oak leviatians, whose huge ribs make
Their creator the vain title take
Of lord of thee, and arbiter of war;
These are thy toys, and, as the snowy flake,
They melt into thy yeast of waves, which mar
Alike the Armada's pride, or spoils of Trafalgar.

Thy shores are empires, changed in all save thee –
Assyria, Greece, Rome, Carthage, what are they?
Thy waters washed them power while they were free,

And many a tyrant since; their shores obey
The stranger, slave or savage; their decay
Has dried up realms to deserts: – not so thou,
Unchangeable save to thy wild waves play –
Time writes no wrinkle on thy azure brow –
Such as creation's dawn beheld, thou rollest now.

Thou glorious mirror, where the Almighty's form
Glasses itself in tempests: in all time,
Calm or convulsed – in breeze or gale, or storm
Icing the pole, or in the torrid clime
Dark-heaving; – boundless, endless and sublime –
The image of Eternity – the throne
Of the invisible; even out thy slime
The monsters of the deep are made; each zone
Obeys thee; thou goest forth, dread, fathomless, alone.

And I have loved thee, Ocean! and my joy
Of youthful sports was on thy breast to be
Borne, like thy bubbles, onward: from a boy
I wantoned with thy breakers – they to me
Were a delight; and if the freshening sea
Made them a terror – 'twas a pleasing fear,
For I was as it were a child of thee,
And trusted to thy billows far and near,
And laid my hand upon thy mane – as I do here.

GEORGE GORDON, LORD BYRON, ~ 1818
From: Canto IV of Childe Harold's Pilgrimage

This piece of a long narrative poem, published by Lord Byron in 1818, considerably moved me when I first read it sometime around 1962; I was then a student at Scripps Institution of Oceanography (SIO). It is worth reading here for another lesson on how both facts and perceptions change with time.

First of all, when I was perhaps 10 years old, I lived near a patch of woods that seemed nearly endless; it had no paths, but it did have a small swamp and a small stream. I explored, and hid in special places that only I knew.

Later, as a graduate student, I would sometimes escape from the lab and in 5 or 10 minutes could walk north to a deserted beach. Then for several miles there was a high cliff on one side, empty sand and beach on the other. Sea gulls flew overhead or chased each other on the sand, speaking harshly to each other; once in a while a sea lion could be seen at some distance, hauled out on the sand. Off shore, just water, occasionally a fishing boat in the distance, waves and the sound of waves breaking on the sand. Sometimes a bit of sea-wrack to catch one's attention.

In front of SIO there was also a good beach, where many students and others would often go to have lunch. I was never a strong swimmer, but I spent half of many such lunch-hours body surfing or, depending on the waves, attempting to body surf. One year I was there 364 days, no doubt at some cost of work to be done. As one learned to appreciate the periods of the waves and guess if the powerful short-period ones were from storms far off in the North Pacific, and even to guess that the very long-period waves or low swells were from much further, powerful storms in the southern Pacific, the image of the ocean: ... *Dark-heaving; – boundless, endless and sublime –* ... captured some of the sense of wonder and awe.

In later years, I looked on that fragment in another way. First of all, that patch of woods is now a housing development. That several-mile stretch of lonely beach is now a clothing-optional place; when I last visited, there were a thousand young Adonises and Dianas sunning, swimming, playing beach volleyball and frisbee, and old men surf fishing, and hang gliders soaring overhead on the updraft along the cliff, and life guards patrolling in jeeps with surfboards on top, and many fishing boats off shore.

My first sudden realization of the global impact of humans, many decades ago, was when I learned that using one of the earliest gas chromatographs it was possible to measure the concentration of the insecticide DDT in Antarctic penguins, which live in places where DDT had never been sprayed. The chemical imprint of man is now certainly measurable everywhere throughout the surface ocean down to a depth of at least 1000 meters and in some places all the way to the bottom. It is now simply a challenge to the analytical chemist to show whether there is yet a single liter of seawater anywhere in which no trace of man's chemicals can be found. Quite probably there is not. In many cases the substances involved may not be especially good for the living things there.

I learned that the Steller's sea cow is extinct, as are the Caribbean monk seal, the great auk, and the grey whale of the North Atlantic. Most populations of most large whales are vastly reduced, as are most of the really large fish. Most populations of fish are over-fished or "optimally" fished. In the latter case, it depends what "optimal" is. A population fished to obtain some calculated maximum long-term yield is still changed by that fishing pressure. If only fish over a certain size are taken, for example, the population is deprived of the often superior reproductive success of the largest fish, and the individuals may on average decrease in size and begin to reproduce at smaller sizes and younger ages. The population is genetically changed, and its role in the ecology of the sea is changed as well. A thousand such changes are continually on-going.

We all learn that the ocean shrank in depth by 120 meters during glacial times; the volume of water frozen as ice on land reached a maximum about 20 000 years ago. And we know that sea level has changed perhaps even more during earlier geological time. Byron knew nothing of the existence of glacial masses covering large areas of Europe and North America some tens of thousands of years ago. He knew nothing of the changing arrangements of continents and oceans over longer geological time. He knew nothing of the mass extinctions that happened tens to hundreds of millions of years ago. If he knew these things, would he have written *Time writes no wrinkle on thy azure brow – Such as creation's dawn beheld, thou rollest now.* ? Nevertheless, in my mind at least, in some sense none of these "natural" changes detract from the romantic view of the ocean as *unchangeable save to thy wild waves play.*

I note with a little surprise that the word "pollution" does not appear anywhere in the text of this book. It seems that I deliberately have avoided such terms, preferring to use the perspective that humans now operate as a geological force, changing the natural geochemical flows of a considerable number of substances, and that there are in many cases observable results of these changes. Such results may be global, and often much more pronounced on a local scale where effects may be concentrated.

We do not see in this text any discussion of the effects of high concentrations of nutrients in causing expansion of anoxic zones in near-coastal waters; such matters are amply considered elsewhere, and this text provides some background for pursuing such investigations. Nor do we see in this text any extended discussion of the numerous organic substances made by chemical industry that are now ubiquitous in the global environment. Whole books could be, and have been, devoted to reporting investigations of these substances.

Another consideration missing from the text is much discussion of the methods for making the measurements that result in the concentrations so simply reported. Indeed, most of the advances in knowledge of the oceans during the last century are due to the many marvelous advances in instrument development and analytical technique. It can be a great thrill to learn the details of how some of the measurements are made. Whole books are devoted to such matters. In this text there is no room for adequate discussion of methods, and here we have simply a brief overview of the whole field of marine chemistry.

I hope that this overview will be helpful to all who care to know something of the marine world.

QUESTIONS FOR CHAPTERS

CHAPTER 2

[1] When a mixture of pure water and pure-water ice in a beaker is progressively frozen the temperature stays constant until all the water is frozen, while when a mixture of seawater and ice in a beaker is progressively frozen the temperature becomes colder and colder. Why is this?

[2] A small body of seawater is isolated from the ocean (perhaps by a growing sandbar in front of a lagoon). It is on average 5 m deep, and has a salinity of 35.00‰. It is winter and now the surface freezes; the ice thickness is 537.14 cm thick. What is the temperature of the bottom water in this lagoon? The density of ice is 0.9167 g mL^{-1}.

[3] The worldwide rise in sea level is about 3 mm per year. Suppose this was all from melting ice, what volume of ice would have to melt each year to cause this rise?

[4] What is the salinity above which a vertically well-mixed body of water must cool all the way to the bottom before it can begin to freeze?

[5] In Table 2.2 there is the statement that 1 kg of seawater contains about 3.225×10^{25} molecules of water. Show how this was calculated.

[6] A sample of water is known to have an absolute ratio $D/H = 120 \times 10^{-6}$. What is the δD of this sample? Where does it seem likely that it could have come from (assuming somewhere in nature)?

CHAPTER 3

[1] Calculate the density at 10 °C of the pure water that constitutes the ice found on the surface in central Antarctica (i.e. near the South Pole).

[2] Suppose an iceberg calves off an Antarctic glacier carrying the isotopic composition the same as SLAP. Further suppose it melts and the melt water mixes with normal seawater at $S = 34.00$‰ carrying the isotopic composition of VSMOW to reach $S = 32$‰ at 0 °C. By how much will the density of this water differ from that simply calculated from the salinity by EOS 1980?

CHAPTER 4

[1] The average concentration of strontium in the world's rivers is known from limited data to be approximately 0.89 μmol L^{-1}.

a. Calculate the residence time of Sr^{2+} in the world ocean.

b. Are there other factors to account for in considering this question?

[2] Suppose that the volume of the deep ocean (below about 1000 m) is 75% of the volume of the whole ocean, that the residence time of this deep water is

1000 years, and that the deep water has a concentration 1.01 times that in the surface water (both salinity normalized). Calculate the dissolution rate, in moles per year, of calcium carried down in sinking particles.

[3] In typical seawater ($S = 35‰$), how many molecules of water are there for each ion of dissolved salt?

[4] You have developed a very quick, precise, and accurate method for measuring the bromide concentration in seawater. The result for one sample is 7.234×10^{-4} mol kg^{-1}. What is the salinity?

CHAPTER 5

[1] In a certain region in the ocean the upper mixed layer is 50 m deep, the salinity is 35‰, the temperature is 20 °C. This region has been made to contain 10 µmol of sulfur hexafluoride, SF_6 in each kilogram of seawater. The wind speed at 10 m is 18 km hr^{-1}.

There is essentially no SF_6 in the atmosphere. How much SF_6 leaves the water each hour (per m^2)?

[2] In a region of the ocean where the salinity is 35‰ and the temperature is 20 °C, the gas exchange velocity is established to be 0.18 m hr^{-1}. The concentration of oxygen in the well-mixed upper 100 m of the water column is observed to be 275.5 µmol kg^{-1}. What is the mass flux (in moles) of the oxygen through 1 km^2 of the air–sea interface, and in what direction is this flux?

[3] Calculate the Henry's law constant for pure oxygen in seawater at $S = 35‰$ and $t = 20$ °C. Express the constant in units of: µmol kg^{-1} kPa^{-1}.

[4] One hundred bubbles of air, each with a volume of 1×10^{-5} L at 15 °C, are carried down to a depth of 20 m in a stormy sea with a mixed-layer temperature of 15 °C.

a. Calculate the initial mass of all the bubbles.
b. Calculate the volume of each bubble at 20 m, assuming no dissolution.
c. Suppose that these bubbles completely dissolve into the water at 20 m, and that the water was initially saturated with atmospheric gases. Calculate the fractional increase of each of the major components of the gas in the water, if the bubbles dissolve in 1 L of sea water.

[5] In a region of the ocean where $S = 34‰$, $t = 20$ °C, the gas exchange velocity is established to be 0.0050 m s^{-1}. The concentration of O_2 in the 100-m thick well-mixed surface layer is measured to be 275.5 µmol kg^{-1}. What is the mass flux in moles of O_2 through 1 km^2 of the air–sea interface?

[6] Suppose that you have 1 kg of seawater ($S = 35‰$) known to be in equilibrium with the atmosphere at $t = 0$ °C. You put this into a container with 37.22 mL of dry air at standard atmospheric pressure and also at 0 °C. You close the container and contrive to decrease the interior volume until all the air has been forced into solution.

a. By how much will the concentration of oxygen in the water be increased?
b. What will be the isotopic composition ($\delta\ ^{18}O$ relative to SMOW) of the oxygen dissolved in the water?

[7] The original solubility data for methane were presented as the Bunsen solubility coefficient. For example, the value at $S = 34‰$ and $t = 5$ °C was given as 0.03918. For your purpose it is convenient to have the solubility expressed as mmol kg^{-1} bar^{-1}. Calculate the numerical value you need.

CHAPTER 6

[1] Gypsum is a common mineral often found in various ancient marine deposits. The formula is $CaSO_4{\cdot}2H_2O$. The solubility product in pure water at 20 °C is reported to be 3.14×10^{-5}. Calculate the saturation state of seawater with this compound.

[2] Glauber's salt ($Na_2SO_4{\cdot}10H_2O$) can precipitate during the freezing of seawater. Calculate the change in ionic strength of seawater that would be caused by the loss of 20 mmol of Glauber's salt from each kilogram of seawater.

[3] Show how the data for sound attenuation were calculated for the situation where the partial pressure is 560 μatm, as in Figure 6.6. This question is best addressed after experience with Chapter 7.

[4] Calculate the sound attenuation at 1 kHz for the situation at a depth of 300 m in the ocean where the salinity is 34‰, the temperature is 10 °C, and the partial pressure of CO_2 is 600 μatm. This question is best addressed after experience with Chapter 7.

CHAPTER 7

[1] In Table 7.10 it is stated that each 1ppmv of CO_2 in the atmosphere corresponds to 2.128 Gt of carbon in the atmosphere. Show how this factor is arrived at.

[2] What is the concentration of borate ion in seawater under the following conditions:

$S = 35‰$, $t = 25$ °C, pH = 8.6 ?

[3] How much CO_2 might have been driven from the ocean into the atmosphere by the net accumulation of $CaCO_3$ on shallow tropical and subtropical regions during the last 10 000 years? How much would this increase the concentration in the atmosphere?

Discuss this result.

[4] Approximately how much fossil carbon (in the form of CO_2) would have to be discharged into the atmosphere after 2010 for the atmosphere to reach a concentration of 500 ppmv? There is no certain answer to this; it requires some assumptions, which should be stated.

[5] You are provided with a sample of deep water from the North Pacific, and you are asked to measure the carbonate alkalinity. The silica concentration is 180 μmol kg^{-1}. Assume that the first dissociation constant for this acid, $K_1^* = 1 \times 10^{-8}$. How much would the dissolved silica contribute to the alkalinity? Should it be taken into account in the evaluation of such water samples?

[6] Surface seawater (the top 200 m) currently has, on average, the capacity to take up carbon dioxide at a rate of very approximately 0.0.84 Gt of carbon for each

rise of 1 ppmv in the atmosphere. What would be the capacity of surface seawater to take up CO_2 if the atmospheric concentration rises to 600 ppmv?

[7] A sample of seawater with living things in it is placed in a closed container in the dark. What happens to the pH, the carbonate alkalinity, the $ppCO_2$, and the $[CO_3^{2-}]$?
Which would be the most sensitive measure of respiration?

[8] What would be the concentration of carbonate ion in seawater (S = 35‰, t = 18°C) when the water is just exactly saturated with calcite?

[9] Three samples of typical seawater (S = 35‰, t = 18 °C, TA = 2.32 mmol kg $^{-1}$) are exposed to different partial pressures of CO_2, with the following results: Sample 1, pH = 8.3; Sample 2, pH = 8.2; Sample 3, pH =8.0.
By what factor is the solubility of calcium carbonate changed between Sample 1 and Sample 2, and between Sample 1 and Sample 3?

The previous questions can be answered without using all the carbonate equations. However, if you set up an Excel spreadsheet (or equivalent) with the equations and algorithms to calculate the constants, you can explore many possibilities. We are not equipped here to carry out these calculations on top of an ocean circulation model, but a lot can be learned by using simple assumptions. Many variations are possible. For example:

[10] Suppose all the estimated resource of fossil carbon is burned, and the ocean, otherwise unchanged, comes to steady state with the result. Calculate the partial pressure of CO_2 in the atmosphere.

[11] After calculating the result in [11], allow the surface average temperature to rise by 1, 2, 3, and 4 °C, and show the new results.

[12] After calculating the result in [11], allow the alkalinity to increase by non-precipitation of the alkalinity entering the sea for 1000 years.

CHAPTER 8

[1] One ton of typical seawater (approximately 1 m^3), obtained from a depth of several hundred meters, is held in a transparent container closed to the atmosphere, and exposed to sunlight. It has a salinity of 35 ‰, a temperature of 18 °C, and an initial pH of 8.2. Phytoplankton are allowed to grow. The plankton are harvested and are found to contain 15.5 mg of P (phosphorus).
   a. How many moles of nitrate may have been removed from the water?
   b. How many moles of carbon dioxide?
   c. What will be the approximate pH of the final solution?
   d. What volume of oxygen may have been produced?

[2] What area of surface seawater, with a mixed layer depth of 50 m, is required for the synthesis of 1 mole of typical Redfield organic substance, if the beginning phosphate concentration is 0.4 μM, and one-half of the phosphorus is used up?

[3] Explore the question of whether, before humans killed off most of them, the sinking of dead whales could account for any significant fraction of the oxygen used in the deep sea, below 1000 m. For the purpose of this calculation assume:

world population of whales, 10^7;
mass of each whale at death, 40 t (whales are 70% water);
mean life span 40 yr.
Note any assumptions necessary.

[4] A sample of water from about 3000 m depth in the middle of the North Atlantic had a salinity of 35‰, a potential temperature of 3.0 °C, a phosphate concentration of 1.3 μmol kg^{-1}, and an oxygen concentration of 270 μmol kg^{-1}. What is the concentration of preformed phosphate?

CHAPTER 9

[1] Suppose that iron is required by phytoplankton in the ratio of one atom of iron to each 2000 atoms of phosphorus. What mass of carbon would be sequestered initially by the growth of phytoplankton, following the addition of 1 ton of dissolved iron to a sufficient area of high-nutrient, low-chlorophyll region in the Southern Ocean?

[2] Calculate the mass of cadmium that would be fixed into living matter during the photosynthetic fixation of 1 gigaton of carbon into biomass, in the surface waters of the North Atlantic. Note any assumptions you might have to make in order to calculate your best estimate.

[3] The rare-earth elements have been much in the news recently, and are the cause of recent international tensions. You are an oceanographer; you are asked how much seawater would have to be processed to obtain a kilogram of neodymium.

[4] We have at present little or no idea what the distribution of lead in the ocean would have been pre-man. Assume that the present concentration below 2500 m characterized the whole North Atlantic. Further assume that the North Atlantic is about one-half the volume of the whole Atlantic, and that the lead profile prepared by Schaule and Patterson was representative of the whole North Atlantic. Estimate roughly the mass of anthropogenic lead dissolved in the water of the North Atlantic in 1979.

CHAPTER 10

[1] A sample of seawater ($S = 35‰$) is acidified and stored in a secure container for a very long time. What will be the maximum concentration (in mol kg^{-1}) of ^{226}Ra that is attained? *Approximately* how long would it take to reach this concentration?

[2] One estimate of the total mass of ^{235}U now contained on Earth gives a value of 9.54×10^{14} kg. How much would have been present 4 billion years ago?

[3] What would have been the ratio of ^{238}U to ^{235}U 4 billion years ago?

[4] A mass of seawater is isolated for long enough for radioactive decay series to come to secular equilibrium. What would be the concentration of ^{210}Pb, in mol kg^{-1}?

[5] A famous geochemist named Devendra Lal investigated the production and fate of cosmic-ray-produced ^{32}P, a radioactive isotope of phosphorus, with a half-life of 14.28 days. He found that this isotope is produced by cosmic-ray-induced

spallation of atmospheric gases, mostly from argon. For the whole Earth he estimated the production rate to be 2.48 mg per day. In surface-water samples from a region in the North Atlantic he found 26.7 atoms of ^{32}P per kilogram. The same water samples had a total inorganic phosphorus concentration of 0.3 μmol kg^{-1}. Then he measured the specific activity (Bq per mole) of the phosphorus in zooplankton collected by net tow from the same region, and found it to be 30 Bq per mole.

a. Calculate the total standing stock of naturally produced ^{32}P on Earth.
b. What is the specific activity, in Bq per mole of phosphorus, of the phosphorus dissolved in this surface seawater?
c. How long on average does it appear that the phosphorus resides in zooplankton?

Answers might require some simplifying assumptions. State what these are, if any.

[6] Some Antarctic penguins collected live in 1950 and stored frozen since then were recently analyzed for carbon-14, and it was found that the readings suggested an age of about 1000 years. How can this be?

[7] Strontium-87 is a daughter of rubidium-87. Both are measurable in seawater. How long would you have to keep a sample of typical seawater (S = 35‰) isolated so that the concentration of ^{87}Sr would be increase by 0.01% ?

[8] Suppose that by 1963 the bomb explosions had resulted in a global inventory of 100 kg of tritium (most of which must be in the ocean). Calculate how much is left now.

[9] The thorium and uranium decay schemes all include isotopes of radon. Why has only one, ^{222}Rn, been used for estimating air–sea exchange rates?

CHAPTER 11

[1] In the popular literature one often reads that the ocean provides half the oxygen we breathe. Show what would happen to the ocean if all marine photosynthesis were to suddenly stop, and stay stopped. Consider timescales of days, years, and a few thousand years.

Make a rough calculation of what would happen to the concentration of oxygen in the atmosphere.

[2] Suppose a 100 km^2 section of the ocean in a central gyre is selected for intense study. The depth of water is 5000 m, and the surface mixed layer is 100 m deep. The daily net photosynthetic production is measured to be 5×10^6 mol per day.

Provide an estimate of the net flux of sinking particles at 100 m, 1000 m, 2000 m, 3000 m, 4000 m, and at the bottom. The sedimentation rate in the region is 5 cm per 1000 years. Estimate the expected oxygen consumption in each 900-m or 1000-m layer of the ocean, and by the benthos at the bottom.

[3] Estimate the mass of organic carbon in the world ocean that is on average about 6200 years old.

[4] Suppose the volume-weighted average concentration of vitamin B_{12} were discovered to be 20 fM. What would be the total mass of this substance in the world ocean?

CHAPTER 12

[1] Some of the most sulfide-rich water anywhere in the world in an arm of the sea is in the upper basin of the Pettaquamscutt River, Rhode Island; the concentration there can sometimes reach 4700 μM. Estimate the temperature rise due to bacterial respiratory activities from the time that all oxygen and nitrate are gone until the above value is reached. Is this a measurable change?

[2] A fish weighing 10 kg was 62.26% water and 37.74% solids. Its composition was basically Redfield, as revised. It died and sank into the depths of an anoxic fjord and was thoroughly digested by bacteria.

a. How many milliliters (at STP) of H_2S were produced?
b. How many grams of H_2S were produced?
c. If this fish were completely oxidized in an oxidizing environment, how many milliliters (at STP) of O_2 would be used up?

[3] Lacking a good hypsographic curve for the Black Sea, you roughly estimate the volume of anoxic water as 465 000 km^3. Given an estimated input of Mediterranean and entrained water to yield a residence time of about 310 years for the anoxic water, and the organic matter remaining for anoxic respiration as estimated in Table 12.3, what should be the average sulfide concentration In the anoxic water? How does this accord with the observed values? Comment on likely sources of error, if any.

CHAPTER 13

[1] In the newspaper you read that the total amount of salt humans obtained last year from mining and from evaporation of sea water amounted to 300 million tons (metric), while from Table 13.2 you learned that the increase of chloride in the global average river water due to human activities amounted to 71 μM.

Are these statements in quantitative agreement. If so why, and if not, what is going on?

[2] Estimate approximately (exact information does not exist) the buoyant force on 1 cubic meter of seawater as it streams out through the hottest of the exit channels in mid-ocean-ridge hydrothermal systems.

[3] Suppose the flow of water that is heated above about 150 °C is discovered to be about 6300 m^3 s^{-1}. What would be rate at which magnesium is removed from the ocean by this process?

[4] Suppose the concentration of Sr in river water is found to be an average of 50 μg L^{-1}. What is the residence time of Sr in seawater? Would you expect Sr to be conservative? In fact is Sr exactly conservative? Is river flow likely to be the only input to the Ocean?

CHAPTER 14

[1] It is estimated that humans annually produce about 300 × 10^6 t of salt, of which about 50 × 10^6 t is from the ocean and the rest from the mining of rock salt. At

this rate, how long would it take for humans to increase the chloride concentration in the sea by 0.001% ?

[2] How much oyster shell would have to be utilized in order to make the chemical reagent needed for the extraction of 1 t of $Mg(OH)_2$ from seawater. Assume 100% efficiency and perfect stoichiometry.

[3] Suppose the rate of evaporation in a hot dry region on the west coast of Mexico is about 2 m per year, and the seawater there has $S = 34‰$. Assuming 90% yield, how much salt (NaCl) can be harvested each year from 1 km^2 of solar-evaporation ponds?

[4] Information on bromine production in the USA is proprietary and not publicly available. The production of bromine by the rest of the world was 460 000 t in 2011. What volume of seawater would have to be processed each year to obtain that bromine from this source? Assume yield of 100%.

CHAPTER 15

[1] In the discussion of oxygen, the statement is made that biological production provided 30 times more oxygen than is present in the atmosphere today. Show how this statement was arrived at.

[2] Suppose the whole ocean, including the deep water, warms by several degrees. Would this lead to any change in the concentration of sulfate ion?

[3] Speculate on whether a doubling of the rate of sea-floor spreading would likely raise or lower the concentration of chloride, magnesium, sulphate, and calcium in the oceans.

[4] Suppose that a dam is placed across the Strait of Gibraltar (and there is no other human intervention). What change would this cause in the rest of the global ocean?

GLOSSARY

AABW Antarctic Bottom Water. Formed in the Weddell and Ross seas.

AAIW Antarctic Intermediate Water.

activity The chemically effective concentration of a substance. The product of the concentration and an activity coefficient appropriate for the conditions.

activity coefficient The quantity that accounts for the non-ideal behavior of a substance in solution.

adiabatic This describes some process where no heat is exchanged with the surroundings.

alkalinity The total concentration of all bases that accept protons during a titration with acid to pH 4.0.

AOU Apparent oxygen utilization.

aragonite One of the common crystalline forms of calcium carbonate. More soluble than calcite.

association constant The numerical value that describes the association of two substances in solution. If $A + B \leftrightarrows AB$, then

$$K_{AB} = \frac{[AB]}{[A] \times [B]}.$$

becquerel One nuclear transformation per second.

beta decay Nuclear transformation by the emission from the nucleus of a single electron or positron.

bitterns The residual liquid after NaCl is precipitated from evaporating seawater.

boiling point The temperature at which the vapor pressure of some substance is equal to standard atmospheric pressure (101 325 Pa). In common usage, the temperature at which the vapor pressure equals whatever is the atmospheric pressure at the time.

Boyle's law The volume of a gas varies inversely with the pressure.

calcite One of the common crystalline forms of calcium carbonate. Less soluble than aragonite, and more abundant.

carbonate alkalinity $CA = [HCO_3^-] + 2[CO_3^{2-}]$

CDIAC Carbon Dioxide Information Analysis Center, Oak Ridge, TN; see http://cdiac.ornl.gov/.

CGPM General Conference on Weights and Measures.

Charles' law The volume of a gas varies directly with the absolute temperature.

chlorinity The number, expressed in parts per thousand (‰), that equals the mass in grams of pure silver (in the form of $AgNO_3$) needed to precipitate the halogens (Cl^-, Br^-, I^-) in 328.5234 g of seawater.

clathrate A cage- or lattice-like arrangement of water molecules that encloses another small molecule, generally a gas molecule.

compensation depth The depth below which $CaCO_3$ is largely absent from marine sediments.

critical temperature The temperature at which the liquid and vapor phases of some substance become equal, and mutually miscible. Above this temperature there can be no liquid phase.

Dalton's law In a mixture of gases the total pressure is equal to the sum of the individual partial pressures.

dB decibels, a measure of relative sound pressure. $dB = 10\log(P_1/P_0)$.

diagenesis The chemical and physical changes that take place in a sediment after it is initially deposited.

DIP Dissolved inorganic phosphorus.

dissociation constant The numerical value that describes the dissociation of some substance in solution. If AB $\leftrightarrows$ A + B, then

$$K_{AB} = \frac{[A] \times [B]}{[AB]}$$

DOC Dissolved organic carbon. Usually defined as organic carbon that passes through a 0.45 μm filter.

Dole effect The observation that atmospheric oxygen is not in isotopic equilibrium with natural waters.

dolomite $CaMg(CO_3)_2$ A common mineral with a crystal structure similar to that of calcite. Probably forms by diagenesis of calcite.

DOM Dissolved organic matter.

DON Dissolved organic nitrogen. See DOC.

DOP Dissolved organic phosphorus. See DOC.

electrostriction The decrease in volume caused by the very close association of water molecules around charged ions in solution.

EOS 1980 Equation of State 1980. The current algorithm to calculate density from temperature and salinity. This may become superseded by TEOS-10.

fatty acids These generally consist of a hydrocarbon chain with a carboxylic acid on one end, $CH_3(CH_2)_nCOOH$, where n is commonly 4 to 22, and, in the marine environment especially, most long-chain fatty acids have one or more unsaturated bonds.

First-order reaction A reaction where the rate depends on the concentration of only one reactant.

free pH scale Defined as $pH_F = -\log [H^+]$, where $[H^+]$ is the concentration of protons.

fugacity Thermodynamic term for the effective pressure or chemical activity of a gas. Most gases dealt with on the surface of the ocean are close enough to ideal behavior that the fugacity is not worth computing. Carbon dioxide departs just enough from ideality that this is taken into account.

GEOSECS Geochemical Ocean Section Study; carried out in the early 1970s.

halocline Vertical gradient in salinity.

Henry's law The concentration of a gas in solution is directly proportional to the partial pressure of the gas.

HNLC High-nutrient, low-chlorophyll.

hydrogen bond A force of attraction between a hydrogen atom, attached by a covalent bond to oxygen or other small atom, and a near-by atom with a partial negative charge (such as the oxygen of a water molecule).

IAEA International Atomic Energy Agency.

IAPSO International Association for the Physical Sciences of the Ocean.

ICES International Council for the Exploration of the Sea.

ideal gas law PV = RT, for one 1 mole of gas.

ionic strength, *I* $I = 0.5 \sum m_i z_i^2$, where m is the molal concentration of the ith ion and z is the charge on the ith ion.

initial nutrient The nutrient in deep water that cannot be accounted for by mineralization of organic matter as estimated by the consumption of oxygen. Similarly: "initial phosphate," "initial nitrate."

IOC Intergovernmental Oceanographic Commission.

IPTS 68 International Practical Temperature Scale 1968.

isopycnal This word means the same (constant) density.

isotope One of two or more kinds of atoms of an element. All have the same number of protons (and thus the same atomic number) and they differ only in the number of neutrons in each nucleus.

ITS 90 International Temperature Scale 1990.

IUPAC International Union of Pure and Applied Chemistry.

JGOFS Joint Global Ocean Flux Study. An international program.

knot Nautical mile per hour. = 0.514 m s^{-1}.

K_{sp} Solubility product constant.

Langmuir circulation Remarkable arrays of rolling vortices developing in the water, parallel to the wind, when the wind blows steadily in one direction.

ligand A molecule that can form a strong bond with a metal ion.

lysocline That depth below which the $CaCO_3$ in sediments begins to become much less abundant.

mass spectrometry The separation of atoms (or small molecules) according to mass, for the purpose of measuring the relative abundances of the different isotopes. Carried out with atoms or small molecules in the vapor state under high vacuum in a mass spectrometer.

Molality (m) Moles of solute per kilogram of solvent.

Molarity (M) Moles of solute per liter of solvent.

NADW North Atlantic Deep Water. Components are formed in the Greenland, Labrador, and Norwegian seas.

NBS United States, National Bureau of Standards, now NIST.

NIST United States, National Institute of Standards and Technology.

NOAA National Oceanic and Atmospheric Administration.

normality (N) The equivalents of a solute per liter of solution.

oligotrophic The situation where the abundance of nutrients is very low.

oxidation The loss of an electron by an atom or molecule.

pH_{NBS} scale $pH = -\log \{H^+\}$, where $\{H^+\}$ is the activity of protons in solution. This scale was developed by Roger Bates at the NBS. The electrodes are calibrated with buffers traceable to NBS, and these are low ionic strength buffers.

POC Particulate organic carbon.

potential temperature The temperature that a mass of deep water would have if raised adiabatically to the surface or to some other specified depth.

ppmv Parts per million by volume. Common concentration unit for gases.

Pratical Salinity Salinity defined by the ratio of conductivity of a seawater sample to that of a solution with a specified concentration of KCl.

μatm Micro-atmosphere, = 10^{-6} of atmospheric pressure.

preformed nutrient Same as initial nutrient (phosphate or nitrate).

primordial element Present at the formation of the Earth.

PSS1978 Practical Salinity Scale of 1978.

PQ Photosynthetic quotient = (O_2 out)/(CO_2 in).

pycnocline A region where the change in density with depth is especially strong.

quartz The predominant crystalline form of pure silia, SiO_2.

reduction The gain of an electron by an atom or molecule.

RQ Respiratory quotient = (O_2 in)/(CO_2 out).

salinity Total concentration of inorganic salts in seawater. Originally specified according to a measurement procedure that resulted in a small loss of salts. Then specified as a function of chlorinity. Currently there are proposals to redefine into several salinities. In no case are other dissolved substances included.

salinity/chlorinity The empirical relationship, $S‰ = 1.80655Cl‰$, was an empirical definition of salinity prior to the introduction of the Practical Salinity Scale.

SCOR Scientific Committee on Oceanic Research.

SI Système International d'Unités (International System of Units).

sigma-t A term expressing the density of ocean water at a given temperature, with fewer digits: $\sigma_t = (\rho - 1)1000$.

SMOW Standard Mean Ocean Water. A calculated isotopic composition, meant to characterize the volume-weighted ocean mean.

specific heat The energy required to raise the temperature 1 g of a substance by 1 °C.

The molar specific heat is the energy required to raise the temperature of 1 mole of substance by 1 °C.

stoichiometry Quantitative relations between concentrations of chemical substances.

Suess effect The dilution of atmospheric CO_2 with ^{14}C-free carbon.

SUN committee Working group on symbols, units and nomenclature in physical oceanography, sponsored by IAPSO.

Sv Sverdrup; unit used for large flows of water. 1 Sv = 10^6 m^3 s^{-1}. Named after Harald U. Sverdrup (1888–1957), physical oceanographer. For a short biography of this remarkable man see: http://scilib.ucsd.edu/sio/biogr/Sverdrup_Biogr.pdf

TEOS-10 International Thermodynamic Equation of Seawater 2010.

thermocline Vertical gradient of temperature. Usually the largest contributor to the gradient in density.

TOC Total organic carbon.

UNESCO United Nations Educational and Cultural Organization.

VSMOW Vienna Standard Mean Ocean Water. The usual standard water used for calibrating mass spectrometers during analysis of samples of water for isotopic composition. Prepared by careful adjustment to be a close as possible to that of the originally defined SMOW.

WOCE World Ocean Circulation Experiment. A major campaign to obtain physical and some chemical data throughout the ocean.

REFERENCES

Abbass, D. K. 1999. *Endeavour* and *Resolution* revisited: Newport and Captain James Cook's vessels. *J. Newport Historical Soc.* **70**: 1–19

Ainslie, M. A. and J. G. McColm. 1998. A simplified formula for viscous and chemical absorption in sea water. *J. Acoust. Soc. Am.* **103**: 1671–1672.

Alduchov, O. A. and R. E. Eskridge. 1996. Improved Magnus form approximation of saturation vapor pressure. *J. Appl. Meterol.* **35**: 601–609.

Alexander, G. B. 1954. The polymerization of monosilicic acid. *J. Am. Chem. Soc.* **76**: 2094–2096.

Alexander, G. B., W. H. Heston, and R. K. Iler. 1954. The solubility of amorphous silica in water. *J. Phys. Chem.* **58**: 453–455.

Alldredge, A. L. and M. W. Silver. 1988. Characteristics, dynamics and significance of marine snow. *Prog. Oceanogr.* **20**: 41–82.

Aluwihare, L. I., D. J. Repeta, and R. F. Chen. 1997. A major biopolymeric component to dissolved organic carbon in surface seawater. *Nature* **387**: 166–169.

Amakawa, H., D. S. Alibo, and Y. Nozaki. 1996. Indium concentration in Pacific seawater. *Geophys. Res. Lett.* **23**: 2473–2476.

AMAP/UNEP. 2008. *Technical Background Report to the Global Atmospheric Mercury Assessment*. Arctic Monitoring and Assessment Programme/UNEP Chemicals Branch, Geneva.

Ambrose, D. and C. H. S. Sprake. 1972. The vapour pressure of mercury. *J. Chem. Thermodynamics* **4**: 603–620.

Anbar, A. D., R. A. Creaser, D. A. Papanastassiou, and G. J. Wasserburg. 1992. Rhenium in seawater: confirmation of generally conservative behavior. *Geochim. Cosmochim. Acta* **56**: 4099–4103.

Anbar, A. D., G. J. Wasserberg, D. A. Papanastassiou, and P. S. Andersson. 1996. Iridium in natural waters. *Science* **273**: 1524–1528.

Anderson, L. A. and J. L. Sarmiento. 1994. Redfield ratios of remineralization determined by nutrient data analysis. *Global Biogeochem. Cycles* **8**: 65–80.

Anderson, M. B., C. H. Stirling, B. Zimmerman, and A. N. Halliday. 2010. Precise determination of the open ocean $^{234}U/^{238}U$ ratio. *Geochem. Geophys. Geosys.* **11**: doi: 1029/2010GC003318.

Andersson, A. J,. F. T. Mackenzie, and A. Lerman. 2005. Coastal ocean and carbonate systems in the high CO_2 world of the Anthropocene. *Am. J. Sci.* **305**: 875–918.

Andersson, M., I Wängberg, K. Gårdfeldt and J. Munthe. 2004. Investigation of the Henry's Law coefficient for elemental mercury. *Proceedings of the 7th Conference "Mercury as a global pollutant". RMZ–Materials and Geoenvironment,* Ljubljana, June, 2004.

Andersson, M. E., J. Sommar, K Gårdfeldt, and O. Lindqvist. 2008. Enhanced concentrations of dissolved gaseous mercury in the surface waters of the Arctic Ocean. *Mar. Chem.* **110**: 190–194.

Andreae, M. O. and R. J. Ferek. 1992. Photochemical production of carbonyl sulfide in seawater and its emission to the atmosphere. *Global Biogeochem. Cycles* **6**: 175–183.

Andreae, M. O. and W. R. Barnard. 1984. The marine chemistry of dimethylsulfide. *Mar. Chem.* **14**: 267–279.

Andrews, A. P. 1983. *Maya Salt Production and Trade*. University of Arizona Press, Tuscon, AZ.

Archer, D. 2005. Fate of fossil fuel CO_2 in geologic time. *J. Geophys. Res.* **110**: C09S05, doi:10.1029/2004JC002625.

Archer, D., M. Eby, V. Brovkin, *et al.* 2009. Atmospheric lifetime of fossil fuel carbon dioxide. *Ann. Rev. Earth Planet. Sci.* **37**: 117–134.

Aristotle. 1931. *The Works of Aristotle*, translated into English under the editorship of W. D. Ross. Vols. III and IV, Oxford, Clarendon Press, p. 351.

Armstrong, F. A. J. and H. W. Harvey. 1950. The cycle of phosphorus in the waters of the English Channel. *J. Mar. Biol. Ass. UK* **29**: 145–162.

Armstrong, F. A. J., P. M. Williams, and J. D. H. Strickland. 1966. Photo-oxidation of organic matter in seawater by ultra-violet radiation, analytical and other applications. *Nature* **211**: 481–483.

Arrhenius, G. and B. R. De H. Alfven. 1974. Origin of the ocean. In *The Sea,* vol. **5**, ed. E. D. Goldberg. John Wiley & Sons, New York, pp. 839–861.

Arrhenius, S. 1896. On the influence of carbonic acid in the air upon the temperature of the ground. *Phil. Mag.* (Ser. 5) **41**: 237–276.

Arrhenius, S. 1908. *Worlds in the Making*. Harper & Brothers, New York. (Translated by Dr. H. Borns from the Swedish: *Världarnas utveckling*, 1906.)

Atlas, E., C, Culberson, and R. M. Pytkowicz. 1976. Phosphate association with Na^+, Ca^{2+}, and Mg^{2+} in seawater. *Mar. Chem.* **4**: 243–254.

Baertschi, P. 1976. Absolute ^{18}O content of standard mean ocean water. *Earth Planet. Sci. Lett.* **31**: 341–344.

Bainbridge, A. E. 1981. *GEOSECS Atlantic Expedition, Hydrographic Data 1972–1973.* US Government Printing Office, Washington, DC.

Baker, E. T. and G. J. Massoth. 1987. Characteristics of hydrothermal plumes from two vent fields on the Juan de Fuca Ridge. *Earth Planet. Sci. Lett.* **85**: 59–73.

Baker, E. T., G. J. Massoth, and R. A. Feely. 1987. Cataclysmic hydrothermal venting on the Juan de Fuca Ridge. *Nature* **329**: 149–151.

Barber, R. T. and J. H. Ryther. 1969. Organic chelators: factors affecting primary production in the Cromwell Current upwelling. *J. Exp. Mar. Biol. Ecol.* **3**: 191–199.

Bareille, G, M. Labracherie, R. A. Mortlock, E. Maier-Reimer, and P. N. Froelich. 1998. A test of $(Ge/Si)_{opal}$ as a paleorecorder of $(Ge/Si)_{seawater}$. *Geology* **26**: 179–182.

Barth, T. F. W. 1952. *Theoretical Petrology*. Wiley, New York,

Baumgartner, A. and E. Reichel. 1975. *The World Water Balance*. R. Olenberg, Munich and Vienna.

Bemis, B. E., H. J. Spero, J. Bijma, and D. W. Lea. 1998. Reevaluation of the oxygen isotopic composition of planktonic foraminifera: experimental results and revised paleotemperature equations. *Paleoceanography* **13**: 150–160.

Bender, M. L., K. A. Fanning, P. N. Froelich, G. R. Heath, and V. Maynard. 1977. Interstitial nitrate profiles and oxidation of sedimentary organic matter in the eastern equatorial Atlantic. *Science* **198**: 605–609.

Bender, M., T. Sowers, and L. Labeyrie. 1994. The Dole effect and its variations during the last 130,000 years as measured in the Vostock ice core. *Global Biogeochem. Cycles* **8**: 363–376.

Benitez-Nelson, C. R. 2000. The biogeochemical cycling of phosphorus in marine systems. *Earth-Sci. Rev.* **51**: 109–135.

Benson, B. B. and D. Krause, Jr. 1980. The concentration and isotopic fractionation of gases dissolved in freshwater in equilibrium with the atmosphere. 1. Oxygen. *Limnol. Oceanogr.* **25**: 662–671.

Berger, W. H., K. Fischer, C. Lai, and G. Wu. 1988. Ocean carbon flux: maps of primary production and export production. In *Biogeochemical Cycling and Fluxes between the Deep Euphotic Zone and Other Oceanic Realms*, ed. C. R. Agegian. NOAA, Washington, DC, pp. 131–176.

Berger, W. H., V. S. Smetacek, and G. Wefer, eds. 1989. *Productivity of the Ocean: Present and Past*. Dahlem Workshop Report, Life Sciences Research Report 44. John Wiley & Sons, New York.

Berman-Frank, I., J. T. Cullen, Y. Shaked, R. M. Sherrell, and P. G. Falkowski. 2001. Iron availability, cellular iron quotas, and nitrogen fixation in *Trichodesmium*. *Limnol. Oceanol.* **46**: 1249–1260.

Berner, E. K. and R. A. Berner. 1987. *The Global Water Cycle: Geochemistry and the Environment*. Prentice-Hall, Englewood Cliffs, NJ.

Berner, R. A. 1982. Burial of organic carbon and pyrite sulfur in the modern ocean: its geochemical and environmental significance. *Am. J. Sci.* **282**: 451–473.

Berner, R. A. and D. E. Canfield. 1989. A new model for atmospheric oxygen over Phanerozoic time. *Am. J. Sci.* **289**: 333–361.

Berner, R. A. and J. W. Morse. 1974. Dissolution kinetics of calcium carbonate in seawater: IV. Theory of calcite dissolution. *Am. J. Sci.* **274**: 108–134.

Bernstein, R. E., P. R. Betzer, R. A. Feely, *et al.* 1987. Acantharian fluxes and strontium-to-chlorinity ratios in the North Pacific Ocean. *Science* **237**: 1490–1494.

Bertine, K. K., M. Koide, and E. D. Goldberg. 1993. Aspects of rhodium marine chemistry. *Mar. Chem.* **42**: 199–210.

Betzer, P. R., W. J. Showers, E. A. Laws, *et al.* 1984. Primary productivity and particle fluxes on a transect of the equator at 153° W in the Pacific Ocean. *Deep-Sea Res.* **31**: 1–11.

Bieri, R. H. 1971. Dissolved noble gases in marine waters. *Earth Planet. Sci. Lett.* **10**: 329–333.

Bigg, P. H. 1967. Density of water in SI units over the range 0–40 °C. *Brit. J. Appl. Phys.* **18**: 521–525.

Bignell, N. 1983. The effect of dissolved air on the density of water. *Metrologia* **19**: 57–59.

Billen, G., C. Lancelot, and M. Meybeck. 1991. N, P, and Si retention along the aquatic continuum from land to ocean. In *Ocean Margin Processes in Global Change*. ed R. F. C. Mantoura, J-M. Martin, and R. Wollast. John Wiley & Sons, New York, pp. 19–44.

Bischoff, J. L. and R. J. Rosenbauer. 1985. An empirical equation of state for hydrothermal seawater (3.2 percent NaCl). *Am. J. Sci.* **285**: 725–763.

Bischoff, J. L. and R. J. Rosenbauer. 1988. Liquid–vapor relations in the critical region of the system $NaCl{\cdot}H_2O$ from 380 to 415°C: a refined determination of the critical point and two-phase boundary of seawater. *Geochim. Cosmochim. Acta* **52**: 2121–2126.

Bischoff, J. L. and R. J. Rosenbauer. 1989. Salinity variations in submarine hydrothermal systems by layered double-diffusive convection. *J. Geol.* **97**: 613–623.

Bishop, J. K. B. 1989. Regional extremes in particulate matter composition and flux: effects on the chemistry of the ocean interior. In *Productivity of the Ocean: Present and Past*, ed. H. Berger, V. S. Smetacek, and G. Wefer. John Wiley & Sons, New York, pp. 117–137.

Bloch, M. R. 1963. The social influence of salt. *Sci. Am.* **209**: 88–98.

Boden, T. A., G. Marland, and R. J. Andres. 2011. *Global, Regional, and National Fossil-Fuel CO2 Emissions*. Carbon Dioxide Information Analysis Center, Oak Ridge National Laboratory, US Department of Energy, Oak Ridge, TN, doi: 10.3334/CDIAC/00001.

Bolin, B. 1960. On the exchange of carbon dioxide between the atmosphere and the sea. *Tellus* **12**: 274–281.

Bolin, B. and C. D. Keeling. 1963. Large scale atmospheric mixing as deduced from the seasonal and meridional variations of carbon dioxide. *J. Geophys. Res.* **68**: 3899–3920.

Borchert, H. 1965. Principles of oceanic salt deposition and metamorphism. In *Chemical Oceanography*, vol. **2**, ed. J. P. Riley and G. Skirrow. Academic Press, New York, pp. 205–276.

Boyd, P. W., A. J. Watson, C. S. Law, *et al.* 2000. A mesoscale phytoplankton bloom in the polar Southern Ocean stimulated by iron fertilization. *Nature* **407**: 695–701.

Boyd, P. W., T. Jickells, C. S. Law *et al.* 2007. Mesoscale iron enrichment experiments 1993–2005: synthesis and future directions. *Science* **315**: 612–617.

Boyle, E. A. 1988. Cadmium: chemical tracer of deepwater paleoceanography. *Paleoceanography* **3**: 471–489.

Boyle, E. and J. M. Edmond. 1975. Copper in surface waters south of New Zealand. *Nature* **253**: 107–109.

Boyle, R. 1673. Observations and experiments about the saltness of the sea. In *The Works of the Honourable Robert Boyle*, vol. **7**, ed. M. Hunter and E. B. Davis. Pickering & Chatto, London, 1999, pp. 389–412.

BP. 2011. *BP Statistical Review of World Energy June **2010***. BP, London.

Bradshaw, A. L., P. G. Brewer, D. K. Shafer, and R. T. Williams. 1981. Measurements of total carbon dioxide and alkalinity by potentiometric titration in the GEOSECS program. *Earth Planet. Sci. Lett.* **55**: 99–115.

Braitsch, O. 1971. *Salt Deposits, Their Origin and Composition.* Translated by P. J. Burek and A. E. M. Nairn. Springer-Verlag, New York.

Brand, L. E., W. G. Sunda, and R. R. L. Guillard. 1986. Reduction of marine phytoplankton growth rates by copper and cadmium. *J. Exp. Mar. Biol. Ecol.* **96**: 224–250.

Brewer, P. 1971. *Hydrographic and Chemical Data from the Black Sea.* WHOI Technical Report 71–65, Woods Hole Oceanographic Institution, Woods Hole, MA.

Brewer, P. G. and D. W. Spencer. 1970. Trace element intercalibration study. Technical Report No. 70–62. Unpublished manuscript. Woods Hole Oceanographic Institution, Woods Hole, MA.

Brewer, P. G. and D. W. Spencer. 1974. Distribution of some trace elements in Black Sea and their flux between dissolved and particulate phases. In *The Black Sea – Geology, Chemistry, and Biology*, ed. E. T. Degens and D. A. Ross. American Association of Petroleum Geologists, Tulsa, OK, pp. 137–143.

Brindley, G. W. and G. Brown, eds. 1980. *Crystal Structures of Clay Minerals and Their X-Ray Identification.* Mineralogical Society, London.

Broadgate, W. J., P. S. Liss, and S. A. Penkett. 1997. Seasonal emissions of isoprene and other reactive hydrocarbon gases from the ocean. *Geophys. Res. Lett.* **24**: 2675–2678.

Broecker, W. S. 1989. The salinity contrast between the Atlantic and Pacific oceans during glacial time. *Paleoceanography* **4**: 207–212.

Broecker, W. S. 1991. The great ocean conveyer. *Oceanography* **4**: 79–89.

Broecker, W. S. and T. -H. Peng. 1974. Gas exchange rates between the air and the sea. *Tellus* **26**: 21–35.

Broecker, W. S. and T. -H. Peng. 1982. *Tracers in the Sea.* ELDIGIO Press, Lamont-Doherty Geological Observatory, Columbia University, New York.

Broecker, W. S. and T. Takahashi. 1978. The relationship between lysocline depth and in-situ carbonate ion concentration. *Deep-Sea Res.* **25**: 65–95.

Broecker, W. S., A. Kaufman, and R. M. Trier. 1973. The residence time of thorium in surface sea water and its implications regarding the rate of reactive pollutants. *Earth Planet. Sci. Lett.* **20**: 35–44.

Broecker, W. S., D. W. Spencer, and H. Craig. 1982. *GEOSECS Pacific Expedition, Hydrographic Data 1973–1974*. US Government Printing Office, Washington, DC.

Broecker, W. S., S. Sutherland, W. Smethie, T. H. Peng, and G. Ostlund. 1995. Oceanic radiocarbon: separation of the natural and bomb components. *Global Biogeochem. Cycles* **9**: 263–288.

Broecker, W. S., T. Takahashi, and T. Takahashi. 1985. Sources and flow patterns of deep-ocean waters as deduced from potential temperature, salinity, and initial phosphate concentration. *J. Geophy. Res.* **90**: 6925–6939.

Broecker, W. S., T.-H. Peng, G. Ostlund, and M. Stuiver. 1985a. The distribution of bomb radiocarbon in the ocean. *J. Geophys. Res.* **90**: 6953–6970.

Brown, C. A. 1942. Liebig and the law of the minimum. In *Liebig and after Liebig*, ed. F. R. Moulton. American Assocation for the Advancement of Science, Washington, DC, pp. 71 82.

Bruland, K. W. 1980. Oceanographic distributions of cadmium, zinc, nickel, and copper in the North Pacific. *Earth Planet. Sci. Lett.* **47**: 176–198.

Bruland, K. W. 1983. Trace elements in seawater. In *Chemical Oceanography*, 2nd edn., vol. **8**, ed. J. P. Riley and R. Chester. Academic Press, New York, pp. 157–220.

Bruland, K. W. and R P. Franks. 1983. Mn, Ni, Cu, Zn and Cd in the western North Atlantic. In *Trace Metals in Sea Water*, ed. C. S. Wong, E. Boyle, K. W. Bruland, J. D. Burton, and E. D. Goldberg. Plenum Press, New York, pp. 395–414.

Bruland, K. W. and E. L. Rue. 2001. Analytical methods for the determination of concentrations and speciation of iron. In *The Biogeochemistry of Iron in Seawater*, ed. D. R. Turner and K. A. Hunter. Wiley, New York,, pp. 255–289.

Buesseler, K. O. 1991. Do upper-ocean sediment traps provide an accurate record of particle flux? *Nature* **353**: 420–423.

Buesseler, K. O., J. E. Bauer, R. F. Chen, *et al.* 1996. An intercomparison of cross-flow filtration techniques used for sampling marine colloids: overview and organic carbon results. *Mar. Chem.* **55**: 1–31.

Burke, W. H., R. E. Denison, E. A. Hetherington, *et al.* 1982. Variation of seawater $^{87}Sr/^{86}Sr$ throughout Phanerozoic time. *Geology* **10**: 516–519.

Burton, J. D. 1975. Radioactive nuclides in the marine environment. In *Chemical Oceanography*, 2nd edn, vol. **3**, ed. J. P. Riley and G. Skirrow. Academic Press, New York, pp. 91–191.

Butler, A. 2005. Marine siderophores and microbial iron utilization. *BioMetals* **18**: 369–374.

Butler, J. N. 1982. *Carbon Dioxide Equilibria and Their Applications*. Addison-Wesley, Reading, MA.

Byrne, R. H. and J. A. Breland. 1989. High precision multiwavelength pH determinations in seawater using cresol red. *Deep-Sea Res.* **36**: 803–10.

Byrne, R. H., L. R. Kump, and K. J. Cantrell. 1988. The influence of temperature and pH on trace metal speciation in seawater. *Mar. Chem.* **25**: 163–181.

Byrne, R. H., Y.-R. Luo, and R. W. Young. 2002. Iron hydrolysis and solubility revisited: observations and comments on iron hydrolysis characterizations. *Mar. Chem.* **70**: 23–35.

Cai, W.-J. and Y. Wang. 1998. The chemistry, fluxes, and sources of carbon dioxide in the estuarine waters of the Satilla and Altamaha Rivers, Georgia. *Limnol. Oceanogr.* **43**: 657–668.

Calvert, S. E. 1983. Sedimentary geochemistry of silicon. In *Silicon Geochemistry and Biogeochemistry*, ed S. R. Aston. Academic Press, London, pp. 143–186.

Calvert, S. E., R. E. Karlin, L. J. Toolin, *et al.* 1991. Low organic carbon accumulation rates in Black Sea sediments. *Nature* **350**: 692–695.

Cann, J. R. and M. R. Strens. 1989. Modeling periodic megaplume emission by black smoker systems. *J. Geophys. Res.* **94**: 12,227–12,237.

Canup, R. M. and E. Asphaug. 2001. Origin of the Moon in a giant impact near the end of the Earth's formation. *Nature* **412**: 708–712.

Capone, D. G., D. A. Bronk, M. R. Mulholand, and E. Carpenter, eds. 2009. *Nitrogen in the Marine Environment*, 2nd edn. Academic Press, San Diego, CA.

Capone, D. G., J. P. Zehr, H. W. Paerl, B. Bergman, and E. J. Carpenter. 1997. *Trichodesmium*, a globally significant marine cyanobacterium. *Nature* **276**: 1221–1229.

Carpenter, E. J. and K. Romans. 1991. Major role of the cyanobacterium *Trichodesmium* in nutrient cycling in the North Atlantic Ocean. *Science* **254**: 1356–1358.

Carpenter, J. H. and M. E. Manella. 1973. Magnesium-to-chlorinity ratios in seawater. *J. Geophys. Res.* **78**: 3621–3626.

Carson, R. 1951. *The Sea Around Us.* Oxford University Press, New York.

Caspers, H. 1957. The Black Sea and Sea of Azov. In *Treatise on Marine Ecology and Paleoecology*, vol. **1**, ed J. W. Hedgpeth. Geological Society of America, Washington, DC, pp. 801–889.

Chappuis, P. 1907. Dilatation de l'eau. *Travaux et Mémoires du Bureau International des Poids et Mesures* **13**: D1–D40 Cited from Menaché in UNESCO 1976.

Charette, M. A. and W. H. F. Smith. 2010. The volume of Earth's ocean. *Oceanography* **23**: 112–114.

Chave, K. E. and E. Suess. 1967. Suspended minerals in sea water. *Trans. New York Acad. Sci.* **29**: 991–1000.

Chen, C., P. N. Sedwick and M. Sharma. 2009. Anthropogenic osmium in rain and snow reveals global-scale atmospheric contamination. *Proc. Natl. Acad. Sci.* **106**: 7724–7728.

Chen, I. H., R. L. Edwards, and G. J. Wasserburg. 1986 ^{238}U, ^{234}U and ^{232}Th in seawater Earth Planet. *Sci. Lett.* **80**: 241–251.

Christian, J. R., M. R. Lewis, and D. M. Karl. 1997. Vertical fluxes of carbon, nitrogen, and phosphorus in the North Pacific Subtropical Gyre near Hawaii. *J. Geophys. Res.* **102**: 15,667–15,677.

Clark, F. W. 1908. *The Data of Geochemistry.* Bulletin No. 330. United States Geological Survey, Washington, DC.

Clark, W. B., M. A. Beg, and H. Craig. 1969. Excess ^{3}He in the sea: evidence for terrestrial primordial helium. *Earth Planet. Sci. Lett.* **6**: 213–220.

Clayton, T. D. and R. H. Byrne. 1993. Spectrophotometric seawater pH measurements: total hydrogen ion concentration scale calibration of *m*-cresol purple and at-sea results. *Deep-Sea Res.* **40**: 2115–2129.

Clever, H. L., S. A. Johnson, and M. E. Derrick. 1985. The solubility of mercury and some sparingly soluble mercury salts in water and aqueous electrolyte solutions. *J. Phys. Chem. Ref. Data* **14**: 631–680.

Coale, K. H. and K. W. Bruland. 1990. Spatial and temporal variability in copper complexation in the North Pacific. *Deep-Sea Res.* **37**: 317–336.

Codispoti, L. A. 2007. An oceanic fixed nitrogen sink exceeding 400 Tg N a^{-1} vs. the concept of homeostasis in the fixed-nitrogen inventory. *Biogeosciences* **4**: 233–253.

Codispoti, L. A., G. E. Friederich, J. W. Murray, and C. M. Sakamoto. 1991. Chemical variability in the Black Sea: implications of continuous vertical profiles that penetrated the oxic/anoxic interface. *Deep-Sea Res.* **38** (Suppl. 2): S691–S710.

Converse, D. R., H. D. Holland, and J. M. Edmond. 1984. Flow rates in the axial hot springs of the East Pacific Rise (21° N): implications for the heat budget and the formation of massive sulfide deposits. *Earth Planet. Sci.* **69**: 159–175.

Coplen, T. B. 1994. Reporting of stable hydrogen, carbon, and oxygen isotopic abundances. *Pure Appl. Chem.* **66**: 273–276.

Copping, A. E. and C. J. Lorenzen. 1980. Carbon budget of a marine phytoplankton-herbivore system with carbon-14 as a tracer. *Limnol. Oceanogr.* **25**: 873–882.

Cowey, C. B. 1956. A preliminary investigation of the variation of vitamin B_{12} in oceanic and coastal waters. *J. Mar. Biol. Ass. UK* **35**: 609–620.

Cox, R. A., F. Culkin, R. Greenhalgh, and J. P. Riley. 1962. Chlorinity, conductivity and density of seawater. *Nature* **193**: 518–520.

Cox, R. A., F. Culkin, and J. P. Riley. 1967. The electrical conductivity/conductivity relationship in natural seawater. *Deep-Sea Res.* **14**: 203–220.

Cox, R. A., M. J. McCartney, and F. Culkin. 1970. The specific gravity/salinity/temperature relationship in natural seawater. *Deep-Sea Res.* **17**: 679–689.

Craig, H. 1957. Isotopic standards for carbon and oxygen and correction factors for mass-spectrometric analysis of carbon dioxide. *Geochim. Cosmochim. Acta* **12**: 133149.

Craig, H. 1961a. Isotopic variations in meteoric waters. *Science* **133**: 1702–1703.

Craig, H. 1961b. Standard for reporting concentrations of deuterium and oxygen-18 in natural waters. *Science* **133**: 1833–1834.

Craig, H. 1966. Isotopic composition and origin of the Red Sea and Salton Sea geothermal brines. *Science* **154**: 1544–1548.

Craig, H. and L. I. Gordon. 1965. Isotopic oceanography: deuterium and oxygen-18 variation in the ocean and marine atmosphere. In *Marine Geochemistry*, ed D. R. Schink and J. T. Corless. Occasasional Publication No. 3, Graduate School of Oceanography, University of Rhode Island, Narragansett, RI, pp. 277–374.

Craig, H. and T. Hayward. 1987. Oxygen supersaturation in the ocean: biological versus physical contributions. *Science* **235**: 199–202.

Craig, H. and R. F. Weiss. 1971. Dissolved gas saturation anomalies and excess helium in the ocean. *Earth Planet. Sci. Lett.* **10**: 289–296.

Crawford, E. 1996. *Arrhenius, From Ionic Theory to the Greenhouse Effect*. Science History Publications, Canton, MA.

Cronan, D. S. 2001. Manganese Nodules. In *Encyclopedia of Ocean Sciences*, ed. J. Steel. Academic Press, San Diego, CA, pp. 1526–1533.

Culberson, C. and R. M. Pytkowicz. 1968. Effect of pressure on carbonic acid, boric acid and the pH in seawater. *Limnol. Oceanogr.* **13**: 403–417.

Culkin, F. 1965. The major constituents of seawater. In *Chemical Oceanography*, vol. **1**, ed J. P. Riley and G. Skirrow. Academic Press, New York, pp. 121–161.

Culkin, F. and R. A. Cox. 1966. Sodium, potassium, magnesium, calcium and strontium in seawater. *Deep-Sea Res.* **13**: 789–804.

Cullen, W. R. and K. J. Reimer. 1989. Arsenic speciation in the environment. *Chem. Rev.* **89**: 713–764.

Cutter, G. A. and K. W. Bruland. 1984. The marine biogeochemistry of selenium: a re-evaluation. *Limnol. Oceanogr.* **29**: 1179–1192.

Damsté, J. S. S., M. Strous, and W. I. C. Rijpstra. 2002. Linearly concatenated cyclobutane lipids form a dense bacterial membrane. *Nature* **419**: 708–712.

Dansgaard, W. 1964. Stable isotopes in precipitation. *Tellus* **16**: 436–468.

Davy, H. 1811. The Bakerian Lecture. On some of the combinations of oxymuriatic gas and oxygene, and on the chemical relations of these principles, to inflammable bodies. *Phil. Trans. R. Soc.* **101**: 1–35.

de Baar, H. J. W., P. M. Saager, R. F. Nolting, and J. van der Meer. 1994. Cadmium versus phosphate in the world ocean. *Mar. Chem.* **46**: 261–281.

de Baar, H. J. W., P. W. Boyd, K. H. Coale, *et al.* 2005. Synthesis of iron fertilization experiments: from the iron age to the age of enlightenment. *J. Geophys. Res.* **110***: C09S16.*

De'ath, G., J. M. Lough, and K. E. Fabricius. 2009. Declining coral calcification on the Great Barrier Reef. *Science* **323**: 116–119.

Deacon, M. 1971. *Scientists and the Sea 1650–1900; A Study of Marine Science*. Academic Press, London.

Degens, E. T. and D. A. Ross, eds. 1974. *The Black Sea – Geology, Chemistry and Biology*. American Association of Petroleum Geologists, Tulsa, OK.

Denman, K. L. G. Brasseur, A. Chidthaisong, *et al.* 2007. Couplings between changes in the climate system and biogeochemistry. In Solomon *et al.* (2007), ch. 7, pp. 499–587.

Deuser, W. G. 1975. Reducing environments. In *Chemical Oceanography*, 2nd edn. vol. **3**, ed. J. P. Riley and G. Skirrow. Academic Press, New York, pp. 1–37.

Deuser, W. G. 1987. Variability of hydrography and particle flux: transient and long-term relationships. *Mitt. Geol-Poläont. Inst. Univ. Hamburg* **62**: 179–193.

Dholabhai, P. D., P. Englezos, N. Kalogerakis, and P. R. Bishnoi. 1991. Equilibrium conditions for methane hydrate formation in aqueous mixed electrolyte solutions. *Can. J. Chem. Eng.* **69**: 800–805.

Dickson, A. G. 1984. pH scales and proton-transfer reactions in saline media such as sea water. *Geochim. Cosmochim. Acta* **48**: 2299–2308.

Dickson, A. G. 1993. The measurement of seawater pH. *Mar. Chem.* **44**: 131–142.

Dickson, A. G. 1990a. Standard potential of the reaction: $AgCl(s) + 1/2\ H_2(g) = Ag(s) + HCl$ (aq), and the standard acidity constant of the ion HSO_4^- in synthetic seawater from 273.15 to 318.15 K. *J. Chem. Thermodynam.* **22**:113–127.

Dickson, A. G. 1990b. Thermodynamics of the dissociation of boric acid in synthetic seawater from 273.15 to 318.15 K. *Deep-Sea Res.* **37**: 755–766.

Dickson, A. G. 1993. pH buffers for sea water media based on the total hydrogen ion concentration scale. *Deep-Sea Res.* **40**:107–118.

Dickson, A. G., C. L. Sabine, and J. R. Christian, eds. 2007. Guide to the best practices for ocean CO_2 measurements. *PICES Special Publication 3*. Available from CDIAC; see http://cdiac.ornl.gov/oceans/Handbook_2007.html.

Dickson, R. R., J. Meincke, S.-A. Malmberg and A. J. Lee. 1988. The great salinity anomaly in the North Atlantic 1968–1982. *Prog. Oceanog.* **20**: 103–151.

Dittmar, W. 1884. Report on researches into the composition of ocean-water collected by H.M.S. Challenger during the years 1873–1876. In *Report of the Scientific Results of the Voyage of H.M.S. Challenger during the years 1873–1876, Physics and Chemistry*, vol. **1**, ed J. Murray. H. M. Stationery Office, London, pp. 1–251.

Doherty, B. T. and D. R. Kester. 1974. Freezing point of seawater. *J. Mar. Res.* **32**: 285–300.

Donat, J. R. and K. W. Bruland. 1995. Trace elements in the oceans. In *Trace Elements in Natural Waters*, ed B. Salbu and E. Steinnes. CRC Press, Boca Raton, FL, pp. 247–281.

Doney, S. C., V. J. Fabry, R. A. Feely, and J. Kleypas. 2009. Ocean acidification: the other CO_2 problem. *Annu. Rev. Mar. Sci.* **1**: 169–192.

Donlon, C. J. and I. S. Robinson. 1997. Observations of the oceanic thermal skin in the Atlantic. *J. Geophys. Res.* **102**: 18,585–18,606

Drake, M. J. and K. Righter. 2002. Determining the composition of the Earth. *Nature* **416**: 39–44.

Druffel, E. M. 1980. *Radiocarbon in annual coral rings of the Pacific and Atlantic Oceans*. Ph.D. Dissertation. University of California, San Diego, CA.

Druffel, E. R. M., P. M. Williams, J. E. Bauer, and J. R. Ertel. 1992. Cycling of dissolved and particulate organic matter in the open ocean. *J. Geophys. Res.* **97**: 15, 639–15,659.

Duarte, C. M., M. Merino, and M. Gallegos. 1995. Evidence of iron deficiency in seagrasses growing above carbonate sediments. *Limnol. Oceanogr.* **40**: 1153–1158.

Duce, R. A. 1989. Exchange of particulate carbon and nutrients across the air/sea interface. In *The Ocean as a Source and Sink for Atmospheric Trace Constituents*. UNESCO, Paris, pp. 30–43.

Duce, R. A. and N. W. Tindale. 1991. Atmospheric transport of iron and its deposition in the ocean. *Limnol. Oceanogr.* **36**: 1715–1726.

Duce, R. A., J. LaRoche, K. Altieri, *et al.* 2008. Impacts of atmospheric anthropogenic nitrogen on the open ocean. *Science* **320**: 893–897.

Duce, R. A., P. S. Liss, J. T. Merrill, *et al* . 1991. The atmospheric input of trace species to the world ocean. *Global Biogeochem. Cycles* **5**:193–259.

Dugdale, R. C. and J. J. Goering. 1967. Uptake of new and regenerated forms of nitrogen in primary productivity. *Limnol. Oceanogr.* **12**: 196–206.

Dunbar, M. J. 1967. Exploring the Arctic Ocean. *Oceanol. Int.* (May–June): 31–35**x**.

Dyni, J. R. 2006. Geology and resources of some world oil-shale deposits. *US Geological Survey Science Investigations Report* **2005–5294**.

Edmond, J. M., C. Measures, R. E. McDuff, L. H. Chan, R. Collier, and B. Grant. 1979. Ridgecrest hydrothermal activity and the balances of the major and minor elements in the ocean; the Galapagos data. *Earth Planet. Sci. Lett.* **46**: 1–18.

Edmonds, J. S., K. A. Francesconi, J. R. Cannon, *et al.* 1977. Isolation, crystal structure and synthesis of arsenobetaine, the arsenical constituent of the western rock lobster *Panulirius longipes cygnus* George. *Tetrahedron Lett.* No. **18**: 1543–1546.

Eglington, T. I. and D. J. Repeta. 2006. Organic matter in the contemporary ocean. In *The Oceans and Marine Geochemitsry*, ed. H. Elderfield (*Treatise on Geochemistry*, vol. 6, ed. H. D. Holland and K. K. Turekian). Elsevier, New York, pp. 145–180.

Elderfield, H. and A. Schultz. 1996. Mid-ocean ridge hydrothermal fluxes and the chemical composition of the ocean. *Ann. Rev. Earth Planet. Sci.* **24**: 191–224.

Emerson, S. 1995. Enhanced transport of carbon dioxide during gas exchange. In *Air–Water Gas Transfer*, ed. B. Jähne and E. C. Monahan. AEON Verlag & Studio, Hanau, pp. 23–35

Emerson, S. and M. Bender. 1981. Carbon fluxes at the sediment–water interface of the deep sea: calcium carbonate preservation. *J. Mar. Res.* **39**: 139–162.

Eppley, R. W. and B. J. Peterson. 1979. Particulate organic matter flux and planktonic new production in the deep ocean. *Nature* **282**: 677–680.

Erickson, D. J., III. 1989. Ocean to atmosphere carbon monoxide flux: global inventory and climate implications. *Global Biogeochem. Cycles* **3**: 305–314.

Eugster, H. P., C. E. Harvie, and J. H. Weare. 1980. Mineral equilibria in a six-component seawater system, Na–K–Mg–Ca–SO_4–Cl–H_2O, at 25 °C. *Geochim. Cosmochim. Acta* **44**: 1335–1347.

Falkner, K. K. and J. M. Edmond. 1990. Gold in seawater. *Earth Planet. Sci. Lett.* **98**: 208–221.

Falkowski, P. G. and J. A. Raven. 2007. Aquatic Photosynthesis, 2nd edn. Blackwell Science, Malden, MA.

Fanning, K. A. 1992. Nutrient provinces in the sea: concentration ratios, reaction rate ratios and ideal covariation. *J. Geophys. Res.* **97**: 5693–5712.

Faraday, M. 1823. On hydrate of chlorine. *Quart. J. Sci.* **15**: 429.

Faulkner, J. and R. J. Anderson. 1974. Natural products chemistry of the marine environment. In *The Sea*, vol. 5, ed. E. D. Goldberg. John Wiley & Sons, New York, pp. 679–714.

Faure, G. 1986. *Principles of Isotope Geology*, 2nd edn. John Wiley & Sons, New York.

Feely, R., C. Sabine, F. Millero, R. Wanninkhof, and D. Hansell. 2006. Carbon dioxide, hydrographic, and chemical data obtained during the R/V Thomas Thompson Cruise in the Pacific Ocean on CLIVAR Repeat Hydrography Sections P16N, 2006. CDIAC, Oak Ridge National Laboratory, US Department of Energy, Oak Ridge, TN. See: http://cdiac.ornl.gov/oceans/RepeatSections/clivar_p16n.html.

Feistel, R. 2003. A new extended Gibbs thermodynamic potential of seawater. *Prog.Oceanogr.* **58**: 43–114.

Field, C. B., M. J. Beheenfeld. J. T. Randerson, and P. Falkowski. 1998. Primary production of the biosphere: integrating terrestrial and oceanic components. *Science* **281**: 237–240.

Firestone, R. B. and V. S. Shirley, eds., 1996. *Table of Isotopes*, vols. **1–2**. John Wiley & Sons, New York.

Fitzgerald, W. F. 1989. Atmospheric and oceanic cycling of mercury. In *Chemical Oceanography*, 2nd edn., vol. **10**, ed. J. P. Riley, R. Chester, and R. A. Duce. Academic Press, New York, pp. 151–186.

Fitzgerald, W. F., C. H. Lamborg, and C. R. Hammerschmidt. 2007. Marine biogeochemical cycling of mercury. *Chem Rev.* **107**: 641–662.

Fitzwater, S. E., G. A. Knauer, and J. H. Martin. 1982. Metal contamination and its effect on primary production measurements. *Limnol. Oceanogr.* **27**: 544–551.

Fofonoff, N. P. 1985. Physical properties of seawater: a new salinity scale and equation of state for seawater. *J. Geophys. Res.* **90**: 3332–3342.

Fofonoff, N. P. and R. C. Millard, Jr. 1983. Algorithms for computation of fundamental properties of seawater. *UNESCO Technical Papers in Marine Science*, vol **44**. UNESCO, Paris.

Fogg, G. E. 1975. Primary productivity. In *Chemical Oceanography*, 2nd edn.,vol. **2**, ed. J. P. Riley and G. Skirrow. Academic Press, New York, pp. 385–453.

Forchhammer, G. 1865. On the composition of seawater in the different parts of the ocean. *Phil. Trans. R. Soc. London* **155**: 203–262.

Fox, D., J. Isaacs, and E. Corcoran. 1952. Marine leptopel, its recovery, measurement and distribution. *J. Mar. Res.* **11**:29–46.

Francesconi, K. A. and J. S. Edmonds. 1998. Arsenic species in marine samples. *Croatica Chemica Acta* **71**: 343–359.

Franck, R. 1694. *Northern Memoirs, Calculated for the Meridian of Scotland....* Henry Mortclock, London. (Written 1658.)

Frankignoulle, M. and A. V. Borges. 2001. Direct and indirect pCO_2 measurements in a wide range of pCO_2 and salinity values (the Schelt estuary). *Aquatic Geochemistry* **7**: 267–273.

Friedlander, G., J. W. Kennedy, and J. M. Miller. 1981. *Nuclear and Radiochemistry, 3rd edn.* John Wiley & Sons, New York.

Froelich, P. N. 1988. Kinetic control of dissolved phosphate in natural rivers and estuaries: a primer on the phosphate buffer mechanism. *Limnol. Oceanogr.* **33**: 649–668.

Froelich, P. N. and M. O. Andreae. 1981. The marine geochemistry of germanium: Ekasilicon. *Science* **213**: 205–207.

Froelich, P. N., G. P. Klinkhammer, M. L. Bender, *et al.* 1979. Early oxidation of organic matter in pelagic sediments of the eastern equatorial Atlantic: suboxic diagenesis. *Geochim. Cosmochim. Acta* **43**: 1075–1090.

Froelich, P. N., M. L. Bender, N. A. Luedtke, G. R. Heath, and T. DeVries. 1982. The marine phosphorus cycle. *Am. J. Sci.* **282**: 474–511.

Froelich, P. N., R. A. Mortlock, and A. Shemesh. 1989. Inorganic germanium and silica in the Indian Ocean: biological fractionation during $(Ge/Si)_{OPAL}$ formation. *Global Biogeochem. Cycles* **3**: 79–88.

Fry, B. 2006. *Stable Isotope Ecology*. Springer, New York.

Fukuyama, H. 1985. Heat of fusion of basaltic magma. *Earth Planet. Sci. Lett.* **73**: 407–414.

Fulweiler, R. W., S. W. Nixon, B. A. Buckley, and S. L. Granger. 2007. Reversal of the net nitrogen gas flux in coastal marine sediments. *Nature* **448**: 180–182.

Galloway, J. N., F. J. Dentener, D. G. Capone, *et al.* 2004. Nitrogen cycles: past, present, and future. *Biogeochemistrry* **70**: 153–226.

Galloway, J. N., A. R. Townsend, J. W. Erisman, *et al* . 2008. Transformation of the nitrogen cycle: Recent trends, questions, and potential solutions. *Science* **320**: 889–892.

Ganachaud, A. and C. Wunsch. 2000. Improved estimates of global ocean circulation, heat transport and mixing from hydrographic data. *Nature* **408**: 453–457.

García, H. E. and L. I. Gordon. 1992. Oxygen solubility in seawater: better fitting equations. *Limnol. Oceanogr.* **37**: 1307–1312.

Garrels, R. M. and M. E. Thompson. 1962. A chemical model for seawater at 25°C and one atmosphere total pressure. *Am. J. Sci.* **260**: 57–66.

Garrels, R. M., M. E. Thompson, and R. Siever. 1961. Control of carbonate solubility by carbonate complexes. *Am. J. Sci.* **259**: 24–45.

Garrett, W. D. 1967. The organic chemical composition of the ocean surface. *Deep-Sea Res.* **14**: 221–227.

Gay-Lussac, J. L. 1817. Note sur la salure de l'Océan atlantique. (Also: Supplement to this note.) *Ann. Chim. Phys., Sér. 2* **6**: 426–436; **7**: 79–83.

Gay-Lussac, J. L. 1819. Premier mémoire sur la dissolubilité des sels dans l'eau. *Ann. Chim. Phys., Sér. 2*, **11**: 296–315.

Gerlach, T. 2011. Volcanic versus anthropogenic carbon dioxide. *EOS* **92** (24): 201–202.

German, C. R. and K. L. Von Damm. 2006. Hydrothermal processes. In *The Oceans and Marine Geochemistry*, ed. H. Elderfield (*Treatise on Geochemistry*, vol. 6, ed. H. D. Holland and K. K. Turekian). Elsevier, New York, pp. 181–222.

Gieskes, J. M. 1983. The chemistry of interstitial waters of deep sea sediments: interpretation of deep sea drilling data. In *Chemical Oceanography*, vol. **8**, ed J. P. Riley and R. Chester. Academic Press, New York.

Gieskes, J. M., H. Elderfield, J. R. Lawrence, and J. LaKind. 1984. Interstitial water studies, Leg 78A. in *Init. Repts. DSDP, 78A*. ed B. Biju-Duval, J. C. Moore, *et al.* U.S. Govt. Printing Office, Washington, D.C., pp. 377–384.

Gill, G. A. and W. F. Fitzgerald. 1988. Vertical mercury distributions in the oceans. *Geochim. Cosmochim. Acta* **52**: 1719–1728.

Girard, G. and M. Menaché. 1971. Variation de la masse volumique de l'eau en fonction de sa composition isotopique. *Metrologia* **7**: 83–87.

Girard, G. and M. Menaché. 1972. Sur le calcul de la masse volumique de l'eau. *C. R. Acad. Sci., Ser. B* **274**: 377–379.

Gist, N. and A. C. Lewis. 2006. Seasonal variations of dissolved alkenes in coastal waters. *Mar. Chem.* **100**: 1–10.

Glueckauf, E. 1951. The composition of atmospheric air. In *Compendium of Meteorology*, ed. Thomas E. Malone. American Meteorological Society, Boston, MA, pp. 3–10.

Goering, J. J. and J. D. Cline. 1970. A note on denitrification in seawater. *Limnol. Oceanogr.* **15**: 306–309.

Goldberg, E. D. 1963. The oceans as a chemical system. In *The Sea; Ideas and Observations on Progress in the Study of the Seas*, vol. **2**, ed. M. N. Hill. Interscience, New York, pp. 3–25.

Goldberg, E. D. 1985. *Black Carbon in the Environment: Properties and Distribution*. Wiley, New York,

Goldberg, E. D. 1987. Heavy metal analyses in the marine environment – approaches to quality control. *Mar. Chem.* **22**: 117–124.

Goldblatt, C., M. W. Claire, and T. M Lenton, *et al.* 2009. Nitrogen-enhanced greenhouse warming on early Earth. *Nature Geoscience* **2**: 891–896.

Goldhammer, T., V. Brüchert, T. G. Ferdelman, and M. Zabel. 2010. Microbial sequestration of phosphorus in anoxic upwelling sediments, *Nature Geoscience* **3**: 557–561.

Goldman, D. T. and R. J. Bell. 1981. *The International System of Units (SI). National Bureau Stds. (U.S.), Spec. Publ.* 330, U.S. Govt. Printing Office, Washington, D.C

Gonfiantini, R. 1978. Standards for stable isotope measurements in natural compounds. *Nature* **271**: 534–536.

Goñi, M. A., D. M. Hartz, R. C. Thunell, and E. Tappa. 2001. Oceanographic considerations for the application of the alkenone-based paleotemperature $U_{37}^{k'}$ index in the Gulf of California. *Geochim. Cosmochim. Acta* **65**: 545–557.

Graedel, T. E. 1978. *Chemical Compounds in the Atmosphere*. Academic Press, New York.

Gran, H. H. 1931. On the conditions for the production of plankton in the sea. *Rapp. Proc. Verb. Cons. Int. Explor. Mer.* **75**: 37–46.

Grasshoff, K. 1975. The hydrogeochemistry of landlocked basins and fjords. In *Chemical Oceanography*, 2nd edn., vol. **2**, ed J. P. Riley and G. Skirrow. Academic Press, New York, pp. 455–597.

Grasshoff, K., K. Kremling, and M. Ehrhardt, eds. 1999. *Methods of Seawater Analysis*, 3rd edn. Wiley-VCH, New York.

Greenwald, I. 1941. The dissociation of Ca and Mg carbonates and bicarbonates. *J. Biol. Chem.* **141**: 789–796.

Grose, T. J., J. A. Johnson, and G. R. Bigg. 1995. A comparison between the FRAM (Fine Resolution Antarctic Model) results and observations in the Drake Passage. *Deep-Sea Res.* **42**: 365–388.

Grosso, P., M. Le Menn, J.-L. D. B. De La Tocnaye, Z. Y. Wu, and D. Malardé. 2010. Practical versus absolute salinity measurement: new advances in high performance seawater salinity sensors. *Deep-Sea Res. I* **57**: 151–156.

Haar, L., J. S. Gallagher, and G. S. Kell. 1984. *NBS/NRC Steam Tables*. Hemisphere Publishing Corporation, Washington, DC.

Hagemann, R., G. Nief, and E. Roth. 1970. Absolute isotopic scale for deuterium analysis of natural waters. Absolute D/H ratio for SMOW. *Tellus* **22**: 712–715.

Halley, E. 1687. An estimate of the quantity of vapour raised out of the sea by the warmth of the sun; derived from an experiment shown before the Royal society, at one of their late meetings. *Phil. Trans. R. Soc.* **16**: 366–370.

Halley, E. 1715. A short account of the cause of the saltness of the ocean, and of several lakes that emit no rivers; with a proposal, by the help thereof, to discover the age of the world. *Phil. Trans. R. Soc.* **29**: 296–300.

Hamme, R. C. and S. R. Emerson. 2002. Mechanisms controllong the global distribution of the inert gases argon, nitrogen and neon, *Geophys. Res. Lett.* **29**: doi: 10.1029/2002GL015273

Hamme, R. C and S. R. Emerson. 2004. The solubility of neon, nitrogen and argon in distilled water and seawater. *Deep-Sea Res. I* **51**: 1517–1528.

Hammond, D. E., J. McManus, W. M. Berelson, T. E. Kilgore, and R. H. Pope. 1996. Early diagenesis of organic material in equatorial Pacific sediments: stoichiometry and kinetics. *Deep-Sea Res. II* **43**: 1365–1412.

Han, M. W. and E. Suess. 1989. Subduction-induced pore fluid venting and the formation of authigenic carbonates along the Cascadia continental margin: implications for the global Ca-cycle. *Palaeogeogr., Palaeoclimat., Palaeoecol.* **71**: 97–118.

Hansen, J., M. Saito, P. Kharecha, *et al.* 2008. Target atmospheric CO_2: where should humanity aim? *The Open Atmos. Sci. J.* **2**: 217–231.

Hansen, J. E. and A. A. Lacis. 1990. Sun and dust versus greenhouse gases: an assessment of their relative roles in global climate change. *Nature* **346**: 713–719.

Hanson, P. J., S. E. Lindberg, T. A. Tabberer, J. G. Owens, and K.-H. Kim. 1995. Foliar exchange of mercury vapor: evidence for a compensation point. *Water, Air, Soil Poll.* **80**: 373–382.

Hardie, L. A. 1996. Secular variation in seawater chemistry: an explanation for the coupled secular variation in the mineralogies of marine limestones and potash evaporites over the past 600 m.y. *Geology* **24**: 279–283.

Harvey, G. R., D. A. Boran, L. A. Chesal, and J. M. Tokar. 1983. The structure of marine fulvic and humic acids. *Mar. Chem.* **12**: 119–132.

Harvey, H. W. 1937. The supply of iron to diatoms. *J. Mar. Biol. Ass. UK* **22**: 205–219.

Harvey, H. W. 1957. *The Chemistry and Fertility of Seawater*. 2nd Ed. Cambridge Univ. Press, London.

Harvie, C. E., J. H. Weare, L. A. Hardie, and H. P. Eugster. 1980. Evaporation of seawater: calculated mineral sequences. *Science* **208**: 498–500.

Hay, W. W. 1985. Potential errors in estimates of carbonate rock accumulating through geologic time. In *The Carbon Cycle and Atmospheric CO_2: Natural Variations Archean to Present*, ed E. T. Sundquist and W. S. Broecker. American Geophysical Union, Washington, DC, pp. 573–583.

Hayes, J. M. 2002. *Practice and Principles of Isotopic Measurements in Organic Geochemistry*. See: http://www.nosams.whoi.edu/docs/IsoNotesAug02.pdf.

Haynes, W. M, ed. 2011. *CRC Handbook of Chemistry and Physics*. CRC Press, Boca Raton, FL.

Heath, G. R. 1974. Dissolved silica and deep-sea sediments. In *Studies in Paleo-Oceanography*, ed. W. W. Hay. Society of. Economic Paleontologists and Mineralogists, Special Publication No. 20, Tulsa, OK, pp. 77–93.

Henry, W. 1803. Experiments on the quantity of gases absorbed by water, at different temperatures, and under different pressures. *Phil. Trans. R. Soc.* **93**: 29–42, 274–276.

Herbert, T. D. 2006. Alkenone paleotemperature determinations. In *The Oceans and Marine Geochemistry*, ed. H. Elderfield (*Treatise on Geochemistry*, vol. 6, ed. H. D. Holland and K. K. Turekian). Elsevier, New York, pp 391–432.

Herrmann, A. G., D. Knake, J. Schneider, and H. Peters. 1973. Geochemistry of modern seawater and brines from salt pans: main components and bromine distribution. *Contr. Mineral. Petrol.* **40**: 1–24.

Hester, K. C., E. T. Peltzer, W. J. Kirkwood, and P. G. Brewer. 2008. Unanticipated consequences of ocean acidification: a noisier ocean. *Geophys. Res. Lett.* **35**: L9601, doi:10.1029/2008GL034913.

Hillis-Colinvaux, L. 1980. Ecology and taxonomy of *Halimeda:* primary producer of coral reefs. *Adv. Mar. Biol.* **17**: 1–327.

Hinga, K. R. 1988. Seasonal predictions for pollutant scavenging in two coastal environments using a model calibration based upon thorium scavenging. *Mar. Environ. Res.* **26**: 97–112.

Hinga, K. R. 2002. Effects of pH on coastal marine phytoplankton. *Mar. Ecol. Prog. Series* **238**: 281–300.

Holland, H. D. 1972. The geologic history of seawater – an attempt to solve the problem. *Geochim. Cosmochim. Acta* **36**: 637–651.

Holland, H. D. 1978. *The Chemistry of the Atmosphere and Oceans.* John Wiley & Sons, New York.

Holland, H. D. 1984. *The Chemical Evolution of the Atmosphere and Oceans.* Princeton University Press, Princeton, NJ.

Holland, H. D. 2006. The geologic history of seawater. In *The Oceans and Marine Geochemistry*, ed. H. Elderfield (*Treatise on Geochemistry*, vol. **6**, ed. H. D. Holland and K. K. Turekian). Elsevier, New York, pp. 583–625.

Holm, N. G. 1996. Hydrothermal activity and the volume of the oceans. *Deep-Sea Res.* **43**: 47–52.

Holmes, A. 1913. *The Age of the Earth.* Harper, New York.

Honjo, S., S. J. Manganini, R. A. Krishfield, and R. Francois. 2008. Particulate organic carbon fluxes to the ocean interior and factors controlling the biological pump: a synthesis of global sediment trap programs since 1983. *Prog. Oceanogr.* **76**: 217–285.

Horibe, Y., E. Keiko, and H. Tsubota. 1974. Calcium in the South Pacific, and its correlation with carbonate alkalinity. *Earth Planet. Sci. Lett.* **23**: 136–140.

Horita, J. and J. R. Gat. 1989. Deuterium in the Dead Sea: remeasurement and implications for the isotopic activity correction in brines. *Geochim. Cosmochim. Acta* **45**: 131–133.

Horiguchi, M. and M. Kandatsu. 1959. Isolation of 2-aminoethane phosphonic acid from rumen protozoa. *Nature* **184**: 901–902.

Houghton, R. A. 2003. Revised estimates of the annual net flux of carbon to the atmosphere from changes in land use and land management 1850–2000. *Tellus B* **55**: 378–390.

Hsü, K. J., M. B. Cita, and W. B. F. Ryan. 1973. *The Origin of the Mediterranean Evaporite. Initial Reports of the Deep Sea Drilling Project*, vol. **XIII**, ed. W. B. F. Ryan and K. J. Hsü. US Government Printing Office, Washington, DC, pp. 1203–1231.

Hu, M., Y. Yang, J. M. Martin, K. Yin, and P. J. Harrison. 1996. Preferential uptake of Se(IV) over Se(VI) and the production of dissolved organic Se by marine phytoplankton. *Mar. Environ. Res.* **44**: 225–231.

Hudson, R. J. M., S. A. Gherini, W. F. Fitzgerald, and D. B. Porcella. 1995. Anthropogenic influences on the global mercury cycle: a model-based approach. *Water, Air, Soil Poll.* **80**: 265–272.

Huizenga, D. L. and D. R. Kester. 1979. Protonation equilibria of marine dissolved organic matter. *Limnol. Oceanogr.* **24**: 145–150.

Humphris, S. E., R. A. Zierenberg, L. S. Mullineaux, and R. E. Thompson, eds. 1995. *Seafloor Hydrothermal Systems.* American Geophysical Union, Washington, DC.

Hunten, D. M. and T. M. Donahue. 1976. Hydrogen loss from terrestrial planets. *Ann. Rev. Earth Planet. Sci.* **4**: 265–292.

Hunter, K. A. and P. S. Liss. 1981. Organic sea surface films. In *Marine Organic Chemistry*, ed. E. K. Duursma and R. Dawson. Elsevier, Amsterdam, pp. 259–298.

Hurd, D. C. 1983. Physical and chemical properties of siliceous skeletons. In *Silicon Geochemistry and Biogeochemistry*, ed S. R. Aston. Academic Press, New York, pp. 187–244.

ICS. 2009. *International Commission on Stratigraphy. International Stratigraphic Chart*. See: www.stratigraphy.org. Accessed June 5, 2011.

Ingle, S. E. 1975. Solubility of calcite in the ocean. *Mar. Chem.* **3**: 301–319.

Ito, E., D. M. Harris, and A. T. Anderson, Jr. 1983. Alteration of oceanic crust and geologic cycling of chlorine and water. *Geochim. Cosmochim. Acta* **47**: 1613–1624.

Jackson, G. A. and P. M. Williams. 1985. The importance of dissolved organic nitrogen and phosphorus to biological nutrient cycling. *Deep-Sea Res.* **32**: 223–235.

Jacobsen, J. P. and M. Knudsen. 1940. Urnormal 1937, or primary standard sea-water 1937. *Int. Union Geodesy Geophys., Assoc. Phys. Oceanogr. Publ. Sci.* **7**: 1–38.

Jacobson, R. L. and D. Langmuir. 1974. Dissociation constants of calcite and $CaHCO_3^+$ from 0 to 50°C. *Geochim. Cosmochim. Acta* **38**: 301–318.

Jähne, B., G. Heinz, and W. Dietrich. 1987a. Measurement of the diffusion coefficients of sparingly soluble gases in water. *J. Geophys. Res.* **92**: 10,767–10,776.

Jähne, B., K. O. Münnich, R. Bösinger, *et al.* 1987b. On the parameters influencing air-water gas exchange. *J. Geophys. Res.* **92**: 1937–1949.

Jensen, P. R., R. A. Gibson, M. M. Littler, and D. S. Littler. 1985. Photosynthesis and calcification in four deepwater *Halimeda* species (Chlorophyceac, Caulerpales). *Deep-Sea Res.* **32**: 451–464.

Jickells, T. D., Z. S. An, K. K. Anderson, *et al.* 2005. Global iron connections between desert dust, ocean biogeochemistry, and climate. *Science* **308**: 67–71.

Johnson, D. L. 1971. Simultaneous determination of arsenate and phosphate in natural waters. *Environ. Sci. Technol.* **5**: 411–414.

Johnson, D. L. 1972. Bacterial reduction in arsenate in seawater. *Nature* **240**: 44–45.

Johnson, K. S. 1982. Carbon dioxide hydration and dehydration kinetics in seawater. *Limnol. Oceanogr.* **27**: 849–855.

Johnson, K. S., F. P. Chavez, and G. E. Friederich. 1999. Continental-shelf sediment as a primary source of iron for coastal phytoplankton. *Nature* **398**: 697–700.

Johnston, R. 1964. Sea water, the natural medium of phytoplankton. I. Trace metals and chelation, and general discussion. *J. Mar. Biol. Ass. UK* **44**: 87–109.

Jones, M. M. and R. M. Pytkowicz. 1973. Solubility of silica in seawater at high pressures. *Bull. Soc. R. Sci. Liege* **42**: 118–120.

Kahane, A., J. Klinger, and M. Philippe. 1969. Dopage selectif de la Glace monocristalline avec de l'helium et du neon. *Solid State Commun.* **7**: 1055–1056.

Kalle, K. 1933. Zum Problem der Meereswasserfarbe. *Annal.Hydrogr. Maritimen Meteorologie* **66**: 1–13.

Kanwisher, J. 1963. Effect of wind on CO_2 exchange across the sea surface. *J. Geophys. Res.* **68**: 3921–3927.

Karl, D., G. M. McMurtry, A. Malahoff, and M. O. Garcia. 1988. Loihi Seamount, Hawaii: a mid-plate volcano with a distinctive hydrothermal system. *Nature* **335**: 532–535.

Karl, D., R. Letellier, L. Tupas, *et al.* 1997. The role of nitrogen fixation in biogeochemical cycling in the subtropical North Pacific Ocean. *Nature* **388**: 533–538.

Karlén, I., I. U. Olsson, P. Kållberg, and S. Kilicci. 1964. Absolute determination of the activity of two C^{14} dating standards. *Arkiv för Geofysik* **4**(22): 465–471.

Kasting, J. F. and N. G. Holm. 1992. What determines the volume of the oceans? *Earth. Planet. Sci. Lett.* **109**: 507–515.

Kastner, M., H. Elderfield, and J. B. Martin. 1991. Fluids in convergent margins: what do we know about their composition, origin, role in diagenesis and importance for oceanic chemical fluxes? *Phil. Trans. R. Soc. London A.* **335**: 243–259.

Keeling, C. D. 1973. Industrial production of carbon dioxide from fossil fuels and limestone. *Tellus* **25**: 174–198.

Keeling, C. D. and T. P. Whorf. 1994. Atmospheric CO_2 records from sites in the SIO sampling network. In *Trends '93: A Compendium of Data on Global Change*, ed T. A. Boden, D. P. Kaiser, R. J. Sepanski, and F. W. Stoss. ORNL/CDIAC-65, Carbon Dioxide Information Analysis Center, Oak Ridge National Laboratory, Oak Ridge, TN, pp. 16–26.

Keeling, R. F. and S. R. Shertz. 1992. Seasonal and interannual variations in atmospheric oxygen and implications for the global carbon cycle. *Nature* **358**: 723–727.

Keeling, R. F., S. C. Piper, and M. Heimann. 1996. Global and hemispheric CO_2 sinks deduced from changes in atmospheric O_2 concentration. *Nature* **381**: 218–221.

Keir, R. S. 1980. The dissolution kinetics of biogenic calcium carbonates in seawater. *Geochim. Cosmochim. Acta* **44**: 241–252.

Kell, G. S. 1972. Effects of isotopic composition, temperature, pressure and dissolved gases on the density of liquid water. *J. Phys. Chem. Ref. Data* **6**: 1109–1131.

Kempe, S., G. Liebezit, A. Diercks, *et al.* 1988. Water column analysis. In *Temporal and Spatial Vairability in Sedimentation in the Black Sea. Cruise Report, R/V Knorr 134–8, Black Sea Leg 1, April 16–May 7,1988*, ed. S. Honjo and B. J. Hay. WHOI Technical Report 88–35, Woods Hole Oceanographic Institution, Woods Hole, MA, pp. 31–58.

Kennett, J. P. 1982. *Marine Geology*. Prentice Hall, Englewood Cliffs, NJ.

Kern, D. M. 1960. The hydration of carbon dioxide. *J. Chem. Educ*. **37**: 14–23.

Kester, D. R. 1975. Dissolved gases other than CO_2. In *Chemical Oceanography*, 2nd edn., vol. **1**, ed J. P. Riley and G. Skirrow. Academic Press, New York, pp. 497–556.

Kim, J. P. and W. F. Fitzgerald. 1986. Sea–air partitioning of mercury in the Equatorial Pacific Ocean. *Science* **231**: 1131–1133.

Kinsey, D. W. and D. Hopley. 1991. The significance of coral reefs as global carbon sinks – response to Greenhouse. *Paleogeogr. Paleoclim. Paleoecol*. **89**: 363–377.

Kittredge, J. S. and E. Roberts. 1969. A carbon–phosphorus bond in nature. *Science* **164**: 37–42.

Knauss, J. A. 1997. *Introduction to Physical Oceanography*, 2nd edn. Prentice-Hall, Upper Saddle River, New Jersey, NJ

Knudsen, M., **ed.** 1902. *Berichte uber die Konstantenbestimmungen zur Aufstellung der hydrographischen Tabellen*. Kon. Danske Videnskab. Selsk. Skrifter, 6 Raekke, Naturvidensk. Mathemat., vol. XII. Includes chapters by C. Forch, M. Knudsen, and S. P. L. Sørensen.

Kolowith, L. C., E. D. Ingall, and R. Benner. 2001. Composition and cycling of marine organic phosphorus. *Limnol. Oceanogr*. **46**: 309–320

Kornitnig, S. 1978. Phosphorus. In *Handbook of Geochemistry*, vol. **2**, ed K. H. Wedephol. Springer-Verlag, New York, pp. 15E1–15E9.

Korson, L., W. Drost-Hansen, and F. J. Millero. 1969. Viscosity of water at various temperatures. *J. Phys. Chem*. **73**: 34–39.

Körtzinger, A., J. I. Hedges, and P. D. Quay. 2001. Redfield ratios revisited: removing the biasing effect of anthropogenic CO_2. *Limnol. Oceanogr*. **46**: 964–970.

Krogh, A. 1934. Conditions of life at great depths in the ocean. *Ecological Monographs* **4**: 431–439.

Kuenen, J. G. 2008. Anammox bacteria: from discovery to application. *Nature Rev. Microbiol*. **6**: 320–326.

Lal, D. and T. Lee. 1988. Cosmogenic ^{32}P and ^{33}P used as tracers to study phosphorus recycling in the upper ocean. *Nature* **333**: 752–754.

Lam, P., G. Lavik, M. M. Jensen, *et al.* 2009. Revising the nitrogen cycle in the Peruvian oxygen minimum zone. *Proc. Natl. Acad. Sci.* **106**: 4752–4757.

Lane, T. W., M. A. Saito, G. N. George, *et al.* (2005). A cadmium enzyme from a marine diatom. *Nature* **435**: 42.

Langdon, C., T. Takahashi, C. Sweeney, *et al.* 2000. Effect of calcium carbonate supersaturation state on the calcification rate of an experimental coral reef. *Global Biogeochem. Cycles* **14**: 639–634.

Langseth, M. G. and J. C. Moore. 1990. Introduction to special section on the role of fluids in sediment accretion, deformation, diagenesis, and metamorphism in subduction zones. *J. Geophys. Res.* **95**: 8737–8741.

Larson, R. L. 1991. Latest pulse of Earth: evidence for a mid-Cretaceous superplume. *Geology* **19**: 547–550.

Latif, M. A., E. Özsoy, T. Oguz, and Ü. Ünlüata. 1991. Observations of the Mediterranean inflow into the Black Sea. *Deep-Sea Res.* **38**(Suppl. 2): S711–S723.

Lavoisier, A. L. 1772. Mémoire sur l'usage de l'esprit-de-vin dans l'analyse des eaux minerals. *Mémoires de l'Académie Royale des Sciences de Paris, année* 1772, pt. **2e**, 555–563. (Published in 1776.)

Laws, E. A. 1991. Photosynthetic quotients, new production and net community production in the open ocean. *Deep-Sea Res.* **38**: 143–167.

Lawson, D. S., D. C. Hurd, and H. S. Pankratz. 1978. Silica dissolution rates of decomposing phytoplankton assemblages at various temperatures. *Am. J. Sci.* **278**: 1373–1393.

Le Pichon, X., K. Kobayashi, and Kaiko–Nankai Scientific Crew. 1992 Fluid venting activity within the eastern Nankai Trough accretionary wedge: a summary of the 1989 Kaiko–Nankai results. *Earth Planet. Sci. Lett.* **109**: 303–318.

Le Pichon, X., P. Henry, and S. Lallemant. 1993. Accretion and erosion in subduction zones: the role of fluids. *Ann. Rev. Earth Planet. Sci.* **21**: 307–331.

Leclercq, N., J.-P. Gattuso, and J. Jaubert. 2000. CO_2 partial pressure controls the calcification rate of a coral community. *Global Change Biology* **6**: 329–334.

Ledwell, J. J. 1984. The variation of the gas transfer coefficient with molecular diffusivity. In *Gas Transfer at Water Surfaces*, ed W. Brutsaert and G. H. Jirka. D. Reidel Publishing Co., Hingham, MA, pp. 293–302.

Lee, C. and J. L. Bada. 1975. Amino acids in equatorial Pacific Ocean water. *Earth Planet. Sci. Lett.* **26**: 61–68.

Lee, C. and S. G. Wakeham. 1989. Organic matter in seawater; biogoechemical processes. In *Chemical Oceanography*, 2nd edn., vol. **9**, ed. J. P. Riley. Academic Press, New York, pp. 1–51.

Lee, C. and S. G. Wakeham. 1992. Organic matter in the water column: future research challenges. *Mar. Chem.* **39**: 95–118.

Lee, D. S. and J. M. Edmond. 1985. Tellurium species in seawater. *Nature* **313**: 782–785.

Lee, D. S., J. M. Edmond, and K. W. Bruland. 1986. Bismuth in the Atlantic and North Pacific: a natural analogue to plutonium and lead? *Earth Planet. Sci. Lett.* **76**: 254–262.

Lee, J. G., S. B. Roberts, and F. M. M. Morel. 1995. Cadmium: a nutrient for the marine diatom *Thalassiosira weissflogii*. *Limnol. Oceanogr.* **40**: 1056–1063.

Levitus, S. 1982. *Climatological Atlas of the World Ocean.* Professional Paper **13**. National Oceanic and Atmospheric Administration, Washington, DC.

Levitus, S. and T. P. Boyer. 1994a. *World Ocean Atlas 1994, Vol. 2: Oxygen.* National Oceanic and Atmospheric Administration, US Department of Commerce, Washington, DC.

Levitus, S. and T. P. Boyer. 1994b. *World Ocean Atlas 1994, Vol. 4: Temperature*. National Oceanic and Atmospheric Administration, US Department of Commerce, Washington, DC.

Levitus. S., R. Burgett, and T. P. Boyer, 1994. *World Ocean Atlas 1994, Vol. 3: Salinity*. National Oceanic and Atmospheric Administration, US Department of Commerce, Washington, DC.

Lewis, B. L., P. N. Froelich, and M. O. Andreae. 1985. Methylgermanium in natural waters. *Nature* **313**: 303–305.

Lewis, B. L., M. O. Andreae, and P. N. Froelich. 1989. Sources and sinks of methylgermanium in natural waters. *Mar. Chem.* **27**: 179–200.

Lewis, E. L. 1980. The practical salinity scale 1978 and its antecedents. *IEEE J. Oceanic Eng.* **OE-5**: 3–21.

Lewis, E. L. and R. G. Perkin. 1978. Salinity: its definition and calculation. *J. Geophys. Res.* **83**: 466–478.

Lewis, G. N. and M. Randall. 1921. The activity coefficient of strong electrolytes. *J. Am. Chem. Soc.* **43**: 1112–1154.

Lhomme, N., G. K. C. Clarke, and C. Ritz. 2005. Global budget of water isotopes inferred from polar ice sheets. *Geophys. Res. Lett.* **32**: L20502.

Lieth, H. and R. H. Whittaker, eds. 1975. *Primary Productivity of the Biosphere*. Springer-Verlag, New York.

Liss, P. S. and L. Merlivat. 1986. Air–sea gas exchange rates: introduction and synthesis. In *The Role of Air-Sea Exchange in Geochemical Cycling*, ed. P. Buat-Ménard. D. Reidel Publishing Co., Hingham, MA, pp. 113–129.

Liss, P. S. and P. G. Slater. 1974. Flux of gases across the air-sea interface. *Nature* **247**: 181–184.

Liu, K., M. G. Brown, J. D. Cruzan, and R. J. Saykally. 1996. Vibration–rotation tunneling spectra of the water pentamer: structure and dynamics. *Science* **271**: 62–64.

Liu, X. and F. J. Millero. 2002. The solubility of iron in seawater. *Mar. Chem.* **77**: 43–54.

Llewellyn, T. O. 1994. *Phosphate Rock 1993*. US Department of the Interior, Bureau of Mines, Washington, DC.

Loring, D. H. and R. T. T. Rantala. 1988. An intercalibration exercise for trace metals in marine sediments. *Mar. Chem.* **24**: 13–28.

Love, S. G. and D. E. Brownlee. 1993. A direct measurement of the terrestrial mass accretion rate of cosmic dust. *Science* **262**: 550–553.

Luther III, G. W., T. M. Church, and D. Powell. 1991. Sulfur speciation and sulfide oxidation in the water column of the Black Sea. *Deep-Sea Res.* **38** (Suppl. 2): S1121–S1137.

Lutz, R. A. and M. J. Kennish. 1993. Ecology of deep-sea hydrothermal vent communities: A review. *Rev. Geophys.* **31**: 211–242.

Lyman, J. 1956. Buffer mechanism of seawater. Ph.D. Thesis, University of California, Los Angeles, CA.

Macdonald, R. W., D. W. Paton, and E. C. Carmack. 1995. The freshwater budget and under-ice spreading of Mackenzie River water in the Canadian Beaufort Sea based on salinity and $^{18}O/^{16}O$ measurements in water and ice. *J. Geophys. Res.* **100**: 895–919.

MacIntyre, F. 1974a. The top millimeter of the ocean. *Sci. Am.* **230**: 62–77.

MacIntyre, F. 1974b. Chemical fractionation and seasurface microlayer processes. In *The Sea*, vol. **5**, ed. E. D. Goldberg. John Wiley & Sons, New York.

Mackenzie, F. T., R. M. Garrels, O. P. Bricker, and F. Bickley. 1967. Silica in seawater: control by silica minerals. *Science* **155**: 1404–1405.

Majoube, M. 1971. Fractionnement en oxygène 18 et en deutérium entre l'eau et sa vapeur. *J. de chim. phys.* **68**: 1423–1436.

Maldonado, M. T. and N. M. Price. 1996. Influence of N substrate on Fe requirements of marine centric diatoms. *Mar. Ecol. Prog. Series* **141**: 161–172.

Manning, A. C. and R. F. Keeling. 2006. Global oceanic and land biotic carbon sinks from the Scripps atmospheric oxygen flask sampling network. *Tellus B* **58**: 95–116.

Mantoura, R. F. C. and E. M. S. Woodward. 1983. Conservative behavior of riverine dissolved organic carbon in the Severn Estuary: chemical and geochemical implications. *Geochim. Cosmochim. Acta* **47**: 1293–1309.

Marcet, A. 1819. On the specific gravity, and temperature, in different parts of the ocean, and in particular seas; with some account of their saline contents. *Phil. Trans. R. Soc. London* **109**: 161–208.

Marsilli, L. F. 1681. *Osservazioni Intorno al Bosforo Tracio overo Canale di Constantinopoli*, Rappresentate in Lettera alla Sacra Real Maestá Cristina Regina di Svecia da Luigi Ferdinando Marsilii. Nicolò Angelo Tinassi, Rome. 108 pp. (See Soffientino and Pilson 2009.)

Marsilli, L. F. 1725. *Histoire Physique de la Mer*. De'pens, Amsterdam. **40** pl. + 173 pp.

Martin, J. H. and S. E. Fitzwater. 1988. Iron deficiency limits phytoplankton growth in the norteast Pacific subarctic. *Nature* **331**: 341–343.

Martin, J. H. and R. M. Gordon. 1988. Northeast Pacific iron distributions in relation to phytoplankton productivity. *Deep-Sea Res.* **35**: 177–196.

Martin, J. H., G. A. Knauer, and R. M. Gordon. 1983. Silver distribution an fluxes in north-east Pacific waters. *Nature* **305**: 306–309.

Martin, J. H., G. A. Knauer, D. M. Karl, and W. W. Broenkow. 1987. VERTEX: carbon cycling in the northeast Pacific. *Deep-Sea Res.* **34**: 267–285.

Martin, J. H., R. M. Gordon, S. Fitzwater, and W. W. Broenkow. 1989. VERTEX – phytoplankton studies in the Gulf of Alaska. *Deep-Sea Res. I* **36**: 649–680.

Masiello, C. A. and E. R. M. Druffel. 1998. Black carbon in deep-sea sediments. *Science* **280**: 1911–1913.

Mason, R. P. and W. F. Fitzgerald. 1990. Alkylmercury species in the equatorial Pacific. *Nature* **347**: 457–459.

Mason, R. P. and W. F. Fitzgerald. 1993. The distribution and biogeochemical cycling of mercury in the equatorial Pacific Ocean. *Deep-Sea Res.* **40**: 1897–1924.

Mason, R. P., W. F. Fitzgerald, J. Hurley, *et al.* 1993. Mercury biogeochemical cycling in a stratified estuary. *Limnol. Oceanogr.* **38**: 1227–1241.

Mason, R. P., W. F. Fitzgerald, and F. M. M. Morel. 1994. The biogeochemical cycling of elemental mercury: anthropogenic influence. *Geochim. Cosmochim. Acta* **58**: 3191–3198.

Mason, R. P., J. R. Reinfelder, and F. M. M. Morel. 1996. Uptake, toxicity, and trophic transfer of mercury in a coastal diatom. *Environ. Sci. Technol.* **30**: 1835–1845.

Mason, R. P., K. R. Rolfhus, and W. F. Fitzgerald. 1998. Mercury in the North Atlantic. *Mar. Chem*, **61**: 37–53.

Maury, M. F. 1855. *The Physical Geography of the Sea*. Harper & Brothers, New York.

McCarthy, M. D., J. I. Hedges, and R. Benner. 1998. Major bacterial contribution to marine dissolved nitrogen. *Science* **281**: 231–234.

McElligott, S. and R. H. Byrne. 1998. Interaction of $B(OH)_3^0$ and HCO_3^- in seawater: formation of $B(OH)_2CO_3^-$. *Aquatic Geochemistry* **3**: 345–356.

McKelvey, B. A. and K. J. Orians. 1993. Dissolved zirconium in the North Pacific Ocean. *Geochim. Cosmochim. Acta* **57**: 3801–3805.

McKillop, H. 2002. *Salt, White Gold of the Ancient Maya*. University Press of Florida, Gainsville, FL.

McManus, J., D. E. Hammond, K. Cummins, G. P. Klinkhammer, and W. M. Berelson. 2003. Diagenetic Ge–Si fractionation in continental margin environments: further evidence for a nonopal Ge sink. *Geochim. Cosmochim. Acta.* **67**: 4545–4557.

McNeil, C. 2006. Undersaturation of inert gases at the ocean surface: a thermal pumping mechanism. *Geophys. Res. Lett.* **33**: LO1607, doi: 10.1029/2005GL024752.

Measures, C. I., B. C. Grant, B. J. Mangum, and J. M. Edmond. 1983. The relationship of the distribution of dissolved selenium IV and VI in three oceans to physical and biological processes. In *Trace Metals in Sea Water*, ed. C. S. Wong, E. Boyle, K. W. Bruland, J. D. Burton, and E. D. Goldberg. Plenum Press, New York, pp. 73–83.

Mehrbach, C., C. H. Culberson, J. E. Hawley, and R. M. Pytkowicz. 1973. Measurement of the apparent dissociation constants of carbonic acid in seawater at atmospheric pressure. *Limnol. Oceanogr.* **18**: 897–907.

Menaché, M. and G. Girard. 1970. Étude de la variation de la masse volumique de l'eau en fonction de sa composition isotopique. *C. R. Acad. Sci., Ser. B* **270**: 1513–1516.

Menaché, M., J. Crease, G. Girard, and R. B. Montgomery. 1979. *SUN Report on the use in physical sciences of the ocean of the Systeme International d'Unites (SI) and related standards for symbols and terminology*. (Adopted at the XVII General Assembly of IAPSO.) IUGG Publishing Office, Paris.

Menard, H. W. and S. M. Smith. 1966. Hypsometry of the ocean basin provinces. *J. Geophys. Res.* **71**: 4305–4325.

Menzel, D. W. and J. P. Spaeth. 1962. Occurrence of vitamin B_{12} in the Sargasso Sea. *Limnol. Oceanogr.* **7**: 151–154.

Menzel, D. W. and R. Vaccaro. 1964. Measurement of dissolved organic and particulate carbon in sea water. *Limnol. Oceanogr.* **9**: 138–142.

Merlivat, L., F. Pineau, and M. Javoy. 1987. Hydrothermal vent waters at 13º N on the East Pacific Rise: isotopic composition and gas concentration. *Earth Planet. Sci Lett.* **84**: 100–108.

Mero, J. L. 1965. *The Mineral Resources of the Sea.* Elsevier, New York.

Meybeck, M. 1979. Concentration des eaux fluviales en éléments majeurs et apports en solution aux océans. *Rev. Geol. Dynam. Geog. Phys.* **21**: 215–246.

Meybeck, M. 1982. Carbon, nitrogen, and phosphorus transport by world rivers. *Am. J. Sci.* **282**: 401–450.

Meybeck, M. and R. Helmer. 1989. The quality of rivers: from pristine stage to global pollution. *Global Planet. Change* **1**: 283–309.

Meyers-Schulte, K. J. and J. I. Hedges. 1986. Molecular evidence for a terrestrial component of organic matter dissolved in ocean water. *Nature* **321**: 61–63.

Middleburg, J. J., D. Hoede, H. A. Van Der Sloot, C. H. Van Der Weijden, and J. Wijkstra. 1988. Arsenic, antimony and vanadium in the North Atlantic Ocean. *Geochem. Cosmochim. Acta.* **52**: 2871–2878.

Milkov, A. V. 2004. Global estimates of hydrate-bound gas in marine sediments: how much is really out there? *Earth-Science Rev.* **66**: 183–197.

Miller, C. E. and L. D. Marinelli. 1956. Gamma-ray activity of contemporary man. *Science* **124**: 122–123.

Miller, S. L. 1974. The nature and occurrence of clathrate hydrates. In *Natural Gases in Marine Sediments*, ed. I. R. Kaplan. Plenum Press, New York, pp 151–177.

Millero, F. J. 1974. Seawater as a multicomponent electrolyte solution. In *The Sea*, vol. **5**, ed. E. D. Goldberg. John Wiley & Sons, New York, pp. 3–90.

Millero, F. J. 1979. The thermodynamics of the carbonate system in seawater. *Geochim. Cosmochim. Acta* **43**: 1651–1661.

Millero, F. J. 1983. Influence of pressure on chemical processes in the sea. In *Chemical Oceanography*, 2nd edn., vol. 8, ed J. P. Riley and R. Chester. Academic Press, New York, pp. 1–88.

Millero, F. J. 1991. The oxidation of H_2S in Black Sea waters. *Deep-Sea Res.* **38** (Suppl. 2): S1139–S1150.

Millero, F. J. 1995. Thermodynamics of the carbon dioxide system in the oceans. *Geochim. Cosmochim. Acta* **59**: 661–677.

Millero, F. J. 2010. Carbonate constants for estuarine waters. *Mar. Freshwater Res.* **61**: 139–142.

Millero, F. J. and W. H. Leung. 1976. The thermodynamics of seawater at one atmosphere. *Amer. J. Sci.* **276**: 1035–1077.

Millero, F. J. and A. Poisson. 1981. International One Atmosphere Equation of State for Seawater. *Deep-Sea Res.* **28A**: 625–629.

Millero, F. J. and D. R. Schreiber. 1982. Use of the ion pairing model to estimate the activity coefficients of the ionic components of natural waters. *Am. J. Sci.* **282**: 1508–1540.

Millero, F. J., G. Perron, and J. F. Desnoyers. 1973. Heat capacity of seawater solutions from 5 to 35 °C and 0.5 to 22‰ chlorinity. *J. Geophys. Res* **78**: 4499–4506.

Millero, F. J., C.-T. Chen, A. Bradshaw, and K. Schleicher. 1980. A new high-pressure equation of state for seawater. *Deep-Sea Res.* **27A**: 255–264.

Millero, F. J., T. Plese, and M. Fernandez. 1988. The dissociation of hydrogen sulfide in seawater. *Limnol. Oceanogr.* 33: 269–274.

Millero, F. J., J.-Z. Zhang, S. Fiol, *et al.* 1993. The use of buffers to measure the pH of seawater. *Mar. Chem.* **44**: 143–152.

Millero, F. J., K. Lee, and M. Roche. 1998. Distribution of alkalinity in the surface waters of the major oceans. *Mar. Chem.* **60**: 111–130.

Millero, F. J., D. Pierrot, K. Lee, *et al.* 2002. Dissociation constants for carbonic acid determined from field measurements. *Deep-Sea Res. I* **49**: 1705–1723.

Millero, F. J., R. Feistel, D. G. Wright, and T. J. McDougall. 2008. The composition of Standard Seawater and the definition of the Reference-Composition Salinity scale. Deep-Sea *Res. I* **55**: 50–72.

Milliman, J. D. 1993. Production and accumulation of calcium carbonate in the ocean: budget of a nonsteady state. *Global Biogeochem. Cycles* **7**: 927–957.

Milliman, J. D. and A. W. Droxler. 1995. Calcium carbonate sedimentation in the global ocean: linkages between the neritic and pelagic environments. *Oceanography* **8**: 92–94.

Milliman, J. D. and R. H. Meade. 1983. World-wide delivery of river sediment to the oceans. *J. Geol.* **91**: 1–21.

Milliman, J. D. and J. P. M. Syvitski. 1992. Geomorphic/tectonic control of sediment discharge to the ocean: the importance of small mountainous rivers. *J. Geol.* **100**: 525–544.

Miyake, Y. and M. Koizumi. 1948. The measurement of the viscosity coefficient of sea water. *J. Mar. Res.* **7**: 63–66.

Monahan, E. C. and I. G. O'Muircheartaigh. 1986. Whitecaps and passive remote sensing of the ocean surface. *Int. J. Remote Sensing* **7**: 627–642.

Montgomery, R. B. 1958. Water characteristics of Atlantic Ocean and of world oceans. *Deep-Sea Res.* **5**: 134–148.

Montoya, J. P., C. M. Holl, J. P. Zehr, *et al.* 2004. High rates of N_2 fixation by unicellular diazotrophs in the oligotrophic Pacific Ocean. *Nature* **430**: 1027–1031.

Moore, J. C. and P. Vrolijk. 1992. Fluids in accretionary prisms. *Rev. Geophys.* **30**: 113–135.

Moore, R. M. and R. Tokarczyk. 1993. Volatile biogenic halocarbons in the Northwest Atlantic. *Global Biogeochem. Cycles* **7**: 195–210.

Mopper, K., R. Dawson, G. Liebezeit, and V. Ittekkot. 1980. The monosaccharide spectra of natural waters. *Mar. Chem.* **10**: 55–66.

Mopper, K., X. Zhou, R. J. Kieber, *et al.* 1991. Photochemical degredation of dissolved organic carbon and its impact on the oceanic carbon cycle. *Nature* **353**: 60–62.

Morey, G. W., R. O. Fournier, and J. J. Rowe. 1962. The solubility of quartz in water in the temperature interval from 25 to 300°C. *Geochim. Cosmochim. Acta.* **26**: 1029–1043.

Morris, A. W. and J. P. Riley. 1966. The bromide/chlorinity and sulfate/chlorinity ratio in seawater. *Deep-Sea Res.* **13**: 699–705.

Morris, J. D., W. P. Leeman, and F. Tera. 1990. The subducted component in island are lavas: constraints from Be isotopes and B–Be systematics. *Nature* **344**: 31–36.

Morse, J. W. and R. A. Berner. 1972. Dissolution kinetics of calcium carbonate in seawater: II. A kinetic origin for the lysocline. *Am. J. Sci.* **272**: 840–851.

Mottl, M. J. and C. G. Wheat. 1994. Hydrothermal circulation through mid-ocean ridge flanks: fluxes of heat and magnesium. *Geochim. Cosmochim. Acta* **58**: 2225–2237.

Mucci, A. 1983. The solubility of calcite and aragonite in seawater at various salinities, temperatures, and one atmosphere total pressure. *Am. J. Sci.* **283**: 780–799.

Mulder, A., A. A. van de Graaf, L. A. Robertson, and J. G. Kuenen. 1995. Anaerobic ammonium oxidation discovered in a denitrifying fluidized bed reactor. *FEMS Microbiol. Ecol.* **16**: 177–184.

Murphy, J. and J. P. Riley. 1962. A modified single solution method for the determination of phosphate in natural waters. *Anal. Chim. Acta.* **27**: 31–36.

Natterer, K. 1892–1894. Chemische Untersuchungen im östlichen Mittelmeer. *Monatsh. Chem.* **13**: 873–890, 897–908; **14**: 624–673; **15**: 530–604.

Nebel, O., E. E. Sherer, and K. Mezger. 2010. Evaluation of the ^{87}Rb decay constant by age comparison against the U–Pb system. *Earth Planet. Sci. Lett.* 301: 1–8.

Nelson, D. M., P. Tréguer, M. A. Brzezinski, A. Leynaert, and B. Quéguiner. 1995. Production and dissolution of biogenic silica in the ocean: revised global estimates, comparison with regional data and relationship to biogenic sedimentation. *Global Biogeochem. Cycles* **9**: 359–372.

Nelson, K. H. and T. G. Thompson. 1954. Deposition of salts from seawater by frigid concentration. *J. Mar. Res.* **13**: 166–182.

Nollet, J. A. 1748. Recherches sur les causes du bouillonnement des liquides. *Mémoires. de Mathématique et de Physique de l'Académie Royal des Sciences*, Paris, **1748**, pp. 57–104 & 2 plates.

Nørby, J. G. 2000. The origin and meaning of the little p in pH. *Trends in Biochem. Sci.* **25**: 36–37.

Nozaki, Y. 1997. A fresh look at element distribution in the North Pacific Ocean. *EOS* **78**(21): 221.

NRC. 2008. *Desalination, a National Perspective*. National Academies Press, Washington, DC.

Oliver, B. M., J. G. Bradley, and H. Farrar. IV. 1984. Helium concentration in the Earth's lower atmosphere. *Geochim. Cosmochim. Acta* **48**: 1759–1767.

Opsahl, S. and R. Benner. 1997. Distribution and cycling of terrigenous dissolved organic matter in the ocean. *Nature* **386**: 480–482.

Oren, A. and F. Rodríguez-Valera. 2001. The contribution of halophilic Bacteria to the red coloration of saltern crystallizer ponds. *FEMS Microbiology and Ecology* **36**: 123–130.

Orians, K. J. and K. W. Bruland. 1985. Dissolved aluminium in the central North Pacific. *Nature* **316**: 427–429.

Orians, K. J. and K. W. Bruland. 1986. The biogeochemistry of aluminum in the Pacific Ocean. *Earth Planet. Sci. Lett.* **78**: 397–410.

Orians, K. J. and K. W. Bruland. 1988a. Dissolved gallium in the open ocean. *Nature* **332**: 717–719.

Orians, K. J. and K. W. Bruland. 1988b. The marine geochemistry of dissolved gallium: A comparison with dissolved aluminum. *Geochim. Cosmochim. Acta* **52**: 2955–2962.

Orians, K. J., E. A. Boyle, and K. W. Bruland. 1990. Dissolved titanium in the open ocean. *Nature* **348**: 322–325.

Ostlund, H. G., H. Craig, W. S. Broecker, and D. Spencer. 1987. *GEOSECS Atlantic, Pacific, and Indian Ocean Expeditions*. Vol. **7**. Shore-based Data and Graphics. National Science Foundation, Washington, DC.

Oudot, C., C. Andrie, and Y. Montel. 1990. Nitrous oxide production in the tropical Atlantic Ocean. *Deep-Sea Res.* **37**: 183–202.

Oxner, M. 1962. *The Determination of Chlorinity by the Knudsen Method. Translation.* G. M. Manufacturing Co., New York.

Özsoy, E., Ü. Ünlüata, and Z. Top. 1993. The evolution of Mediterranean water in the Black Sea: interior mixing and material transport by double diffusive intrusions. *Prog. Oceanogr.* **31**: 275–320.

PACODF. 1986. *Transient Tracers in the Ocean. North Atlantic Study. 1 April–19 Oct. 1981.* SIO Ref. No. 86–15. Scripps Institution of Oceanography, University California, San Diego, CA.

Pacyna, E. G., J. M. Pacyna, F. Steenhuisen, and S. Wilson. 2006. Global anthropogenic mercury emission inventory for the year 2000. *Atmospheric Environment* **40**: 4048–4063.

Palmer, M. R. and J. M. Edmond. 1989. The strontium isotope budget of the modern ocean. *Earth Planet. Sci. Lett.* **92**: 11–26.

Parai, R. and S. Mukhopadhyay. 2012. How large is the subducted water flux? New constraints on mantle regassing rates. *Earth Planet. Sci. Lett.* **317–318**: 396–406.

Park, K. 1969. Oceanic CO_2 system: an evaluation of ten methods of investigation. *Limnol. Oceanogr.* **14**: 179–186.

Parsons, B. 1981. The rates of plate creation and consumption. *Geophys. J. R. Astr. Soc.* **67**: 437–448.

Parsons, B. 1982. Causes and consequences of the relation between area and age of the ocean floor. *J. Geophys. Res.* **87**: 289–302.

Patterson, C. 1974. Lead in seawater. *Science* **183**: 553–554.

Paytan, A. and K. McLaughlin. 2007. The oceanic phosphorus cycle. *Chem. Rev.* **107**: 563–576.

Paytan, A., M. Kastner, E. E. Martin, J. D. Macdougall, and T. Herbert. 1993. Marine barite as a monitor of seawater strontium isotope composition. *Nature* **366**: 445–449.

Peltzer, E. T., B. Fry, P. H. Doering, *et al.* 1996. A comparison of methods for the measurement of dissolved organic carbon in natural waters. *Mar. Chem.* **54**: 85–96.

Perez, F. F. and F. Fraga. 1987. Association constant of fluoride and hydrogen ions in seawater. *Mar. Chem.* **21**: 161–168.

Perkin, R. G. and E. L. Lewis. 1980. The practical salinity scale 1978: fitting the data. *IEEE J. Oceanic Eng.* **OE-5**: 9–16.

Peterson, M. N. A. 1966. Calcite: rates of dissolution in a vertical profile in the Central Pacific. *Science* **154**: 1542–1544.

Petit, G. R., J. Jouzel, D. Ragnoud *et al.* 1999. Climate and atmospheric history of the past 420,000 years from the Vostoc ice core, Antarctica. *Nature* **399**: 429–436.

Petley, B. W. 1996. The mole and the unified atomic mass unit. *Metrologia* **33**: 261–264.

Phillips, O. L. and A. H. Gentry. 1994. Increasing turnover through time in tropical forests. *Science* **263**: 954–958.

Pickard, G. L. and W. J. Emery. 1990. *Descriptive Physical Oceanography,* 5th edn. Pergamon Press, New York.

Pilson, M. E. Q. 1974. Arsenate uptake and reduction by *Pocillopora verrucosa. Limnol. Oceanogr.* **19**: 339–341.

Pilson, M. E. Q. 1985. Annual cycles of nutrients and chlorophyll in Narragansett Bay, Rhode Island. *J. Mar. Res.* **43**: 849–873.

Pilson, M. E. Q. 1998. *Introduction to the Chemistry of the Sea.* Prentice Hall, Upper Saddle River, NJ.

Pilson, M. E. Q. 2006. We are evaporating our coal mines into the air. *Ambio* **35**: 130–133.

Pilson, M. E. Q. 2010. *Blue is the desert color of the sea.* Where does this sentence come from? *Limnol. Oceanol. Bull.* **19(3)**: 62–63.

Pimentel, G. C. and A. L. McClellan. 1960. *The Hydrogen Bond.* Reinhold Publishing. Company, New York.

Piña-Ochoa, E., S. Høgslund, E. Geslin, *et al.* 2010. Widespread occurrence of nitrate storage and denitrification among Foraminifera and *Gromiida. Proc. Natl. Acad. Sci.* **107**: 1148–1153.

Plass-Dülmer, C., A. Khedim, R. Koppmann, F. J. Johnen, and J. Rudolph. 1993. Emissions of light nonmethane hydrocarbons from the Atlantic into the atmosphere. *Global Biogeochem. Cycles* **7**: 211–228.

Plath, D. C. 1979. The solubility of $CaCO_3$ in seawater and the determination of activity coefficients in electrolyte solutions. M.Sc. Thesis, Oregon State University. 92 pp. (Cited from UNESCO, 1983.)

Plath, D. C., K. S. Johnson, and R. M. Pytkowicz. 1980. The solubility of calcite – probably containing magnesium – in seawater. *Mar. Chem.* **10**: 9–29.

Plummer, L. N. and E. Busenberg. 1982. The solubilities of calcite, aragonite and vaterite in CO_2–H_2O solutions between 0 and 90 °C, and an evaluation of the aqueous model for the system $CaCO_3$–CO_2–H_2O. *Geochim. Cosmochim. Acta* **46**: 1011–1040.

Plunkett, M. A. and N. W. Rakestraw. 1955. Dissolved organic matter in the sea. *Deep-Sea Res.* **3**(Suppl.): 12–14.

Poisson, A., C. Brunet, and J. C. Brun-Cotton. 1980. Density of standard seawater solutions at atmospheric pressure. *Deep-Sea Res.* **27A**: 1013–1028.

Pollack, H. N., S. J. Hurter, and J. R. Johnson. 1993. Heat flow from the Earth's interior: analysis of the global data set. *Rev. Geophys.* **31**: 267–280.

Postgate, J. 1998. *Nitrogen Fixation*, 3rd edn, Cambridge, Cambridge University Press.

Prahl, F. G. and S. G. Wakeham. 1987. Calibration of unsaturation patterns in long-chain ketone compositions for paleoceanography assessment. *Nature* **330**: 367–369.

Prentice, I. C., G. D. Farquhar, M. J. R. Fasham *et al.* 2001. The carbon cycle and atmospheric carbon dioxide. In Houghton and Ding *et al.* (2001), ch. 3, pp 182–237.

Pütter, A. 1909. *Die Ernährung der Wassertiere und der Stoffhaushalt der Gewässer*. Fischer, Jena (cited from Stephens 1982).

Pytkowicz, R. M. 1962. Effect of gravity on the distribution of salts in sea water. *Limnol. Oceanogr.* **7**: 434–435.

Pytkowicz, R. M. 1973. Calcium carbonate retention in supersaturated seawater. *Am. J. Sci.* **273**: 515–522.

Pytkowicz, R. M., ed. 1979. *Activity Coefficients in Electrolyte Solutions*, vols. 1 & 2. CRC Press, Boca Raton, FL.

Quay, P. D., B. Tilbrook, and C. S. Wong. 1992. Oceanic uptake of fossil fuel CO_2: carbon-13 evidence. *Science* **256**: 74–79.

Raven, J. A. 1988. The iron and molybdenum use efficiencies of plant growth with different energy, carbon and nitrogen sources. *New Phyto.* **109**: 279–287.

Raynaud, D., J. Jouzel, J. M. Barnola, *et al.* 1993. The ice record of greenhouse gases. *Science* **259**: 926–934.

Redfield, A. C. 1934. On the proportions of organic derivatives in seawater and their relation to the composition of plankton. In *James Johnston Memorial Volume*, ed R. J. Daniel. University Press of Liverpool, Liverpool, pp. 176–192.

Redfield, A. C. 1958. The biological control of chemical factors in the environment. *Am. Sci.* **46**: 205–221.

Redfield, A. C., H. P. Smith, and B. Ketchum. 1937. The cycle of organic phosphorus in the Gulf of Maine. *Biol. Bull.* **73**: 421–443.

Redfield, A. C., B. H. Ketchum, and F. A. Richards. 1963. The influence of organisms on thc composition of seawater. In *The Sea*, vol. **2**, ed. M. N. Hill. Interscience, New York, pp. 26–87.

Reid, R. T., D. H. Live, D. J. Faulkner, and A. Butler. 1993. A siderophore from a marine bacterium with an exceptional ferric ion affinity constant. *Nature* **366**: 455–458.

Repeta, D. J., T. M. Quan, L. I. Aluwihare, and A. M. Accardi. 2002. Chemical characterization of high molecular weight dissolved organic matter in fresh and marine waters. *Geochim. Cosmochim. Acta* **66**: 955–962.

Revelle, R. 1956. Testimony to the House Appropriations Committee. Cited from: Shor, E. N. 1978. *Scripps Institution of Oceanography: Probing the Oceans 1936 to 1976*. Tofua Press, San Diego. 502 pp.

Revelle, R. and H. E. Suess. 1957. Carbon dioxide exchange between atmosphere and ocean and the question of an increase of atmospheric CO_2 during the past decades. *Tellus* **9**: 18–27.

Richards, F. A. 1965. Anoxic basins and fjords. In *Chemical Oceanography*, vol. **1**, ed. J. P. Riley and J. Skirrow. Academic Press, New York, pp. 611–645.

Richards, F. A. 1971. Comments on the effects of denitrification on the budget of combined nitrogen in the ocean. *Mer. Bull. Soc. Franco-Japanise d'Oceanogr.* **9**: 68–77.

Riebesell, U., Fabry, V. J., Hansson, L., Gattuso, J.-P., eds. 2010. Guide to best practices for ocean acidification research and data reporting. EUR24328 EN, Publication Office of the European Union.

Riley, G. 1951. Oxygen, phosphate and nitrate in the Atlantic Ocean. *Bull. Bingham Oceanogr. Coll.* **13**: 1–126.

Riley, G. A., D. Von Hemert, and P. J. Wangersky. 1965. Organic aggregates in surface and deep waters of the Sargasso Sea. *Limnol. Oceanogr.* **10**: 354–363.

Riley, J. P. 1965. Historical introduction. In *Chemical Oceanography*, vol. **1**. ed. J. P. Riley and G. Skirrow. Academic Press, New York, pp. 1–41.

Riley, J. P. and G. Skirrow, eds. 1975. *Chemical Oceanography*, 2nd edn., vol. **1**. Academic Press, New York.

Riley, J. P. and M. Tongudai. 1967. The major cation/chlorinity ratios in seawater. *Chem. Geol.* **2**: 263–269.

Risgaard-Petersen, N., A. M. Langezaal, S. Ingvardsen, *et al.* 2006. Evidence for complete denitrification in a benthic foraminifer. *Nature* **443**: 93–96.

Robbins, L. L. and P. L. Blackwelder. 1992. Biochemical and ultrastructural evidence for the origin of whitings: a biologically induced calcium carbonate precipitation mechanism. *Geology* **20**: 464–468.

Robinson, R. A. 1954. The vapor pressure and osmotic equivalence of sea water. *J. Mar. Biol. Assoc. UK* **33**: 449–455.

Robinson, R. A. and R. H. Stokes. 1970. *Electrolyte Solutions, 2nd edn.* Butterworths, London.

Romankevich, E. A. 1984. *Geochemistry of Organic Matter in the Ocean.* Springer-Verlag, New York.

Ronov, A. B. and A. A. Yaroshevsky. 1969. Chemical composition of the Earth's crust. In *The Earth's Crust and Upper Mantle*, ed P. J. Hart. American Geophysical Union, Geophysical Monograph 13, Washington, DC, pp. 37–57.

Ross, D. A., E. Uchupi, K. E. Prada, and J. C. MacIlvaine. 1972. Bathymetry and microtopography of the Black Sea. In *The Black Sea – Geology, Chemistry, and Biology*, ed E. T. Degens and D. A. Ross. American. Association of Petroleum Geologists, Tulsa, OK, pp. 1–10.

Roy, R. N., L. N. Roy, K. M. Vogel, *et al.* 1993. The dissociation constants of carbonic acid in seawater at salinities 5 to 45 and temperatures 0 to 45 °C. *Mar. Chem.* **44**: 249–267.

Rudnick, R. L. and S. Gao. 2004. Composition of the continental crust. In *The Crust*, ed. R. L. Rudnick (*Treatise on Geochemistry*, vol. 3, ed. H. D. Holland and K. K. Turekian), Elsevier, Oxford, pp. 1–64.

Rumpler, A., J. S. Edmonds, M. Katsu, *et al.* 2008. Arsenic-containing long-chain fatty acids in cod-liver oil: a result of biosynthetic infidelity? *Angew. Chem., Int. Ed. Engl.* **47**: 2665–2667.

Ryan, W. B. F. 1973. Geodynamic implications of the Messinian crisis of salinity. In *Messinian Events in the Mediterranean*, ed. C. W. Drooger. North-Holland, Amsterdam, pp. 26–38.

Sabine, C. L., R. A. Feely, N. Gruber, *et al.* 2004. The oceanic sink for anthropogenic CO_2. *Science* **305**: 367–371.

Saito, M. A. and J. W. Moffett. 2002. Temporal and spatial variability of cobalt in the Atlantic Ocean. *Geochim. Cosmochim. Acta* **66**: 1943–1953.

Sanemasa, I. 1975. The solubility of elemental mercury vapor in water. *Bull. Chem. Soc. Jpn.* **48**: 1795–1798.

Santschi, P. H., D. Adler, M. Amdurer, Y. H. Li, and J. J. Bell. 1980a. Thorium isotopes as analogues for 'particle-reactive' pollutants in coastal marine environments. *Earth Planet. Sci. Lett.* **47**: 327–335.

Santschi, P. H., Y. H. Li, and S. R. Carson. 1980b. The fate of trace metals in Narragansett Bay, Rhode Island: radiotracer experiments in microcosms. *Estuar. Coast. Mar. Sci.* **10**: 635–654.

Sarmiento, J. L. and N. Gruber. 2006. *Ocean Biogeochemical Dynamics.* Princeton University Press, Princeton, NJ.

Sayles, F. L. 1979. The composition and diagenesis of interstitial solutions – I. Fluxes across the sediment–water interface in the Atlantic Ocean. *Geochim. Cosmochim. Acta* **43**: 527–545.

Sayles, F. L. 1981. The composition and diagenesis of interstitial solutions – II. Fluxes and diagenesis at the sediment–water interface in the high latitude North Atlantic and South Atlantic. *Geochim. Cosmochim. Acta* **45**: 1061–1086.

Schaule, B. K. and C. C. Patterson. 1981. Lead concentrations in the northeast Pacific: evidence for global anthropogenic perturbations. *Earth Planet. Sci. Lett.* **54**: 97–116.

Schaule, B. K. and C. C. Patterson. 1983. Perturbations of the natural lead depth profile in the Sargasso Sea by industrial lead. In *Trace Metals in Sea Water,* ed. C. S. Wong, E. Boyle, K. W. Bruland, J. D. Burton, and E. D. Goldberg. Plenum Press, New York, pp. 487–503.

Schink, D. R., K. A. Fanning, and M. E. Q. Pilson. 1974. Dissolved silica in the upper pore waters of the Atlantic Ocean floor. *J. Geophys. Res.* **79**: 2243–2250.

Schink, D. R., N. L. Guinasso Jr., and K. A. Fanning. 1975. Processes affecting the concentration of silica at the sediment–water interface of the Atlantic Ocean. *J. Geophys. Res.* **80**: 3013–3031.

Schlesinger, W. H. 1997. *Biogeochemistry; An Analysis of Global Change*, 2nd edn. Academic Press, New York.

Schopf, J. W., ed. 1983. *Earth's Earliest Biosphere: Its Origin and Evolution.* Princeton University Press, Princeton, NJ.

Schrag, D. P., G. Hampt, and D. W. Murray. 1996. Pore fluid constraints on the temperature and isotopic composition of the glacial ocean. *Science* **272**: 1930–1932.

Schubert, G. and A. P. S. Reymer. 1985. Continental volume and freeboard through geological time. *Nature* **316**: 336–339.

Schuffert, J. D., R. A. Jahnke, M. Kastner, *et al.* 1994. Rates of formation of modern phosphorite off western Mexico. *Geochim. Cosmochim. Acta* **58**: 5001–5010.

Schütt, F. 1892. Das Pflanzenleben der Hochsee. In *Ergebnisse der Plankton-Expedition der Humboldt-Stiftung. Band 1*, ed. V. Hensen. Lipsius & Tischer, Kiel und Leipzig, pp. 243–314

Seitzinger, S. P. and A. E. Giblin. 1996. Estimating denitrification in North Atlantic continental shelf sediments. *Biogeochemistry* **35**: 235–260.

Seitzinger, S. P., S. Nixon, M. E. Q. Pilson, and S. Burke. 1980. Denitrification and N_2O production in nearshore marine sediments. *Geochim. Cosmochim. Acta* **44**: 1853–1860.

Seitzinger, S. P., M. E. Q. Pilson, and S. Nixon. 1983. N_2O production in near-shore marine sediments. *Science* **222**: 1244–1246.

Seitzinger, S. P., S. W. Nixon, and M. E. Q. Pilson. 1984. Denitrification and nitrous oxide production in a coastal marine ecosystem. *Limnol. Oceanogr.* **29**: 73–83.

Seitzinger, S. P., J. A. Harrison, J. K. Bohlke, *et al.* 2006. Denitrification across landscapes and waterscapes: a synthesis. *Ecol. Appl.* **16**: 2064–2090.

Sharp, J. H. 1973. Total organic carbon in seawater – comparison of measurements using persulfate oxidation and high-temperature combustion. *Mar. Chem.* **1**: 211–229.

Sharp, J. H. 1993. The dissolved organic carbon controversy: an update. *Oceanography* **6**: 45–50.

Shi, D., Y. Xu, B. M. Hopkinson, and F. M. M. Morel. 2010. Effect of ocean acidification on iron availability to marine phytoplankton. *Science* **327**: 676–679.

Shigley, C. M. 1951. Minerals from the sea. *J. Met.* **3**: 25–29.

Shoemaker, E. M. 1983. Asteroid and comet bombardment of the Earth. *Ann. Rev. Earth Planet. Sci.* **11**: 461–494.

Siever, R. 1968. Sedimentological consequences of a steady-state ocean–atmosphere. *Sedimentology* **11**: 5–59.

Sillén, L. G. 1961. The physical chemistry of seawater. In *Oceanography*, ed M. Sears. American Association for the Advancement of Science, Washington, DC, pp. 549–581.

Simpson, T. L. and B. E. Volcani, eds. 1981. *Silicon and Siliceous Structures in Biological Systems.* Springer-Verlag, New York.

Skei, J. M. 1988. Framvaren – environmental setting. *Mar. Chem.* **23**: 209–218.

Sloan, E. D. 2003. Fundamental principles and applications of natural gas hydrates. *Nature* **246**: 353–359.

Sloan, E. D. and C. A. Koh. 2008. *Clathrate Hydrates of Natural Gases*, 3rd edn. CRC Press. Boca Raton, FL.

Smil, V. 1997. Global population and the nitrogen cycle. *Sci. Am.* **277**: 76–81.

Smil, V. 2001. *Enriching the Earth.* MIT Press, Cambridge, MA.

Soffientino, B. and M. E. Q. Pilson. 2005. The Bosporus Strait, a special place in the history of oceanography. *Oceanography* **18**: 16–23.

Soffientino, B. and M. E. Q. Pilson. 2009. *Osservazioni intorno al Bosforo Tracio overo Canale di Constantinopoli.* Presented in a letter to Her Sacred Royal Majesty Queen Christina of Sweden in 1681 by Luigi Ferdinando Marsilii: English translation with footnotes. *Earth Sci. Hist.* **28**: 57–83.

Sohrin, Y., K. Isshiki, and T. Kuwamoto. 1987. Tungsten in North Pacific waters. *Mar. Chem.* **22**: 95–103.

Sohrin, Y., Y. Fujishima, K. Ueda, *et al.* 1998. Dissolved niobium and tantalum in the North Pacific. *Geophys. Res. Lett.* **25**: 999–1002.

Soli, A. L. and R. H. Byrne, 2002. CO_2 System hydration and dehydration kinetics and the equilibrium CO_2/H_2CO_3 ratio in aqueous NaCl solution. *Mar. Chem.* **78**: 65–73.

Solomon, S., D. Qin, M. Manning *et al.* 2007. *Climate Change 2007. The Physical Science Basis.* Cambridge, Cambridge University Press.

Solomon, S., G.-K. Plattner, R. Knutti, and P. Friedlingstein. 2009. Irreversible climate change due to carbon dioxide emissions. *Proc. Natl. Acad. Sci.* **106**: 1704–1709.

Sonstadt, E. 1872. On the presence of gold in seawater. *Chem. News* **26**: 159–161. Cited from Falkner and Edmond (1990).

Sorokin, Y. I. 1983. The Black Sea. In *Ecosystems of the World 26. Estuaries and Enclosed Seas*, ed B. H. Ketchum. Elsevier, New York, pp. 253–292.

Sourirajan, S. and G. C. Kennedy. 1962. The system H_2O–NaCl at elevated temperatures and pressures. *Am. J. Sci.* **260**: 115–141.

Spencer, D. W. and P. G. Brewer. 1971. Vertical advection, diffusion and redox potentials as controls on the distribution of manganese and other trace metals dissolved in the Black Sea. *J. Geophys. Res.* **76**: 5877–5892.

Spencer, R. J. and L. A. Hardie. 1990. Control of seawater composition by mixing of river waters and midocean ridge hydrothermal brines. In *Fluid-Mineral Interactions: A Tribute to H. P. Eugster*, ed. R. J. Spencer and I.-M. Chou. Special Publication No. **2**, The Geochemical Society, San Antonio, TX, pp. 409–419.

Spinrad, R. 2008. The salt of the sea. *Oceanography* **21(1)**: 7.

Stacey, F. D. 1992. *Physics of the Earth, 3rd edn.* Brookfield Press, Brisbane.

Starikova, N. D. 1970. Vertical distribution patterns of dissolved organic carbon in sea water and interstitial solutions. *Oceanol. Acad. Sci. USSR* **10**: 796–807. (Eng. transl.).

Staudigel, H. 2004. Hydrothermal alteration processes in the oceanic crust. In *The Crust*, ed. R. L. Rudnick (*Treatise on Geochemistry*, vol. 3, ed. H. D. Holland and K. K. Turekian), Elsevier, Oxford, pp. 511–535.

Steemann Nielsen, E. 1952. The use of radioactive carbon (C^{14}) for measuring organic production in the sea. *J. Cons. Int. Explor. Mer* **18**: 117–140.

Stein, C. A. and S. Stein. 1994. Constraints on hydrothermal heat flux through the oceanic lithosphere from global heat flow. *J. Geophys. Res.* **99**: 3081–3095.

Stephens, G. C. 1982. Recent progress in the study of Die Ernährung der Wassertiere und der Stoffhaushalt der Gewässer. *Amer. Zool.* **22**: 611–619.

Stevenson, D. J. 1983. The nature of the Earth prior to the oldest known rock record: the Hadean Earth. In *Earth's Earliest Biosphere: Its Origin and Evolution*, ed. J. W. Schopf. Princeton University Press, Princeton, NJ, pp. 32–40.

Stoffers, P., M. Hannington, I. Wright *et al.* 1999. *Elemental mercury at submarine hydrothermal vents in the Bay of Plenty*, Taupo volcanic zone, New Zealand.

Stoye, J. 1994. *Marsigli's Europe*. Yale University Press, New Haven, CT.

Strous, M. and M. S. M. Jetten. 2004. Anaerobic oxidation of methane and ammonium. *Ann. Rev. Microbiol.* **58**: 99117.

Strous, M., J. A. Fuerst, E. H. M. Kramer, *et al.* 1999. Missing lithotroph identified as a new planctomycete. *Nature* **400**: 446–449.

Stuiver, M. and H. A. Polach. 1977. Discussion: reporting of ^{14}C data. *Radiocarbon* **19**: 355–363.

Stull, R. B. 1988. *An Introduction to Boundary Layer Meteorology*. Kluwer Academic Publishers, Boston, MA.

Suess, E. 1980. Particulate organic carbon flux in the oceans – surface productivity and oxygen utilization. *Nature* **288**: 260–263.

Suess, E. and M. J. Whiticar. 1989. Methane-derived CO_2 in pore fluids expelled from the Oregon subduction zone. *Palaeogegr., Palaeoclimat. Palaeoecol.* **71**: 119–136.

Sugimura, N. and Y. Suzuki. 1988. A high-temperature catalytic oxidation method for the determination of non-volatile dissolved organic carbon in seawater by direct injection of a liquid sample. *Mar. Chem.* **24**: 105–131.

Sunda, W. G. 2001. Bioavailability and bioaccumulation of iron in the sea. In *The Biogeochemistry of Iron in Seawater*, ed. D. R. Turner and K. A. Hunter. Wiley, New York, pp. 41–84.

Sunda, W. G., P. A. Tester, and S. A. Huntsman. 1987. Effects of cupric and zinc ion activities on the survival and reproduction of marine copepods. *Mar. Biol.* **94**: 202–210.

Sundquist, E. T. 1993. The global carbon dioxide budget. *Science* **259**: 934–941.

Sundquist, E. T., L. N. Plummer, and T. M. L. Wigley. 1979. Carbon dioxide in the ocean surface: the homogeneous buffer factor. *Science* **204**: 1203–1205.

Sutcliff, W. H., E. R. Baylor, and D. W. Menzel. 1963. Sea surface chemistry and Langmuir circulation. *Deep-Sea Res.* **10**: 233–243.

Suzuki, N. and K. Kato. 1953. Studies on suspended materials – marine snow in the sea. Part 1, Sources of marine snow. *Bull. Fac. Fish. Hokkaido Univ.* **4**: 132–137.

Sverdrup, H. U., M. W. Johnson, and R. Fleming. 1942. *The Oceans, Their Physics, Chemistry, and General Biology*. Prentice-Hall, Englewood Cliffs, NJ.

Swift, D. 1980. Vitamins and phytoplankton growth. In *The Physiological Ecology of Phytoplankton*, ed. I. Morris. Blackwell, Boston, MA, pp. 329–368.

Swinnerton, J. W. and V. J. Linnenbom. 1967. Gaseous hydrocarbons in seawater. Determination. *Science* **156**: 1119–1120.

Takahashi, T., W. S. Broecker, and S. Langer. 1985. Redfield ratio based on chemical data from isopycnal surfaces. *J. Geophys. Res.* **90**: 6907–6924.

Taleshi, M. S., K. B. Jensen, G. Raber, *et al.* 2008. Arsenic-containing hydrocarbons: natural compounds in oil from the fish capelin, *Mallotus villosus*. *Chem. Commun.* **39**: 4706–4707 DOI: 10.1039/b808049f.

Tanaka, M., G. Girard, R. Davis, A. Peuto, and N. Bignell. 2001. Recommended table for the density of water between 0 °C and 40 °C based on recent experimental reports. *Metrologia* **38**: 301–309. TEOS-10. See www.teos-10.org.

The Royal Society. 2005. Ocean acidification due to increasing atmospheric carbon dioxide. Policy Document 12/05. See: www.royalsoc.ac.uk.

Thiesen, M. 1900. Untersuchungen über die thermische Ausdehnung von festen and tropfbarflüssigen Körpern, ausgeführt durch M. Thiesen, K. Scheel und H. Diesselhorst. *Wissenschaftl.Abhandl. Physik.-Tech. Reichanstalt* **3**: 1–70. Cited in Menaché in UNESCO 1976.

Thompson, C. W. 1873. *The Depths of the Sea*. MacMillan, London.

Thurman, E. M. 1985. *Organic Geochemistry of Natural Waters*. Martinus Nijhoff, Dordrecht.

Tilton, L. W. and J. K. Taylor. 1937. Accurate representation of the refractivity and density of distilled water as a function of temperature. *J. Res. Nat. Bur. Stand. Wash.* **18**: 205–214.

Tortell, P. D., M. T. Maldonado, and N. M. Price. 1996. The role of heterotrophic bacteria in iron-limited ocean ecosystems. *Nature* **383**: 330–332.

Toy, A. D. F. 1973. Phosphorus. In *Comprehensive Inorganic Chemistry*, vol. **2**, ed. J. C. Bailar, H. J. Emeleus, R. S. Nyholm, and A. F. Trotman-Dickenson Pergamon Press, Oxford, pp. 389–545.

Tréguer, P., D. M. Nelson, A. J. Van Bennekom, *et al.* 1995. The silica balance in the world ocean: a reestimate. *Science* **268**: 375–379.

Trenberth, K. E. and C. J. Guillemot. 1994. The total mass of the atmosphere. *J. Geophys. Res.* **99**: 23,079–23,088.

Turner, D R. and K. A. Hunter, eds., 2001. *The Biogeochemistry of Iron in Seawater*. Wiley, New York..

Turner, R. E. and N. N. Rabalais. 1991. Changes in Mississippi River water quality in this century. *Bioscience* **41**: 140–147.

Tyndall, J. 1861. On the absorption and radiation of heat by gases and vapors, and on the physical connection of radiation, absorption, and conduction. *Phil. Mag.* (Ser. 4) **22**: 169–194, 273–285.

Tyndall, J. 1896. *The Forms of Water in Clouds and Rivers, Ice and Glaciers*. Appleton & Co., New York.

UNEP Chemicals Branch, 2008. The Global Atmospheric Mercury Assessment: Sources, Emissions and Transport. UNEP-Chemicals, Geneva.

UNESCO. 1962. Joint Panel on the Equation of State of Sea Water. UNESCO, Paris Cited in Wallace 1974.

UNESCO. 1966. International Oceanographic Tables. *UNESCO Technical Papers in Marine Science*, vol. **39**. UNESCO, Paris.

UNESCO. 1976. Seventh report of the Joint Panel on Oceanographic Tables and Standards. *UNESCO Technical Papers in Marine Science, vol.* 24. UNESCO, Paris.

UNESCO. 1981a. Background papers and supporting data on the Practical Salinity Scale 1978. *UNESCO Technical Papers in Marine Science* **37**. UNESCO, Paris.

UNESCO. 1981b. Background papers and supporting data on the International Equation of State of Sea-water 1980. *UNESCO Technical papers in Marine Science* 38. UNESCO, Paris.

UNESCO. 1983a. Carbon dioxide sub-group of the joint panel on oceanographic tables and standards. *UNESCO Technical Papers in Marine Science, vol.* **42**. UNESCO, Paris.

UNESCO. 1983b. Algorithms for computation of fundamental properties of seawater. *UNESCO Technical Papers in Marine Science, vol.* 44. UNESCO, Paris.

UNESCO. 1985. The international system of units (SI) in oceanography. *UNESCO Technical Papers in Marine Science, vol.* 45. UNESCO, Paris.

UNESCO. 1987. Thermodynamics of the carbon dioxide system in seawater. UNESCO Technical Papers in Marine Science, vol. 51. UNESCO, Paris.

Uppström, L. R. 1974. The boron/chlorinity ratio of deep-sea water from the Pacific Ocean. *Deep-Sea Res.* **21**: 161–162.

USGS. 2012. Commodity Statistics and Information. See http://minerals.er.usgs.gov/minerals/pubs/commodity.

Usiglio, M. J. 1849. Études sur la composition de l'eau de la Mediterranée et sur l'exploitation des sels qu'elle contient. *Ann. Chim. Phys.* (Ser. 3) **27**: 172–191.

Usoskin, I. G. and B. Kromer. 2005. Reconstruction of the ^{14}C production rate from measured relative abundance. *Radiocarbon* **47**: 31–37.

Valladares, C aptain J., W. Fennel, and E. G. Morozov. 2011. Replacement of EOS-80 with the International Thermodynamic Equation of Seawater – 2010 (TEOS-10). *Deep-Sea Research I* **58**: 978.

von Bibra, E. 1851. Untersuchung von Seewasser des stillen Meeres und des atlantischen Oceans. *Ann. Chem. Pharm.* **77**: 90–102.

von Brand, T., N. W. Rakestraw *et al.* 1937–1942. Decomposition and regeneration of nitrogeneous organic matter in seawater. *Biol. Bull.* **72**: 165; **77**: 285; **79**:231; **81**:63; **83**:273

Von Damm, K. L. 1990. Seafloor hydrothermal activity: black smoker chemistry and chimneys. *Ann. Rev. Earth Planet. Sci.* **18**: 173–204.

Von Damm, K. L. and J. L. Bischoff. 1987. Chemistry of hydrothermal solutions from the southern Juan de Fuca Ridge. *J. Geophys. Res.* **92**: 11,334–11,**346**.

von Huene, R. and D. W. Scholl. 1991. Observations at convergent margins concerning sediment subduction, subduction erosion, and the growth of continental crust. *Rev. Geophys.* **29**: 279–316.

Vraspir, J. M. and A. Butler. 2009. Chemistry of marine ligands and siderophores. *Ann. Rev. Marine Sci.* **1**: 43–63.

Walker, J. C. G. 1977. *Evolution of the Atmosphere*. MacMillan, New York.

Wallace, W. J. 1974. *The Development of the Chlorinity/Salinity Concept in Oceanography*. Elsevier, New York.

Walsh, J. J. 1989. How much shelf production reaches the deep sea? In *Productivity of the Ocean: Present and Past*, ed. W. H. Berger, V. S. Smetacek, and G. Wefer. John Wiley & Sons, New York, pp. 175–191.

Wangersky, P. J. 2000. Intercomparisons and intercalibrations. In *The Handbook of Environmental Chemistry*, vol. **5D**. Springer-Verlag, Berlin, ch. 7, pp. 167–191.

Wanninkhof, R. 1992. Relationship between wind speed and gas exchange over the ocean. *J. Geophys. Res.* **97**: 7373–7382.

Wanninkhof, R. and M. Knox. 1996. Chemical enhancement of CO_2 exchange in natural waters. *Limnol. Oceanogr.* **41**: 689–697.

Wanninkhof. R. K. F., Sullivan, and Z. Top. 2004. Air–sea gas transfer in the Southern Ocean. *J. Geophys. Res.* **109**: CO8S19, doi: 10.1029/2003JC001767.

Warneck, P. 1988. *Chemistry of the Natural Atmosphere*. Academic Press, New York.

Warner, M. J. and G. I. Roden. 1995. Chlorofluorocarbon evidence for recent ventilation of the deep Bering Sea. *Nature* **373**: 409–412.

Wedepohl, K. H., ed. 1969. *Handbook of Geochemistry*. Springer-Verlag, New York.

Wedepohl, K. H. 1995. The composition of the continental crust. *Geochim. Cosmochim. Acta* **59**: 1217–1232.

Wefer, G., G. Fischer, D. Fuetterer, and R. Gersonde. 1988. Seasonal particle flux in the Bransfield Strait, Antarctica. *Deep-Sea Res.* **35**: 891–898.

Weiss, R. F. 1970. The solubility of nitrogen, oxygen and argon in water and seawater. *Deep-Sea Res.* **17**: 721–735.

Weiss, R. F. 1971. Solubility of helium and neon in water and seawater. *J. Chem. Eng. Data* **16**: 235–241.

Weiss, R. F. 1974. Carbon dioxide in water and seawater: the solubility of non-ideal gas. *Mar. Chem.* **2**: 203–215.

Weiss, R. F. and T. K. Kyser. 1978. Solubility of krypton in water and seawater. *J. Chem. Eng. Data* **23**: 69–72.

Weiss, R. F and B. A. Price. 1980. Nitrous oxide solubility in water and seawater. *Mar. Chem.* **8**: 347–359.

Weiss, R. F., H. G. Östlund, and H. Craig. 1979. Geochemical studies of the weddell sea. *Deep-Sea Res.* **26A**: 1093–1120.

Weiss, R. F., W. S. Broecker, H. Craig, and D. Spencer. 1983. *GEOSECS Indian Ocean Expedition, Hydrographic Data 1977–1978*. US Government Printing Office, Washington, DC.

Wells, M. L. and E. D. Goldberg. 1992. Marine submicron particles. *Mar. Chem.* **40**: 5–18.

Westheimer, F. H. 1987. Why nature chose phosphates. *Science* **235**:1173–1178.

Whitfield, M. 1975a. An improved specific interaction model for seawater at 25 °C and 1 atmosphere total pressure. *Mar. Chem.* **3**: 197–213.

Whitfield, M. 1975b. The extension of chemical models for sea water to include trace components at 25 °C and 1 atm pressure. *Geochim. Cosmochim. Acta* **39**: 1545–1557.

Whitfield, M., R. A. Butler, and A. K. Covington. 1985. The determination of pH in estuarine waters I. Definition of pH scales and the selection of buffers. *Oceanologica Acta* **8**: 423–432.

Wiebe, W. J., R. E. Johannes, and K. L. Webb. 1975. Nitrogen fixation by a coral reef community. *Science* **188**: 257–259.

Wiesenberg, D. A. and N. M. Guinasso, Jr. 1979. Equilibrium solubilities of methane, carbon monoxide and hydrogen in water and seawater. *J. Chem. Eng. Data* **24**: 356–360.

Wieser, M. E. 2006. Atomic weights of the elements 2005 (IUPAC Technical Report). *Pure Appl. Chem.* **78**: 2051–2066.

Williams, P. M. 1961. Organic acids in Pacific Ocean waters. *Nature* **189**: 219–220.

Williams, P. M. 1965. Fatty acids derived from lipids of marine origin. *J Fish. Res. Bd. Can.* **22**: 1107–1122.

Williams, P. M. and E. R. M. Druffel. 1987. Radiocarbon in dissolved organic matter in the central North Pacific Ocean. *Nature* **330**: 246–248.

Williams, P. M. and E. R. M. Druffel. 1988. Dissolved organic matter in the ocean: comments on a controversy. *Oceanography* **1**: 14–17.

Williams, P. M., H. Oeschger, and P. Kinney. 1969. Natural radiocarbon activity of the dissolved organic carbon in the North-East Pacific Ocean. *Nature* **224**: 256–258.

Wilson, T. R. S. 1975. Salinity and the major elements of sea water. In *Chemical Oceanography*, 2nd edn, vol. 1, ed. Riley, J. P, and G. Skirrow. Academic Press, New York, ch. 7, pp. 365–413.

Wilson, T. R. S. 1978. Evidence for denitrification in aerobic pelagic sediments. *Nature* **274**: 354–356.

Wise, D. L. and H. G. Houghton. 1966. The diffusion coefficients of ten slightly soluble gases in water at 10–60°C. *Chem. Eng. Sci.* **21**: 999–1010.

Wollast, R. 1974. The silica problem. In *The Sea*, vol. **5**, ed E. Goldberg. John Wiley & Sons, New York, pp. 359–392.

Wong, G. T. F. and C. E. Grosch. 1978. A mathematical model for the distribution of dissolved silicon in interstitial waters – an analytical approach. *J. Mar. Res.* **36**: 735–750.

Wright, D. G., R. Pawlowicz, T. J. McDougall, R. Feistel, and G. M. Marion. 2011. Absolute salinity, "Density Salinity" and the Reference-Composition Salinity Scale: present and future use in the seawater standard TEOS-10. *Ocean Sci.* **7**: 1–26.

Wu, J, and E. A. Boyle. 1997. Lead in the western North Atlantic Ocean: Completed response to leaded gasoline phaseout. *Geochim. Cosmochim. Acta* **61**: 3279–3283.

Yamamoto, S., J. B. Alcauskas, and T. E. Crozier. 1976. Solubility of methane in distilled water and seawater. *J. Chem. Eng. Data* **21**: 78–80.

Yonge, S. M. 1972. The inception and significance of the Challenger Expedition. *Proc. Roy. Soc. Edinburgh B* **72**: 1–13. (And subsequent articles in the same volume.)

Zeebe, R. E. and D. Wolf-Gladrow. 2001. *CO_2 in Seawater: Equilibrium, Kinetics, Isotopes.* Elsevier, New York.

Zehr, J. P., J. B. Waterbury, P. J. Turner, *et al.* 2001. Unicellular cyanobacteria fix N_2 in the subtropical North Pacific Ocean. *Nature* **412**: 635–638.

Zehr. J. P. 2009. New twist on nitrogen cycling in oceanic oxygen minimum zones. *Proc. Natl. Acad. Sci.* **106**: 4575–4576.

Zhang, J., H. Amakawa, and Y. Nozaki. 1994. The comparative behaviors of yttrium and lanthanides in seawater of the North Pacific. *Geophys. Res. Lett.* **21**: 2677–2680.

Zhang, J., P. D. Quay, and D. O. Wilbur. 1995. Carbon isotope fractionation during gas-water exchange and dissolution of CO_2. *Geochim. Cosmochim. Acta* **59**: 107–114.

Zumft, W. G. 1997. Cell biology and molecular basis of denitrification. *Microbiol. Molec. Biol. Rev.* **61**: 533–616.

INDEX

CPSIA information can be obtained
at www.ICGtesting.com
Printed in the USA
LVHW102034031218
599101LV00018B/339/P

9 780521 887076